Animals

Animals

R.McNEILL ALEXANDER
Professor of Zoology, University of Leeds

CAMBRIDGE
UNIVERSITY PRESS

Published by the Press Syndicate of the University of Cambridge
The Pitt Building, Trumpington Street, Cambridge CB2 1RP
40 West 20th Street, New York NY 10011–4211 USA
10 Stamford Road, Oakleigh, Melbourne 3166, Australia

First published 1990
Reprinted 1995

Printed in Great Britain by Bell and Bain Ltd, Glasgow

British Library cataloguing in publication data

Alexander, R. McNeill (Robert McNeill)
Animals.
1. Animals
I. Title
591

Library of Congress cataloguing in publication data

Alexander, R. McNeill.
Animals / R. McNeill Alexander.
   p.   cm.
Includes index.
ISBN 0 521 34391 7. – ISBN 0 521 34865 X (paperback)
1. Zoology.   I. Title
QL 47.2.A44   1990
591—dc20   89-9766   CIP

ISBN 0 521 34865 X paperback

SE

Contents

Contents

Preface

I have built this book from two of my previous ones, *The Invertebrates* and *The Chordates*. Together, they surveyed the animal kingdom (including the protozoans) in 1072 pages. This book covers the same ground in much less space. I believe that this shorter treatment will be found more suitable for modern university courses, which devote less time than older courses did to the structure and diversity of animals. I expect the book will be used mainly by undergraduates in their first or second year.

This book is about the major groups of animals, about their structure, physiology and ways of life. Each chapter, except the first, deals with a taxonomic group of animals, usually a phylum or class. Brief descriptions of a few examples are followed by more detailed discussion of selected topics. Some of these topics are peculiarities of the groups (for instance, the shells of molluscs and the flight of insects). Others are more widespread features or properties of animals which can be illustrated particularly well by reference to the group. Thus I have used jellyfish in chapter 3 to illustrate the workings of simple nervous systems, and have explained some of the basics of muscle physiology in my account of molluscs, in chapter 6. I have described many experiments because I think it as important and interesting to know how zoological information is obtained, as to know the information itself.

The diversity of animals is a major problem in writing about them. Any attempt at encyclopaedic coverage results in an enormous quantity of indigestible morphological and taxonomic information. I have tried to overcome this difficulty by omitting minor groups of animals, and describing only a few examples of each major group. I would rather give students a good understanding of typical examples of familiar groups of animals than tell them about obscure groups they may seldom or never see.

Though I have described rather few species, I want students to appreciate the extraordinary diversity of animals. That appreciation is better gained in the field and the laboratory, by studying specimens found locally, than by reading about animal diversity in books.

One of the main aims of zoology is to explain the structure and physiology of animals in terms of physical science. For instance, explanations of nerve conduction, swimming and skeletal strength depend on physical chemistry, hydrodynamics and materials science, respectively. I have used many branches of physics and physical chemistry but have tried to explain them in simple terms. I have also used a lot of simple calculations to check whether explanations are plausible.

I have benefitted greatly from the advice of colleagues who read chapters in manuscript: Professor Donald Lee and Drs Jeff Bale, John Grahame, Joe Jennings, Peter Mill, Judith Smith and Stephen Sutton.

University of Leeds   R. MᶜNEILL ALEXANDER

1 *Introduction*

Subsequent chapters describe a great many observations on animals, some of them anatomical and some physiological. This chapter explains some of the techniques used by zoologists to make these observations. It also explains how zoologists classify animals.

Many people seem still to think of a zoologist as someone with a microscope, a scalpel and a butterfly net. These are still important tools but a modern zoological laboratory requires a far more varied (and far more expensive) range of equipment. Some of the most important tools which will be mentioned repeatedly in later chapters are described in this one.

1.1. Microscopy

It is convenient to start with the conventional light microscope (Fig. 1.1a). Light from a lamp passes through a condenser lens which brings it to a focus on the specimen S which is to be examined. The light travels on through the objective lens which forms an enlarged image of the specimen at I_1. This is viewed through an eyepiece used as a magnifying glass so that a greatly enlarged virtual image is seen at I_2. Each of the lenses shown in this simple diagram is multiple in real microscopes, especially the objective, which may consist of as many as 14 lenses. This complexity is necessary in good microscopes to reduce to an acceptable level the distortions and other image faults which are known as aberrations.

Microscrope technology has long been at the state at which the capacity of the best microscopes to reveal fine detail is limited by the properties of light rather than by any imperfections of design. It would be easy to build microscopes with greater magnification than is generally used but this would not make finer detail visible any more than magnification will reveal finer detail in a newspaper photograph. If light of wavelength λ and glass lenses of refractive index 1.5 are used, objects less than $\lambda/3$ apart cannot be seen separately, however great the magnification and however perfect the lenses. Since visible light has wavelengths around 0.5 μm, details less than about 0.2 μm apart cannot be distinguished. This is expressed by saying that light microscopes are incapable of resolutions better than about 0.2 μm.

Even this resolution is only possible with an oil-immersion lens of high numerical aperture. An oil-immersion lens is one designed to have the space between it and the specimen filled by a drop of oil of high refractive index. High numerical aperture implies that light from a single point on the specimen may enter the objective at a wide range of angles, i.e. that the angle θ (Fig. 1.1a) is large.

Fig. 1.1(b) shows a projection microscope. The image I_1 is just outside the focal length of the projector lens whereas in a conventional microscope it is just inside the focal length of the eyepiece. Consequently the final image I_2 is real instead of virtual

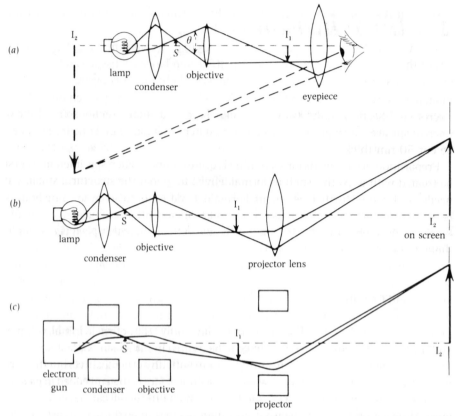

Fig. 1.1. Diagrams of (*a*) a conventional light microscope, (*b*) a projecting light microscope and (*c*) a transmission electron microscope. The paths of a few rays (of light or of electrons) are indicated.

and can be projected onto a screen. The projection microscope has little use in zoology except in teaching, and is illustrated solely for comparison with the transmission electron microscope (Fig. 1.1*c*). This uses a beam of electrons instead of a beam of light. Magnetic fields set up by electric currents in coils of wire serve as lenses, refracting the rays of electrons in the same way as convex glass lenses refract rays of light. Electrons from the electron gun are focussed on the specimen by the condenser, and other magnetic lenses produce a greatly magnified final image in similar fashion to the lenses of a light projection microscope. The image is thrown onto a fluorescent screen, which can be viewed directly, or onto photographic film.

The electron beam has a wavelength which depends on the potential difference used to accelerate it. Electron microscopes use large potential differences which make the wavelength exceedingly small, so that extremely fine resolution is possible in principle. The resolution actually achieved is much less good because even the best magnetic lenses are far less free from aberrations than the lenses used in light microscopes. Very small numerical apertures have to be used to reduce the aberrations to an acceptable level, making the resolution less good than is theoretically possible. It is still far better than for the light microscope. Resolutions aound 1 nm (0.001 μm) are achieved in biological work.

To reveal fine internal detail, tissues must be cut into very thin slices. They cannot be cut thin enough without prior treatment. Frozen blocks of tissue can be cut thin enough for some purposes but most specimens for light microscopy are embedded in molten wax which is allowed to solidify and then sliced in a machine called a microtome. Sections only 2 μm thick can be cut. Thinner sections for electron microscopy are cut from specimens embedded in plastics such as Araldite. They can be cut 50 nm thick.

Preparation of specimens for sectioning is quite complex. Each specimen must first be treated with a fixative such as formaldehyde to give it the structural stability it needs to withstand further treatment. Formaldehyde seems to act by forming bridges between protein molecules (this is a process like vulcanization, which converts liquid latex to solid rubber by inserting sulphur between the molecules). Next the water in the specimen must be removed and replaced by a liquid miscible with molten wax, so that the wax can permeate the specimen. This is usually done by transferring the specimen from water, through a series of water–ethyl alcohol mixtures, to pure alcohol and then to xylene (which is miscible both with alcohol and with wax, but not with water). This unfortunately tends to shrink the specimen. For instance, sea urchin eggs in xylene have been found to have only 48% of their initial volume. Specimens for electron microscopy have to be treated in rather similar fashion. They are more often fixed with osmium tetroxide than with any other fixative, but they are passed through a series of concentrations of alcohol before being embedded in plastic.

Even all this processing produces sections in which little detail can be seen because most tissue constituents are transparent, especially in thin sections. Dyes are usually used in light microscopy, to colour different constituents different colours. The ways in which most of them work are not fully understood, but some of them are marvellously effective. For instance the Mallory technique colours muscle red, collagen blue and nerves lilac. Contrast in electron microscopy depends on different parts of the specimen scattering electrons to different extents. The greater the mass per unit area of section, the more electrons are scattered and the darker that part of the section appears in the image. Dyeing would be ineffective but contrast can be enhanced by treatment with compounds of heavy metals, such as phosphotungstic acid and uranyl acetate. These 'electron stains' attach preferentially to certain cell constituents.

How does all this treatment alter the structure of the specimens? Is the structure seen through the microscope more or less as in the living animal, or is it largely new structure produced by chemical treatment? Fig. 1.2 shows that the same material prepared by different but accepted techniques may look a little different. However, fairly similar appearances are obtained by grossly different techniques and it seems more likely that the structure which is seen was there initially than that it is formed independently by several different treatments.

The technique of serial sectioning is often used. A specimen is cut into sections all of which are examined, to discover the three-dimensional structure of the original specimen.

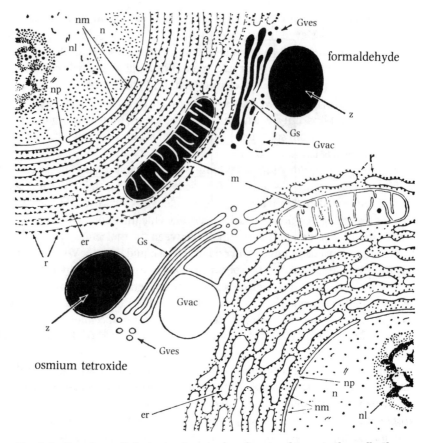

Fig. 1.2. Drawings of electron micrographs of parts of two similar cells (from a mouse pancreas). The one shown above and to the left was fixed by formaldehyde and the other (below and to the right) by osmium tetroxide. Abbreviations: er, endoplasmic reticulum; Gs, Golgi saccule; Gvac, Golgi vacuole; Gves, Golgi vesicle; m, mitochondrion; n, nucleus; nl, nucleolus; nm, nuclear membranes; np, nuclear pore; r, ribosome; z, zymogen granule. From J.R. Baker (1966). *Cytological technique*, 5th edn. Methuen, London.

Protozoans and isolated cells can be examined alive and intact at high magnification, but little detail can be seen in them by ordinary light microscopy. Most of their parts are more or less equally colourless and transparent so details in them are as hard to see as glass beads in a tumbler of water. Fortunately there are differences of refractive index, and these are used to reveal more detail in the techniques of phase contrast and interference microscopy.

Ordinary (transmission) electron microscopes are used for examining thin sections. Scanning electron microscopes are used for examining solid objects. They have been found particularly convenient for studying the hard parts of animals and have been used, for instance, to examine broken pieces of mollusc shell and find out how the crystals in them are arranged (Fig. 6.6). Fig. 1.3 shows how they work. The beam of electrons is focussed to a tiny spot on the specimen. One of its effects on the

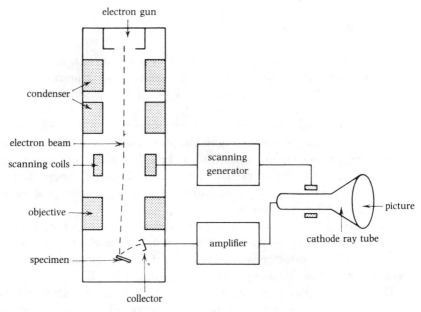

Fig. 1.3. A diagram of a scanning electron microscope.

specimen is to release other electrons which are drawn to the positively charged collector, where they are detected. The beam is deflected by the scanning coils which are used to move the spot systematically backwards and forwards over the specimen. The signal from the collector is used to control the brightness of a spot on a cathode ray tube (like the tube of a television set). This spot is moved backwards and forwards over the screen in precisely the same way as the spot is moving over the specimen, so it builds up a picture of the specimen in the same way as a television picture is built up by a bright, rapidly scanning spot. The pictures give an excellent three-dimensional effect: hollows look dark and projections throw shadows.

The scanning electron microscope cannot resolve detail finer than the diameter of the spot, which is typically 10 nm. Its resolution is therefore much less good than that of the transmission electron microscope, though very much better than that of the light microscope. It is sometimes used at very low magnifications to observe details which are easily visible by light microscopy, because it has much greater depth of field than the light microscope. When a thick specimen is examined by light microscopy it is necessary to focus up and down to see the detail at different levels. When a scanning electron microscope is used to view the same specimen, often everything can be seen simultaneously, all crisply in focus.

Though specimens do not need sectioning for scanning electron microscopy, they need some processing. Since a beam of electrons in air is scattered before it has travelled far, there has to be a vacuum in the microscope. In the vacuum, water would evaporate rapidly from a specimen which had not been previously dried. The specimen must be dried out without shrivelling it. This presents no problem when the specimen is a rigid one, like a piece of mollusc shell, but soft tissues need the special technique of critical point drying.

1.2.   Chemical analysis

Zoologists often want to know the chemical composition of a structure or fluid. If large enough samples are available ordinary analytical techniques can be used. Two instruments deserve special mention. One of them is the automatic amino acid analyser, which separates automatically the constituent amino acids of proteins and measures how much of each is present. The other is the X-ray diffraction spectrometer, which measures the spacings of the repeating patterns in crystals. These two techniques have been used, for instance, to show that the horny skeletons of sea fans consist largely of a protein extremely like the collagen of which the tendons of vertebrates are made (section 3.5). Further analysis was needed to show why they are so much stiffer than tendons.

Often zoologists have only tiny samples available. This was the case, for instance, in an investigation of the contractile vacuole of an amoeba which is described in section 2.7. The contractile vacuole is a drop of fluid of diameter about 50 μm. The contents of contractile vacuoles were analysed individually.

The first problem in an investigation like this is to obtain the sample. This is often done by sucking it into a micropipette, a glass capillary drawn out to a very fine point. The tips of the micropipettes used in the contractile vacuole investigation had diameters less than 5 μm. They were stuck into the amoeba under a microscope. This would have needed a phenomenally steady hand if the micropipette had not been held in a micromanipulator, a device in which very fine movements in three dimensions are produced by quite coarse movements of knobs.

The zoologist often wants to know the osmotic concentration of a sample. It is usually most convenient to discover this by measuring the freezing point. If the osmotic concentration of the sample is X mol 1^{-1} its freezing point is $-1.86\,X°$C. A tiny sample is sucked into a fine glass capillary and frozen. It is watched through a microscope while it is warmed again in a specially designed bath, until the last of the ice disappears. The temperature at which this happens is measured by a sensitive thermometer, capable of registering hundredths of a degree.

The most plentiful cations in animals are sodium and potassium. Their concentrations in small samples are usually measured by flame photometry. If you heat a little of a sodium salt in a hot flame it emits yellow light of wavelength 589 nm. This is because electrons in the vaporized sodium atoms are temporarily excited to a higher energy level. When one falls back to its initial level it releases just enough energy to produce a photon of light of wavelength 589 nm. Potassium is similarly affected but the energy change is smaller so the wavelength of the light is longer, 766 nm. The intensity of the emitted light is used in flame photometry to measure the concentrations of the elements.

Fig. 1.4 shows a simple flame photometer. Air and fuel (often acetylene) are blown into the spray chamber. The air, entering through the nebulizer, draws the sample in through the long tube on the left and makes it into a spray which mixes with the fuel and so gets heated in the flame. Light from the flame is made to pass through a filter (to cut out wavelengths emitted by elements other than the one being measured) before falling on a photocell. The reading of the microammeter indicates the intensity

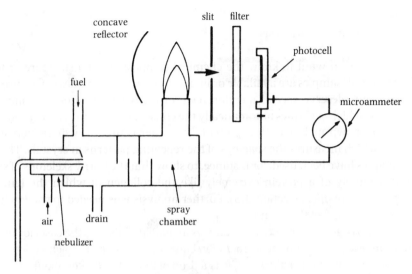

Fig. 1.4. A diagram of a simple flame photometer. From R. Ralph (1975). *Methods in experimental biology.* Blackie, Glasgow.

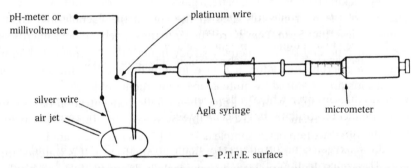

Fig. 1.5. Apparatus for titration of chloride in small samples. From R. Ralph (1975). *Methods in experimental biology.* Blackie, Glasgow.

of the light reaching the cell and the concentration of the sample can be calculated from it.

The most plentiful anion in animals is chloride. It is measured by titration with silver nitrate, which produces a precipitate of silver chloride. If the sample is large the titration can be done with a burette in the conventional way. If it is small the technique must be modified, for instance in the way shown in Fig. 1.5. The sample is a single drop which is stirred by a jet of air. Silver nitrate is added to it from a syringe operated by a micrometer. The endpoint is sensed electrically: the potential of the silver electrode changes rapidly at the endpoint as the concentration of silver ions in the drop rises.

Histochemistry is used to find out which cells in a microscope section contain a particular chemical compound. Substances are added which react with the compound to produce a coloured product which can be seen under the microscope. There is an account in section 4.4 of an investigation in which histochemical techniques

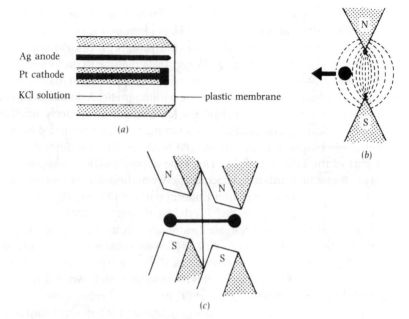

Fig. 1.6. Diagrams of equipment for measuring partial pressures of oxygen. Part (*a*) represents a Clark oxygen electrode; (*b*) shows the force which acts on a diamagnetic body in a non-uniform magnetic field in the presence of oxygen, and (*c*) shows a paramagnetic oxygen analyser which utilizes this effect.

were used to discover which cells produce which digestive enzymes in flatworms. Histochemical techniques have been devised for electron microscopy, as well as for light microscopy.

Data on oxygen concentrations are often wanted, particularly in studies of respiration. The concentration of oxygen in a gas mixture, or dissolved in water, can be measured by chemical methods, but there are other methods which are often more convenient. One uses the Clark oxygen electrode (Fig. 1.6*a*). This can be used to measure the partial pressure of oxygen, either in a gas mixture or in solution. It actually contains two electrodes, a platinum cathode and a chloride-coated silver anode. The cathode is kept around 0.5 V negative to the anode. Cathode and anode are in a potassium chloride solution, behind a plastic membrane. The device senses the partial pressure of oxygen in the gas or solution immediately *outside* the membrane. Oxygen diffuses through the membrane to the cathode where it is immediately reduced:

$$O_2 + 2H_2O + 4e = 4OH^-$$

or

$$O_2 + 2H_2O + 2e = 2OH^- + H_2O_2.$$

The current depends on the rate at which the oxygen diffuses in, which in turn depends on its partial pressure outside the membrane. Thus the current indicates the partial pressure. Oxygen electrodes have been used to record the rate at which swimming animals remove oxygen from the water (Fig. 6.19) and to find out how fast oxygen can diffuse out of fish swimbladders (section 14.3).

Paramagnetic oxygen analysers are devices that measure the proportion of oxygen in gas mixtures by quite different means. We are all familiar with the forces which act on ferromagnetic materials (such as iron) in a magnetic field. Most materials are not ferromagnetic but diamagnetic, and some are paramagnetic. A diamagnetic body in a non-uniform magnetic field experiences a force which is relatively small if it is surrounded by a diamagnetic gas but much stronger if it is surrounded by oxygen, which is strongly paramagnetic. The oxygen tends to displace the body from the strongest part of the field (Fig. 1.6*b*). One type of paramagnetic oxygen analyser contains a diamagnetic dumb-bell suspended between the poles of a powerful magnet in such a way that its ends are in non-uniform parts of the field (Fig. 1.6*c*). When oxygen is present forces act on the dumb-bell tending to make it turn on its suspension. Electrostatic forces are applied automatically to balance these forces, and the potential required is proportional to the oxygen concentration. Oxides of nitrogen are paramagnetic and would affect readings if they were present, but all other common gases are diamagnetic and have very little effect on the instrument. Paramagnetic oxygen analysers have been used in many experiments with vertebrates, for instance to measure the oxygen consumption of running lizards (section 16.4).

1.3.   Uses of radioactivity

Many zoological experiments exploit the properties of radioactive isotopes. The atoms of an element are not all identical. For instance, carbon consists mainly of atoms containing six protons and six neutrons, giving a mass number of $6+6=12$. However, it includes about 1% of atoms with six protons and seven neutrons (mass number 13) and a tiny proportion with six protons and eight neutrons (mass number 14). These three types of atom are described as isotopes of carbon and are represented by the symbols ^{12}C, ^{13}C and ^{14}C. Their chemical properties are identical, apart from small differences in rates of reaction.

^{12}C and ^{13}C are stable but ^{14}C is not. One of the neutrons in its nucleus disintegrates, becoming a proton and an electron. This leaves the nucleus with seven protons and seven neutrons so that it is no longer carbon but the common isotope of nitrogen.

$$^{14}C \quad \rightarrow {}^{14}N \quad + \quad \beta^-$$

| 6 protons | 7 protons | 1 electron |
| 8 neutrons | 7 neutrons | |

The electron leaves at high velocity because the change releases energy. Fast-moving electrons emitted like this by radioactive materials are known as β-rays.

Breakdown is a random process so the number of ^{14}C atoms in a sample falls exponentially. Half would vanish in the course of 5700 years but only a tiny proportion vanish during an ordinary experiment. The presence of the radioactive isotope is nevertheless easily detected because the β-rays can be detected individually, and counted.

^{14}C is present in normal carbon because ^{14}N atoms in the upper atmosphere, bombarded by neutrons in cosmic radiation, are converted to ^{14}C.

$$^{14}\text{N} \qquad + \text{neutron} \rightarrow {}^{14}\text{C} \qquad + \text{proton.}$$

| 7 protons | 6 protons |
| 7 neutrons | 8 neutrons |

^{14}C is manufactured by bombarding nitrogen compounds with neutrons, in atomic reactors.

The special value of radioactive isotopes in research is due to their being detectable in tiny quantities, and distinguishable from stable isotopes. For instance, in experiments described in section 3.6, radioactive sodium bicarbonate ($NaH^{14}CO_3$) was put in the water with corals, and radioactive glycerol was later detected. The glycerol must have been made from the bicarbonate.

Radioactive isotopes of many other elements as well as carbon are available. A particularly useful one is the hydrogen isotope tritium, ^{3}H, which has a proton and two neutrons in its nucleus instead of the lone proton of the common isotope.

Some isotopes emit radiation which is easily detected by means of a Geiger counter. Others, notably ^{14}C and ^{3}H, emit low-energy β-radiation which is best measured by the instrument called a liquid scintillation counter. The sample is mixed in solution with an organic compound which emits flashes of light when bombarded by β-radiation. Each β-ray causes a flash, and the instrument counts the flashes. The intensity of each flash corresponds to the energy of the β-ray which caused it. This makes it possible to count separately the dim flashes caused by the very low-energy β-rays from ^{3}H and the brighter flashes caused by β-rays from ^{14}C.

Radioactive isotopes can be detected in microscope sections by autoradiography. A photographic emulsion is laid over the section and left for a while in darkness before being developed and fixed. The β-rays from radioactive isotopes in the section affect the emulsion in the same way as light would. Clusters of silver crystals in the developed emulsion show where the radioactive isotopes are in the section. The emulsion is left in place on the section making it possible to see under a microscope which tissues contain the radioactive material. The technique has been used, for instance, to show that foodstuffs produced by photosynthesis by algal cells in corals pass into the cells of the coral itself.

The radiation emitted by radioactive isotopes is harmful to health, so appropriate precautions have to be taken in experiments. The stringency of the precautions which are needed depends on the isotope. ^{14}C and ^{3}H are the most useful isotopes in biology and are also, conveniently, among the least hazardous.

1.4.   Recording events

Zoologists often want a permanent record of the movements of an animal, of the forces it exerts or of changes of temperature or pressure within it.

The most versatile instrument for recording movements is the cinematograph camera. Films are normally taken and shown at 18 or 24 frames (pictures) per second, which is just fast enough to avoid a flickering effect. The wings of flying insects would be blurred in films taken at this rate and no details of their movements could be made out, even if the frames were examined one by one. Films have been taken at rates up to at least 7000 frames per second, to show insect wing movements clearly (the outlines shown in Fig. 9.7*b* are traced from selected frames of a film taken at about 3500 frames per second). When films taken at high framing rates are projected at normal rates, the motion is seen slowed down. Conversely, films are sometimes taken at exceptionally low framing rates so that when they are projected, slow movements are seen speeded up. This is called time-lapse photography. It was used in an investigation of the development of sea urchins so that events which take 48 hours could be seen happening in a few minutes (Fig. 11.5).

X-ray cinematography is sometimes much more useful than light cinematography. It has been used to show how the limb bones of mammals move when they walk (Fig. 18.15) and how teeth move over each other in chewing (Fig. 18.20). A visible image can be obtained by projecting a beam of X-rays through the specimen on to a fluorescent screen. A film of the screen can be taken with an ordinary cine camera, but this is not generally a satisfactory procedure because the image is rather dim. It is usual to obtain a very much brighter image by means of an electronic device known as an image intensifier, and to take a film of that. Framing rates up to about 150 frames per second are practicable.

The opacity of materials to X-rays depends on their density and on the atomic number of the atoms they contain. Dense materials, and elements of high atomic number, are relatively opaque. Bone contains a high proportion of calcium phosphate, which makes it denser than other tissues. Calcium and phosphorus have higher atomic numbers than the elements which are most plentiful in other tissues. Hence bone is relatively opaque and can be distinguished in X-ray images. Air, on the other hand, has a low density and is relatively transparent. X-ray films which have been taken of lungfish breathing show both the X-ray-opaque bones, and the X-ray-transparent air in the lung (Fig. 15.9). Structures which would not normally be distinguishable in X-ray images can be made visible by injecting fluids containing elements of high atomic number. This method has been used, for instance, in investigations of the flow of blood through the hearts of frogs.

Movements and forces are often recorded by means of transducers, devices that translate one type of signal into another. For instance, a microphone is a transducer that translates sound into electrical signals and a loudspeaker is a transducer that has the reverse effect. Experimental zoologists have made a great deal of use, in recent years, of transducers which produce an electrical signal that can be displayed on an oscilloscope screen or recorded by a pen recorder or ultra-violet recorder.

Fig. 1.7*a* shows a transducer which can be used (in different circumstances) to sense forces or movements. It is simply a flexible beam fixed at one end, with devices called strain gauges glued to its upper and lower surfaces. When its free end is pulled down the upper strain gauge is stretched a little and the lower one is compressed, and this alters their electrical resistances. There are two types of strain gauge. One type consists of a thin wire or a strip of metal foil. Stretching makes it longer and thinner, so that its electrical resistance increases. The other type is a slice of semiconductor

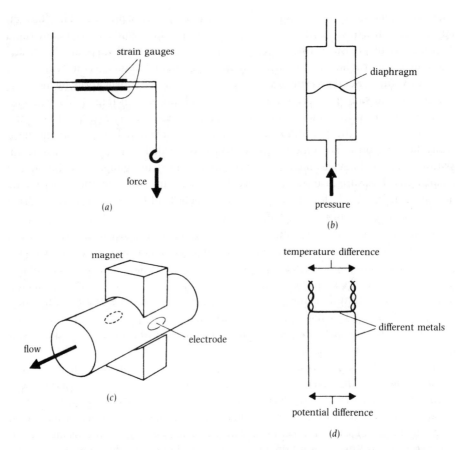

Fig. 1.7. **Diagrams illustrating (*a*) a force–displacement transducer, (*b*) a pressure transducer, (*c*) an electromagnetic flowmeter and (*d*) a thermocouple.**

material, which undergoes a *decrease* in resistance when it is stretched, and is much more sensitive than the wire or foil type. By incorporating strain gauges in a Wheatstone bridge circuit, an electrical output proportional to their extension or compression can be obtained. Thus a record of any bending of the beam shown in Fig. 1.7*a* can be obtained. This transducer can be used to sense forces if its beam is relatively stiff, so that the forces which act on it bend it only a little. A transducer of this type was used as a force transducer to measure the drag force on hydroid colonies in a water current (section 3.5). Alternatively, if the beam is so flexible that it does not limit the movements of the organ to which it is attached, it can be used as a displacement transducer, to sense movement.

The forces which animals exert on the ground as they walk or run can be recorded by means of force platforms. These are plates set into the floor, mounted on force transducers. Separate transducers can be used to sense the vertical, longitudinal and transverse components of any force which acts on the platform. Force platforms have been used in investigations of walking, running and jumping (Fig. 18.12).

There are also transducers designed to sense pressure. Steady pressures, or pressures which change only slowly, can be measured with U-tube manometers, but such manometers do not respond very quickly to a change of pressure. They take an appreciable time to settle at the new equilibrium position, and cannot follow rapid fluctuations of pressure. Often, the pressures in which zoologists are interested fluctuate rather rapidly: the pressures in arteries, which fluctuate as the heart beats, are one example. Such pressures can generally be recorded satisfactorily by means of a pressure transducer incorporating a stiff metal diaphragm (Fig. 1.7*b*). The diaphragm is forced into a domed shape by the pressure to be measured. The higher the pressure, the more the diaphragm is distorted. The diaphragm has to be stiff to give a really fast response, so some sensitive device is needed to register its distortion. Various electrical devices have been used, including strain gauges attached to the diaphragm. Pressure transducers have been used in many of the experiments described in this book, for instance to obtain records of the pressures which drive water over the gills of fishes (section 13.4).

Velocity of air flow has sometimes to be measured, for instance in an investigation of flow in the lungs and air passages of birds (Fig. 17.10). The hot-wire anemometer is a convenient instrument for the purpose. The electrical resistance of a wire increases as its temperature rises. The anemometer has an electrically heated filament which is held in the air current. The moving air tends to cool it, so the faster the flow the cooler the wire and the lower its resistance.

It is often useful to be able to record the velocity of flow of blood in a blood vessel. It is undesirable to introduce any instrument into the vessel where it might interfere with the flow. There is no need to do this because flowmeters are available which fit like a cuff around the vessel, without any need to cut or pierce it. One type is the electromagnetic flowmeter. It depends on the principle of electromagnetic induction: blood is a conductor of electricity so blood moving across the lines of force of a magnetic field will set up a potential difference. The flowmeter contains an electromagnet which is fitted around the blood vessel, and potential differences across the vessel are recorded (Fig. 1.7*c*). The ultrasonic flowmeter is another instrument which can be used in the same way but depends on the Doppler effect. An electromagnetic flowmeter has been used to record flow in the ventral aorta of a shark (Fig. 13.8).

It is often useful to have small devices to measure temperature changes, because small devices are easiest to fit into animals and because they heat and cool quickly. One such device is the thermocouple. If a conductor runs through a gradient of temperature, a potential difference is set up between its ends. This cannot be measured in a circuit made entirely of one metal, because the effect on one wire running down the gradient is cancelled out by the effect on the other wire running up the gradient to complete the circuit. This difficulty can be overcome by using two metals which are affected to different extents, as shown in Fig. 1.7*d*. A potential difference is developed proportional to the difference in temperature between the two junctions where one metal joins the other. Thermocouples are incorporated in the device for studying air flow in bird lungs which is illustrated in Fig. 17.10.

Another small device for measuring temperature is the thermistor, which is simply a small bead of a metal oxide or oxide mixture with electrical leads attached. The electrical resistance of a thermistor decreases as its temperature rises. (Metals also change their resistance as temperature changes, but not nearly as much and, as it happens, in the opposite direction.) Thermistors have been used to measure the

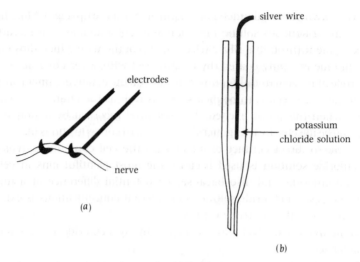

silver wire

electrodes

potassium
chloride solution

nerve

(*a*)

(*b*)

Fig. 1.8. Diagrams of (*a*) metal hooks being used as electrodes to record from a nerve; and (*b*) a micropipette electrode.

temperatures of the muscles of freshly caught tuna (Fig. 14.12). Electrically heated thermistors are sometimes used to measure velocities of moving air or water. The moving fluid tends to cool the thermistor, so the faster the flow, the lower the temperature, and the electrical output indicates the rate of flow. Water flow through sponges has been measured in this way (Fig. 3.2).

1.5.   Electrodes

Messages are transmitted around the bodies of animals as electrical signals in nerve cells, and zoologists often want to record these signals.

The most important type of signal is the action potential, a brief reversal of the potential difference across the cell membrane. While an action potential is travelling along a nerve cell, there are potential differences in the surrounding extracellular fluid. These are often detected by means of fine tungsten or platinum wires, connected through an amplifier to a cathode ray oscilloscope. In some experiments a fine nerve or a strand dissected from a nerve is picked up on a pair of tungsten or platinum hooks, which serve as electrodes (Fig. 1.8*a*). In others an electrode is pushed into a brain or ganglion to record the electrical activity within. Electrodes to be used in this way are insulated (often by varnish) with only the tip left bare. Recording electrodes are usually held in micromanipulators.

The potential differences recorded in nerves by extracellular electrodes are small, usually of the order of 1 mV. The changes of electrical potential which occur in nerve cells while an action potential is passing are much larger, of the order of 0.1 V, but they cannot be recorded unless an electrode is pushed into the cell. This electrode must be fine enough not to damage the cell too much, for the cell must remain alive and active with the electrode tip inside it. The electrodes most often used are glass micropipettes, drawn out to a diameter of 0.5 μm or less at the tip (Fig. 1.8*b*). They are filled with a strong aqueous solution of potassium chloride and connected to the recording equipment through a silver wire. The tip of the wire which dips into the potassium chloride is lightly coated (by electrolysis) with silver chloride.

This electrode may seem unduly complicated but all the details are important if the resting potential in the cell is to be measured, as well as the changes of potential which occur when the cell is excited. The potential recorded by a simple metal electrode would be affected by the (generally unknown) concentrations of salts in the fluid in the cell, so direct contact of metal with the cell contents is avoided. A potassium chloride solution is used because the most plentiful ions in cells are potassium and chloride, and also because the potential difference at a junction between two potassium chloride solutions of different concentrations is extremely small. The solution in the electrode is made strong, so that it has reasonably high electrical conductivity. The silver wire is coated with silver chloride to prevent it from becoming polarized.

A zoologist who has struck an electrode into a brain or ganglion and recorded from a cell may want to identify the cell and investigate its anatomy. This can be done if the electrode is filled with a substance which can be injected and later seen in the cell. The recording is made, the substance is injected, the animal is killed and the ganglion is prepared for microscopy. The injected substance can be seen in the cell from which the records were made. A technique of this sort was used to show that certain action potentials recorded from jelly fish were in giant neurites (Fig. 3.16).

In many experiments, muscles or nerves are given small electric shocks to stimulate activity. Metal electrodes are often used.

Zoologists often want to know which muscles are responsible for particular movements, and the sequence in which they act. This information can be obtained by the technique of electromyography, which takes advantage of the electrical changes that occur within muscles when they are active. Vertebrate striated muscle fibres contract in response to action potentials in their motor nerves. Most have only one motor nerve ending, from which the stimulus to contract spreads as an action potential along the muscle fibre. These muscle action potentials can be detected by electrodes in the muscles. To be sure that any potentials that are recorded actually arise in the muscle being investigated, two electrodes should be placed in the muscle and the potential difference between them recorded. Concentric electrodes are often used, made from hypodermic needles. The needle is filled with an insulating material, with a wire running down its centre. The bare tip of the wire is exposed at the point of the needle, and the potential difference between wire and needle is recorded. Alternatively, two fine wire electrodes, each insulated except at the tip, are placed in the muscle. Very fine, flexible electrodes can be inserted with the help of a hypodermic needle. The needle is slipped over the electrode, pushed into the muscle and then withdrawn leaving the electrode in place. The potentials detected by the electrodes must be amplified and can then be displayed on a cathode-ray oscilloscope screen (which can be photographed) or can be recorded directly on sensitive paper by an ultraviolet recorder. Electromyography shows when a muscle is active and gives some indication of how active it is, but does not give any quantitative indication of the force that is being exerted.

1.6.   Experiments on animals

Many of the experiments described in this book were performed on living animals. The information they provide could not have been obtained in any other way. It is important that zoologists should take all reasonable precautions to avoid causing pain or distress to experimental animals. Some information cannot be obtained without risk of causing pain or discomfort, and in such cases zoologists should consider carefully whether an experiment is justified before performing it. In Britain there are legal restrictions as well as moral ones. Potentially painful experiments on living vertebrates may only be performed by holders of a licence from the Home Office. This applies even if the entire experiment is performed under anaesthetic and the animal is killed before it comes round.

1.7.   Fossils

Most of the animals described in this book are modern but a few are extinct, known only as fossils. The next few paragraphs explain briefly how fossils are formed, and how their relative ages can be discovered.

The surface of the earth is continually being crinkled by the processes which produce mountains, and levelled again by processes of erosion and sedimentation. A variety of processes break down rocks, particularly on land. Heating of their surfaces by day and cooling by night sets up stresses due to thermal expansion and breaks fragments off. Water expands when it freezes, so water freezing in cracks in rocks is apt to split them. Streams carrying abrasive particles such as sand scour and erode the rocks over which they run. Water containing dissolved carbon dioxide removes calcium and other elements from rocks, carrying them away as a solution of carbonates and bicarbonates.

Materials removed from rocks in these ways are deposited as sediments in other places. Particles which are carried along by fast-flowing water settle out where flow is slower, for instance on the flood plains of rivers and around their mouths. The smaller the particles, the slower the flow must be before they will settle, so relatively large particles settle as gravel or sand in different places from the small particles which form mud. Dry sand may be blown by the wind, and accumulate as dunes. Dissolved calcium carbonate is apt to be precipitated out of water in areas where algae are removing carbon dioxide from the water by photosynthesis. There are also sediments which are not formed from products of erosion, but from animal or plant remains. Shell gravel and the ooze formed on the ocean floor by accumulation of the shells of planktonic Foraminifera are two exampes of calcium carbonate sediments of animal origin. Peat is a deposit of incompletely decomposed plant material (in modern times mainly mosses).

Sediments are generally soft when they are formed but if they are not disturbed they tend in time to become rocks. Mud becomes shale, sand becomes sandstone and deposits of calcium carbonate become limestone. The change is partly due to the particles becoming more tightly packed and partly to processes which cement them together. Mud is a mixture of fine mineral particles and water. Initially up to 90% of its volume may be water, the particles are not in contact and it is sloppy. As more mud accumulates on top of it, it is subjected to pressure and water is squeezed out. Adjacent particles make contact when the water content is about 45% by volume and further compaction involves rearrangement and crushing of particles. Note that complete compaction of a layer of mud which initially contained 90% water involves reduction to one tenth of its initial thickness. Settled sand contains only about 37% water by volume, so compaction cannot reduce its thickness much. The grains in sandstones do not generally seem to have been crushed, but to have dissolved at the points where they touch other grains, so that the grains fit more closely together. Generally silica (perhaps from the dissolved corners of the grains) or a deposit of calcium carbonate cements sandstone together.

An animal which dies where a sediment is being deposited, or is carried there by currents after death, is liable to become embedded in the sediment. If it does not decay completely, it becomes a fossil. Usually only hard parts such as the shells of molluscs and the bones of vertebrates survive. Occasionally traces of soft tissues are also found, as stains or impressions in the rock.

It is always interesting to know the age of the rock in which fossils are found. Sediments are formed in layers. Layers of the same material may be distinct because sedimentation was interrupted. Layers of different sediments may be formed on top of one another because local conditions changed. The layers mark time intervals. Successive layers sometimes followed one another immediately, but sometimes there were extremely long intervals of time while no sediment was formed, and erosion may even have occurred, at the locality in question. The order in which the sediments were formed at any particular locality is generally obvious since later sediments are on top of earlier ones, though later earth movements may fold sediments so that part of the sequence is upside-down. Sediments formed simultaneously at different places can often be matched, particularly if they contain similar fossils. Thus the relative ages of fossils which are being studied can generally be established.

Fossils are very rare in the oldest sedimentary rocks. The time spanned by rocks in which fossils are reasonably plentiful is divided into three eras and eleven periods, as shown in Table 1.1. It is usually possible to determine the period in which a particular sedimentary rock was formed, and the approximate position within the period.

1.8. Concepts in morphology

There are some concepts considered basic by most zoologists concerned with the structure of animals, which are used in this book and had better be explained. One is the concept of homology. It is seldom mentioned explicitly in this book, but it is often implied. For instance, Figs 16.3 and 18.3 show the skulls of a python and an opossum, with many of the bones labelled. In many cases, the same name is given to a bone in each skull. As a general rule, bones in comparable positions are given the same name: for instance, the bones labelled maxilla both run along the side of the mouth and bear the main teeth of the upper jaw. There are, however, exceptions. The bone next posterior to the maxilla is called the ectopterygoid in the python, but the

Table 1.1. *The main divisions of time since the beginning of the Palaeozoic era* The Present is at the top of the table. Age is the approximate time since the *beginning* of the period, estimated mainly from the decay of radioactive elements.

Era	Period	Age (million years)
Cenozoic	Quaternary	2
	Tertiary	70
Mesozoic	Cretaceous	140
	Jurassic	190
	Triassic	230
Palaeozoic	Permian	280
	Carboniferous	350
	Devonian	400
	Silurian	440
	Ordovician	500
	Cambrian	570

jugal in the opossum. This implies that the maxilla of the python and the maxilla of the opossum are considered homologous, but the ectopterygoid of the python and the jugal of the opossum are not.

What does this mean? It means that the python and the opossum are believed to have had a common ancestor that had a structure identifiable as a maxilla, and that their maxillae have been derived from this ancestral maxilla by the process of evolution. Further, there was no single structure in any ancestor that gave rise to both the ectopterygoid of the python and the jugal of the opossum. The most recent common ancestor of the python and the opossum was probably an early reptile with a skull only a little different from those of primitive amphibians (Fig. 15.5). More distant ancestors are believed to have included fish with skulls like the one shown in Fig. 15.3. Notice that bones labelled maxilla, ectopterygoid and jugal appear in this illustration.

Identical names do not always imply homology. Insects and cows both have structures called legs, but no modern zoologist would suggest that they had a common ancestor with legs. On the other hand, structures given different names are sometimes claimed to be homologous. The hyomandibular cartilage of sharks, the hyomandibular bone of teleosts and the tiny bone called the stapes in the ears of mammals are considered homologous. The reasons will become apparent later in this book.

There is a severe difficulty implicit in this evolutionary concept of homology. There is no complete and unambiguous record of the course of evolution. The fossil record is fragmentary. For instance, to refer back to the example which has just been used, there is no known fossil which seems at all likely to resemble a common ancestor of sharks and teleosts, and which has any structure identifiable as a hyomandibular cartilage or bone. The conclusion that the hyomandibular cartilage of sharks and the hyomandibular bone of teleosts are homologous has not been reached by tracing their ancestry, but by comparing the structure of both adult and embryo sharks and teleosts. Indeed the concept of homology was introduced before it was generally accepted that evolution had occurred, and it did not at first have evolutionary overtones. It implied correspondence of a rather abstract type.

Structures are often described as adaptations. For instance, the very strong forelimbs of moles may be described as adaptations for burrowing. This means that they are believed to have evolved because ancestral moles with stronger forelimbs were better able to burrow and were consequently favoured by natural selection. They were more likely to leave offspring than their fellows with weaker forelimbs, and so forelimbs in the population as a whole became progressively stronger.

Zoologists discussing adaptations are apt to make statements such as 'Moles have strong forelimbs for burrowing'. Statements like this, which indicate that something has a purpose or function, are called teleological. Many teachers of biology condemn teleological statements, perhaps because they feel they imply *conscious* purpose; that they imply for instance, that God provided moles with strong forelimbs so that they could burrow. Others contend that teleological statements are proper and useful in biology, and have no metaphysical implications. If strong forelimbs evolved because they gave a selective advantage attributable to their use in burrowing, surely they are *for* burrowing. I support this view, and have made no conscious effort to avoid teleological language in this book.

Animals, or features of animals, are often described as primitive or advanced. The opossum is a primitive mammal, and its molar teeth are among its primitive features. This means that it has many features in which it resembles the ancestral mammals rather than many present-day mammals, and that its molar teeth are among these features. Though it is a primitive mammal it is not a primitive vertebrate, for the mammals are more different from the ancestral vertebrates than any other vertebrates except perhaps the birds: mammals and birds are advanced vertebrates.

1.9. Classification

This book is full of generalizations, some about large groups of animals (such as the molluscs) and some about small ones (such as the members of a single species). Names are needed for all these groups. A system of classification, if it is well designed, helps zoologists to marshal their knowledge of animals and provides them with names for most of the groups about which they wish to make generalizations.

The smallest unit of classification with which we are concerned is the species. This unit is notoriously difficult to define, and most readers will have at least an intuitive understanding of its scope. The garden snail is an example of a species. It is a snail with a brown and black shell, common in Britain and also in North America (where it has been introduced). Like other species it is assigned to a genus and has a Latin name (*Helix aspersa*) which consists of two words, of which the first is the name of the genus (*Helix*). Several other very similar snails are assigned to the same genus, including

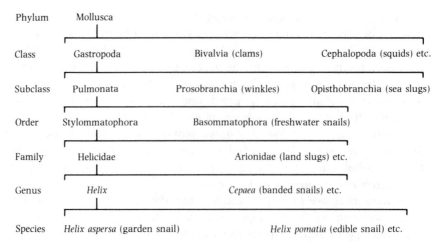

Fig. 1.9. **A diagram showing the position of the garden snail (*Helix aspersa*) in a system of zoological classification.**

the edible snail *Helix pomatia*. This is the snail which is sold as food. It is larger than *Helix aspersa* and has a cream-coloured shell. Though the differences are quite small they are distinct, and *Helix aspersa* does not breed with *Helix pomatia*. The banded snail (*Cepaea nemoralis*, another common European species) is more different. Its shell is much flatter in shape as well as being smaller and differently coloured. It is universally judged too different from the species of *Helix* to be included in that genus and is therefore put in a separate genus *Cepaea*.

Precise rules are needed to ensure (as far as possible) that all zoologists call the garden snail *Helix aspersa* and that none use this name for other animals. These rules are incorporated in the *International Code of Zoological Nomenclature* and are administered by an international commission. If there is no other compelling reason to prefer one of two rival names, the one which was introduced first has priority.

Genera are grouped in families, families in orders, orders in classes and classes in phyla. Thus *Helix* and *Cepaea* belong (with various other similar snails) to the family Helicidae but the terrestrial slugs are put in a separate family Arionidae (Fig. 1.9). These two families and several other families of terrestrial snails are grouped together in an order Stylommatophora while most freshwater snails are put in a separate order Basommatophora. The Stylommatophora have four tentacles with eyes at the tips of the second pair (see *Helix* in Fig. 6.23) but the Basommatophora have only two tentacles with eyes at their bases (see *Planorbis* in Fig. 6.3*d*). The Stylommatophora and Basommatophora form the subclass Pulmonata while winkles and sea slugs are put in separate subclasses. All these subclasses are grouped together in the class Gastropoda. Finally the Gastropoda are grouped with other classes (which include the clams and squids) in the phylum Mollusca. A snail is not very obviously like a clam and still less like a squid but there are basic similarities of body plan which justify putting this strange assemblage of animals in a single phylum. This should be apparent from chapter 6.

Every animal is assigned to a species, genus, family, order, class and phylum, and additional groupings are used when convenient. For instance, it has been found

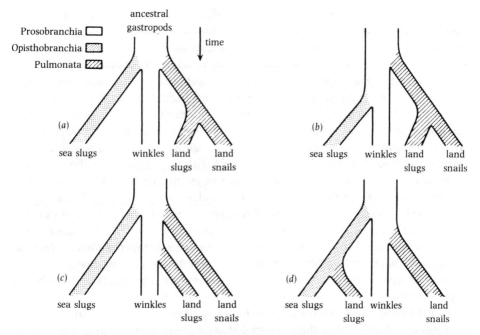

Fig. 1.10. **Diagrams illustrating hypotheses about the evolution of gastropod molluscs.**

convenient to divide the large class Gastropoda into three subclasses, as shown in Fig. 1.9. Putting two species in the same genus implies close similarity between them, putting them in the same family implies rather less similarity and putting them in the same order, class or phylum imply successively lower degrees of similarity. There are no formal definitions of the degrees of similarity required, but experienced zoologists get a feel for them.

Diligent students will quickly find discrepancies between the classifications used in different books. Some of the discrepancies are mere differences of name: for instance, the class of molluscs which includes the clams is called the Bivalvia by some zoologists and the Pelecypoda by others. Such inconsistencies are more or less inevitable because the names of groups above the level of family are not subject to the *International Code*. Other discrepancies reflect differences or changes of opinion. For instances the Polyplacophora (the chitons, or coat-of-mail shells) and the Aplacophora (an obscure group of worm-like molluscs) are listed by some zoologists as separate classes of mollusc but by others, who are more impressed by the similarities between them, as subclasses of a united class Amphineura.

A classification is not right or wrong, but it may be good or bad. The best classification is generally the most useful one. It helps zoologists to marshal their thoughts in the most profitable way, it provides the names they need when they wish to make generalizations, and it does not unnecessarily change old names which they have grown accustomed to use.

It is generally agreed that a classification should not be inconsistent with existing knowledge of the course of evolution. In practice this means that if group *a* is believed to have evolved directly from class (or order or family) *A* and group *b* from class (or order or family) *B*, *a* and *b* will not be put together in the same class (or order or family) *C*. However, if *a* is believed to have evolved from order *x* of class *A* and *b* from order *y* of the same class, it would be permissible to include *a* and *b* in a single class *C*.

Fig. 1.10 will be used to illustrate the way in which these rules are applied. It represents four possible hypotheses about the evolution of the gastropods. Hypothesis (*a*) is the one immediately suggested by the classification of Fig. 1.9: it shows the ancestral gastropods (which are assumed to be members of the Prosobranchia) giving rise at about the same time to the Opisthobranchia and Pulmonata. However, the classification is also consistent with (*b*), which shows the Pulmonata arising before the Opisthobranchia. Further, it is consistent with (*c*) which shows the terrestrial slugs and snails arising separately from different groups of Prosobranchia. However, it is not consistent with (*d*), which shows the snails evolving from Prosobranchia and the terrestrial slugs from the Opisthobranchia. A zoologist who believed that the gastropods had evolved as shown in Fig. 1.10(*d*) ought not to use the classification shown in Fig. 1.9, because this puts together in one subclass the snails (which he believes to have evolved from members of the subclass Prosobranchia) and the terrestrial slugs (which he believes to have evolved from members of the subclass Opisthobranchia).

Our fictitious zoologist who believes that the gastropods evolved as shown in Fig. 1.10(*d*) might argue that the terrestrial slugs and sea slugs (with inconspicuous shells or no shell at all) are so similar to each other that they must be more closely related to each other than either is to the other gastropods (with well-developed shells). Most zoologists would disagree. They would argue that there are a great many close similarities between terrestrial slugs and snails, for instance in the arrangement of eyes and tentacles and in the possession of a lung instead of gills. There are so many features in which terrestrial slugs and snails resemble each other and differ from other gastropods that they must be closely related. Reduction of the shell in slugs and sea slugs does not necessarily indicate relationship; such a simple change could have occurred independently in two lines of evolution, as implied by Figs. 1.10(*a*)–(*c*).

The following list includes all the phyla and subphyla that are mentioned in this book, and also the classes of the subphylum Vertebrata (the vertebrates). More detailed classifications are given at the beginnings of the chapters in which the groups are described. Minor groups that are not described in this book are omitted from the list.

KINGDOM PROTISTA

Phylum Sarcomastigophora,   chapter 2
    Subphylum Mastigophora   *flagellates*
    Subphylum Opalinata
    Subphylum Sarcodina   *amoebas*
Phylum Apicomplexa   chapter 2
Phylum Ciliophora   *ciliates*   chapter 2

KINGDOM ANIMALIA

Phylum Porifera,	chapter 3
Subphylum Symplasma　*glass sponges*	
Subphylum Cellularia　*other sponges*	
Phylum Cnidaria　*corals, jellyfishes etc.*	chapter 3
Phylum Platyhelminthes　*flatworms*	chapter 4
Phylum Nemertea	chapter 4
Phylum Rotifera　*rotifers*	chapter 5
Phylum Nematoda　*roundworms*	chapter 5
Phylum Mollusca　*molluscs*	chapter 6
Phylum Annelida　*segmented worms*	chapter 7
Phylum Chelicerata　*spiders etc.*	chapter 7
Phylum Crustacea　*crabs etc.*	chapter 8
Phylum Uniramia,	
Subphylum Onychophora	chapter 7
Subphylum Myriapoda　*centipedes etc.*	chapter 7
Subphylum Hexapoda　*insects*	chapter 9
Phylum Bryozoa	chapter 10
Phylum Brachiopoda	chapter 10
Phylum Echinodermata　*starfish etc.*	chapter 11
Phylum Chordata,	
Subphylum Urochordata　*sea squirts etc.*	chapter 12
Subphylum Cephalochordata　*amphioxus*	chapter 12
Subphylum Vertebrata,	
Class Agnatha　*lampreys etc.*	chapter 12
Class Acanthodii	chapter 13
Class Placodermi	chapter 13
Class Selachii　*sharks and rays*	chapter 13
Class Osteichthyes　*bony fish*	chapters 14 & 15
Class Amphibia　*frogs, newts etc.*	chapter 15
Class Reptilia　*reptiles*	chapters 16 & 18
Class Aves　*birds*	chapter 17
Class Mammalia　*mammals*	chapter 18

Further Reading

General books on animals

Alexander, R.McN. (1983). *Animal mechanics*, 2nd edn. Blackwell, Oxford.

Schmidt-Nielsen, K. (1983). *Animal physiology: adaptation and environment*, 3rd edn. Cambridge University Press.

Sibly, R.M. & Calow, P. (1986). *Physiological ecology of animals*. Blackwell, Oxford.

Welsch, U. & Storch, V. (1976). *Comparative animal cytology and histology*. Sidgwick & Jackson, London.

General books on invertebrates

Barnes, R.D. (1987). *Invertebrate Zoology*, 5th edn. Saunders, Philadelphia.
Barnes, R.S.K., Calow, P. & Olive, P.J.W. (1988). *The invertebrates: a new synthesis.* Blackwell, Oxford.
Barrington, E.J.W. (1979). *Invertebrate structure and function*, 2nd edn. Nelson, London.

General books on vertebrates

Hildebrand, M., Bramble, D.M., Liem, K.F. & Wake, D.B. (1985). *Functional vertebrate morphology.* Harvard University Press, Cambridge, Massachusetts.
Romer, A.S. & Parsons, T.S. (1986). *The vertebrate body*, 6th edn. Saunders, Philadelphia.
Young, J.Z. (1981). *The life of vertebrates*, 3rd edn. Clarendon Press, Oxford.

Microscopy

Grimstone, A.V. (1976). *The electron microscope in biology*, 2nd edn. Arnold, London.
Kiernan, J.A. (1981). *Histological and histochemical methods: theory and practice.* Pergamon, Oxford.

Chemical analysis

Ralph, R. (1975). *Methods in experimental biology.* Blackie, Glasgow.

Uses of radioactivity

Wolf, G. (1964). *Isotopes in biology.* Academic Press, New York.

Recording events

De Marre, D.A. & Michaels, D. (1983). *Bioelectronic measurements.* Prentice-Hall, Englewood Cliffs, New Jersey.
Turner, A.P.F. (1987). *Biosensors: fundamentals and applications.* Oxford University Press.

Electrodes

Bures, J., Petrán, M. & Zachar, J. (1987). *Electrophysiological methods in biological research*, 3rd edn. Academic Press, New York.
Loeb, G.E. & Gans, C. (1986). *Electromyography for experimentalists.* University of Chicago Press.

Fossils

Carroll, R.L. (1987). *Vertebrate palaeontology and evolution.* Freeman, New York.
Raup, D.M. & Stanley, S.M. (1978). *Principles of palaeontology*, 2nd edn. Freeman, San Francisco.

Concepts in morphology

Alexander, R.McN. (1988). The scope and aims of functional and ecological morphology. *Neth. J. Zool.* **38**, 3–22.

O'Grady, R.T. (1986). Historical processes, evolutionary explanations and problems with teleology, *Can. J. Zool.* **64**, 1010–1020.

van Valen, L. (1982). Homology and causes. *J. Morph.* **173**, 305–312.

Classification

Joysey, K.A. & Friday, A.E. (eds) (1982). *Problems of phylogenetic reconstruction.* Academic Press, London.

Parker, S.P. (ed.) (1982). *Synopsis and classification of living organisms.* McGraw Hill, New York.

2   *Single-celled animals*

Kingdom Protista,
  Phylum Sarcomastigophora,
    Subphylum Mastigophora (flagellates)
    Subphylum Opalinata
    Subphylum Sarcodina (amoebas, foraminiferans etc.)
  Phylum Apicomplexa
  Phylum Ciliophora (ciliates)
  and other phyla

2.1.   Introduction

The characteristic that distinguishes the protozoans from the animals described in later chapters can be expressed in two different ways. Some zoologists like to say that each individual protozoan has just one cell in its body. Others prefer to say that their bodies are not divided into cells, which means much the same thing.

Protozoans are eukaryotes, as also are the many-celled animals and plants: that is to say, each cell has a nucleus and other organelles enclosed by membranes within it. In contrast, prokaryotes such as bacteria have no membrane-enclosed organelles.

Zoologists study protozoans as well as many-celled animals, so I have included them in this book, but they are often excluded from the animal kingdom in formal classifications. The fashion is to form a Kingdom Protista for all the single-celled eukaryotes, whether they have traditionally been regarded as animals (like *Amoeba*) or as plants (like the diatoms). The Kingdom Animalia then consists only of many-celled animals. This arrangement avoids the difficulty that would otherwise arise, in deciding which protistans should be regarded as animals and which as plants. The division may seem obvious in the cases of *Amoeba* and diatoms but there are other, much less obvious, cases.

The Kingdom Protista is divided into phyla, of which three are listed at the beginning of this chapter. These are the groups of protistans traditionally studied by zoologists, who tend to think of them as animals whether or not they are included formally in the Kingdom Animalia. They are known informally as the protozoans, and are the subject of this chapter.

All of them are small. Two of the largest are the giant amoeba *Pelomyxa palustris* and the dinoflagellate *Noctiluca milaris*, each about 2 mm long. One of the smallest is a marine flagellate, *Micromonas pusilla*, which is only 1–1.5 μm long, about the size of a typical bacterium. This is of course a very wide range of sizes. *Pelomyxa* is over 1000 times as long as *Micromonas* and must be at least $(1000)^3 = 10^9$ times as heavy. A 100 t whale is only 5×10^7 times as heavy as a 2 g shrew.

Micromonas is one of the smallest known eukaryotes, but it is not by any means the smallest known organism. Viruses are much smaller but are incapable of independent life: they can grow and multiply only inside a living cell and they depend on the biochemical processes of that cell. However, mycoplasmas only 0.3 μm in diameter have been grown on non-living media, and so shown to be capable of independent life.

The first protistan phylum that I want to introduce is the Sarcomastigophora. Nearly all its members belong either to the subphylum Mastigophora (the flagellates) or to the subphylum Sarcodina (amoebas etc.). The third subphylum, Opalinata, is needed for some peculiar protozoans found in the guts of frogs.

Fig. 2.1 shows a selection of flagellates. Among them, (*a*)–(*c*) are capable of phytosynthesis. Look first at *Chlamydomonas* (Fig. 2.1*a*), which is found in ponds and ditches. The drawing is based on electron microscope sections. The two long projections are the flagella, whose movements propel the organism through the water: section 2.8 of this chapter explains how flagella move. The rounded body of the cell has a rigid polysaccharide cell wall outside the cell membrane. The largest organelle in the cell is the plastid, which is green because it contains chlorophyll, which is used in photosynthesis. A food store of starch is laid down in it. The stigma is a cluster of lipid globules which contrasts with the rest of the plastid because dissolved pigments give it a red colour. The protoplasm in the hollow of the cup-shaped plastid contains the nucleus, mitochondria, Golgi bodies and contractile vacuole. The nucleus of course contains the chromosomes, with the coded genetic information which enables the cell to synthesize its enzymes and other proteins. The mitochondria contain the enzymes required for the Krebs cycle, which completes the oxidation of glucose to carbon dioxide and water and provides energy for vital processes in the cell. The Golgi bodies probably secrete the cell wall, just as the Golgi bodies of the cells of higher plants secrete their cell walls. The contractile vacuole is an organelle which pumps excess water (drawn in osmotically) out of the cell. Its working is discussed in section 2.7.

Euglena (Fig. 2.1*b*) is another green flagellate. It flourishes in pools visited by cattle, or polluted with organic matter in other ways. It has numerous plastids, each with a store of polysaccharide (paramylon, not starch). There are two flagella which have their bases in a pocket at one end of the body but one is very short and does not protrude from the pocket. There is no polysaccharide cell wall but a pellicle which is flexible enough to allow limited changes of shape of the body. It consists of long, narrow strips of a material which is mainly protein. Each strip is linked to its neighbours by a tongue-and-groove arrangement, like floor boards. The strips are not external to the cell membrane like the cell wall of *Chlamydomonas*, but internal to it.

Gymnodinium (Fig. 2.1*c*) is one of the dinoflagellates, a group of flagellates which is abundant in the plankton of seas and lakes. The body is protected by plates of cellulose which are internal to the cell membrane, enclosed in membrane-lined cavities. There are two flagella, of which one lies in a groove running round the equator of the body and the other points posteriorly. There are numerous brown plastids. *Gymnodinium amphora* is a free-living member of the plankton but another species of *Gymnodinium* lives symbiotically within the cells of corals (see section 3.6).

Trypanosoma brucei (Fig. 2.1*d*) is a parasite which causes a dangerous disease,

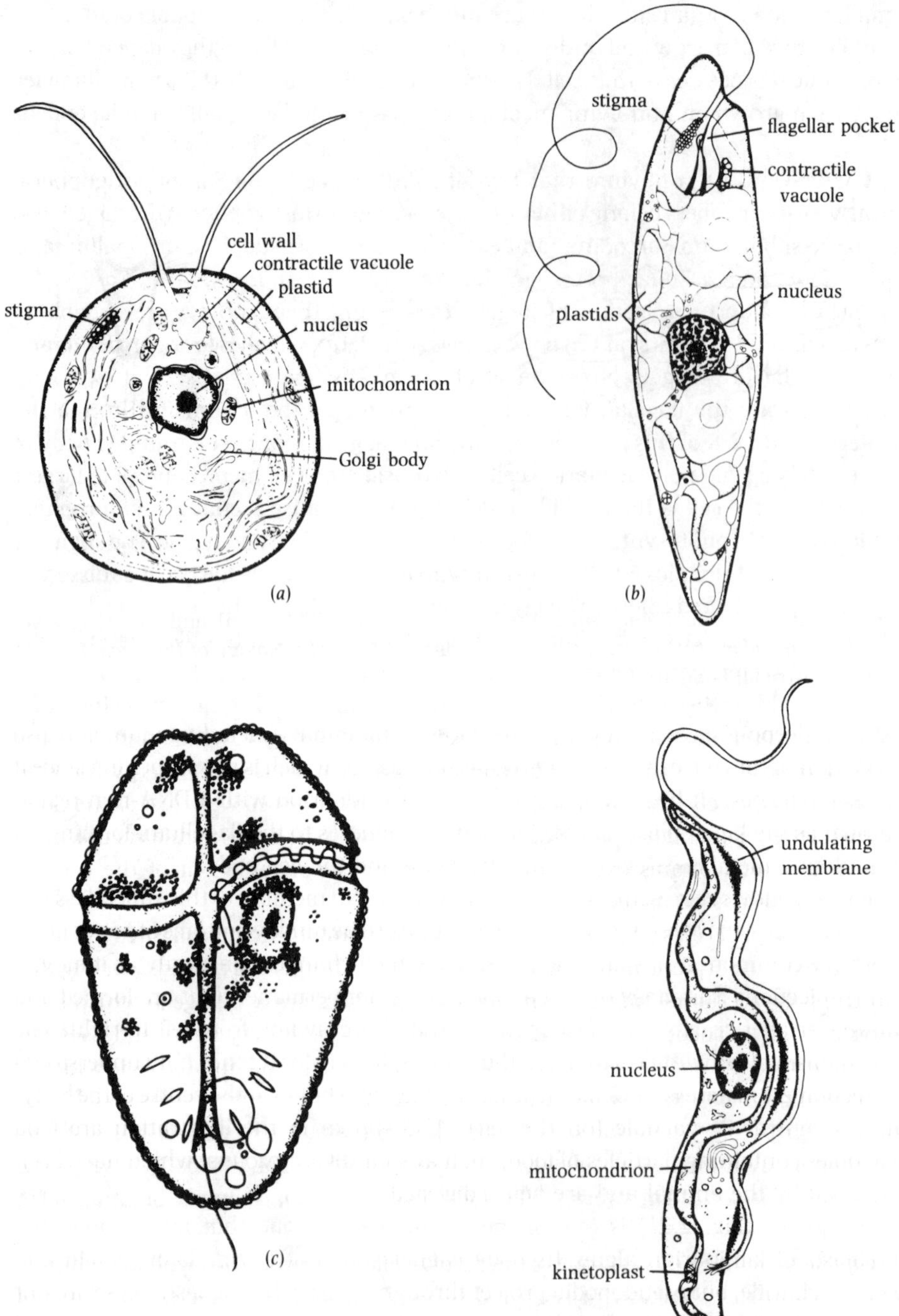

Fig. 2.1. A selection of flagellates. (a) *Chlamydomonoas reinhardi* (length of cell body 10 μm); (b) *Euglena gracilis* (50 μm); (c) *Gymnodinium amphora* (30 μm); and (d) *Trypanosoma brucei* (20 μm). From M.A. Sleigh (1973). *The biology of Protozoa.* Edward Arnold, London.

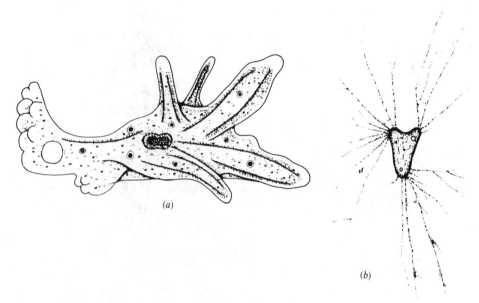

Fig. 2.2. Sarcodina. (*a*) *Amoeba proteus* (length 500 μm) and (*b*) *Allogromia laticollaris* (600 μm). From M.A. Sleigh (1973). *The biology of Protozoa.* Edward Arnold, London.

African sleeping sickness. It lives in the blood of mammals (including man) and also in the gut of the tsetse fly (*Glossina*). It has no plastids, which is not surprising since it lives in darkness. It has a single very long mitochondrion with a DNA-rich region known as the kinetoplast. The cell membrane adheres to the flagellum, forming an undulating membrane between the flagellum and the cell body.

Fig. 2.2 shows representatives of the Sarcodina. No member of this superclass has plastids. *Amoeba proteus* (Fig. 2.2*a*) can be found on the muddy bottoms of ponds but is not very common. It has an irreguar shape which changes constantly as it moves. The projections from its body (pseudopodia) are not permanent but are formed and disappear as it crawls. The specimen illustrated is crawling from left to right. The mechanism of crawling is discussed (but not explained, for it is not fully understood) in section 2.8. There is a nucleus (the large organelle shown at the centre of the body) and a contractile vacuole (on the left). Also shown in the illustration are food vacuoles containing particles of food, such as smaller protozoans, which have been engulfed by the animal and are being digested.

Allogromia (Fig. 2.2*b*) is one of the foraminiferans, one that lives among the holdfasts of laminarian algae. Its body is enclosed in a test made of protein and polysaccharide, but pseudopodia project through openings in the test. These are not stubby lobes like the pseudopodia of *Amoeba*, but slender strands that branch and rejoin to form a network. Protozoans and other small animals, and bacteria, get caught on the network and are transported to the main body for digestion, as described in section 2.8. This particular foraminiferan has an organic test, but many others have their tests reinforced by sand grains or sponge spicules, or by incorporation of calcium carbonate. About one third of the area of the ocean floor is covered

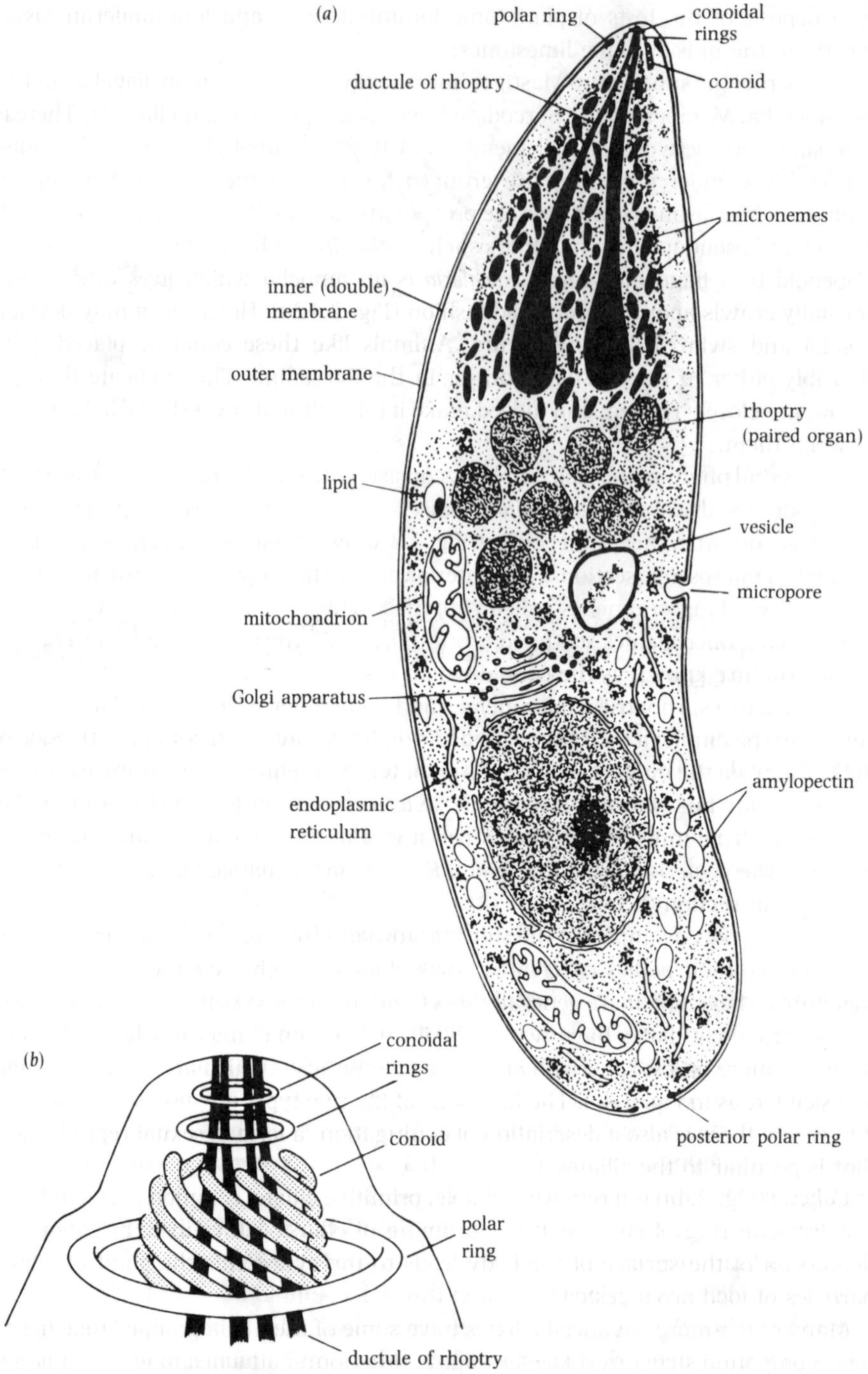

Fig. 2.3. (*a*) A diagrammatic longitudinal section of *Eimeria perforans*, a sporozoan parasite of rabbits. This is the stage in the life cycle known as the merozoite (see Fig. 2.9). From E. Scholtyseck (1973). In *The Coccidia* (ed. D.M. Hammond & P.L. Long). University Park Press, Baltimore. (*b*) A diagram of a conoid in the protruded position.

by a deposit of the tests of planktonic foraminiferans, and foraminiferan fossils constitute the bulk of some limestones.

Members of the subphylum Mastigophora mostly have one or more flagella, and no pseudopodia. Members of the Sarcodina have pseudopodia and no flagella. There is little similarity between typical members of the two subphyla, and readers may reasonably wonder why they are grouped together in the phylum Sarcomastigophora. The reason is that there are protozoans such as *Pedinella* which have both flagella and pseudopodia, and others such as *Naegleria* which may change from an amoeboid to a flagellated form. *Naegleria* is an amoeba which lives in soil, and normally crawls about in amoeboid fashion (Fig. 2.12*a*). However, it may develop flagella and swim if flooding occurs. Animals like these could be placed quite plausibly either in the Mastigophora or in the Sarcodina. They indicate that the groups are closely related, and indeed make it difficult to draw a sharp dividing line between them.

The second phylum, the Apicomplexa, consists entirely of parasites, some of which cause serious diseases. *Plasmodium* causes malaria and *Eimeria* species cause coccidiosis of cattle, sheep and poultry. Fig. 2.3 shows the structure of *Eimeria* as seen in electron microscope sections. The anterior part of the body (at the top of the figure) contains two long organelles called rhoptries, which look like glands, with small bodies called micronemes scattered around them. At the extreme anterior end is a small structure known as the conoid which is shown in more detail in Fig. 2.3(*b*). These structures, rhoptries, micronemes and conoid, form the apical complex that gives the Apicomplexa their name. In addition, they have a depression in the side of the body called the micropore. There is an outer cell membrane enclosing the cell in the usual way but within it is a double layer formed from flattened sacs of similar membrane. This inner membrane is perforated at the conoid, the micropore and (usually) the posterior end of the body. *Eimeria* has a complex life cycle, which is described in section 2.5.

Fig. 2.4 shows examples of the third protozoan phylum, the Ciliophora. Typical ciliates have their bodies covered by rows of cilia, which have the same internal structure as flagella but are generally short and numerous (as in Fig. 2.4*a*) and beat asymmetrically. Ciliates have two types of nucleus, small micronuclei (which are generally more or less spherical) and large macronuclei (which are sometimes long and slender, as in Fig. 2.4*c*). The functions of the two types are discussed in section 2.6, where there is also a description of conjugation, a form of sexual reproduction that is peculiar to the ciliates.

Colpoda (Fig. 2.4*a*) is a relatively simple, primitive ciliate, common in soil. It has a uniform covering of cilia, with no grouping of cilia together in tight clumps. A depression of the surface of the body leads to the cytostome, the point at which particles of food are ingested.

Many of the more advanced ciliates have some of their cilia grouped together to form compound structures. *Stentor* (Fig. 2.4*b*) is found attached to weeds in ponds and slow streams. It is trumpet-shaped, with its cytostome in the bell of the trumpet. There are ordinary cilia on the general body surface and in addition groups of about 70 long (25 μm) cilia form membranelles around the rim of the trumpet. *Stentor* is very large and individual cilia are not shown in the figure: each line in the fringe

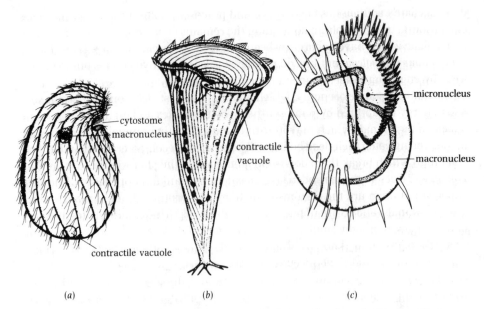

Fig. 2.4. A selection of ciliates. (*a*) *Colpoda cucullus* (length 75 μm); (*b*) *Stentor coeruleus* (1.5 mm); (*c*) *Euplotes patella* (100 μm). From M.A. Sleigh (1973). *The biology of Protozoa*. Edward Arnold, London.

around the rim represents a membranelle. Each membranelle beats as a unit but it is not known what prevents the cilia from beating separately or fraying apart. Electron micrographs of sections of membranelles show that the cilia are packed closely, almost touching each other, but do not show anything holding them together. *Stentor* can contract rapidly, shortening its body to as little as a quarter of its extended length.

Euplotes (Fig. 2.4*c*), which is common in various marine and freshwater habitats, has no ordinary cilia on its ventral surface. All of its ventral cilia are grouped together in membranelles or similar structures. Each membranelle consists of two or three closely packed rows of cilia. Much of the surface of the body is bare.

The rest of this chapter is about various aspects of the Protista that seem interesting and important. First, a series of sections describe how they live and feed: there are sections about free-living protozoans and about parasites that live in the gut, in the blood and in the cells of their hosts. They are followed by sections on sex and reproduction, on osmotic regulation and on locomotion.

2.2.   Photosynthesis and feeding

Many flagellates have plastids and produce foodstuffs by photosynthesis. They thus resemble plants, and indeed the higher plants are believed to have evolved from ancestors resembling *Chlamydomonas*. Some other flagellates get their energy as animals do, by eating other organisms. Amoebas and ciliates also feed like animals.

The distinction between plant-like and animal-like flagellates is blurred. The dinoflagellate *Ceratium* has chlorophyll and practises photosynthesis, but vacuoles can be found in its cytoplasm containing the remains of bacteria, diatoms and blue-green algae which have apparently been engulfed. *Peranema* is a flagellate which is very similar to *Euglena*, but has no plastids. *Euglena* photosynthesizes but *Peranema* feeds by engulfing other organisms (including *Euglena*).

The ultimate source of energy, which maintains almost all life on earth, is the sun. Solar energy is captured by photosynthesis, by flagellates and higher plants. These are eaten by animals which may in turn be eaten by other animals, and if they die without being eaten their dead bodies will nourish the decomposers (such as bacteria and fungi) which break them down. On land, most of the photosynthesis is done by grasses and trees. In the sea and in lakes, it is done by planktonic organisms. The most important of them are the diatoms (which are always thought of as plants rather than as animals, and so are not described in this book), the dinoflagellates and very small flagellates broadly similar to *Chlamydomonas*. These photosynthesizing members of the plankton are known collectively as the phytoplankton, to distinguish them from the zooplankton which feed in animal fashion.

Some of the energy captured by photosynthesis is used by the photosynthesizing organisms for their own metabolism, to supply the energy they need for the maintenance of life. The remainder is used for growth or to build up food stores. The chemical energy accumulated in this way is referred to as the net primary production, and is energy potentially available to herbivores. It can be expressed in joules per square metre per year ($J\, m^{-2}\, a^{-1}$; the abbreviation a^{-1} means 'per annum'). Marine plankton and freshwater plankton in unpolluted waters achieve in favourable conditions a net primary production of about $4\, MJ\, m^{-2}\, a^{-1}$ (representing about 200 g dry organic matter $m^{-2}\, a^{-1}$). However, polluted waters produce more, and as much as $90\, MJ\, m^{-2}\, a^{-1}$ has been achieved by cultivating algae in sewage ponds in California. This is comparable to the highest net primary production which can be obtained on land. Good agricultural grassland in New Zealand can produce $60\, MJ$ $m^{-2}\, a^{-1}$ and tropical rain forest and sugar cane plantations can each produce about $120\, MJ\, m^{-2}\, a^{-1}$.

If the energy which reaches the earth's atmosphere passed through it undepleted, around $10\, GJ$ ($10\,000\, MJ$) would fall on each square metre of level ground or water in the course of a year (rather more near the equator, and less near the poles). Much of this energy is absorbed or scattered by the atmosphere but far more reaches the ground than can be captured by photosynthesis. Even in the most favourable circumstances crops and plankton seem unable to capture more than about 3% of the radiation energy which falls on them.

The net primary production by plankton in unpolluted waters is much less than is possible in sewage ponds, with the same supply of light. It seems generally to be limited in unpolluted waters by the supplies of inorganic salts. Nitrate or an alternative nitrogen source is required to supply the nitrogen for protein synthesis. Phosphate is required for the synthesis of phospholipids for cell membranes and for many other essential constituents of cells. Diatoms need silicate, to build their silica tests. In temperate and polar lakes and seas nitrate, phosphate and silicate concentrations may fall from relatively high winter values to very low summer values.

The annual cycle of plankton population changes is closely linked to the phenom-

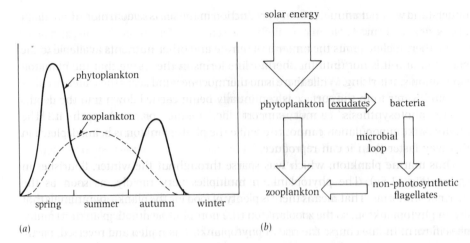

Fig. 2.5.(*a*) A schematic graph showing how phytoplankton and zooplankton biomass fluctuates in the course of a year in temperate seas. (*b*) A diagram representing energy flow through the plankton in summer.

enon of thermal stratification. In spring, the temperature of surface waters is rising. They become warmer than the deeper water below them, and less dense than it. The difference in density tends to prevent mixing. Winds cause mixing down to a certain depth but at that depth there is a rather sharp change of temperature (the thermocline). The depth of the thermocline depends on the place and the season. It is often about 20 m from the surface in the English Channel, but may be as deep as 100 m in the North Atlantic. The thermocline persists through the summer but in the autumn the surface waters become cooler and denser, and are mixed with the deep water by convection.

The photosynthetic members of the plankton (the phytoplankton) can only grow and multiply where the light intensity is high enough for them to photosynthesize faster than they use energy in respiration. Thus they cannot survive for long in the perpetual darkness of the depths, and need to be reasonably near the surface. They can live deeper in clear oceanic water than in turbid coastal water, but it seems to be a general rule that little phytoplankton is found anywhere below the thermocline.

Consider an area where the water at the beginning of spring contains 0.1 mg nitrogen (as nitrate) per litre and where a thermocline forms at a depth of 50 m. These data are reasonably typical for many areas in the sea. There would be 100 mg available nitrogen in each cubic metre of water, and the total mass of available nitrogen between the surface and the thermocline would be 5 g m^{-2}. This is enough to make about 30 g protein m^{-2} or perhaps about 100 g (dry mass of plankton) m^{-2}. If no mixing occurred across the thermocline there could never be more than about this amount of plankton. The limit would be different if the initial concentration of available nitrogen, or the depth of the thermocline, were different. This is a limit to the amount of plankton which could be present at any one time. Total primary production during the season may be more than this because some mixing of water occurs across the thermocline and because decay of dead plankton makes its nitrogen available again. Mixing and decay are rather slow. This calculation helps us to understand why net annual primary production in the sea is seldom more than about 200 g dry mass m^{-2}.

The thermocline limits the amount of nitrate and other nutrients available to the plankton, but it is not until the thermocline forms in the spring that the plankton population starts rising. While there is no thermocline wind keeps the water circulating, and water from the surface is continually being carried down into the depths where photosynthesis cannot support life, taking plankton with it. The phytoplankton population cannot rise while the phytoplankton is being depleted in this way faster than it can reproduce.

Thus marine plankton, which was sparse throughout the winter, flourishes in spring (Fig. 2.5a). The phytoplankton multiplies very rapidly, as soon as the thermocline forms. That means there is plenty of food for the planktonic animals that fed on phytoplankton, so the zooplankton (the non-photosynthetic plankton) multiplies in turn. In due course the rise in phytoplankton is halted and reversed, partly because the phytoplankton are being eaten by zooplankton but largely because nitrate, phosphate and silicate have become scarce. Later still the zooplankton declines for lack of food. In temperate seas there is often a brief second peak of phytoplankton population, but the plankton declines again, as the thermocline breaks down at the beginning of winter.

In summer, phytoplankton is still fairly plentiful but nutrients are scarce. The plankton produces organic compounds by photosynthesis but cannot use them for growth because of the lack of nitrates to make proteins and of phosphates to make cell membranes etc. Excess organic compounds, including glycollic acid ($CH_2OH.COOH$), are exuded by the phytoplankton cells and serve as food for bacteria in the surrounding water. These bacteria multiply and are eaten by small non-photosynthetic flagellates which in turn are eaten by other members of the zooplankton, especially by ciliate protozoans. Thus as well as feeding on the phytoplankton, the zooplankton get food from them indirectly via the 'microbial loop' (Fig. 2.5b).

It is difficult to measure directly the rate at which organic matter is exuded by the phytoplankton, because it gets used up by the bacteria, but it has been estimated from the rate of multiplication of the bacteria that in summer up to 60% of the phytoplankton production may be lost as exudates.

Thus three categories of protozoan are important in marine plankton: the photosynthetic flagellates, the small bacteria-eating flagellates and the ciliates that eat both. Other important members of the plankton are discussed in sections 6.12 and 8.5.

We have not yet asked how the non-photosynthetic protozoans swallow and digest their prey. Fig. 2.6(a) shows how amoebas feed. A hollow forms in the protozoan's surface, next to the food, and its edges close in so that the food (if small enough) is enclosed in a vacuole. If the food is too large to be engulfed whole, the 'lips' closing round it may bite a chunk off. This simple method of catching food is astonishingly effective. Add some fast-swimming ciliates to a dish containing *Amoeba* and within 10 minutes or so there are ciliates in vacuoles inside the amoebas. Granules called lysosomes coalesce with the food vacuole, releasing digestive enzymes into it. The prey is digested and eventually its indigestible remains are discarded by bringing the vacuole to the cell surface and reversing the process which formed it. Amoebas may

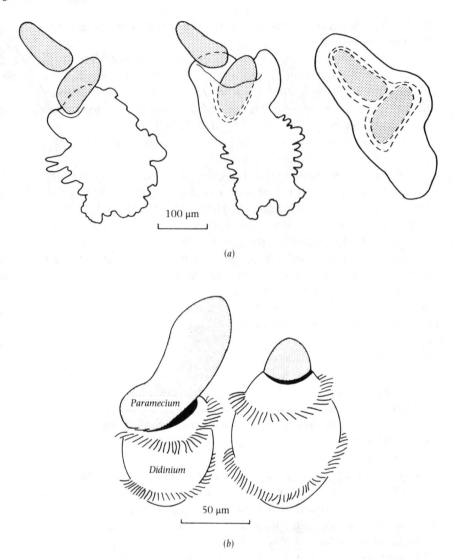

Fig. 2.6.(*a*) Outlines from a film of *Amoeba* ingesting two *Paramecium*. After K.G. Grell (1973) *Protozoology*; Springer, Berlin. (*b*) Sketches of *Didinium* ingesting *Paramecium*, based on scanning electron micrographs by H. Wessenberg & G. Antipa (1970). *J. Protozool.* 17, 250–270.

form food vacuoles at any point on the cell surface but ciliates such as *Colpoda* (Fig. 2.4) have a cytostome which serves as a mouth.

Fig. 2.6(*b*) shows a ciliate feeding. This particular one (*Didinium*) eats other ciliates (such as *Paramecium*) which may be larger than itself. It collides with a *Paramecium*, bumping into it with its pointed snout. During the moment of contact its snout discharges trichocysts. These are threads which penetrate the *Paramecium* but remain firmly fixed at the other end to the *Didinium*: they thus attach the animals to each other. Once this has happened the *Paramecium* stops swimming, and it dies in a few minutes even if it is not ingested. A cavity opens around the point of attachment

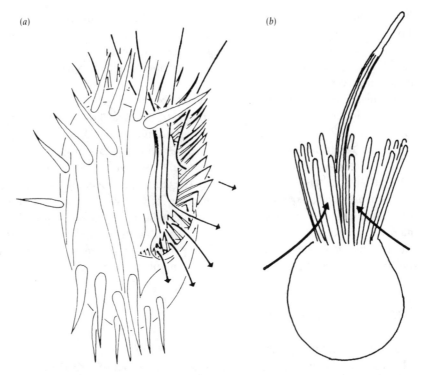

Fig. 2.7. Two filter-feeding protozoans with arrows showing the water currents they produce. (*a*) is the ciliate *Euplotes* (length 100 μm) and (*b*) is the choanoflagellate *Monosiga* (diameter 2 μm). Part (*a*) is from T. Fenchel (1986). *Prog. Protistol.* 1, 65–113.

of the *Paramecium*, which is drawn down into it. Though the victim is initially larger than the predator it is believed that its volume may be reduced during ingestion by removing water, which the predator gets rid of by means of its contractile vacuole. Ingestion may be completed in a minute but two hours are needed for digestion.

The movements that protozoans use to ingest prey are probably powered in the same sort of way as the creeping movements of some protozoans. Electron micrographs of amoebas show a network of thin filaments in the protoplasm of 'lips' closing around prey. Some ciliates have structures, built of bundles of microtubules, around their cytostomes. Section 2.8 of this chapter tells how microfilaments may be involved in amoeboid crawling and microtubules in the 'gliding' of sporozoans such as *Eimeria*.

Didinium captures large prey but many other ciliates eat bacteria and other food items that are tiny compared to their own bodies. *Stentor* and *Euplotes* (Fig. 2.4*b*,*c*) both do this. They set up currents in the water by beating their membranelles, and filter out small particles of food from the flowing water. Fig. 2.7(*a*) shows how the water flows over *Euplotes*: the information was obtained by studying video recordings, made through a microscope, of *Euplotes* in a suspension of tiny plastic spheres. The water passes through the gaps between successive membranelles at speeds of 0.5–1.0 mm s^{-1}. The membranelles are about 2 μm apart, so particles larger than this get caught. (Experiments with suspensions of plastic spheres of different sizes show that some 2 μm spheres are caught, but 4–5 μm spheres are caught more effectively.) Once caught, particles are moved into the cytostome (it is not clear how) and ingested. Food particles and indigestible plastic spheres are ingested indiscriminately.

The membranelles beat at a frequency of about 50 Hz (cycles per second) but the details of their movements can be seen under stroboscopic illumination, using a light flashing at a slightly lower frequency. They undulate in such a way as to squeeze water through the gaps between them, as if each gap was a tiny peristaltic pump.

Fig. 2.7(*b*) shows a flagellate that feeds in similar fashion, by straining out particles of food from a water current. In this case the current is driven by the beating of the flagellum and food is strained out by the 'collar' of microvilli that surrounds the base of the flagellum. Water flows between the microvilli at only about 30 μm s⁻², but the microvilli are so close together (about 0.3 μm) that very tiny food particles can be caught. The choanoflagellates, of which this is an example, are the most important of the small, bacteria-eating flagellates in the plankton (Fig. 2.5*b*).

2.3.   Gut-living protozoans

This and the next few sections are about protozoans that live in or on the bodies of other animals. They are parasites, in the broad sense of the word. However, the term 'parasite' is often restricted to organisms which harm their host by feeding on its tissues or by robbing it of food without giving a compensating advantage. Protozoans such as *Trypanosoma* and *Eimeria*, that cause disease, are parasites in this narrow sense, but the ciliates that live in the stomachs of cattle are not: the cattle benefit greatly from having them, as will be explained.

It is customary to distinguish between ectoparasites, which live on the outer surfaces of their hosts, and endoparasites, which live inside their hosts' bodies. There are many ectoparasitic protozoans. One example is *Brooklynella*, a ciliate that lives on the gills of fishes. It crawls over the gills but has a cavity in its body, opening by a pore, from which it secretes a substance which is suspected of being an adhesive. Perhaps it sticks itself temporarily to the gills and feeds there for a while, before releasing itself and moving on. It feeds on the epithelial cells of the gills, and may cause inflammation, bleeding and the death of the host. It has been a very troublesome pest in at least one public aquarium where it infested a variety of marine fish. The other parasitic protozoans described in this book are endoparasites.

The rest of this section is about protozoans that live in the guts of vertebrates, starting with the ciliates such as *Entodinium* which live in the stomachs of ruminant animals (cattle, deer, etc.). They live in the rumen, the largest of the several chambers of the stomach, where bacteria are also plentiful. No hydrochloric acid or gastric enzymes are secreted into the rumen, so the pH is close to neutrality and the microorganisms are not digested while they remain there.

The rumen is continually being filled with food from the mouth and emptied into the more posterior parts of the gut. Bacteria and ciliates carried out with the food are digested in the intestines, so the populations in the rumen must reproduce fast enough to compensate for the losses. Suppose that the volume of the rumen is *V* and that it contains C ciliates per unit volume, so that the total number of ciliates is VC. These ciliates reproduce at a rate r, so if there were no losses the population would increase at a rate rVC. These rates have been measured in cultures of rumen ciliates kept in appropriate conditions. Different rates have been found for different species but a typical value of r is 0.04 h^{-1}. That means that, without losses, there would be 1.04 times as many ciliates at the end of an hour as at the beginning, and numbers would double in 18 hours ($1.04^{18} = 2$).

However, rumen contents leave at a rate R (that is the volume leaving in unit time) and you might suppose that they might carry ciliates away at a rate RC. In that case, for the gains to balance the losses,

$$rVC = RC,$$
$$r = R/V.$$

The relative rate of loss of rumen contents R/V has been measured in various ways, and values around 0.07 h^{-1} have been obtained for cattle on normal rations. This is greater than the typical value of r (0.04 h^{-1}) so you might suppose that the ciliate population in the rumen would dwindle. However, most of the ciliates are probably closely associated with the solid matter in the rumen, which passes through much more slowly: it is changed at rates around 0.02 h^{-1} when the cattle are eating hay. Ground and pelleted food passes through the rumen faster than this, and cattle that are fed on it are liable to lose their ciliates.

The ruminants feed almost entirely on plants which contain a lot of the cell-wall materials cellulose and hemicellulose. For example, these materials constitute about 65% of the organic content of mature ryegrass. The ruminants produce no enzymes capable of digesting them, but the ciliates and bacteria do.

Foodstuffs which are used by the rumen bacteria and ciliates for growth become available to their host in due course, when the micro-organisms are carried through to more posterior parts of the gut and digested. Energy which they use for metabolism is lost to the host, but they can only use a small fraction of the energy of the food passing through the rumen because the partial pressure of oxygen in the rumen is extremely low. They convert cellulose and other carbohydrates to fatty acids such as acetic acid by reactions such as

$$C_6H_{12}O_6 = 2CH_3COOH + CO_2 + CH_4.$$

Note that this requires no oxygen, but produces carbon dioxide and methane. The heat of combustion of one mole of glucose is 2.9 MJ; that of the two moles of acetic acid which would be formed from it in this reaction is 1.8 MJ. This acetic acid is available to the ruminant. Thus the ruminant loses only a fraction of the energy content of its food to the micro-organisms. It even gets most of the energy from the cellulose and hemicellulose which it could not itself digest. Both the micro-organisms and the ruminant benefit from their association.

Ruminants have no bacteria or ciliates in the rumen when they are born, but they quickly acquire them. The habit of chewing the cud brings rumen contents (including the micro-organisms) into the mother's mouth. When she licks and grooms her young it is likely to swallow a little of her saliva and so acquire enough bacteria and ciliates to start a population in the rumen. Also, it may swallow small quantities of another animal's saliva and micro-organisms accidentally by grazing beside it. Rumen ciliates have been found on the grass in a sheep pasture.

Some other protozoans are capable of passing through the guts of mammals without being digested. They are transmitted via faeces which contaminate the food of new hosts. An example is *Entamoeba histolytica*, which causes amoebic dysentery in man. It lives in the large intestine, which is also inhabited by bacteria. The bacteria feed on such remnants of food as reach the large intestine without having been digested, and the *Entamoeba* feed on them. While it lives in the lumen of the intestine *Entamoeba* does no harm but it invades the gut wall and sometimes other tissues, feeding on them and causing disease. The response of the large intestine to damage is diarrhoea. This may seem inconvenient to the sufferer, but is an appropriate response because it increases the relative rate of loss of intestine contents (R/V). If this exceeds the rate of reproduction of the amoebas (r) the population of amoebas in the lumen of the intestine must fall, though parasites may remain in the gut wall.

When the host is healthy he or she forms normal faeces, removing water from them in the large intestine. *Entamoeba* in the faeces respond to dehydration by enclosing themselves in protective cysts. These survive in the faeces and may enter another person who eats contaminated food or drinks contaminated water. Food may be contaminated by handlers who take insufficient care over the cleanliness of their hands, or by flies which land on food after visiting faeces. The *Entamoeba* emerge from their cysts in the small intestine of their new host.

2.4. Blood parasites

Many parasites spend at least part of their life cycle in the blood of vertebrates. Blood-sucking insects carry many of them from one host to another. The malaria parasites (*Plasmodium*) are transmitted by mosquitoes. African sleeping sickness is transmitted by the tsetse flies (*Glossina*) in essentially the same way. It is caused by *Trypanosoma brucei* (Fig. 2.1d) which lives in the blood plasma of people, antelopes and other mammals. The trypanosomes are swallowed by tsetse flies when they take meals of blood, and multiply in the flies' guts. They infect the salivary glands and so are likely to be injected with saliva into a new host. Another trypanosome, *Trypanosoma cruzi*, causes Chagas' disease, which is important in S. America. It parasitizes dogs, cats, opossums and monkeys as well as people and is transmitted by blood-sucking bugs including *Rhodnius prolixus*. It establishes itself in the gut of the bug and is passed in the faeces. If the bug defecates while feeding on a man the faeces may get into the bite or other abrasions and infect him with the disease.

Trypanosoma brucei occupies very different habitats, in the blood of mammals and in the gut of the tsetse fly. It has to adapt itself to each, as it is transferred from one to the other. The bloodstream form (Fig. 2.1d) has long but unbranched mitochondria, and the cell membrane is covered by an external coat about 15 nm thick. In the tsetse fly the mitochondria are larger and more elaborate in shape, and the surface coat is lost. The trypanosomes change back to the bloodstream form in the insect's salivary glands. Trypanosomes can be kept in culture as well as in host animals. They swim free in the blood, but attach themselves by their flagella to the wall of the tsetse fly's gut. The attachment is by close apposition, depending on forces like those that hold cells together in tissues (section 3.3).

There is a difference in metabolism as well as in appearance between the bloodstream and tsetse fly forms of trypanosomes. This has been demonstrated by experiments with suspensions of trypanosomes from the blood of rats and from cultures of the tsetse fly form. The bloodstream forms were separated from blood corpuscles by centrifugation. The rates of oxygen consumption of both forms were measured in solutions of various foodstuffs, and the solutions were analysed afterwards to discover how much of each foodstuff was used and what substances were produced.

Both forms ceased moving and stopped using oxygen in solutions containing no foodstuffs; they seem to have no food reserves in their bodies. They could use glycerol, glucose and certain other sugars. The tsetse form could also use succinate and other substances involved in the Krebs cycle, but the bloodstream form could not. The tsetse form oxidizes most of the glucose it uses completely, to carbon dioxide and water:

$$C_6H_{12}O_6 + 6O_2 = 6CO_2 + 6H_2O + 38(\sim),$$

where $(\sim)$ represents an energy-rich phosphate bond. However, the bloodstream form cannot oxidize glucose completely, but even in the presence of ample oxygen converts it mainly to pyruvic acid

$$C_6H_{12}O_6 + O_2 = 2CH_3CO.COOH + 2H_2O + 8(\sim),$$

It apparently lacks the enzymes of the Krebs cycle, which oxidize pyruvic acid to carbon dioxide and are normally located in mitochondria. This may explain why the mitochondria of the bloodstream form are relatively small.

When glucose is oxidized to carbon dioxide, 6 molecules of oxygen yield 38 energy-rich bonds, 6.3 bonds per molecule. When it is oxidized to pyruvic acid each molecule of oxygen yields 8 energy-rich bonds, so slightly less oxygen is needed to obtain a given amount of energy (though very much more glucose is used). The partial pressure of oxygen is high in arteries but may be quite low in veins, so there may be some advantage to the bloodstream form in using oxygen economically. In any case, glucose is very plentiful in mammal blood so there is no need to use it economically. In the gut of the tsetse fly glucose and alternative foods are probably not as constantly plentiful, so there is an advantage to the parasite in using them economically. The partial pressure of oxygen in the guts of tsetse flies is not known, but is probably much higher than in the guts of vertebrates. Otherwise the culture (and tsetse fly) form of the trypanosome would presumably not have the ability to oxidize glucose completely.

Parasites such as trypanosomes that live in the blood (or other tissues) of vertebrates have to withstand the immune responses of the host, which tend to destroy them and any other foreign cells which enter the body. The vertebrate host produces proteins known as antibodies which react specifically with particular molecules on the surfaces of the foreign cells, but not with the surfaces of host cells. The foreign molecules against which antibodies are produced are known as antigens. Antibodies attack foreign cells in various ways. Some agglutinate them (stick them together in clumps). Others make them swell and burst. These antibodies attach complement (a group of proteins) to the cell membrane, and the complement apparently makes small holes in the membrane by enzyme action. Ions can diffuse through these holes but the large organic molecules of the cell cannot. This destroys the osmotic control mechanism which will be described in section 2.7, and the cell swells up and bursts. Foreign cells may also be engulfed and digested by macrophages, much as *Amoeba* engulfs food. The macrophages seem to 'recognize' the foreign cells by antibody attached to them.

All these processes involve antibodies synthesized by the host to attack the particular foreign cells which have invaded it. They are produced by the reticuloendothelial cells in the spleen, lymph nodes and bone marrow and can be produced quite quickly, within a few days. The host also responds to foreign cells by producing more macrophages in the spleen and lymph nodes, and releasing them into the blood. Parasites that flourish and multiply rapidly when they first infect a host may be eliminated suddenly when the immune response becomes effective.

In African sleeping sickness, instead of simply rising and then falling, the parasite population fluctuates at intervals of a few days. The patient's symptoms also fluctuate. Partial recoveries, during which the number of trypanosomes in the blood is quite low, alternate with relapses during which the number is much higher. It seems that the immune responses of the host are relatively effective during the recovery phases and ineffective during the relapses. The host's antibodies act against the macromolecule of the external coat of the trypanosomes. The trypanosomes alter the chemical constitution of this coat periodically, so that the antibodies produced by the host against previous coats are no longer effective. A relapse occurs in the interval between adoption of a new coat and production of new antibody to attack it. Each change of coat gives the parasite temporary respite from the attacks of host antibodies.

Evidence for this comes from experiments with goats and rabbits. The animals (which were initially uninfected) were infected by a single bite of a tsetse fly. Blood samples were taken at intervals of a few days and used to prepare samples of serum and trypanosomes. The trypanosomes were rather sparse in the blood of the goats and rabbits, so larger numbers were obtained by injecting samples of the blood into mice. The trypanosomes multiplied in the mice. Blood from the first mouse was used to infect a second one after 2–3 days, a third mouse was infected from the second 2–3 days after that, and so on. In this way blood containing very large numbers of trypanosomes was obtained, without allowing enough time in any one mouse for that mouse to produce antibodies or for the trypanosome to change its coat. Trypanosomes were separated from the mouse blood by centrifugation and suspended in a saline solution. The goat and rabbit serum and the suspensions of trypanosomes were stored deep frozen until they were required for agglutination tests.

In each test, a drop of trypanosome suspension and a drop of diluted serum were mixed on a microscope slide. They were examined half an hour later to see whether the trypanosomes had formed clumps. If they had, the serum must have contained antibodies effective against the surface coats of those particular trypanosomes. The tests were repeated with more and more dilute sera, to assess the concentration of the antibodies, which was expressed as the titre. If the trypanosomes were clumped by serum diluted to 10 times its initial volume, but not by serum diluted 20 times, the titre was 10. If another sample of serum agglutinated the trypanosomes when diluted 1280 times but not 2560 times, its titre was 1280. A high titre indicates a high concentration of antibodies.

Fig. 2.8 shows some of the results of these experiments. In (*a*) are the results of an

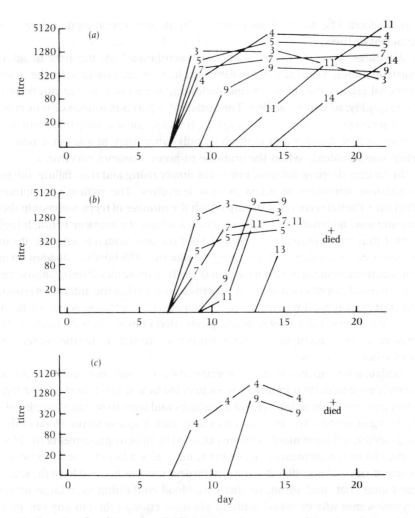

Fig. 2.8. Graphs of titre of antibodies against *Trypanosoma brucei*, against number of days since initial infection. (*a*) Titres of antibodies in the serum of a rabbit to trypanosomes from the same rabbit; (*b*) titres in a goat to trypanosomes from the goat; and (*c*) titres in the goat to trypanosomes from the rabbit. In each case separate lines show titres to trypanosomes taken from the host on the days indicated by numbers on the lines; for instance, the line 3–3 refers to trypanosomes taken on day 3. Re-drawn from A.R. Gray (1965). *J. gen Microbiol.* **41**, 195–214.

experiment with a rabbit. Samples of blood were taken every few days and serum from each sample was tested for antibodies against samples of trypanosomes taken from the same animal on the same or different days. Up to 7 days after infection, no evidence of antibodies against any of the trypanosomes could be found. Serum from day 9 had high titres of antibody effective against trypanosomes from days 3, 4, 5 and 7 but failed to agglutinate trypanosomes from day 9 or later days. By day 11, antibodies against day 9 trypanosomes had appeared. Antibodies against day 11 and day 14 trypanosomes first appeared on days 14 and 18, respectively.

These results can be explained as follows. The trypanosomes injected by the tsetse fly had antigen A in their surface coats. No antibodies were produced by the rabbit until between days 7 and 9 when antibody A (effective against antigen A) was produced. By day 9, however, the trypanosomes had changed their antigen to B so that antibody A was no longer effective against them. By day 11 the blood had antibody B as well as antibody A, but the trypanosomes had changed to antigen C. On day 14 the blood had antibodies A, B and C but the trypanosome had changed again to antigen D.

Fig. 2.8(*b*) shows very similar results from an experiment with a goat. Successive antibodies were detected on days 8, 11 and 15. Fig. 2.8(*c*) shows the results of testing serum from the goat for antibodies against trypanosomes from the rabbit. The antibody which was first detected on day 8 was effective against day 4 trypanosomes from the rabbit and the one detected on day 11 against day 9 trypanosomes from the rabbit. It seems that the trypanosomes injected into the rabbit and goat (by different tsetse flies) both had the same antigen A, and that the first change of antigen in each case was the same antigen B. Further experiments with rabbits and goats confirmed what this one suggests, that *Trypanosoma brucei* generally goes through the same sequence of antigen changes (A, B, C, etc.) in mammal hosts but reverts to A when ingested by a tsetse fly.

In more recent research it has been possible to isolate from trypanosomes the RNA responsible for the synthesis of their protective coats. The corresponding DNA has been prepared and sequenced and used to locate the matching DNA in the chromosomes. It has been shown that the variable part of the coat molecule is a peptide chain, 360 amino acids long, so the scope for variation is enormous. Trypanosomes carry genes for at least 100 different surface coats, but they may have many more. Each individual trypanosome has genes for all the alternatives, but it has only one of them in its surface coat at any time: only one of the genes is expressed.

We do not yet know how just one of the genes is selected for expression, but the research has shown that in many cases an extra copy is made of that gene and incorporated in a chromosome near one of its ends. The RNA is made from the extra copy, not from the original one.

2.5.   Intracellular parasites

Apicomplexa spend part of their life cycle inside the cells of their host. *Plasmodium* (the cause of malaria) get into red blood corpuscles. Once inside they feed partly on haemoglobin (which they take in through the micropore) and partly on glucose and other foodstuffs which diffuse in from the blood plasma.

Infected corpuscles are less dense than uninfected ones and so can be separated almost completely from them by repeated centrifugation of blood from an infected animal. When they are separated in this way it can be shown that infected corpuscles use oxygen and glucose many times faster than uninfected ones, presumably owing to the metabolism of the parasites.

The parasites apparently increase the permeability of the red cell membrane, making it easier for foodstuffs to diffuse in. This has been demonstrated by experiments with infected and uninfected cells from the same animal, separated by

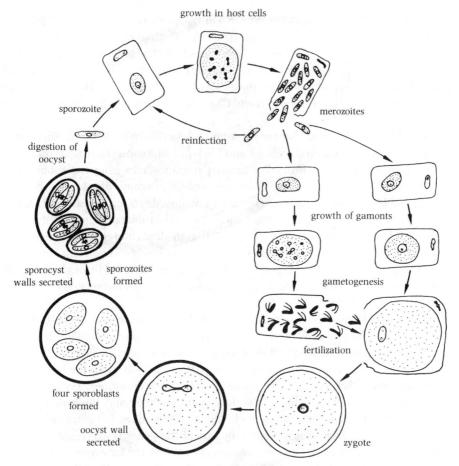

growth in host cells

sporozoite

merozoites

digestion of oocyst

reinfection

sporocyst walls secreted

sporozoites formed

growth of gamonts

gametogenesis

four sporoblasts formed

fertilization

oocyst wall secreted

zygote

Fig. 2.9. Stages in the life cycle of *Eimeria steidae*, a parasite of rabbits. From M.A. Sleigh (1973). *The biology of Protozoa*. Edward Arnold, London.

centrifugation. The cells were suspended for a short time in a solution of L-glucose labelled with ^{14}C. They were centrifuged out again, and the radioactivity of the pellet was measured. The pellet retained a little of the solution between the cells, but the amount was measured and corrected for. It was found that much more L-glucose had entered the infected cells than the uninfected ones. The parasites had apparently made the cell membrane much more permeable to L-glucose and presumably also to D-glucose and other small molecules (D-glucose is the form which occurs naturally in blood, and is metabolized).

Plasmodium is transmitted from host to host by mosquitoes. *Eimeria*, the cause of coccidiosis, is transmitted in food contaminated by faeces. The life cycle of a species that infects rabbits is shown in Fig. 2.9. The host ingests accidentally the stage known as the oocyst, which is shown on the left of the diagram. This contains eight individual cells ('sporozoites') enclosed in a double protective covering: pairs of sporozoites are enclosed in the four sporocysts, which in turn are enclosed by the outer oocyst wall. The protection is so effective that the sporozoites are unharmed

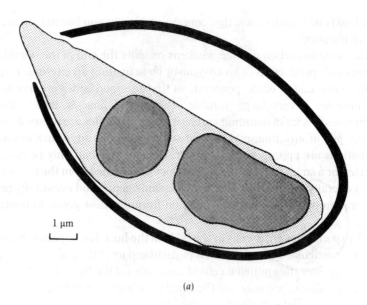

1 µm

(a)

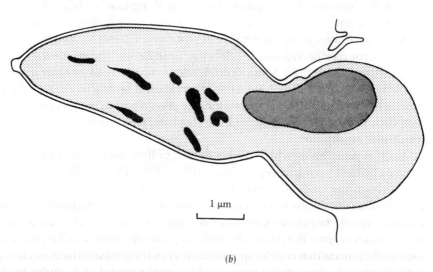

1 µm

(b)

Fig. 2.10. Sporozoites of *Eimeria larimerensis* (*a*) escaping from a sporocyst and (*b*) entering a host cell. Based on electron microscope sections by W.L. Roberts, C.A. Speer & D.M. Hammond: (1970) *J. Parasitol.* 56, 918–926; and (1971) *J. Parasitol.* 57, 615–625, respectively.

when the oocyst is put in strong disinfectants or acids. However, when the oocyst reaches the small intestine of the host, openings are formed in its wall and in the walls of the sporocysts, and the sporozoites escape (Fig. 2.10*a*). Oocysts can be made to release their sporozoites outside a host by exposing them first for several hours to a high partial pressure of carbon dioxide (such as would be found in all parts of the gut of a host) and then to a solution containing trypsin and bile (which are found in the small intestine).

Exposure to carbon dioxide weakens or splits the end of the oocyst, probably by stimulating production of an enzyme or by activating an enzyme. Trypsin digests a plug in one end of each sporocyst, so that the sporozoites are free to escape.

There are a very large number of species of *Eimeria*, which are parasites of numerous species of mammal and bird. The sporozoites are released from the oocyst in the gut of any mammal or bird (or presumably any other animal where the conditions are appropriate). However, the life cycle can only be completed in one species, or a small number of species. Oocysts obtained from the faeces of one species can generally infect other species of the same genus and eventually produce a new generation of oocysts, but transmission from one host genus to another generally fails.

The sporozoites enter cells in the body of the host. Each species of *Eimeria* tends to enter a particular type of cell, in a particular part of the gut or liver. *Eimeria steidae* (Fig. 2.9) enter the epithelial cells of tributaries of the bile duct. They apparently get there by making their way into the blood vessels of the intestine and getting carried in the hepatic portal circulation to the liver.

Various species of *Eimeria* have been watched and filmed entering cells in tissue culture. The sporozoites were generally moving along immediately before entering cells. Only a few seconds were needed to get into the cell, and their motion was scarcely interrupted. Sometimes a slender protuberance from the anterior end of the body is thrust repeatedly into the cell before penetration. The protuberance apparently contains the conoid. Some electron microscope sections of *Eimeria* show the conoid retracted (as in Fig. 2.3a) while others show it protruded through the polar ring (as in Fig. 2.3b). The conoid presumably plays a part in the process of penetration. Both *Eimeria* and *Plasmodium* have rhoptries, which may secrete a substance which aids penetration.

Details of penetration, which cannot be seen by light microscopy and in films, are revealed by electron microscope sections (Fig. 2.10b). The parasite does not simply pass through the host cell membrane: rather, the membrane caves in so that the parasite is eventually enclosed in a membrane-lined vacuole within the cell. While entering the cell, the parasite is constricted at the point of entry. (This can also be seen by light microscopy.) The process by which it pulls itself into the cell is probably the same as the process that enables sporozoites to crawl over solid surfaces (section 2.8).

Inside the cell, the parasite grows and becomes rounded or irregular in shape, instead of spindle-shaped. The conoid, inner membrane, micronemes and rhoptries gradually disappear. The parasite feeds on the host cell, using the micropore as a cytostome.

As the parasite grows, its nucleus divides repeatedly. Duplicate copies of the chromosomes are made before each division and the two nuclei that are produced have the same genetic composition as the original one: this is the process called mitosis. Eventually the whole parasite divides into a large number of small merozoites, each with its own nucleus. This is the stage of the life cycle shown in Fig. 2.3. The merozoites escape from the host cell and enter other cells, and repeat the process of growth and division. This division of a relatively large intracellular parasite into many small merozoites is a characteristic of most sporozoan life cycles, and is called schizogony.

Schizogony is an asexual process, a process of reproduction simply by division. A few generations of it are followed by a different, sexual proces of reproduction. At this stage, the parasites are called gamonts. They grow within the host cell but do not divide to form merozoites again. Instead they form gametes, either a single large (female) gamete, comparable to the eggs of multicellular animals, or many small (male) ones. The male gametes consist of a nucleus, a mitochondrion, two or three flagella and very little else: they are very similar to vertebrate spermatozoa (which, however, have only one flagellum). They escape from the host cells and fertilize any female gametes that may be around, forming zygotes which secrete oocyst walls around themselves, are discharged from the host cells and eventually leave the host in the faeces.

Sporozoites, merozoites, gamonts and gametes are haploid: that is to say, each nucleus contains only a single set of chromosomes. When a zygote is formed, the male and female nuclei fuse so the zygote nucleus is diploid, with a double set of chromosomes. The subsequent divisions that produce the sporozoites involve a meiosis: the nuclei divide one more time than the chromosomes so the sporozoites are haploid, each with only a single set of chromosomes.

The likelihood that a particular oocyst will be eaten by the host species may be very low, but if it is eaten it may produce an enormous number of oocysts of the next generation. *Eimeria bovis*, which causes a troublesome disease of cattle, provides a striking example. Each oocyst contains eight sporozoites. Each of these may produce, in the first asexual generation, 100 000 merozoites. There is a second asexual generation in which each of these merozoites may produce about 30 merozoites, which produce gamonts. If no parasite cells died, one oocyst could produce $8 \times 100\ 000 \times 30 = 24$ million gamonts. Only 50% of gamonts can be expected to be female so n oocysts could, in principle, give rise to 12 million n oocysts of the next generation. The number actually produced is likely to be a lot less than this but it seems clear that each oocyst needs only a small probability of entering an appropriate host, for the species to survive. *E. bovis* grows exceptionally large in the host cells and produces exceptionally large numbers of merozoites, but many other species of *Eimeria* have more than two asexual generations and may have as high a potential for multiplication within the host.

2.6.   Sex and reproduction

In the previous section I described the reproduction of sporozoan parasites, and in this one I discuss protozoan reproduction more generally.

Amoebas reproduce by dividing. Some species divide by multiple fission, producing a large number of tiny offspring, as in sporozoan schizogony. Others perform binary fission, dividing their cytoplasm into two approximately equal halves. In either case the process is asexual and the necessary divisions of the nucleus are mitoses.

Flagellates and ciliates often perform binary fission, but they also reproduce sexually. When that happens meiosis must occur, because sexual reproduction involves fusion of the haploid nuclei of two gametes to form the diploid nucleus of the zygote. Many flagellates are haploid while reproducing asexually, and the zygote divides by meiosis to get back to the haploid condition; sporozoans do the same, as we have seen. Ciliates, however, are diploid while reproducing asexually, and meiosis occurs in the divisions that produce the gamete nuclei. Similarly, metazoan animals are diploid and meiosis occurs in their gonads to produce haploid eggs and sperm.

Many protozoans have complicated organelles that have to be duplicated in binary fission. For example, the ciliate *Tetrahymena* divides in such a way that one of the offspring retains the original cytostome and the other has the original contractile vacuole. Each develops a replacement for the missing organelle.

The flagellate *Chlamydomonas reinhardi* can be cultured either in water or on agar jelly in petri dishes. They reproduce asexually so long as there is an adequate supply of nitrogen (as nitrate or ammonium ions) for protein synthesis. They become ready for sexual reproduction when transferred to a medium in which the concentrations of these ions are low.

A culture of genetically identical *Chlamydomonas* (a clone) can be produced by allowing a single individual to reproduce asexually. Members of the same clone will not join in sexual reproduction, and they will only join with members of 50% of other clones. It appears that there are two sexes and that all members of a clone have the same sex (as they should, being genetically identical). In some species of *Chlamydomonas* the sexes are similar in size but in others they are very different: in such cases the small sex may be regarded as male and the large one as female.

The genetics of *Chlamydomonas* have been studied in detail. Members of two clones of opposite sex are mixed. Zygotes are taken from the mixture and set individually on agar jelly. If they are kept in suitable conditions meiosis occurs and the diploid zygote produces four haploid individuals in the course of a week. These four are genetically different. They can be separated by a fine glass loop under a dissecting microscope (a practised experimenter can do this in less than a minute) and clones can be grown from them. Many mutations of *Chlamydomonas* are known, involving peculiarities of colour, requirements for nutrients, drug resistance, paralysis of flagella, etc. Clones showing these mutations have been bred together and it has been possible to demonstrate crossing over and to work out the relative positions of the mutant genes on the chromosomes.

Sexual reproduction raises interesting questions. Why has it evolved? Why are ova large and spermatozoa small?

The question of why sex has evolved is particularly difficult because sex brings an appalling evolutionary penalty. Imagine a species with some members whose genes make them reproduce asexually and others (otherwise identical) that reproduce sexually. Energy and materials are needed to produce offspring, and each animal has the resources to produce just n asexual offspring. Assume that in sexual reproduction the male contributes only a tiny sperm (as is the case, for many species) so that sexually reproducing mothers can produce no more than n offspring, $n/2$ females and $n/2$ males. The asexual members of the species have the potential to multiply their number by n in each generation but the sexual ones can multiply only by $n/2$. It is most unlikely that the full potential for multiplication will be realized, but the proportion of asexual members will increase very rapidly, in successive generations, unless sexual reproduction brings some huge advantage that we have not yet noticed.

That argument assumed that sexual reproduction produced equal numbers of sons and daughters. The genetic mechanism that decides the sex of offspring in most animals tends to produce equal numbers, but it is easy to imagine other mechanisms that would produce different proportions of the two sexes. A single male can produce enough sperm to fertilize a great many females, so it might seem best to produce mainly daughters, with just a tiny proportion of sons.

That is not the logic of evolution. Imagine a population of N_m males and N_f females that produce N_o offspring. Each of these offspring has a father and a mother so the males produce on average N_o/N_m offspring and the females on average N_o/N_f offspring. Assume for simplicity that the cost, in energy and materials, of producing a son is the same as the cost of producing a daughter (as it probably will be, if the sexes are equal in size). The cost of producing them is the same but the number of grandchildren they can be expected to yield is different, unless N_m equals N_f. If males are rarer than females, natural selection will favour genes that make animals produce male offspring because N_o/N_m will be greater than N_o/N_f. If females are rarer, it will favour production of females. In either case, the balance of the sexes will tend to change until there are equal numbers of males and females. (If, however, the costs of producing the two sexes were different, the evolutionary outcome would be different: the population would come to devote equal proportions of its resources to producing sons and daughters.) Thus the penalty of having sexual rather than asexual reproduction cannot be escaped by producing mainly daughters. Sexual reproduction must indeed have a huge advantage that we have not yet noticed.

The advantage seems to be that sexual reproduction brings the possibility of rapid evolutionary change. Asexual reproduction produces offspring that are genetically identical to their parents (apart from rare mutations). Sexual reproduction produces offspring that have half their mother's genes and half their father's genes, so each generation brings together new combinations of genes. Imagine a mixed population of sexually and asexually reproducing members of the same species, living in an environment in which the temperature changes at intervals of a few generations. Selection will occur in a cold period so that only cold-tolerant members of the species survive. When the environment becomes warmer the asexual reproducers will still produce only cold-tolerant offspring (except in the rare event of an appropriate mutation), and these offspring will probably be ill-adapted to the warm conditions. However, recombination of genes will probably produce some heat-tolerant individuals among the offspring of the sexual reproducers. In a varying environment the genes of sexual reproducers are more likely to be transmitted to subsequent generations than the genes of asexual reproducers.

The problem with that explanation is that the changes in the environment would have to be very frequent and very extreme for the capacity for quick change, brought by sexual reproduction, to balance the huge disadvantage of the lower potential rate of multiplication.

Professor W.D. Hamilton has suggested that the important advantage of sexual reproduction may show itself in the interaction between hosts and parasites. Suppose that a host with genes *ABCD* is particularly susceptible to a parasite with genes *abcd*. If *ABCD* is the commonest genotype in the host population, the parasite population will come to be predominantly *abcd*. In that case, different host genotypes will be advantageous and if *ABED* (for example) emerges, it may be favoured by natural selection. It may become predominant and the corresponding parasite genotype, *abed*, will be favoured. If *abed* becomes common the host population will tend to change again, perhaps to *AFED* or even back to *ABCD*. Evolution will become a kind of chase with the parasite 'trying' to match the host's current genotype and the host 'trying' to keep a step ahead. In that situation, the capacity for rapid change given by sexual reproduction may be a huge advantage.

The only empirical evidence I know to support this idea comes from observations of a New Zealand freshwater snail, *Potamopyrgus*. This snail reproduces in two ways, either sexually or by the asexual process of parthenogenesis, whereby a female produces female offspring genetically identical to herself (see section 5.1). The number of males in a local population shows how prevalent sexual reproduction is. Samples were collected from many sites in New Zealand and examined for sex and for infection by parasitic trematode worms. Males were rare in populations where there was little infection, and common where there was a lot. Sexual reproduction seemed to be favoured where parasites were prevalent.

To explain the evolution of sex, we must show its value in protozoans. The protozoans that have been mentioned as involved in parasitic interactions have been the parasites, not the hosts. Sexual reproduction will generally be less advantageous for parasite than for host because parasites generally reproduce much faster than the larger organisms that serve as their hosts, which gives them an initial advantage in the kind of chase that I have described. However, protozoans themselves serve as hosts for parasites. For example, bacteria of the genus *Holospora* infect the nuclei of *Paramecium*, causing fatal diseases.

That gives us a possible explanation for the evolution of sex, but why should one sex produce large gametes and the other small ones, as happens in many protozoans and all sexually reproducing metazoans?

I will try to explain this indirectly by showing why the size difference is unlikely to be lost, once it has evolved. Imagine a protozoan species in which female gametes have volume V and male gametes have much smaller volume v, so zygotes have volume $V + v$. Suppose a mutant appears that makes one division fewer when producing male gametes: it produces only half the normal number of male gametes but their volume is $2v$ and they form zygotes of volume $V + 2v$. These slightly larger zygotes will probably survive slightly better than normal ones, but this advantage is unlikely to offset the huge disadvantage of halving the number of male gametes. (I am assuming that every male gamete has the same chance of fertilizing a female one). Mutants with larger male gametes are unlikely to be favoured by natural selection: indeed, smaller male gametes will be favoured until they are so small that further reduction in size would seriously impair their chance of fertilizing a female gamete.

This does not mean that smaller female gametes will be favoured. If a mutant halved the size of female gametes, to $V/2$, its zygotes would be much smaller than normal ones and might survive much less well. This effect might be serious enough to offset the advantage of doubling the number of female gametes.

Ciliate protozoans have a peculiar form of sexual reproduction, called conjugation, but before describing it I must explain the difference between the micronucleus and the macronucleus. All known ciliates (except the primitive *Stephanopogon*) have both kinds of nucleus, usually one of each. The micronucleus is diploid. It contains DNA but cytochemical tests show no RNA in it. The macronucleus is generally much larger and contains both DNA and RNA. Only the micronucleus takes part in sexual reproduction (the process of conjugation): the old macronucleus disappears and is replaced by a new one formed by division of the zygote micronucleus. Thus the micronucleus alone is responsible for transmission of genetic information at sexual reproduction. However, it is presumably not involved in the ordinary processes of running the cell, in providing templates for protein synthesis, etc., since it produces no RNA. The macronucleus, which does produce RNA, presumably has this responsibility.

Since the macronucleus is formed from the micronucleus it cannot carry more genetic information than the micronucleus. However, it contains far more DNA. This has been shown by staining the DNA by the Feulgen method and measuring photometrically the amount of stain taken up by each nucleus. Various species have been investigated. In most of them the macronucleus contains at least 8 times as much DNA as the micronucleus, and in some over 6000 times as much. Either the macronucleus contains a lot of random DNA which carries no genetic information or else it contains many duplicate sets of chromosomes (i.e. it is polyploid). The latter alternative seems the more probable, and there is further evidence for it. It is possible to dissect out parts of the macronucleus of *Stentor*. Even when 90–95% of the nucleus is removed the remaining fragment can regenerate an apparently normal macronucleus and the animal regains a normal appearance. It seems that any reasonably sized piece of macronucleus contains a complete set of genes.

Ciliates usually reproduce asexually, but conjugation occurs from time to time. In wild populations it often occurs regularly at a particular season. In laboratory cultures several asexual generations usually occur between successive sexual ones, and clones can survive indefinitely without conjugation. Clones of *Tetrahymena* with no micronucleus reproduced asexually for over 30 years and showed no other sign of abnormality.

The details of conjugation vary between species, but here is what happens in the typical case of *Paramecium caudatum*. Two individuals meet and adhere to each other. Their micronuclei then undergo meiosis, each producing four haploid nuclei of which three disappear. The remaining haploid nucleus in each individual divides again and one of the resulting nuclei stays where it is while the other moves across into the other individual's cytoplasm to fuse with the stationary nucleus there and produce a diploid zygote nucleus. At this stage the two *Paramecium* separate and the zygote nucleus makes three mitotic divisions, producing eight nuclei. The original macronucleus (which has remained intact so far) breaks up. One of the eight newly formed nuclei forms the new macronucleus, three disappear and four become micronuclei. Two of these four are passed to each of the offspring of the next (asexual) division, and one to each of the offspring of the next division after that.

Notice that the old macronucleus is discarded and replaced: it has to be, to ensure that the macronucleus carries the same genetic information as the micronucleus.

Notice also that each conjugating individual supplies a migrating nucleus (like a sperm) and a nucleus surrounded by cytoplasm (like an ovum). Each of the partners can be considered to play both a male and a female role in the same reproductive act.

2.7. Osmotic regulation

A protozoan must contain organic molecules which are not present in its environment. Its cell membrane must prevent these molecules from escaping, but it must also allow small molecules such as carbon dioxide and oxygen to pass through, to satisfy the needs of the cell. It seems inevitable that the cell membrane will be semipermeable, permeable to water and other small molecules but not to large organic molecules. The organism will thus tend to take up water from the environment and swell, owing to the osmotic pressure of the organic molecules, unless there is an adaptation to prevent this.

The situation is rather more complicated than this because many of the organic molecules are ionized and because there are inorganic ions both in the cell and in the environment. Amino acids have molecules of the form $NH_2CHRCOOH$, where the part R of the molecule is different in different amino acids. At low (acid) pH they form positively charged ions (cations) while at high (alkaline) pH they form negatively charged ions (anions). Amino acids (and proteins and peptides) are mostly negatively charged at the values of pH (around 7) which are usual in cells. There must be an interaction between these anions which are trapped in the cell and any inorganic ions which pass freely through the cell membrane.

An uncharged molecule which passed freely through the membrane would reach equilibrium when its activity was the same on both sides of the membrane. (Activity is effective concentration. It may be less than the concentration determined by chemical analysis because of interaction between molecules.) In the case of an ion, however, the activities may not be the same at equilibrium; a difference in activity may be maintained by an electrical potential difference because positive ions are repelled by positive charges and negative by negative. The Nernst equation tells us how big a potential difference is needed:

$$E_i - E_o = (58/n) \log_{10} (X_o/X_i) \text{ millivolts.} \tag{2.1}$$

In this equation, X_o and X_i are activities of the ion outside and inside the cell, E_o and E_i are the electrical potentials in millivolts outside and inside, and n is the valency of the ion (positive for cations and negative for anions). The equation tells us that a membrane potential of 58 mV can maintain a ten to one ratio of activity of a univalent ion, or a hundred to one ratio for a divalent ion.

Fig. 2.11 is a simplified representation of a cell in sea water. Sodium and chloride are much the most common ions in sea water and each has a concentration of about $0.5 \text{ mol } l^{-1}$. We will ignore the other inorganic ions. The organic particles trapped in the cell will vary in nature and in charge but we will suppose they are all univalent anions A^-, and that their concentration is $0.1 \text{ mol } l^{-1}$. This concentration has been chosen arbitrarily but lies within the range of concentrations of organic solutes which occur in living cells. What will the concentrations of sodium, $[Na^+]_i$, and chloride, $[Cl^-]_i$, in the cell be, when equilibrium has been reached? Assume that the

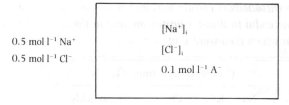

Fig. 2.11. A diagram of a cell in seawater, showing the concentrations of some ions.

activities of the ions are about equal to their concentrations. (This assumption is probably fairly accurate but could be badly wrong if ions were bound to organic compounds in the cell.) The membrane potential is the same for both sodium (which has a valency of $+1$) and chloride (valency -1) so

$$E_i - E_o = 58 \log_{10} (0.5/[Na^+]_i)$$
$$= -58 \log_{10} (0.5/[Cl^-]_i).$$

Also, since the concentrations of positive and negative charges in the cell must be (very nearly) equal,

$$[Na^+]_i = [Cl^-]_i + 0.1.$$

By solving these simultaneous equations we find $[Na^+]_i = 0.55 \text{ mol } l^{-1}$, $[Cl^-]_i = 0.45$ mol l^{-1} and $(E_i - E_o) = -2.4$ mV. A very small negative membrane potential would develop, which would keep the sodium ions a little more concentrated inside the cell than outside and the chloride ions a little less concentrated. An equilibrium of this sort, involving some ions which pass freely through the membrane and others which are trapped on one side of it, is called a Donnan equilibrium.

This particular equilibrium would involve a total concentration of $1.1 \text{ mol } l^{-1}$ ions inside the cell and $1.0 \text{ mol } l^{-1}$ outside. There would be a difference in osmotic concentration of $0.1 \text{ mol } l^{-1}$ and the cell would tend to take up water. The osmotic pressure Π of a solution containing c moles per unit volume is given by the equation

$$\Pi = RTc, \tag{2.2}$$

where R is the universal gas constant ($8.3 \text{ J K}^{-1}\text{mol}^{-1}$) and T is the absolute temperature (about 300 K for the environment of a living cell). The osmotic concentration difference of $0.1 \text{ mol } l^{-1}$ is equivalent to 100 mol m^{-3} (we have to express it this way, in basic SI units, to get our units right), so the corresponding osmotic pressure is $8.3 \times 300 \times 100 = 2.5 \times 10^5 \text{ N m}^{-2}$ or 2.5 atmospheres. If there were no mechanism to prevent it, the cell would take up water, swell and burst, like a parasite cell attacked by complement (section 2.4).

This necessity could be avoided by making the cell membrane relatively impermeable to one of the inorganic ions and pumping that ion out of the cell to keep its concentration below the equilibrium concentration. Energy would be needed to pump the ion out, but the more impermeable the cell membrane was made to the ion, the less would leak in and the less energy would be needed. It is probably not too difficult to make a membrane less permeable to sodium ions than to potassium and chloride ions, as hydrated sodium ions have the largest radius of the three.

All that was theory. What are the concentrations of ions in real marine protozoans? Here is some information about *Miamiensis*, a marine ciliate. It was grown in culture, in sea water with added nutrients. Samples of the culture were

Table 2.1. *Concentrations of certain ions and of amino acids in a marine and a freshwater protozoan, and in the culture media in which they were kept*

	Concentrations (mmol kg^{-1})			
	Na$^+$	K$^+$	Cl$^-$	Amino acids
Miamiensis avidus	88	74	61	317
Seawater medium	372	12	443	21
Amoeba proteus	1	25	10	—
Freshwater medium	0	0.08	0.08	—

Data from E.S. Kaneshiro, P.B. Dunham & G.G. Holz (1969). *Biol. Bull.* **136**, 63–75; E.S. Kaneshiro, G.G. Holz & P.B. Dunham (1969). *Biol. Bull.* **137**, 161–9; and R.D. Prusch & P.B. Dunham (1972). *J. exp. Biol.* **56**, 551–63.

centrifuged to obtain pellets of ciliates for analysis. The ciliates did not pack tightly enough to squeeze all the culture medium out from between them, but the amount of medium left in the pellets was measured and a correction was applied. The pellets were analysed by flame photometry (for cations), by titration (for chlorides) and in an automatic amino acid analyser (for amino acids). Some of the results are shown in Table 2.1. This table does not show a complete analysis, and there was presumably a substantial concentration of other organic molecules inside the ciliates. Notice that the concentration of potassium ions was about six times as high in the ciliates as in the medium, and the concentration of chloride ions was about seven times as high in the medium as in the ciliates. As $58 \log_{10} (1/7)$ is -49, then if the membrane potential was -49 mV the potassium and chloride ions were both more or less in equilibrium. If so, the sodium ions (which were much less concentrated in the ciliates than in the medium) were plainly not in equilibrium and must have been being pumped out of the ciliates. The membrane potential would have to be measured to confirm this interpretation, and it is certainly not the whole story. Calcium and magnesium ions would have to be excluded as well as sodium, for otherwise they would have reached very high concentrations in the cells, which they did not do.

Table 2.1 also gives data for *Amoeba proteus*, which lives in fresh water. The concentration of potassium ions in the cell is much higher than in the medium, but so is the concentration of chloride ions. The membrane potential was measured by inserting a microelectrode into the amoeba and found to be -90 mV, enough to keep potassium 35 times as concentrated in the amoeba as outside. The concentration in the amoeba was much higher than this so potassium as well as chloride must have been being pumped into the cell.

Marine protozoans may be able to prevent their osmotic concentration rising above that of the surrounding water, but freshwater protozoans cannot do so. The osmotic concentration inside them is inevitably greater than the very low osmotic concentration of the water. Water will diffuse in, and must be pumped out if the animal is not to swell. It is pumped out by one or more contractile vacuoles.

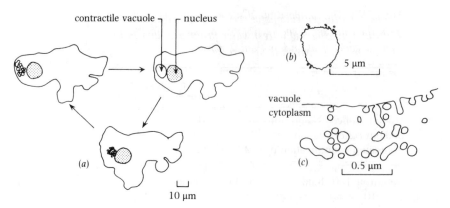

Fig. 2.12.(*a*) Outlines traced from a film of *Naegleria*, showing the contractile vacuole at three stages in its cycle. After K.G. Grell (1973). *Protozoology.* Springer, Berlin. (*b*), (*c*) A section through the contractile vacuole of *Acanthamoeba*, and part of the edge of the vacuole at a higher magnification. After B. Bowers & E.D. Korn (1968). *J. cell Biol.* **39**, 95–111.

Fig. 2.12(*a*) shows the action of the contractile vacuole of *Naegleria*. The vacuole attains a maximum diameter and then discharges its contents through the cell membrane, leaving only a cluster of much smaller vacuoles. These enlarge and coalesce and the cycle is repeated. Electron microscope sections of another amoeba (*Acanthamoeba*, Fig. 2.12*b,c*) show fine convoluted tubules around the vacuole and it is believed that the water which is excreted first enters the tubules and then drains from them into the vacuole. Microtubules have been found around the contractile vacuole of another amoeba (though not in *Acanthamoeba*) and may cause the contraction which drives the water out.

The action of the contractile vacuole of *Acanthamoeba* has been watched under the microscope. Measurements with a micrometer eyepiece showed that it reached an average diameter of 6 μm before discharging. It discharged on average once every 50 s in fresh water, and it is easy to calculate (since the vacuole was roughly spherical) that 8000 μm³ water were being pumped out every hour. Most of this must have entered the amoeba by diffusion. Food vacuoles totalling 500 μm³ were formed every hour. Water must also have been produced as a byproduct of metabolism (which converts foodstuffs and oxygen to carbon dioxide and water) but the rate at which it was produced can be calculated, from the oxygen consumption, to be very small. Hence about 7500 μm³ of water must have diffused into the amoeba every hour. The volume of the amoebas was determined by sucking them in and out of a pipette until they drew in their pseudopodia, measuring their diameter and calculating the volume. It was found to be 3000 μm³, so the water diffusing into the amoebas every hour was 2.5 times the volume of the body.

This may seem remarkable, but it must be remembered that *Acanthamoeba* is very small and so has a large ratio of surface area to volume. A real *Acanthamoeba* is about 50 μm long, but imagine one the size of a small fish, 50 mm long. It would be 10^3 times as long so it would have 10^6 times the surface area and so would presumably take up water 10^6 times as fast. However, it would be 10^9 times as heavy so the rate of

Table 2.2. *The composition of the cytoplasm, and of fluid from the contractile vacuole, of* Pelomyxa *in a medium of osmotic concentration less than 2* mmol l^{-1}

	Cytoplasm	Vacuole fluid
Osmotic concentration (mmol l^{-1})	117	51
Sodium concentration (mmol l^{-1})	6	20
Potassium concentration (mmol l^{-1})	30	5

Data from D.H. Riddick (1968). *Am. J. Physiol.* **215**, 736–40.

uptake of water expressed in body volumes per hour would be only 10^{-3} times the value for the small amoeba. The fish-sized amoeba would take up 2.5×10^{-3} body volumes per hour or 2.5 g water (kg body mass)$^{-1}$h^{-1}. Small freshwater fish produce urine as fast as this, or faster.

Samples of fluid from the contractile vacuoles of the giant amoeba, *Pelomyxa*, have been analysed. Micropipettes, 2–5 μm in diameter, were thrust into contractile vacuoles, and the fluid was sucked out. The freezing points of the tiny samples were determined, and used to calculate the osmotic concentration. The sodium and potassium concentrations were determined by flame photometry. Samples of cytoplasm were analysed in the same way. The results are shown in Table 2.2. The fluid in the vacuole has a lower osmotic concentration than the cytoplasm, and contains less potassium but more sodium.

Nearly all freshwater protozoans have contractile vacuoles. Some marine forms and many which live as internal parasites do not; a contractile vacuole is unnecessary for a protozoan living in a medium of reasonably high osmotic concentration. However, some marine and parasitic protozoans have contractile vacuoles, which generally beat rather slowly. Presumably their cytoplasm has a higher osmotic concentration than the medium they live in.

2.8.   Locomotion

Amoebas crawl slowly, at speeds up to about 5 μm s^{-1} (2 cm h^{-1}). Fig. 2.13(*a*) shows what seems to happen when an amoeba crawls. The outer layer of protoplasm (ectoplasm) appears to be relatively stiff and jelly-like, and anchored to the ground. Granules which can be seen in it under the microscope remain stationary relative to each other and to the ground. The inner core (endoplasm) can be seen from the movement of granules in it to be fluid, flowing forwards. Ectoplasm must be converted to fluid endoplasm at A and endoplasm must be converted to jelly-like ectoplasm at B. A particular particle remains stationary while it is in the ectoplasm but in time it finds itself at the rear end of the animal, becomes endoplasm and flows forward.

Sections of amoebas examined by electron microscopy are found to contain two

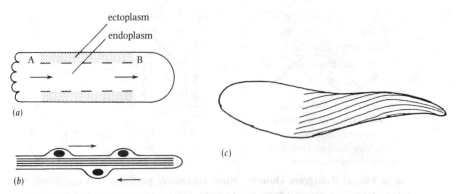

Fig. 2.13. Diagrams showing (*a*) how *Amoeba* crawls; (*b*) how particles in the cytoplasm move along the pseudopodia of foraminiferans; and (*c*) how the microtubules are arranged in an *Eimeria* sporozoite.

types of filament which are suspected of playing a part in locomotion. There are relatively short, thick filaments about 16 nm in diameter and long, thin filaments about 7 nm in diameter, scattered in the cytoplasm. These two types of filament are similar in diameter to the thick myosin filaments and the thin actin filaments of muscle (section 6.3), and there is good evidence that the thin microfilaments are actin. Treatment with heavy meromyosin gives the thin microfilaments a characteristic fringed appearance in electron micrographs, just as it does with actin filaments in muscle. (Heavy meromyosin is the part of the myosin molecule which forms the cross-bridges in muscle).

The cytoplasm of amoebas can be separated from the nuclei and cell membranes by high-speed centrifugation. Cytoplasm obtained in this way and kept at a low temperature (0.4°C) was fixed, sectioned and examined by electron microscopy. It contained thick microfilaments, but very few thin ones. When the cytoplasm was warmed to 22°C and ATP was added, it started moving in a most dramatic way. Streams of particles could be seen flowing in various directions through it. Apparently ATP can provide energy for amoeboid movement, as it does for muscle contraction. Cytoplasm fixed while in motion and examined by electron microscopy contained dense networks of thin microfilaments. It is suspected that the ectoplasm of living amoebas may be stiffened by a dense network of thin microfilaments which disappear when it is converted to endoplasm. If so, most of the thin microfilaments also disappear when the amoeba is prepared for electron microscopy, for no difference between ectoplasm and endoplasm is apparent in sections of intact amoebas. The isolated cytoplasm, with no cell membrane, may be less sensitive than the amoeba to the chemicals used for fixation, and this may be why the networks of thin microfilaments survive in sections of it.

The thick and thin microfilaments and the effect of ATP suggest that amoeboid movement may have a good deal in common with muscle contraction. Muscle works by the sliding of filaments along each other. Could amoeboid movement work in the same way? It seems likely but the details of the mechanism are still unknown, despite a great deal of research.

Foraminiferans such as *Allogromia* (Fig. 2.2*b*) use their pseudopodia for feeding

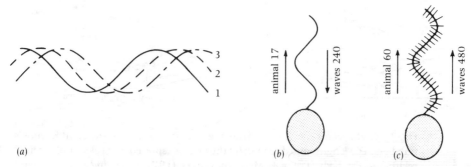

Fig. 2.14.(*a*) A diagram showing three successive positions of a flagellum, with waves travelling along it from left to right. (*b*), (*c*) Diagrams of *Strigomonas* and *Ochromonas*, respectively, swimming. The direction of swimming, and the direction in which waves travel along the flagellum, are shown. Typical speeds of swimming, and of movement of waves along the flagellum, are given in μm s^{-1}. These speeds were obtained from high-speed cinematograph films.

rather than locomotion, but it seems appropriate to discuss their movements here. Granules in the cytoplasm can be seen moving along them, often towards the tip on one side of the pseudopod and towards the base on the other (Fig. 2.13*b*). Bacteria and other food particles that collide with the pseudopodia stick to them, and are carried towards the cell body in this way.

The fine structure of foraminiferan pseudopodia is hard to study because fixatives distort them or even make them break up into droplets of cytoplasm. One cannot be sure that structures seen in sections of distorted or fragmented pseudopodia were present in life. However, it has been found that sea water with a high concentration of magnesium chloride immobilizes *Allogromia* without distorting it. It can then be fixed, embedded and sectioned for electron microscopy. Sections of undistorted pseudopodia obtained in this way have bundles of microtubules running along them. These are hollow tubes, not solid filaments, of diameter about 20 nm. The cytoplasm moving along the pseudopodia is probably made to slide along the microtubules by a mechanism like the one that makes microtubules slide along each other in cilia and flagella, as will be described.

Eimeria sporozoites glide along on solid surfaces at speeds of 4–8 μm s^{-1}, rotating as they glide. Three observations help to explain this mysterious-looking movement. First, particles labelled with a fluorescent substance (to make them conspicuous in ultra-violet radiation) adhere to the body and move slowly along it, accumulating at the posterior end. Secondly, there are microtubules, helically arranged, close under the cell membrane (Fig. 2.13*c*). Thirdly, motion is stopped by a substance that inhibits microfilament action but not by one that inhibits microtubules. It seems likely that molecules in the cell membrane attach to the substrate and that microfilaments then propel the cell by sliding these molecules posteriorly, parallel to the microtubules.

Most flagellates swim with their flagella, usually at speeds between 20 and 200 μm s^{-1}. *Chlamydomonas* beats its flagella in a manner rather like the arm action of a human breast stroke, but most other flagellates make symmetrical waves travel along their flagella as shown in Fig. 2.14(*a*). The frequency of beating is in some cases

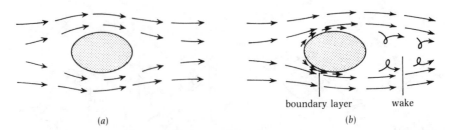

Fig. 2.15. Diagrams showing fluid moving past an object at two different Reynolds numbers. The lengths and directions of the arrows represent the velocity of the fluid relative to the body. From R. McN. Alexander (1971). *Size and shape*. Edward Arnold, London.

as high as 70 Hz, which is far too high to be observed by eye and has to be recorded by high-speed cinematography. Many flagellates swim with their flagella pulling them from in front, as in Fig. 2.14(b) and (c), but some use their flagella to push them from behind. Some move their flagella in a single plane like the tail of an eel but others throw them into helical (corkscrew) waves.

The flagella of some flagellates are fringed by fine hairs called flimmer filaments. Fig. 2.14(b, c) shows two flagellates which have been studied by filming, one with flimmer filaments and one without. Each has only one flagellum, which is held in front and beaten with planar waves. *Strigomonas* has a smooth flagellum and waves travelling down it from tip to base pull it forwards; the animal travels in the opposite direction to the waves. *Ochromonas* has flimmer filaments and waves travelling from base to tip of the flagellum pull it forwards; the animal travels in the same direction as the waves. The flimmer filaments explain the difference, as will be seen.

The movements of flagella are very like the movements made by the tails of swimming fish. It is tempting to conclude that forces of the same nature propel flagellates and fish, but such a conclusion would be false. Water flows around a flagellum in quite a different way from its flow around a fish's tail, because the flagellum is so much smaller and moves at so much lower a speed. This requires some explanation.

Fig. 2.15 shows two possible patterns of flow of fluid around a stationary object. The pattern of flow of fluid *relative to the object* would be the same in each case if the object were moving and the bulk of the fluid were stationary. In (a) the fluid is flowing smoothly around the body, parting in front of it and closing up behind it. The fluid in contact with the stationary body is itself stationary and the velocity increases gradually with distance from the body. Thus there are gradients of fluid velocity, and forces must act to overcome the viscosity of the fluid. The fluid exerts a force on the body in the direction of flow. This force is called drag. In the case shown in Fig. 2.15(a) nearly all the drag is due to viscosity.

Fig. 2.15(b) shows a different pattern of flow. Fluid in contact with the body is stationary but fluid quite close to it is moving almost as fast as the bulk of the fluid so steep gradients of velocity are confined to a thin boundary layer. The fluid does not close up smoothly behind the body but forms a wake of swirling eddies. Drag acts, partly owing to the viscosity of the fluid in the boundary layer and partly owing to the

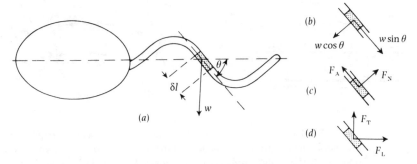

Fig. 2.16. Diagrams of a flagellum, showing components of velocity and of force.

changed momentum of the fluid in the swirling wake (a force is needed to change momentum, according to Newton's second law of motion).

Which pattern of flow occurs depends on the Reynolds number, which (for an object moving in water) is

$$10^6 \text{ (length in metres)} \times \text{(speed in metres per second)}.$$

The pattern shown in Fig. 2.15(a) occurs if the Reynolds number is less than 1, as it is for all protozoans. The pattern in Fig. 2.15(b) occurs if the pattern is well over 1, as it is for fish swimming at their normal speeds.

In this chapter we are only concerned with Reynolds numbers less than 1 and flow of the type shown in Fig. 2.8(a). In such situations the drag on a body of length l moving at velocity u through a fluid of viscosity η is given by

$$\text{Drag} = \eta k l u, \tag{2.3}$$

where k is a constant which depends on the shape and orientation of the body. It has a value k_A for a cylinder moving lengthwise along its own axis and a different value k_N for the same cylinder moving broadside on, normal to its axis. It is found that $k_N \approx 2k_A$. (At higher Reynolds numbers, drag is proportional to the square of speed.)

Fig. 2.16(a) shows a flagellate passing waves along its flagellum from tip to base, that is towards the left of the diagram. Waves like this would propel *Strigomonas* (Fig. 2.14b) towards the right. Can we explain why?

We will suppose that the body of the flagellate is initially stationary, and discover the direction of the force which the flagellum exerts on it. Consider the short segment of flagellum δl which at the instant illustrated is inclined at an angle θ to the longitudinal axis of the flagellate and is moving transversely with velocity w. This velocity can be resolved into a component $w \sin \theta$ along the axis of the segment and a component $w \cos \theta$ at right angles to the axis (Fig. 2.16b). Hence the force exerted by the water on the segment has components F_A, F_N (Fig. 2.16c) where, from equation (2.3)

$$F_A = \eta k_A w \, \sin\theta \cdot \delta l,$$
$$F_N = \eta k_N w \, \cos\theta \cdot \delta l,$$

This force can be resolved along different axes, into a transverse component F_T and

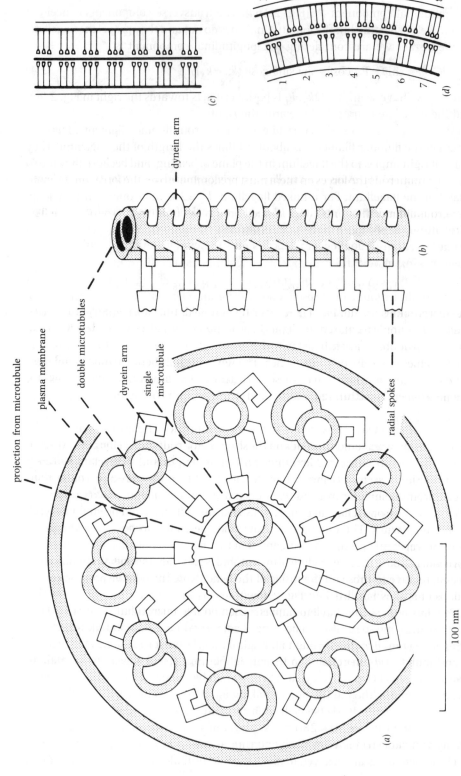

Fig. 2.17. **Diagrams of the structure of flagella and cilia, based on electron microscope sections. (a) A transverse section of a flagellum or cilium. (b) One of the double microtubules, with dynein arms and radial spokes projecting from it. (c), (d) Longitudinal sections of straight and bent portions of cilia.**

projection from microtubule

plasma membrane

double microtubules

dynein arm

single microtubule

radial spokes

dynein arm

100 nm

(a)

(b)

(c)

(d)

a longitudinal component F_L (Fig. 2.16d). The transverse components cancel out over a cycle of beating but the longitudinal ones do not and they propel the flagellate. At the instant we are considering the longitudinal component is

$$F_L = F_N \sin\theta - F_A \cos\theta = \eta w \sin\theta \cos\theta \, (k_N - k_A)\delta l. \tag{2.4}$$

Since, as we have seen, $k_N \approx 2k_A$, F_L is positive: it acts towards the right in Fig. 2.9(a) and the flagellate is propelled towards the right.

Ochromonas (Fig. 2.14c) has very fine, very numerous flimmer filaments. The total length of the flimmer filaments is about 20 times the length of the flagellum. They stand at right angles to the flagellum in the plane of beating, and because they are so long and numerous the forces on them must predominate over the forces on the main strand of the flagellum. For a naked flagellum, $k_N \approx 2 \, k_A$. For one with very long, numerous flimmer filaments $k_A \approx 2 \, k_N$ and F_L must (by equation 2.4) be negative. The flagellate will be propelled in the same direction as the waves on the flagellum, as is in fact observed.

Fig. 2.17(a) shows the structure of a flagellum, as seen in transverse section by electron microscopy. Note the microtubules, which have about the same diameter as the microtubules found in the pseudopodia of foraminiferans. Here, however, most of the microtubules are double. There is a ring of nine double microtubules, with two single microtubules in the centre. Pairs of projections, dynein arms, project from one side of each double microtubule towards the next. Radial spokes run from the double microtubules towards the central ones. The whole flagellum is enclosed in an outer membrane. All flagella have these structures but some, including those of dinoflagellates, have an additional rod running parallel with the bundle of microtubules.

There are two ways in which a flagellum could bend. The microtubules on the inside of each bend could shorten and those on the outside could elongate. Alternatively, the microtubules could remain constant in length and slide relative to each other. The latter seems to be what happens. Fig. 2.17(c) and (d) are based on sections of cilia from freshwater mussels, examined at very high magnification under the electron microscope (cilia have the same structure as flagella and presumably work in the same way). These diagrams show that the radial spokes are arranged in regularly repeating groups of three, and that the spacing is not altered on either side of a bend. Bending involves sliding of microtubules and radial spokes on the outside of a bend, relative to those on the inside of the bend. Note the relative positions of the groups of spokes labelled 7 in Fig. 2.17(d).

How fast does sliding have to happen? Each double microtubule is about 70 nm from the next one so the radii of bending of adjacent double microtubules may differ by up to 70 nm. Bending through an angle θ radians must slide some of the double microtubules 70θ nm relative to their neighbours. Typical flagella bend through about 2 radians (114°), so a cycle of beating involves about 140 nm sliding, first in one direction and then in the other, making a total of 280 nm. A typical flagellum might beat at about 30 Hz (some beat faster, and some more slowly) which would require sliding rates of 30×280 nm s$^{-1} = 8$ μm s^{-1}. This is similar to the rates of sliding which occur between thick and thin filaments in moderately fast muscles.

The dynein arms are believed to make the microtubules slide, in the same way as

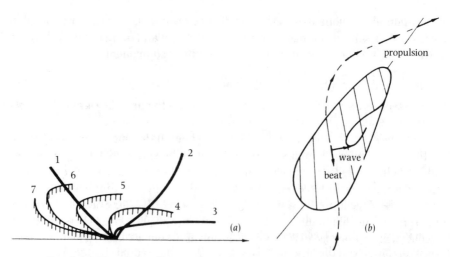

Fig. 2.18. Diagrams showing how the cilia of *Paramecium* beat. (*a*) Successive positions of a cilium. (*b*) The directions of the effective stroke of the ciliary beat, of metachronal waves and of propulsion. From J.R. Blake & M.A. Sleigh (1974). *Biol. Rev.* 49, 85–125.

the cross-bridges cause sliding in muscle. As in muscle, the energy is supplied by ATP. This has been demonstrated by experiments with flagella treated with glycerol, which disrupts the outer membrane. Any ATP that is in the flagella is thus allowed to escape, but when the treated flagella are put in suitable solutions containing ATP, this ATP can reach the microtubules, and the flagella beat. The experiment has been done, for instance, with flagella of *Polytoma* treated with glycerol and then broken off the body of the parent cell by centrifugation. In suitable solutions containing ATP these flagella beat more or less normally with waves of bending starting at the base of the flagellum as in the intact flagellate. They swim along, base leading. Magnesium ions are needed in the medium, as well as ATP. Abnormal or irregular beating occurs at ATP concentrations below about 10^{-5} mol l^{-1}.

Cilia have the same internal structure as flagella and seem like them to be driven by the action of the dyein arms. However, they beat asymmetrically, as Fig. 2.18(*a*) shows. The cilium is held fairly straight as it beats to the right (positions 1–3). It is bent as it returns more slowly to the left (positions 4–7). The effective stroke (1–3) drives water to the right and exerts relatively large forces because the movement is fast and because the cilium is moving at right angles to its long axis ($k_N \approx 2k_A$). The recovery stroke (4–7) exerts forces to the left but they are relatively small because the cilium is moving more slowly and because much of it is moving more or less parallel to its length. The net effect of the cycle is to drive water to the right, parallel to the surface of the body. It will tend to drive the animal to the left.

In the effective stroke most of the cilium is held straight and only the base bends. In the recovery stroke a bend formed at the base travels out to the tip of the cilium, rather like a bending wave travelling along a flagellum. Many cilia do not bend in a single plane. In Fig. 2.18(*a*) the effective stroke is in the plane of the paper but in the recovery stroke the cilium bends out of this plane, away from the reader.

Fig. 2.18(*b*) shows the direction in which the cilia of *Paramecium* beat. This ciliate has a long groove (leading to the cytosome) along one side of the body. The cilia in the groove must be less effective than the others in propulsion. The grooved side thus travels more slowly than the other and the animal swims along a helix, with the groove always facing the axis of the helix.

Ciliates swim faster than flagellates. Speeds of 0.4–2 mm s^{-1} are usual, while flagellates can only achieve 20–200 μm s^{-1}. The reason seems clear. Compare *Euglena* (Fig. 2.1*b*) with *Tetrahymena*, a ciliate of about the same size. *Euglena* has a single flagellum about 100 μm long. *Tetrahymena* has over 500 cilia. Each of them is only 5 μm long but their total length is several millimetres. Since the flagellum of *Euglena* and the cilia of *Tetrahymena* have the same diameter and internal structure, the ciliate has many times as much propulsive machinery as the flagellate.

The individual cilia do not beat at random, but in a highly organized way. In *Paramecium*, for instance, cilia along a line parallel to the direction of beating beat in phase with each other. To either side of this line, the cilia are more and more out of phase with the cilia on the original line (Fig. 2.19*a*). The gradual change of phase in a direction at right angles to the beat makes the ciliated surface look as though waves were passing over it in this direction (Fig. 2.18*b*). The pattern of beating is often described by speaking of metachronal waves travelling over the surface. There are usually about ten complete waves on a *Paramecium*, at a given instant.

The pattern is difficult to elucidate in detail, because the waves look different from different angles. One investigator found it useful to make models from wires stuck into a plastic base (Fig. 2.19). These were made to match electronic flash photographs of living animals, and could be viewed from different angles to check the interpretation of the pattern on parts of the animal which were seen obliquely. Fig. 2.19(*a*) shows the pattern normally observed all over the body of *Paramecium*. Note how the cilia bend to the side in their recovery stroke. Figs. 2.19(*b*), (*c*) and (*d*) show what would happen if the waves travelled in other directions. In each case there are places (indicated by arrows) where the cilia get in each other's way. The pattern which occurs seems to be the one which involves least interference between cilia, when the cilia bend to their left side in the recovery stroke. In some other animals cilia beat in a single plane, or bend to the right in the recovery stroke, and the metachronal waves move in other directions. The mechanism which controls metachrony has not been fully explained, but seems to depend on hydrodynamic interaction between adjacent cilia.

When *Paramecium* collides with an obstacle it reverses its cilia and swims backwards a little, before swimming forward again. This manoeuvre often gets it clear of the obstacle. How is it done, by an animal which has no nervous system? Poking the rear end of an animal does not cause reversal but (appropriately) faster forward swimming.

Paramecium has a negative membrane potential like *Amoeba* (section 2.7). This can be destroyed by sticking a microelectrode into *Paramecium* and applying a positive potential through it. When this is done, the beat of the cilia reverses. The negative membrane potential can also be destroyed by putting *Paramecium* in a solution containing a high concentration of potassium: this also reverses the beat. However, neither electrodes nor a changed solution will reverse the beat if there is no calcium in the water.

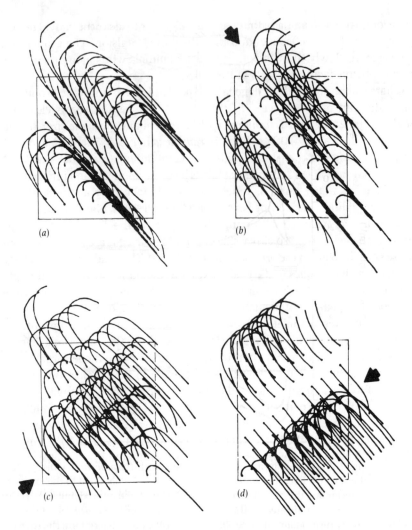

Fig. 2.19. Models of cilia beating (in the effective stroke) from top left to bottom right, showing different patterns of metachrony. Part (*a*) shows how the cilia of *Paramecium* beat. Metachronal waves are travelling from bottom left to top right. Parts (*b*), (*c*) and (*d*) are hypothetical: they represent cilia beating in the same way, but with different patterns of metachrony. In (*b*) the waves are travelling towards the bottom left, in (*c*) towards bottom right and in (*d*) towards top left. The arrows point to places where cilia get in each other's way. From H. Machemer (1972). *J. exp. Biol.* **57**, 239–59.

The role of calcium is shown more precisely by experiments on *Paramecium* treated with a detergent, Triton X. This has the same effect as the glycerol used in experiments with flagella. It disrupts the membranes of the cell, making them freely permeable to inorganic ions and other small molecules. Treated *Paramecium* still swim, if they are kept in a suitable solution containing ATP. If the solution contains very little calcium, for instance 10^{-8} mol l^{-1}, the ciliates swim forwards. If the calcium concentration is increased above about 10^{-6} mol l^{-1} the ciliates swim

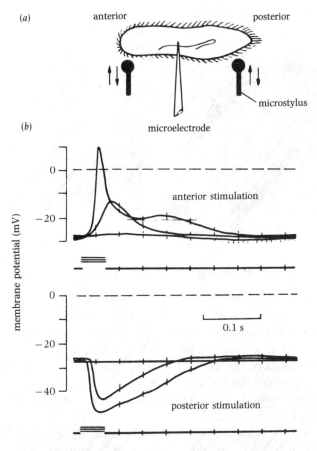

Fig. 2.20.(*a*) A diagram of an experiment with *Paramecium*, which is described in the text. (*b*) Records of membrane potential (mV) obtained in the experiment. The anterior and the posterior end were each stimulated by prodding with a microstylus, at the time indicated by the lines below the records. Records of three experiments, involving prodding at three different intensities, are superimposed in each case. From R. Eckert (1972). *Science* 176, 473–481.

backwards. These calcium concentrations are low: calcium concentrations in fresh water are generally 10^{-4} mol l^{-1} or higher. The experiment suggests that the concentration of calcium inside *Paramecium* is normally kept very low, but that calcium is admitted at appropriate times to reverse the cilia.

Another experiment is illustrated in Fig. 2.20. A *Paramecium* is impaled on a microelectrode, which is used to record its membrane potential. When the animal is undisturbed the membrane potential is -30 mV. The animal is prodded with an electrically driven microstylus which can be made to prod harder or less hard. When it is prodded very gently at the anterior end there is no effect. A moderate prod makes the membrane potential less negative and a harder one causes a bigger change and may even make the membrane potential positive. Conversely, prodding at the posterior end makes the membrane potential more negative. The harder the prod, the bigger the effect.

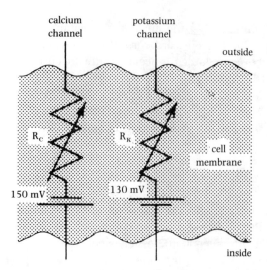

Fig. 2.21. A diagram representing the cell membrane of *Paramecium*.

These phenomena are particularly interesting, because they have a lot in common with important processes in nerve cells. They have been partially explained by Dr Roger Eckert, but the explanation requires a little more physical chemistry than has been used so far. The Nernst equation (section 2.7) refers to a situation where equilibrium has been reached. The ions which can penetrate the membrane have diffused through and achieved equilibrium, while the other ions cannot penetrate it at all. Now we have to consider a situation where ions penetrate the membrane, but not freely, and are not in equilibrium.

We have obviously got to consider calcium. It must have a very low concentration inside *Paramecium* (since *Paramecium* does not normally swim backwards!). It is probably usually about 10^5 times as concentrated outside the animal as inside it, and if so a membrane potential $V_C \approx (58/2) \log_{10} (10^5) \approx 150$ mV would be required to halt diffusion of calcium into the cell completely. If the membrane potential is less than this, a current of calcium ions will flow into the cell. We can represent this by an equivalent electrical circuit, a resistance and battery in series penetrating the cell membrane (Fig. 2.21). The battery produces an electromotive force V_C (≈ 150 mV) which will make current flow into the animal unless the membrane potential exceeds 150 mV. The resistance R_C represents the resistance of the membrane to passage of calcium ions through it. It is shown as a variable resistance, for reasons which will be explained.

Potassium is likely to be important as well as calcium: it is almost certainly the most plentiful ion in *Paramecium*. There is evidence that the potassium concentration in *Paramecium* is about 20 mmol l^{-1}, a little less than in *Amoeba* (Table 2.1). The potassium concentration of fresh water varies, but we will assume a value of 0.1 mmol l^{-1}, so that the concentration is 200 times as high in *Paramecium* as outside it. A membrane potential of $V_K = 58 \log_{10} (1/200) \approx -130$ mV would be required to halt diffusion of potassium out of the cell. A potassium channel is shown in Fig. 2.21 with a 130 mV battery and a variable resistance R_K in it. Other channels could be drawn for other ions but these two are enough to explain the observed phenomena.

When the membrane potential is V_m the current of calcium ions out of the *Paramecium* will be $(V_m - V_C)/R_C$. The outward current of potassium ions will be $(V_m - V_K)/R_K$. (Inward currents are regarded as negative outward currents.) The net current out of the cell must be zero, so

$$(V_m - V_C)/R_C + (V_m - V_K)/R_K = 0.$$

By putting $V_C = 150\,\text{mV}$, $V_K = -130\,\text{mV}$ in this equation, and solving it, we find that V_m is

$$\frac{(150\,R_K - 130\,R_C)}{(R_K + R_C)}\text{mV}.$$

If the calcium resistance R_C is much greater than the potassium resistance R_K, V_m will be approximately equal to the Nernst potential for potassium, $-130\,\text{mV}$. If R_K is much greater than R_C, the membrane potential will be approximately equal to the Nernst potential for calcium, $+150\,\text{mV}$. If neither R_K nor R_C is overwhelmingly greater than the other, the membrane potential will lie somewhere between these extremes. This seems to be the normal situation. In the experiment shown in Fig. 2.20 the membrane potential of the unstimulated *Paramecium* was $-30\,\text{mV}$. Poking at the anterior end made the membrane potential less negative and poking at the posterior end made it more negative. Dr Eckert suggested that distortion of the cell membrane at the anterior end reduced R_C temporarily, while distortion at the posterior end reduced R_K.

The initial effect is presumably local, reducing R_C and allowing calcium ions to flow in faster at the anterior end only. This inward current must be balanced by a current flowing outward through other parts of the cell membrane. The latter current, flowing outwards through the resistance of the cell membrane, must raise V_m (make it less negative) and tend to stimulate a fall in R_C over the whole cell surface: we have already seen that when electrodes are used to raise V_m the beat of the cilia is reversed, presumably by an influx of calcium. Thus a fall in R_C tends to increase V_m and an increase in V_m tends to reduce R_C. This is a positive feedback situation which could explain the effect of anterior stimulation spreading over the whole cell surface. A similar phenomenon (with sodium playing the part of calcium) is responsible for nerve conduction.

Calcium flows in until the calcium concentration in the *Paramecium* is high enough to reverse the cilia and make the animal swim backwards. Presumably the extra calcium is then pumped out of the animal, so that forward swimming starts again.

Further reading

General

Fenchel, T. (1987). *Ecology of Protozoa*. Sci Tech, Madison.
Grell, K.G. (1973). *Protozoology*. Springer, Berlin.
Kreier, J.P. & Baker, J.R. (1987). *Parasitic Protozoa*. Allen & Unwin, Boston.

Levine, N.D. *et al.* (1980). A newly revised classification of the Protozoa. *J. Protozool.* **27**, 37–58.

Long, P.L. (1982). *The biology of the Coccidia.* University Park Press, Baltimore.

Sleigh, M.A. (1973). *The biology of Protozoa.* Arnold, London.

Taylor, F.J.R. (ed.) (1987). *The biology of dinoflagellates.* Blackwell, Oxford.

Westphal, A. (1977). *Protozoa.* Blackie, Glasgow.

Wichterman, R. (1986) *The biology of Paramecium,* 2nd edn. Plenum, New York.

Photosynthesis and feeding

Azam, F., Fenchel, T., Field, J.G., Gray, J.S., Meyer-Reil, L.A. & Thingstad, F. (1983). The ecological role of water-column microbes in the sea. *Mar. Ecol. Prog. Ser.* **10**, 257–263.

Fenchel, T. (1986). Protozoan filter feeding. *Prog. Protistol.* **1**, 65–113.

Grahame, J. (1987). *Plankton and Fisheries.* Arnold, London.

Jeon, K.W. & Jeon, M.S. (1983). Generation of mechanical forces in phagocytosing amoebae. *J. Protozool.* **30**, 536–538.

Newell, R.C. & Linley, E.A.S. (1984). Significance of microheterotrophs in the decomposition of phytoplankton. *Mar. Ecol. Prog. Ser.* **16**, 105–119.

Nisbet, B. (1984). *Nutrition and feeding strategies in Protozoa.* Croom Helm, London.

Gut-living protozoans

Hungate, R.E. (1966). *The rumen and its microbes.* Academic Press, New York.

Martinez-Palomo, A. (ed.) (1986). *Amebiasis.* Elsevier, Amsterdam.

Shirley, R.L. (1986). *Nitrogen and energy nutrition of ruminants.* Academic Press, Orlando.

Blood parasites

Donelson, J.E. & Turner, M.J. (1985). How the trypanosome changes its coat. *Scient. Am.* **252**(2), 32–39.

Hudson, L. (ed.) (1985). *The biology of trypanosomes.* Springer, Berlin.

Intracellular parasites

Russell, D.G. (1983). Host cell invasion by Apicomplexa. *Parasitology* **87**, 199–209.

Sex and reproduction

Hamilton, W.D. (1980). Sex versus non-sex versus parasite. *Oikos* **35**, 282–290.

Preer, J.R. (1969). Genetics of the Protozoa. In *Research in Protozoology* (ed. T.T. Chen), vol. 3, pp. 129–278. Pergamon, Oxford.

Raikov, I.B. (1972). Nuclear phenomena during conjugation and autogamy in ciliates. In *Research in Protozoology* (ed. T.T. Chen), vol. 4, pp. 147–290. Pergamon, Oxford.

Treisman, M. (1976). The evolution of sexual reproduction: a model which assumes individual selection. *J. theor. Biol.* **60**, 421–431.

Osmotic regulation

Patterson, D.J. (1980). Contractile vacuoles and associated structures: their organization and function. *Biol. Rev.* **55**, 1–46.

Locomotion

Amos, W.B. & Duckett, J.G. (eds) (1982). Prokaryotic and eukaryotic flagella. *Symp. Soc. exp. Biol.* **35**, 1–632.

Eckert, R. (1972). Bioelectric control of ciliary activity. *Science* **176**, 473–481.

Russell, D.G. & Sinden, R.E. (1982). Three-dimensional study of the intact cytoskeleton of coccidian sporozoites. *Int. J. Parasitol.* **12**, 221–226.

3 Animals with mesogloea

Kingdom Animalia
   Phylum Porifera,
      Subphylum Symplasma (glass sponges)
      Subphylum Cellularia (other sponges)
   Phylum Cnidaria
     Class Anthozoa,
       Subclass Alcyonaria (soft corals, sea fans, etc.)
       Subclass Zoantharia (stony corals, sea anemones, etc.)
     Class Hydrozoa (hydroids)
     Class Scyphozoa (jellyfishes)

3.1.   Sponges

This chapter is about two phyla of simple multicellular animals. Their cells do not form thick tissues like those of the animals described in later chapters. Instead, they form sheets that are generally just one cell thick, covering all the animals' surfaces. These sheets of cells are supported by an underlying structure of jelly-like material, the mesogloea, which is described in more detail in section 3.5.

This arrangement is illustrated by the sponge *Leucosolenia*, shown in Fig. 3.1. It is white, about 2 cm high, and common in pools on British shores, attached to the rocks. It has a central cavity connected to the surrounding water by many small pores (ostia) and one large one (the osculum). Its outer surface is covered by an epithelium of cells that are flat like paving stones, and the inner surface by cells called choanocytes that are very like the entire body of a choanoflagellate (Fig. 2.7b). Each has a collar of microvilli around the base of its single flagellum. They filter food particles from the water, just as choanoflagellates do. Cells called porocytes, shaped like napkin rings, form walls for the ostia. Between them, these cells cover all the surfaces of the mesogloea, but there are also some cells in the mesogloea.

Spicules of calcium carbonate are embedded in the mesogloea, some completely embedded and others projecting from the animal's surface. They stiffen the animal (see section 3.5) and probably help to protect it: any animal that eats *Leucosolenia* will get an unpleasantly spiky mouthful.

Leucosolenia is a simple sac with a wall consisting of two layers of cells separated by mesogloea. This structure is suitable enough for a small sponge, 1 or 2 cm high. It would not be suitable for large sponges which grow up to 1 m in diameter. Large sponges (and some small ones) are more complicated than *Leucosolenia*. They have large numbers of small chambers lined with choanocytes, with branching tubes carrying water to them from the ostia and from them to one or several oscula.

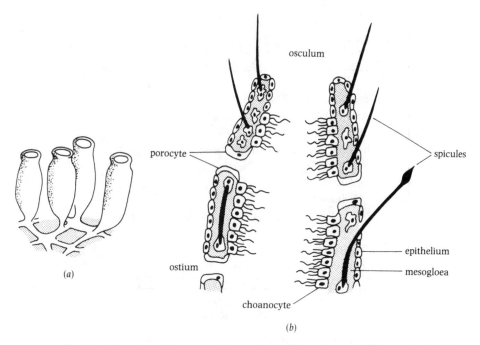

Fig. 3.1. Structure of the sponge *Leucosolenia*: (*a*) a colony and (*b*) a vertical section.

Sponges in still water (for instance in bowls or aquaria) keep water moving constantly through them, in at the ostia and out at the oscula. It is presumably propelled by the flagella of the choanocytes. If small particles are suspended in the water, they are filtered out by the collars of the choanocytes. Suspensions of graphite seem to be filtered as assiduously as suspensions of particles which can serve as food, and it has been shown that graphite particles of diameter 1 μm are filtered out effectively.

This happens even in still water. However, sponges do not seem to flourish in still water: they are found on shores where they are exposed to waves and tidal currents, and in deeper waters only where currents flow. One zoologist judged from his observations of sponges in their natural habitats that current speeds of $0.5 - 1 \mathrm{~m~s}^{-1}$ were most favourable to them. In such currents, water will flow through the sponges whether the flagella beat or not, as has been demonstrated by experiments. The flagella may still be needed to make water flow through the collars; since the collars do not completely block its passage, water can flow from ostia to osculum without being filtered.

Experiments on the effect of external currents on flow through sponges have been performed in a flow table. This is an apparatus designed to produce smooth, even currents of water across a shallow tank. Specimens of the sponge *Halichondria* were used: they were finger-shaped, about 3 cm tall with an osculum of diameter 5 mm at the top. A fine rod with a heated bead thermistor on its end was slipped into the osculum, so that the velocity of flow of water out of the osculum could be measured.

Fig. 3.2(*a*) shows the results of these experiments. In still water, the velocity of flow through the osculum was less than 3 cm s^{-1}. It increased when there was a current

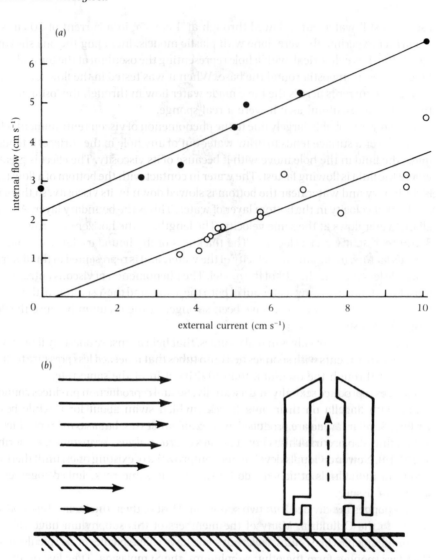

Fig. 3.2.(*a*) A graph of velocity of flow of water through the oscula of sponges (*Halichondria*) against the velocity of the current flowing over them in a flow table. Separate graphs are shown for living sponges (●) and for ones treated with fresh water to stop the flagella (○). From S. Vogel (1974). *Biol. Bull.* 147, 443–456. (*b*) A vertical section through a plastic model, about 2 cm high, which was used to demonstrate the principle of viscous entrainment. At the base of the model, inner and outer rings of radial holes connect with an annular slot. From S. Vogel & W.L. Bretz (1972). *Science* 175, 210–211.

across the flow table and was nearly 7 cm s⁻¹ when the current was 10 cm s⁻¹. Faster currents would presumably have caused faster flow through the sponge but could not be obtained with this particular flow table. The choanocyte flagella could be stopped by putting the sponge into fresh water for a while. Sponges treated in this way were returned to the flow table. It was found that no water flowed through their oscula in still water but it flowed through at 4 cm s^{-1} in a current of 10 cm s^{-1}.

Further experiments were done with plastic models, including the one shown in Fig. 3.2(*b*). It is cylindrical, with a hole representing the osculum at the top and radial holes representing ostia round the base. When it was tested in the flow table it was found that currents across the table made water flow in through the 'ostia' and out through the 'osculum' as if it were a real sponge.

This flow is probably largely due to the phenomenon of viscous entrainment. Fluid flowing over a surface tends to draw water out of any hole in the surface; it tends to make the fluid in the hole move with it because of its viscosity. The effect is greatest where the fluid is flowing fastest. The water in contact with the bottom of a flow table is stationary and water near the bottom is slowed down by its viscosity, so there is a gradient of velocity in the bottom layer of water. This is the boundary layer. Above it all the water flows at the same velocity. The lengths of the horizontal arrows in Fig. 3.2(*b*) represent water velocities. The thickness of the boundary layer depends on various factors, including the velocity of the water, and is represented in this diagram as a little less than the height of the model. The phenomenon of viscous entrainment must have tended to draw water out of both the 'ostia' and the 'osculum' of the model. However, the tendency must have been stronger at the osculum because the flow past it was faster.

There is evidence of valves in real sponges, that help to ensure one-way flow. It was found in experiments with sponges tied onto tubes that it needed less pressure to draw water out through the osculum, than to drive it in at the same rate.

Sponges reproduce sexually and asexually. Sexual reproduction produces rounded larvae with flagella on their outside cells, which swim about for a while before settling. Similar larvae are produced by asexual processes, but asexual reproduction by budding also occurs. This is how colonies of *Leucosolenia*, for instance, are formed (Fig. 3.1*a*). New individuals develop from outgrowths of existing ones, until there is a colony of individuals attached side by side to the same rock, linked together by strands of tissue.

The sponges are divided into two subphyla. Most of them (including *Leucosolenia*) belong to the Cellularia. Many of the members of this subphylum have calcium carbonate spicules like *Leucosolenia*, but some have silica spicules or no spicules at all. The glass sponges form the other subphylum, the Symplasma. They live on the sea bottom, most of them at depths of 100 m or more. They have spicules of silica which consist of three bars, joined mutually at right angles. These sometimes fuse together as a three-dimensional network, in which case the skeleton may remain intact when the soft tissues have decayed: the beautiful objects called Venus's flower basket are skeletons of the sponge *Euplectella*. The most remarkable feature of the Symplasma is that much of their tissue is not divided into separate cells, though the cytoplasm contains many nuclei. Tissues like this are called syncytia. A three-dimensional syncytial network, the trabecular tissue, permeates the mesogloea.

3.2.   Cnidarians

Cnidarians (also called coelenterates) have two principal body forms, polyps and

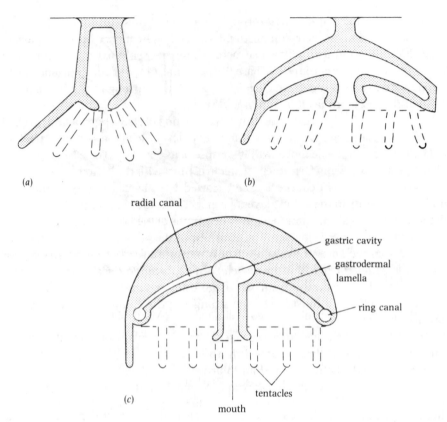

Fig. 3.3. **Diagrammatic sections of (*a*) a polyp, (*b*) an imaginary intermediate form and (*c*) a medusa.**

medusae. Polyps are hollow cylindrical animals with a mouth at one end, surrounded by tentacles: for example, sea anemones (Fig. 3.14). Medusae are shaped like a bell or an upside-down saucer, with the mouth underneath at the end of a tube which hangs down like the clapper of the bell. The large jellyfishes (Scyphozoa) are the most familiar medusae but a much smaller one is shown in Fig. 3.8 (bottom right). Fig. 3.3 represents the two basic forms, polyp and medusa, with an intermediate to show how they are related.

Both polyp and medusa have single layers of cells covering their surfaces, inside and out. The mesogloea is generally quite thin in polyps but thick in medusae, in which it generally occupies most of the volume of the body. Polyps have a simple gastric cavity but in medusae the central gastric cavity is connected by radial canals to a ring canal round the edge of the bell (Fig. 3.8, bottom right). A single sheet of cells, the gastrodermal lamella, fills the gaps between the radial canals.

Fig. 3.3(*b*) is wholly imaginary but helps to stress a basic similarity between the polyp and the medusa. A soft clay model of a polyp could be distorted to the medusa form in the manner indicated by the series of diagrams. The gastrodermis of opposite faces of the gastric cavity could be made to meet and merge in places, forming a gastrodermal lamella but leaving radial and ring canals.

Cnidaria have nerve and muscle cells, which sponges do not. These enable them to

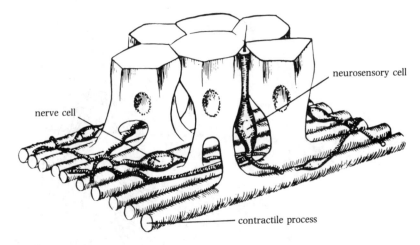

nerve cell

neurosensory cell

contractile process

Fig. 3.4. A diagram of a group of musculo-epithelial cells, with nerve cells and a sensory cell. From G.O. Mackie & L.M. Passano (1968). *J. gen. Physiol.* **52, 600– 621.**

move the parts of their bodies in a co-ordinated way. Many of the muscle cells have the curious and characteristic form shown in Fig. 3.4. Where they occur, they are the principal cells of the epidermis or gastrodermis. Their bodies are packed neatly together in a single layer but their bases (next to the mesogloea) are drawn out into long muscle fibres (contractile processes). The length of these fibres is not obvious in ordinary microscope sections but can be seen when the cells are separated by chemical treatment. In the sea anemone *Metridium* the fibres are up to 1 mm long though only about 1 μm in diameter. Networks of nerve cells run among the bases of the epithelial cells. The muscles and nerve cells and the range of behaviour they make possible are described in section 3.4.

Interspersed in places among the epithelial cells are cells called cnidoblasts which contain nematocysts (Fig. 3.5). These are rather like the trichocysts of ciliates. They contain hollow coiled threads which can be extruded, entangling prey or enemies or penetrating them and injecting toxin into them. They are plentiful in the epidermis of the tentacles and in other parts of some cnidarians.

It is difficult to discover how nematocysts work, because they discharge very rapidly and because important details of their structure are too small to be seen by optical microscopy. However, it is possible to obtain large nematocysts that have stopped short in the process of discharge by wiping the anemone *Corynactis* over a sheet of plastic. Electron microscope sections of these show that the process of discharge involves the hollow thread turning inside-out, bringing barbs that were initially inside it to the outside (Fig. 3.5*b*). Notice that the barbs emerging from the tip of the thread, as it turns inside out, are well placed to pierce prey. Notice also that the barbs move much further apart, as they come to the outside: the undischarged thread has deep helical pleats in its wall, which straighten out during discharge.

The nematocyst capsule does not collapse as the thread emerges from it, so the total volume of the system increases greatly. Fluid must enter the capsule during discharge, probably by an osmotic mechanism.

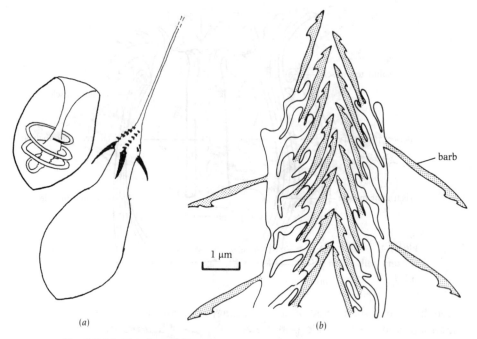

Fig. 3.5.(*a*) Sketches of a nematocyst before and after discharge. (*b*) A longitudinal electron microscope section of the tip of an incompletely discharged *Corynactis* nematocyst. The external barbs had broken off the original specimen but have been drawn in approximately their original positions. Drawn from a photograph in R.J. Skaer & L.E.R. Picken (1965). *Phil. Trans. R. Soc. Lond.* B250, 131–164.

To investigate this possibility, the freezing point of the fluid in undischarged nematocysts was determined. The nematocysts were cut open under liquid paraffin with a splinter of silica, and the contents were sucked into micropipettes. The freezing point of fluid from undischarged capsules was about − 5°C. This is well below the freezing point of sea water (about − 1.9°C) so the undischarged capsules have an osmotic concentration well above that of sea water. It would be feasible for them to draw water in osmotically as they discharge.

When a nematocyst has discharged, droplets of fluid emerge from the open tip of the thread. Such droplets were collected and found to have a freezing point of − 3°C. This confirms that the osmotic concentration of the nematocyst contents falls during discharge, as it would if water were being taken up osmotically. The droplets contain a mixture of venoms that serve to kill or immobilize prey.

The process of discharge cannot be reversed. Nematocysts can be used only once, and must then be discarded.

The Cnidaria are divided into three classes. The class Anthozoa consists of the sea anemones and corals, all of which have the polyp body form. The section of a sea anemone (Fig. 3.6) shows that the body wall is turned in at the mouth to form a tubular pharynx, and that the gastric cavity is incompletely partitioned by mesenteries which extend radially inwards from the outer body wall. The diagram shows just one mesentery which is 'perfect' in the sense that it reaches the pharynx,

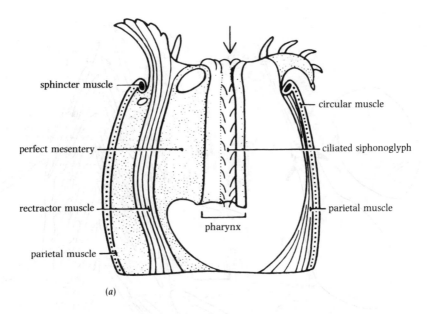

(a)

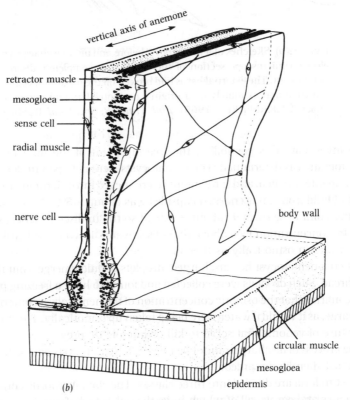

(b)

Fig. 3.6.(*a*) A diagram of a sea anemone cut in half to show the principal muscles. From E.J. Batham & C.F.A. Pantin (1950). *J. exp. Biol.* 27, 264–288. (*b*) A diagram of part of the outer body wall and a mesentery of a sea anemone, showing muscle fibres and nerve cells. From T.H. Bullock & G.A. Horridge (1965). *Structure and function of the nervous system of invertebrates.* Freeman, San Francisco.

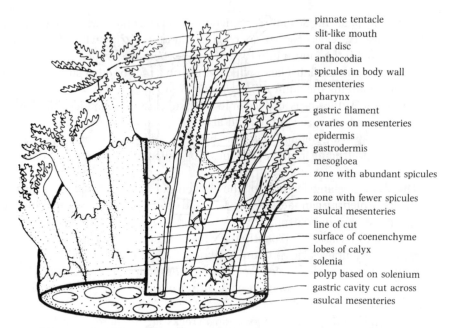

pinnate tentacle
slit-like mouth
oral disc
anthocodia
spicules in body wall
mesenteries
pharynx
gastric filament
ovaries on mesenteries
epidermis
gastrodermis
mesogloea
zone with abundant spicules

zone with fewer spicules
asulcal mesenteries
line of cut
surface of coenenchyme
lobes of calyx
solenia
polyp based on solenium
gastric cavity cut across
asulcal mesenteries

Fig. 3.7. Part of a colony of *Alcyonium*. The external surface is shown on the left but on the right the colony has been cut to show internal features. The polyps project about 1 cm from the surface of the colony when extended. From W.S. Bullough (1950). *Practical invertebrate anatomy*. Macmillan, London.

but a sea anemone may have dozens of mesenteries, only some of them perfect. The siphonoglyph is a ciliated groove in the pharynx. All the musculoepithelial cells are in the gastrodermis.

Sea anemones are solitary polyps, but many anthozoans are colonies of large numbers of polyps connected by living tissue. Some have the polyps connected by stolons, like those of the sponge *Leucosolenia* (Fig. 3.1a), but many form more massive colonies. An example is *Alcyonium digitatum*, which is called dead men's fingers because its colonies can look unpleasantly like a swollen hand, drained of blood. Fig. 3.7 shows the tip of a single branch ('finger') of a colony. *Alcyonium* is found attached to rocks near and below the low tide mark. The bulk of the colony is mesogloea which has spicules of calcium carbonate embedded in it but still has a soft, fleshy texture. The gastric cavities of the polyps extend well down into this mesogloea, and are connected by branching tubes. The polyps normally project from the surface of the colony as shown in Fig. 3.7 but they can withdraw into their cavities where they are well protected.

Most stony corals form colonies of large numbers of polyps, connected by a sheet of tissue and covering a massive structure made of calcium carbonate which they themselves have secreted. Many corals have their polyps standing in depressions in the stony material, into which they can retract when danger threatens.

Anthozoans reproduce both sexually and asexually. Asexual reproduction is by splitting in half or by forming buds which develop into polyps, which may either remain attached to the parent polyp as part of the same colony or may split off to become independent. Sexual reproduction involves gonads which develop on the mesenteries. The eggs or sperm are released into the gastric cavity whence they can escape through the mouth into the open sea, where fertilization generally occurs. The zygote forms a ciliated larva which settles on a rock or other suitable site and in due course attains the adult form.

The class Anthozoa is divided into two subclasses. The Alcyonaria includes *Alcyonium* and other soft corals, and also the gorgonians or sea fans which form much more delicate colonies which are branched or form networks (Fig. 3.19) with tiny polyps protruding from the branches. Gorgonians are common on Atlantic coral reefs.

The distinguishing features of the Alcyonaria are that they have just eight tentacles, which are pennate (see Fig. 3.7), and eight mesenteries in each polyp. They have only one siphonoglyph.

The other subclass is the Zoantharia, which includes the sea anemones and stony corals. Their tentacles vary in number (but are often a multiple of six) and are not pennate. There are at least twelve mesenteries and often many more, and the muscles on the mesenteries are arranged differently from those of Alcyonaria. There are usually two siphonoglyphs.

Sea anemones (Fig. 3.14) are common on rocks, on the shore and in deeper waters. Some of them are brightly coloured, and are perhaps the most spectacular inhabitants of rock pools. As well as anemones which attach to rocks there are others which burrow. *Cerianthus*, for instance, lives with its body buried in sand and only the tentacles projecting into the water above.

Some stony corals form massive reefs, which are discussed in the final section of this chapter. Most of them live in the tropics and few are found deeper than 45 m. The others which do not form reefs are found in all latitudes, mainly at depths of 180–550 m but sometimes in shallower water. *Astrangia danae*, for example, forms small encrusting colonies about 10 cm across on stones in shallow water on the Atlantic coast of the United States.

The Hydrozoa, the second class of Cnidaria, have both polyp and medusa forms. *Bougainvillea* (Fig. 3.8) is a typical marine example. Its polyps grow in bushy colonies on stones and shells and on the surfaces of sponges, sea squirts, etc. Many colonies may grow alongside one another forming quite a thick carpet. Each colony consists of a branching root-like hydrorhiza and a branching stem-like hydrocaulus which bears the polyps. The hydrocaulus has a tubular outer casing, the perisarc, which is discussed in section 3.5. The polyps differ from those of Anthozoa in having no mesenteries, and also in having musculoepithelial cells in the epidermis (where the fibres run longitudinally) as well as in the gastrodermis (where they run circularly, around the polyp's circumference).

The medusae of *Bougainvillea* develop as buds on the colony but detach from it and swim freely. Fig. 3.8 shows that they have the basic structure already described and in addition that they have ocelli and a velum. The ocelli are believed to function as very simple eyes, and are described in section 3.4. The velum is a shelf around the inside edge of the bell. Medusae swim by alternate expansion and contraction of the bell.

The medusa shown in Fig. 3.8 is immature. An adult medusa would have four

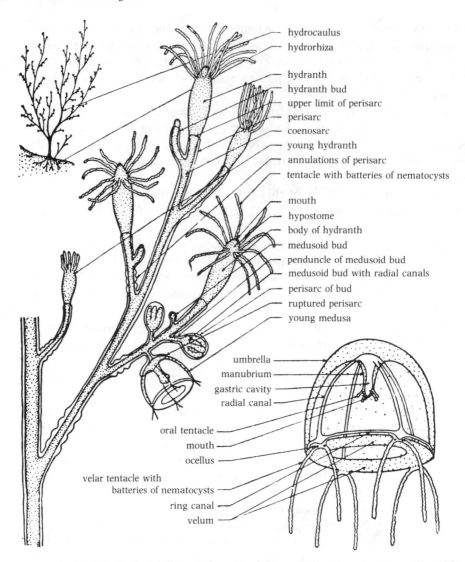

Fig. 3.8. *Bougainvillea*: a colony (top left), part of a colony (centre) and a medusa (bottom right). The colony was probably about 4 cm high and the diameter of the medusa about 3 mm. From W.S. Bullough (1951). *Practical invertebrate anatomy*. Macmillan, London.

gonads (all of the same sex) on the sides of the manubrium. Eggs released into the sea by female medusae are fertilized by spermatozoa released by male ones, giving rise to larvae which settle and develop into colonies of polyps. Thus polyps and medusae alternate in the life cycle; polyps give rise to medusae by the asexual process of budding and medusae give rise to polyps by sexual reproduction.

Hydra is a very simple member of the Hydrozoa, one of the few genera of Cnidaria which live in fresh water. Its polyps are solitary, not colonial. It does not produce medusae: instead, gonads develop on the sides of the polyps which can reproduce both asexually (by budding) and sexually.

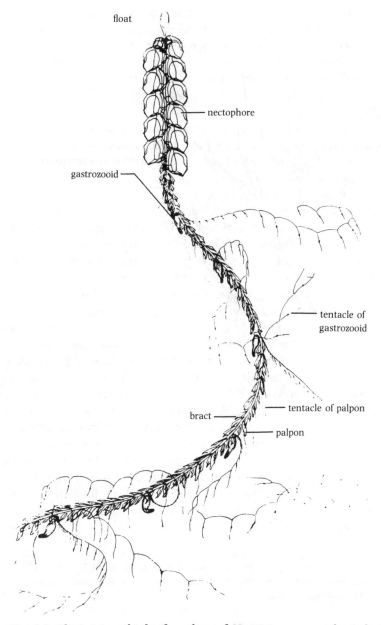

float

nectophore

gastrozooid

tentacle of
gastrozooid

tentacle of palpon

bract

palpon

**Fig. 3.9. The top two thirds of a colony of *Nanomia cara*, a siphonophore. The
overall length of this colony was probably about 25 cm. From G.O. Mackie (1964).
Proc. R. Soc. Lond. B159, 366–391.**

Unlike the Hydrozoa mentioned so far, the siphonophores have no sessile polyp
stage. Some, such as *Physalia*, the Portuguese man-o'-war, float at the surface of the
sea. Others such as *Nanomia* (Fig. 3.9) swim below the surface. Siphonophores form
polymorphic colonies: that is, colonies of which the individual members have
different forms. The nectophores are bell-shaped and seem to be modified medusae.

They contract and expand like free-swimming medusae, propelling the colony through the water. Further down the colony the gastrozooids are polyps, each with a single long tentacle. They are the only members of the colony capable of swallowing food. The palpons are polyps with a tentacle but no mouth. The bracts are leaf-shaped, but seem to be modified medusae. Yet other members of the colony bear gonads. At the top of the colony there is a gas-filled float containing (mainly) carbon monoxide.

Nanomia and related siphonophores are very common in the oceans, but their commonness was not realized until the 1960s. It was discovered in a curiously indirect way, through the use of echo-sounding. This technique was originally developed to measure the depth of the sea bottom, by emitting a sound pulse downwards from a ship and recording the time taken for its echo to return. Other objects, as well as the bottom, may return echoes. Submerged bubbles of gas are particularly effective, and this makes it possible to use echo-sounding for finding shoals of fish with gas-filled swimbladders. Echo-sounding in the oceans often detects echoes from a 'deep scattering layer' which is continuous, at the same depth, over large areas. This layer moves up near the surface at dusk each evening, and down at dawn to a depth of several hundred metres. Observations from submersible craft ('diving saucers') show that siphonophores and lantern fishes (Myctophidae, up to 10 cm long) are common in the deep scattering layer. The gas-filled floats of the siphonophores and the swimbladders of some of the fishes (some have no swimbladder) are presumably mainly responsible for the echoes. Population densities of *Nanomia* in the deep scattering layer have been estimated as up to 0.3 per cubic metre, a very high density for colonies that may be as much as 75 cm long.

The third class of Cnidaria, the Scyphozoa, includes the large jellyfishes that are familiar to sea bathers: the nematocysts of some of them inflict painful stings. Though their medusae are large (*Cyanea arctica* reaches diameters of two metres) their polyps are small and inconspicuous. The radial canals of the medusa are more numerous than those of hydrozoan medusae, and are branched. There is no velum. The gonads develop internally, on the floor of the gastric cavity. Fertilized eggs form ciliated, swimming larvae which settle on stones or algae and become polyps, which may form new polyps by budding (Fig. 3.10). These polyps seldom grow more than 1 cm high. They live for several years and then split up into ephyra larvae which are small medusae. Each cylindrical polyp forms a pile of saucer-shaped ephyrae which break free and grow to become adult medusae.

Colonial cnidarians are modular organisms. That means that (like modular furniture) they are built from a limited number of kinds of unit. Similarly, flowering plants are modular organisms: the modules are the leaves, flowers, branches and roots. In colonies of *Alcyonium* (Fig. 3.7) and *Bougainvillea* (Fig. 3.8) the polyps are the most obvious modules. In *Nanomia* (Fig. 3.9) the nectophores, gastrozooids, palpons and bracts are different kinds of modules. The term 'modular organism' is applied to organisms like these, but at a different level of organization all many-celled animals are modular: the cells can be regarded as modules.

These colonial cnidarians raise the question, what is an individual animal? The polyps might be regarded as individuals, because they resemble solitary individuals such as sea anemones or *Hydra*. Alternatively, the colony might be regarded as an

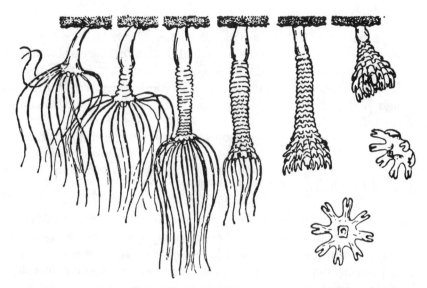

Fig. 3.10. Stages in the development of the scyphozoan *Aurelia* from the polyp (scyphistoma) to the young medusa (ephyra). From A.C. Hardy (1956). *The open sea, its natural history* (part 1, *The world of plankton*). Collins, London.

individual, because the modules are connected together and may share resources and behave in a co-ordinated way. The gastric cavities of the polyps interconnect and it has been shown in experiments with *Pennaria* (a hydroid) that radioactively labelled foodstuffs fed to one polyp spread through the colony. A stimulus applied to one polyp of a coral colony may make not only it contract, but may make all the polyps contract in a large part of the colony. The most impressive integration is seen in siphonophore colonies. Only the gastrozooids can feed, but food from them spreads through the colony to all the other modules. Only the nectophores can swim, but they pull the rest of the colony along with them. *Nanomia* disturbed by a bathyscaphe may dart away suddenly, by simultaneous contraction of all the nectophores. Is a *Nanomia* colony a single individual, or dozens of individuals connected together? Fortunately, the answer to the question does not matter: it is merely a question of how we define our terms.

Colonial forms presumably evolved from solitary ones, by incomplete separation of the products of asexual reproduction. Another consequence of asexual reproduction, in animals that live attached to the bottom, is that patches of neighbouring individuals or colonies tend to be genetically identical clones. There were 291 colonies of the coral *Porites compressa* on a 20 m² area of reef in Hawaii, but they belonged to only 15 clones. The clones were distinguishable by slight differences of appearance and also, more objectively, by having different forms of five enzymes. The different forms of each enzyme were separated by electrophoresis.

Sea anemones do not copulate, but release their gametes into the sea. They are nevertheless more likely to succeed in sexual reproduction if there is a member of the opposite sex nearby. They can move about only very slowly, crawling like slugs at 0.01 to 0.3 mm s⁻¹, and are apt to be surrounded by asexually produced members of the same clone, which have the same sex as themselves. This is apparent in *Metridium*

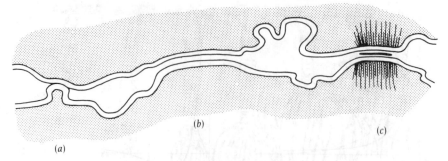

(b)

(c)

(a)

Fig. 3.11. A diagrammatic section through two adjacent cells, showing three types of attachment which are described in the text. This diagram is based on electron microscope sections.

senile, whose clones often differ in colour. They tolerate the presence nearby of members of their own clone and also of opposite-sexed members of other clones (which are potential mates) but they attack same-sexed members of other clones with their nematocysts and often succeed in driving them off.

3.3.   Multicellular structure

The nucleus synthesizes the messenger RNA which moves out into the cytoplasm and serves as a template for the synthesis of enzymes and other proteins. It seems reasonable to suppose that there must be a limit to the size of cell that can be served by an ordinary diploid (or haploid) nucleus. If the cell is too large the nucleus may be unable to produce RNA fast enough, and parts of the cell may be a long way from the nucleus so that RNA has to travel long distances. Study of protozoans seems to confirm that there is indeed a practical limit for most purposes. The giant amoeba *Pelomyxa* has a length of about 2 mm but has many nuclei. The dinoflagellate *Noctiluca* is about the same size and has only one nucleus, but much of its volume is occupied by fluid-filled vacuoles. Ciliates such as *Stentor* (Fig. 2.4*b*) and *Spirostomum* have lengths of 1 mm or more but their RNA is produced by macronuclei which have many copies of each gene, which can produce RNA simultaneously. Also, the macronucleus tends to be long and to wind through the body so that none of the cytoplasm is very far from it.

Multicellular animals can be very large and nevertheless have all parts of their cytoplasm close to a nucleus, if the individual cells are small. They tend to consist of cells of 20–30 μm diameter, similar in size to many protozoans, but some types of cell are much larger. For example, some human striated muscle cells are as much as 20 mm long, but their diameters are only 10–100 μm and they have nuclei spaced at short intervals along them.

Even quite small multicellular animals may consist of very large numbers of cells. For example, the simple hydroid *Hydra attenuata* is about 15 mm long (excluding its tentacles) and consists of about 130 000 cells. The cells of a multicellular organism must obviously be attached to each other, but it is not so obvious how they are attached. Fig. 3.11 shows three types of attachment between cells which are found in

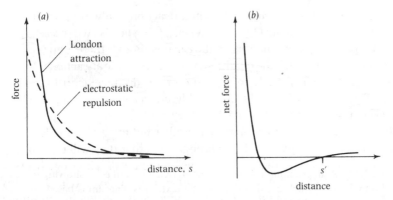

Fig. 3.12. Diagrammatic graphs of forces acting between two cells, against their distance apart. (*a*) The London–van der Waals attraction and the electrostatic repulsion shown separately; (*b*) the net attraction.

different tissues. Cell membranes are shown in this diagram as double black lines, which is how they appear in electron microscope sections of material treated with osmic acid. (The osmic acid attaches to the polar groups which are concentrated near the two faces of the membrane.) They are about 8 nm thick. At junction (*a*) the cell membranes of the two cells seem to be in close contact but at (*b*) and (*c*) they run parallel to each other, with a gap of 15 nm between them. This seems to be a genuine gap, not just a band of unstained material: it has been shown that molecules of proteins such as haemoglobin percolate quite freely into the gaps. The type of attachment shown in (*c*) is known as a desmosome. There are filaments in each cell, ending at the desmosome, and there seems to be some material in the gap between the cell membranes. Nevertheless the gap is not blocked, for protein molecules can enter it. Cells with attachments of type (*a*) between them are damaged if they are pulled apart but attachments of types (*b*) and (*c*) can be broken quite easily without damaging the cells. Most attachments of sponge cells to each other are of types (*b*) and (*c*), and sponges can be broken up into their component cells by squeezing them through cloth.

Attachments of types (*b*) and (*c*) seem rather puzzling. It is easy to envisage mechanisms which could hold two membranes in contact, but how can membranes be held 15 nm apart? The answer seems to involve interaction between two opposing forces, one of mutual attraction between the cells and one of mutual repulsion. The force of attraction is the London–van der Waals force, which should be familiar to readers who have studied chemistry. The force of repulsion is an electrostatic one, due to the outer surfaces of the cell membranes being negatively charged. The charge can be demonstrated by applying a potential across a suspension of cells, which move towards the anode. It is due to neuraminic acids (which are acid sugars) being attached to the membrane and being partly ionized. This negative charge on the outside of the cell should not be confused with the potential difference (inside negative) which occurs across the cell membrane because of the distribution of ions.

Both the London–van der Waals attraction and the electrostatic repulsion diminish as the distance s between the cell membranes increases (Fig. 3.12*a*). The London attraction is roughly proportional to $1/s^3$ while the electrostatic repulsion is proportional to e^{-ks} (where k is a constant). The attraction is always greater than the repulsion when s is very small ($1/s^3 = \infty$ when $s = 0$) and also when s is large, but if there is enough charge per unit area on the cell membranes there is an intermediate range of values of s for which the repulsion is greater than the attraction. Thus the net force between two cells is likely to depend on their distance apart in the manner shown in Fig. 3.12 (*b*). The cells are weakly attracted to each other when well separated but there is a distance s' at which they neither attract nor repel each other. At slightly shorter distances they repel each other but at very short distances they attract each other strongly. If cells are brought together they will tend to stop and stick at a distance s' apart. If they are somehow brought really close together they will adhere very strongly in close contact. Cell junctions of type (*a*) (Fig. 3.11) involve close contact while those of types (*b*) and (*c*) involve separation by, presumably, a distance s'.

If a suspension of separated sponge cells in sea water is allowed to settle the cells move about, clump together and eventually form a new sponge. Large clumps of cells form in suspensions which are prevented from settling by shaking. An experiment which has often been tried is to mix suspensions of sponge cells of two species. Clumps form and in some cases it can be shown that there is a strong tendency for cells of the same species to clump together. For instance, *Microciona prolifera* has red cells and *Haliclona oculata* has brown ones. The clumps which form in a mixed suspension of cells of these two species consist either of red cells alone or of brown ones alone. Some other pairs of species form mixed clumps initially but sort themselves out later.

Animals are assemblies of different types of cell. The cell types have been counted in *Hydra attenuata* after separating the cells by soaking and then shaking the animal in a mixture of glycerol and acetic acid. Of the 130 000 or so cells, about 25 000 were found to be epidermal and gastrodermal cells. These are the largest cells in the body and make up most of the total volume. There are also about 35 000 cnidoblasts, 5000 nerve cells and 5000 gland cells, and many small, undifferentiated interstitial cells.

A culture of well-fed *Hydra attenuata*, kept in uncrowded conditions at 20°C, reproduces by the asexual process of budding and thereby doubles its numbers every 3 days. The average size of the individuals remains constant so the numbers of each type of cell must double every 3 days. Additional cells are needed to replace used nematocysts and to replace cells which are sloughed off the pedal disc and the ends of the tentacles.

Rates of cell division have been investigated by experiments with radioactive thymidine. Division of a cell into two daughter cells involves doubling the quantity of DNA in its nucleus. Thymidine is a constituent of DNA but not of RNA and is taken up by cells in the period prior to division when they are synthesizing DNA (the so-called S phase). A dilute solution of radioactive thymidine is injected through the mouth of a living *Hydra*, into the gastric cavity. Any cells that are in the S phase at the time or within a few minutes thereafter take up some of the radioactive thymidine and incorporate it into their DNA. The *Hydra* is broken up then or later, and samples of its cells are spread out on slides. The cells which have radioactive DNA in their nuclei can be identified by autoradiography.

If the cells are examined immediately, a proportion of the epidermal, gastrodermal, gland and interstitial cells are found to be radioactive. All these types of cell divide and some of each type are in the S phase at any instant. No cnidoblasts or nerve cells are found to be radioactive so it seems that these types of cell do not divide. However, if a long enough interval is left between injection and autoradiography, radioactive cnidoblasts and nerve cells are found. The parents of these cells must have been in the S phase at the time of injection.

Further experiments with radioactive thymidine showed how fast each type of cell was dividing. It was found that epidermal and gastrodermal cells divide (and so double their numbers) every 3.5 days, or about once for every doubling of the population. Interstitial cells divide about once every 24 h, or even more frequently in the case of the clumps of developing cnidoblasts. They give rise to the new cnidoblasts and nerve cells in addition to doubling their own numbers every 3.5 days: this is why they have to divide so frequently.

Cells of different types have the same genes but nevertheless differ in structure and behave in different ways. The genes responsible for the differences are expressed in cells of one type, but not in cells of other types – a situation rather like the one we met in trypanosomes (section 2.4), where only one of the many surface coat genes is expressed at a time.

Hydra is much simpler in structure than most metazoans but it has, nevertheless, a clear pattern, with a mouth and tentacles at one end and a pedal disc at the other. The manner in which this pattern is maintained has been investigated by experiments in which it has been disturbed by surgery. The simplest experiments consist simply of cutting pieces off *Hydra*, which have remarkable powers of regeneration. They do not simply re-grow lost parts as an earthworm, for instance, replaces an amputated posterior end, but re-form the remaining parts of the body to form a complete but smaller animal. Even when a piece is cut with neither an oral (mouth) end nor a pedal disc, the end which was originally nearer the oral end becomes the new oral end.

More subtle experiments have been performed by grafting. If a small piece cut from one *Hydra* is slipped into an incision in the side of another it generally becomes incorporated in it. Fig. 3.13 shows the results of some such grafts. A piece from the extreme oral end of one *Hydra* grafted into the side of another forms a new oral end with a mouth and tentacles (*a*). A piece from just below the tentacles is simply absorbed into the existing structure of the host animal (*b*) unless the oral end of the host is removed (*c*) or the graft is placed near the pedal disc (*e*). It appears that pieces from below the tentacles have the capacity to form a new oral end but are normally inhibited from doing so by the oral end nearby. In experiment (*d*) the oral end of the donor animal was cut off 4 h before the piece was taken for the graft. No new tentacles or mouth were formed in this time but the processes of regeneration must have started, for the cut end acquired the ability to overcome the inhibitory effect of the host's oral end.

Attempts are being made to explain the results of these and other grafting experiments but they have not so far been wholly successful. There is evidence that the oral end exerts its inhibiting influence by releasing a substance which diffuses rapidly along the animal and is destroyed. Thus a gradient of inhibitor concentration is set up, highest at the oral end. It also appears that individual cells must have some

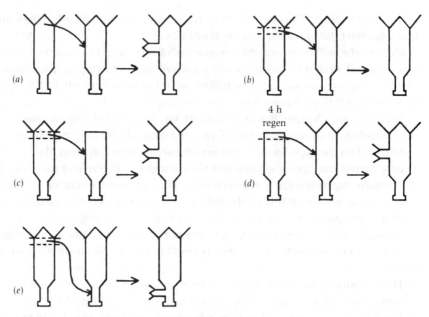

Fig. 3.13. Grafting experiments on *Hydra*, which are explained in the text. From L. Wolpert, J. Hicklin & A. Hornbruch (1971). *Symp. Soc. exp. Biol.* 25, 391–415.

more stable property which has been called positional information, which determines their aptitude to form a new oral end if the inhibitor concentration is not too high.

3.4.   Muscles and nerves

Sponges have no muscles, but some of their cells are capable of contracting: indeed, the whole body may contract noticeably if the animal is poked.

Cnidarians have musculoepithelial cells, as already described (Fig. 3.4). We have little direct knowledge of how they work, but they contain filaments of the proteins actin and myosin and presumably work like the muscles of more advanced animals (section 6.3).

Fig. 3.6 shows how the muscle cells are arranged in a typical sea anemone. The epidermis of the column is not muscular but the gastrodermis consists largely of musculoepithelial cells. The gastrodermis of the body wall has muscle fibres which run circumferentially round the column, forming the circular muscle. The sphincter is a large bundle of these fibres. Each mesentery has longitudinal muscle fibres on both sides, close to the body wall. These form the parietal muscles. There is another strip of longitudinal fibres forming the retractor muscle on one face of each mesentery, and there are radial muscle fibres (fibres running at right angles to the body wall) on the other face. Under the retractor muscles, the interface between the gastrodermis and the mesogloea is deeply folded (Fig. 3.6*b*). This greatly increases the number of musculoepithelial cells which can be fitted as a single layer onto a given area of mesentery. The oral disc has radial and circular muscle fibres and the tentacles have longitudinal and circular ones.

Most of the muscle fibres are outgrowths of typical musculoepithelial cells but some have their cell bodies reduced to a small mass of cytoplasm containing the nucleus, and so resemble the muscle fibres found in most other phyla. The sphincter muscle is like this, and so are some of the muscle fibres of the oral disc and tentacles.

Muscles can shorten, but they cannot forcibly extend: they cannot make a contracted anemone expand again by pushing it up. The movements of anemones do not depend on the muscles alone but on interaction between them, the liquid in the gastric cavity, the siphonoglyphs and the mesogloea. The role of the mesogloea is discussed in the next section of this chapter. The siphonoglyphs drive water into the anemone and are apparently used to inflate it to large sizes. This would not work if the mouth were a gaping hole: water could leave by the mouth as fast as it entered by the siphonoglyphs. However, the flexible pharynx must act as a valve, which tends to be closed if the pressure in the gastric cavity rises above the pressure of the surrounding water.

The pressure in the gastric cavity has been measured, by means of a manometer or a pressure transducer connected to the cavity by a glass tube pushed through a hole in the body wall. In one investigation the anemone *Tealia* was found to maintain pressures up to 3.5 cm water (350 N m^{-2}) while inactive. Higher pressures up to 10 cm water occurred during natural movement and up to 15 cm water when the anemone was made to contract by poking it.

When an anemone is inflated, contraction of the circular muscle of the column must tend to make it tall and thin, while contraction of the parietal muscles must tend to make it squat and fat (Fig. 3.14*e, f*). These movements will occur with no change of volume if the mouth remains closed. The parietal and circular muscles thus have opposite effects (they are antagonistic to one another) so long as the volume of water in the gastric cavity remains constant. If both contract together water must be driven out of the mouth and siphonoglyphs and the size of the anemone must be reduced (Fig. 3.14*d*). Anemones can also bend to one side (presumably by contracting the parietal muscles of that side only) and they sometimes contract only part of the circular muscle, producing a waisted effect. Contraction of the retractor muscles pulls the oral disc and tentacles down into the column, and if the sphincter then contracts it closes over the tentacles (Fig. 3.14*a, c*). An anemone contracted like this is much less vulnerable than when it has its tentacles spread. Extreme contraction of all the muscles buckles the body wall (Fig. 3.14*c*).

The retractor and sphincter muscles of at least some sea anemones have short myosin filaments (about 2 μm long) but the parietal and pharynx muscles have long filaments (about 7 μm). These are characteristics of fast and slow muscles, respectively, in other phyla (section 6.3).

Medusae also have musculoepithelial cells, but only on the inner surface of the bell. The fibres of some run circularly, and others radially. The swimming movements that they drive are illustrated in Fig. 3.15 (*a, b*). Scyphozoan medusae such as *Cyanea* have bands of radial and circular muscle which are thickened by folding of the epidermis, like the retractor muscles of sea anemones (Fig. 3.6*b*). The muscle fibres are about 0.3 μm in diameter, both in small hydrozoan medusae such as *Obelia* and in

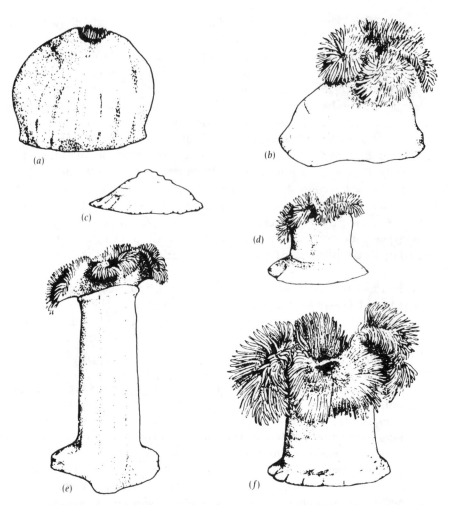

Fig. 3.14. Outlines traced from photographs to the same scale of the same individual *Metridium* on different occasions. Its diameter in (*a*) is 4 cm. From R.B. Clark (1964). *Dynamics in metazoan evolution*. Oxford University Press.

large scyphozoan medusae such as *Cyanea*. If large medusae had only a single unfolded layer of musculoepithelial cells, the proportion of muscle in the body would be very small.

The muscles of cnidarians are controlled by nerve cells, as are the muscles of more complex animals. Before examining these nerve cells, however, we must return briefly to the sponges. They have no nerve cells, but glass sponges have a system that behaves like a nervous system in that it carried messages around the body. When a glass sponge is poked, pinched or given an electric shock, water stops flowing out of the oscula near the point of stimulation. This has been demonstrated by holding tiny thermistor flowmeters over the oscula. (It is presumably due to the flagella of the choanocytes stopping beating.) It has been shown, by using several flowmeters held over oscula at different distances from the point of stimulation, that the stoppage spreads over the surface of the sponge at a speed of about 2.5 mm s^{-1}. It is believed

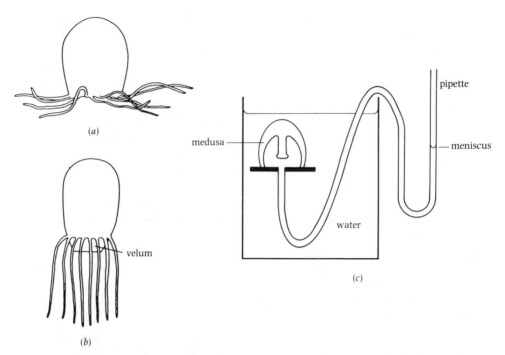

Fig. 3.15.(*a*), (*b*) A medusa of *Polyorchis montereyensis* swimming. Its diameter is about 4 cm. Based on photographs in W.B. Gladfelter (1972). *Helgolander wiss. Meeresunters.* 23, 38–79. (*c*) Apparatus used to investigate the elastic properties of *Polyorchis.*

that the signals responsible for this are transmitted through the trabecular tissue, the syncytial network of cells in the mesogloea.

Cnidarian neurons (nerve cells) were shown schematically in Fig. 3.4. Each has a central body containing the nucleus and two (or more) long processes which are called neurites. (The more familiar term 'axon' is generally reserved for a single long process.) The simplest type of nervous system, found in *Hydra*, is simply a network of neurons in the epidermis, all over the surface of the body. However, the following introduction to cnidarian nervous systems is based on the more obviously organized system of the medusa of *Polyorchis*, a hydrozoan (Fig. 3.15). Its main advantage for study is that some of the neurites are thick enough for an electrode to be put into them, to record changes of electrical potential.

The most obvious feature of the nervous system of *Polyorchis* (and of other hydrozoan medusae) is a double ring of neurons around the edge of the bell, close to the ring canal. The outer and inner nerve rings lie on opposite sides of the mesogloea, but neurites run through the mesogloea, connecting them. Sections through the inner nerve ring show several hundred neurites, most of them no more than a micrometre in diameter, but eight to ten of them are much larger with diameters of 10–25 μm. These neurites are big enough to stick an electrode into. If the electrode is hollow it can be used to inject dye into the neurite, to mark it and so enable the experimenter to check that the electrode had been where he or she intended. A fluorescent yellow dye which showed up brilliantly under ultra-violet light was used in the experiments on *Polyorchis*. This dye, injected into one of the giant neurites, spread within a few minutes into all the others, showing that they formed a continuous network. There were many cells in the network, each with its own nucleus, but there must be some sort of connection between the cells that allows the dye to pass through. This is an unusual arrangement. The cells in most nervous systems are effectively partitioned off from each other.

To obtain the recordings shown in Fig. 3.16, a piece cut from a medusa was pinned down in a dish of sea water. An electrode stuck into the inner nerve ring gave the upper record in (*b*), and subsequent injection of dye confirmed that the electrode was in a giant neurite. This record shows a fairly steady horizontal line with seven upward spikes: the electrical potential in the neurite was fairly constant for most of the time but seven brief electrical events ('action potentials') occurred in the ten seconds represented by the record. The membrane potential was about -60 mV for most of the time but reversed, reaching about $+30$ mV, during each action potential. Thus the action potentials are rather like the electrical events that occur in *Paramecium*, when it is prodded hard at the anterior end. There is, however, an important difference. The electrical events in *Paramecium* are due to a brief change in the permeability of the cell membrane that allows calcium ions to diffuse in. They cannot be produced when the ciliate is in calcium-free water. In contrast, the giant neurites of *Polyorchis* continue to produce action potentials in calcium-free seawater, but not in sodium-free seawater (which was prepared from water and salts, using the chloride of choline, an organic compound, instead of sodium chloride). Their action potentials, and those of other neurons, depend on movements of sodium rather than calcium. The mechanism has been worked out in considerable detail in experiments on the giant axons of squid.

Action potentials serve as signals, conveying messages along neurites. The giant neurites of *Polyorchis* make contact with the musculoepithelial cells that power swimming, and every action potential in the neurites is followed by a contraction of the swimming muscles. Fig. 3.16(*b*) (below) shows a record from an electrode in a musculoepithelial cell and shows that action potentials occur in them, as well as in the neurites, but there is a delay of about 3 ms between each nerve action potential and the one in nearby musculoepithelial cells.

Electron microscope sections show regions of contact between giant neurites and musculoepithelial cells with a group of vesicles in the neurite cytoplasm at the point of contact. These are presumably the 'synapses' at which excitation is transmitted from neuron to muscle. The vesicles presumably contain a transmitter substance that is released when an action potential arrives and stimulates the action potential that occurs (3 ms later) in the muscle. That is how synapses work in other animals (cat nerve–muscle synapses have been studied particularly thoroughly) and at present we can only presume that cnidarian synapses work the same way. There is some evidence that a small peptide found in the nerve cells of anthozoans may function as a transmitter substance in them. There are synapses between neurons, that transmit excitation from one neuron to another, as well as between neurons and muscle cells.

Action potentials in the giant neurites of *Polyorchis* always produce action potentials in the muscle cells, but there are other synapses where an incoming action

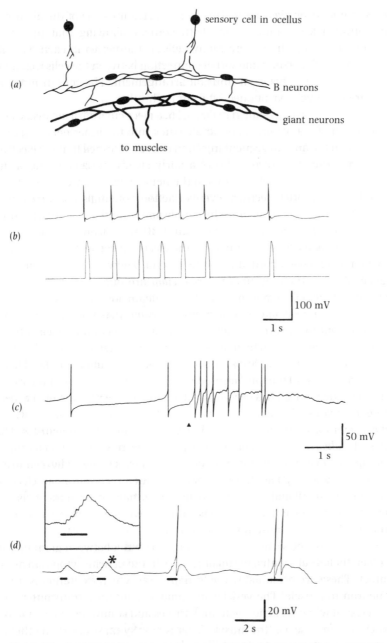

Fig. 3.16. Activity in the nervous system of *Polyorchis penicillatus*. (*a*) A diagram showing some of the connections in the nervous system. (*b*) Action potentials recorded simultaneously in (above) a giant neuron and (below) a musculoepithelial cell. (*c*) Action potentials recorded from a B neuron. The light was turned off at the arrow. (*d*) Electrical activity in a giant neuron. The bars show when the light was off. The asterisked response is shown enlarged in the inset. Parts (*b*) and (*d*) are from A.N. Spencer & W.E. Schwab (1982). In *Electrical conduction and behaviour in 'simple' invertebrates* (ed. G.A.B. Shelton), pp. 73–148. Clarendon Press, Oxford. (*c*) is from A.N. Spencer & S.A. Arkett (1984). *J. exp. Biol.* 110, 69–90.

potential does not always stimulate an outgoing one, as we shall see later.

Polyorchis has ocelli (simple eyes) at the bases of its tentacles. They are cup-shaped and contain two types of cell mingled together, sensory cells and pigment cells. The latter are filled with dark pigment, so the sensory cells only receive light that enters by the mouth of the cup. The sensory cells have neurites that synapse with neurites from neurons (called B neurons) in the outer nerve ring. The B neurons are smaller than the giant neurons of the inner nerve ring, but it has nevertheless been possible to place recording electrodes in them. Action potentials occur in them in the absence of stimulation but there is a particularly fast burst of action potentials whenever the light is turned out (Fig. 3.16c). This is presumably due to the connection with the ocelli.

The B neurons in turn connect with the giant neurons, presumably through the neurites that connect the outer to the inner nerve ring. Fig. 3.16(d) shows what happens in the giant neurons when the light is turned off. Short periods of darkness produce small increases in giant neuron membrane potential, but no action potentials. The enlarged inset shows that the membrane potential rises in small steps, each step probably the result of an action potential in the B neurons. Each of these B neuron action potentials probably releases a little transmitter substance, but only enough to stimulate a small rise (called an excitatory post-synaptic potential, EPSP) in the giant neuron membrane potential. If a second action potential arrives before the first EPSP dies away, the membrane potential builds up. If it builds up high enough, an action potential occurs in the giant neuron: this happens during the two longest periods of darkness in Fig. 3.16(d).

Thus messages pass from ocelli to B neurons to giant neurons to swimming muscles. This particular series of connections is responsible for an easily observed pattern of behaviour: an abrupt fall in light intensity stimulates a burst of swimming movements. The fall in light intensity might be the shadow of a predator and the response may have evolved as an escape mechanism, but swimming upwards does not seem the ideal response to the shadow of a predator overhead.

Much of the swimming that *Polyorchis* does seems to be spontaneous, due to rhythmic activity of the giant neurons rather than to any external stimulus.

Here is another example of cnidarian behaviour. This time we will not examine the nervous mechanisms (which are partly known) but simply enquire what stimuli evoke the behaviour. Many sea anemones eat quite large crustaceans, such as shrimps, and even small fish. Others, and other members of the Anthozoa, eat smaller members of the zooplankton. When potential prey touches the tentacles nematocysts discharge, attaching the prey to the tentacles and injecting venom into it. Next, the tentacles and the edge of the oral disc bend inward, carrying the food towards the mouth which is opened to receive it. How does the anemone recognize the food? We will enquire first about the stimulus for nematocyst discharge and then about the stimulus for passing food to the mouth.

Fig. 3.17 shows some experiments designed to discover the natural stimuli for nematocyst discharge. When a tentacle of *Anemonia* was touched with a human hair, nematocysts at the point of contact discharged (a). Touching with a clean glass rod did not cause discharge (b, c), and adding saliva to the water stimulated only a few nematocysts to discharge (d). However, large numbers were discharged by touching

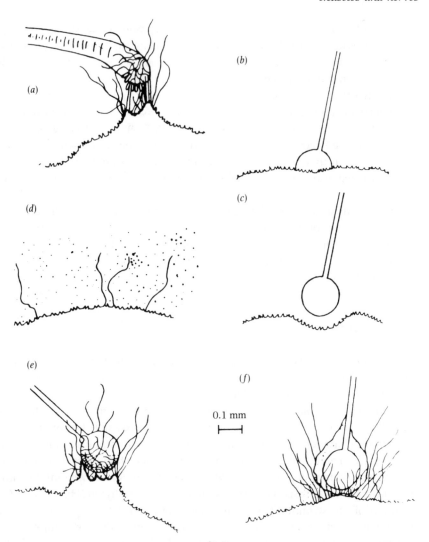

Fig. 3.17. Responses of nematocysts on the tentacles of the sea anemone *Anemonia* to various stimuli which are described in the text. From C.F.A. Pantin (1942). *J. exp. Biol.* 19, 294–310.

with the glass rod when there was saliva in the water (*e*) or when the rod had been smeared with a preparation of fatty material from the scallop, *Pecten* (*f*). These experiments seem to show that a combination of mechanical and chemical stimulation is normally required to cause discharge. The nematocysts (or at least the epidermis near them) must be touched, and some substance or substances must be present. The substances have not been identified but since they are present on hair, in saliva and in scallop tissues, they are probably widespread in and on animals. If the nematocysts discharged in response to touch or to chemical stimulation alone they would be likely to be wasted by being discharged at inappropriate times. As it is, they are likely to discharge and capture any potential food object which happens to touch them.

Experiments to discover the stimulus for passing food to the mouth were performed on *Palythoa*, which looks like a colonial sea anemone but belongs to a different group of Anthozoa. Pieces of filter paper were placed on the tentacles of *Palythoa* to see whether they were swallowed or not. Clean pieces were not swallowed, but pieces impregnated with an extract of brine shrimps (*Artemia*) were. The extract was prepared by shaking ground brine shrimps with alcohol. It was of course a complex mixture but it was separated into components by paper chromatography. This process yielded a sheet of filter paper impregnated at different places with different compounds. Pieces were cut from the sheet and offered to *Palythoa*. It was found that the part of the sheet which was impregnated with the amino acid proline was always swallowed, the part containing alanine was sometimes swallowed and other parts were seldom swallowed. Further experiments showed that paper impregnated with pure proline from other sources was swallowed, as was paper impregnated with glutathione (a peptide consisting of one residue each of glutamine, cysteine and glycine). Glutathione is present in crustacean body fluids and so is likely to leak or diffuse out of potential prey.

3.5.   Skeletons

Sponges, medusae and many cnidarial polyps are given shape by their mesogloea. In many of them the mesogloea is highly deformable, but it recoils elastically to its original shape after being deformed. To that extent, it serves as a skeleton. Other cnidarian structures such as the gorgonin of the stems of sea fans or the hard parts of corals are more obviously skeletal. This section is about all these kinds of skeleton.

Mesogloeas are jelly-like materials, containing protein and polysaccharide. For example, the mesogloea of the sea anemone *Metridium* contains about 8% protein and 1% polysaccharide: the rest is water and dissolved salts. That of the scyphozoan medusa *Aurelia* is even more dilute, with only 1% organic matter. For comparison, table jelly needs at least 5% gelatin to be reasonably stiff.

To understand the functions of mesogloea and other skeletal materials, we need to be clear about the difference between two kinds of mechanical property. Elasticity is the property that makes rubber bands spring back to their original length after being stretched. Viscosity is the property that makes treacle (molasses) hard to stir. Both properties make materials resist deformation but the resisting force depends on the amount of deformation (in the case of elasticity) or the rate of deformation (in the case of viscosity).

We need to know something about the properties of high polymers, materials with molecules made of long chains of more or less similar units. Proteins are high polymers built of amino acid residues and polysaccharides are high polymers of sugar units. Rubber and plastics are also high polymers. The mechanical properties of these materials depend on how their long, flexible molecules are linked together.

Some of the possibilities are shown in Fig. 3.18. Here (*a*) represents a material like latex, in which the molecules are not attached to each other. When it is stretched, the molecules slide past each other. The material may have a high viscosity but it is not elastic (except for transient elastic effects due to molecules getting tangled with each

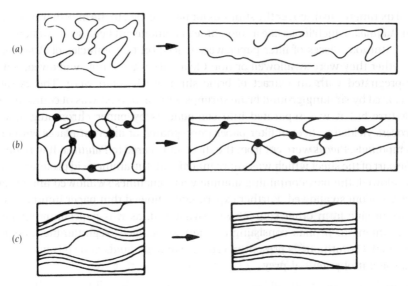

Fig. 3.18. A diagram showing the arrangement of molecules in three types of high polymer, and the effects of stretching. (*a*) An amorphous polymer which is not cross-linked; (*b*) an amorphous cross-linked polymer; and (*c*) a fibre. From R. McN. Alexander (1975). *Biomechanics.* Chapman & Hall, London.

other). Fig. 3.18(*b*) shows a similar material, but with the molecules connected together in a three-dimensional network. Rubber is like this: the process of vulcanization inserts sulphur bridges between the latex molecules. The molecules cannot slide past each other but a great deal of stretching is possible, unravelling the molecules and drawing them out in the direction of stretch. When the material is released it recoils elastically, for the following reason. The long molecules are continually coiling and uncoiling, in Brownian motion. A given molecule will sometimes be rather extended and sometimes be rolled up into quite a tight ball. The distance between its ends fluctuates, but will generally not be too far from a certain most probable value. Stretching draws the molecules out in the direction of stretch but Brownian motion tends to restore them to their most probable lengths when the stretching force is removed, causing elastic recoil. Very large extensions are possible: for instance, a rubber band can be stretched to about four times its initial length. Elastic moduli are fairly low (i.e. quite small stresses can produce large distortions). Young's modulus* is commonly of the order of 1 MN m^{-2} but may be very much lower, particularly for dilute gels. The more cross-links there are, the higher the modulus. In (*b*) the material is amorphous but in (*c*) parts of the molecules are lined up parallel with each other in a crystalline array. This severely limits the amount of stretching that is possible without breaking molecules. Some fibres, including collagen, seem to have this sort of structure. They cannot generally be stretched by more than 20% and Young's modulus is quite high, generally of the order of 1 or 10 GN m^{-2}. That means that the stress needed to produce a given distortion is often 1000 or more times as much as for amorphous polymers.

* Young's modulus measures how hard a material is to stretch: it is the tensile stress (the stretching force, per unit cross-sectional area) divided by the strain (the extension as a fraction of the initial length).

The mechanical properties of sea anemone mesogloea have been investigated by stretching strips of body wall. Cutting a sea anemone makes its muscles contract violently, and the mechanical properties of strips cut from untreated anemones owe as much to the muscle as to the mesogloea. However, the muscles can be made inactive by leaving the anemone for several hours in a mixture of equal volumes of sea water and magnesium chloride of the same osmotic concentration, or in a solution of menthol in sea water (many other marine invertebrates can be immobilized in the same ways). The anemone becomes almost entirely unresponsive to mechanical stimulation. Strips of body wall from anemones narcotized in these ways have been used in tests of mechanical properties. When such strips are stretched by a constant load, they stretch at a gradually diminishing rate. In a typical experiment with *Metridium* body wall, the specimen doubled in length in the first half hour but was still stretching at a perceptible rate after 10 hours and eventually (after 20 hours) reached three times its initial length. When the load was removed the strip returned gradually, over many hours, to its initial length.

These properties represent a combination of elasticity and viscosity: the viscous component prevents the mesogloea from stretching fast but the elastic component sets a limit to the amount of stretching (for any given load) and makes the specimen recoil when the load is removed. The viscous component makes the animal rather resistant to brief forces, though it is easily deformed by forces lasting several hours. The animal resists waves, for instance, as though it were rather rigid, but a small pressure exerted for an hour or more by the siphonoglyph may inflate it enormously. (Unlike many other sea anemones, *Metridium* has only one siphonoglyph.) *Metridiur.* seems unable to inflate itself quickly: inflation from the condition shown in Fig. 3.14 (*d*) to that shown in Fig. 3.14(*e*) takes an hour or so. Contraction is effected by muscles, assisted by elastic forces in the stretched mesogloea, and can be much faster. A fully expanded anemone can be stimulated by vigorous poking to contract to the condition shown in Fig. 3.14(*c*) in a few seconds.

Since the mesogloea is elastic it will always return (slowly) to the same size, when it is unstressed. This is presumably the size eventually reached when the animal is narcotized in magnesium chloride or menthol solution. Narcotized *Metridium* generally look slightly more inflated than the specimen shown in Fig. 31.4(*d*). A fully inflated *Metridium* which halts the cilia of its siphonoglyph will tend to shrink to this size and a fully contracted one which relaxes its muscles will spring up to this size.

Sea anemone mesogloea has fibres running through it in various directions. These fibres are believed to be collagen, the protein that forms the tendons of vertebrate animals. (Part of the evidence is that the mesogloea gives an X-ray diffraction pattern like that of rat tendon.) This presents a puzzle, because tendon collagen can only be stretched about 8% before breaking. How can mesogloea with collagen fibres running through it stretch to three times its unstressed length?

The most plausible explanation seems to be that the collagen fibres are not continuous but are relatively short. Only the dilute protein–polysaccharide gel between them forms a continuous network of linked molecules extending throughout the mesogloea. This has not been demonstrated directly. No method has been devised for measuring the length of the fibres and it is not at all obvious, when they are examined under a microscope, that individual fibres do not run right through the mesogloea.

Some other cnidarians and many sponges have their mesogloea reinforced by crystalline spicules. For example, the mesogloea of the 'fingers' of *Alcyonium* contains irregularly star-shaped spicules of calcium carbonate, up to 0.3 mm in diameter, making up 12% of its volume. That of the sponge *Suberites* has needle-like crystals of silica, up to 0.3 mm long, amounting to 14% of its volume. These reinforced mesogloeas are stiffer than ones without spicules: identical stretching tests on *Metridium* and *Alcyonium* mesogloea showed that the latter was almost ten times stiffer. Similarly, many man-made materials are stiffened by embedded particles or fibres. The rubber used for making automobile tyres is greatly stiffened by mixing in carbon particles (which also make it black).

A simple experiment seems to confirm that spicules have a stiffening effect on mesogloea. Cnidarian and sponge spicules were mixed with melted gelatine (as sold for making table jelly) and a little water. The jelly was allowed to set and then strips were cut from it and subjected to stretching tests. Specimens containing 15% by volume of *Alcyonium* spicules or 5% by volume of *Suberites* crystals were about ten times stiffer than similar jelly without spicules.

Animals with spicules in their mesogloea seem unable to make big changes of size and shape, like those of *Metridium* (Fig. 3.14). However, *Alcyonium* does inflate itself and contract a little, varying the height of the colony by about 30%.

Gorgonians have a delicate fan-like structure (Fig. 3.19) but grow upright on coral reefs and have to withstand wave action. Their blades are stiffened only by mesogloea with calcareous spicules in it, and are quite flexible. Their slender stems have a core of material called gorgonin, which is not part of the mesogloea. This makes the stems stiff, and remarkably strong. To collect a large gorgonian from a coral reef it is far easier to hack away the reef limestone at its base than to break or cut the stem.

Gorgonin is fibrous, with fibres running predominantly parallel to the axis of the stem. X-ray diffraction patterns and amino acid analysis show that it consists largely of collagen, but its Young's modulus is several times higher than that of ordinary tendon collagen. Also, ordinary collagen can be dissolved by autoclaving (heating with water under pressure) but gorgonin cannot. The high modulus and the insolubility suggest that gorgonin has additional cross-links which are more resistant to heat than the cross-links of tendon collagen. The only known way of dissolving gorgonin (except by really drastic action with concentrated acids or alkalis) is by putting it in a 10% solution of sodium hypochlorite. It shares this property with some other proteins (including the protein of insect cuticle) which are known to be cross-linked by a process called quinone tanning. It is therefore presumed that gorgonin is a quinone-tanned collagen.

Ortho quinones have structures like this

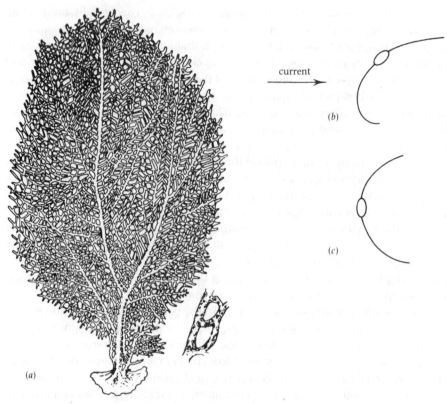

Fig. 3.19.(*a*) *Gorgonia*, a sea fan (height about 50 cm), and a small portion of it enlarged to show the holes for the polyps. From L.H. Hyman (1940). *The invertebrates*, vol. 1. McGraw-Hill, New York. (*b*), (*c*) Diagrammatic horizontal sections of a sea fan bending in a water current.

(there are various possible side chains R). In the process of quinone tanning an *ortho* quinone reacts with free NH_2 groups in adjacent peptide chains, linking them together. For example, it may attach to the second NH_2 group of lysine, thus

Clumps of sea fans are common on Atlantic reefs. They grow upright, and the fans in any clump tend to be parallel to each other. Divers on reefs off Florida measured the compass bearings of individual fans and found that large fans (0.5 m or more high) seldom deviated by more than 20° from the average direction of the clump. Smaller fans tended to deviate more, and sections of the stems of some large fans seemed to show that they had twisted as they grew. The large fans generally stand at right angles to the prevailing wave motion. Thus sea fans apparently start their life oriented at random and gradually twist at right angles to the direction of water movement.

Living sea fans are flexible, unlike the dried specimens kept in museums. Fig. 3.19 (*b,c*) shows sections of a flexible blade in a current. When the blade is set obliquely in the current, it bends in such a way that its upstream edge is more nearly at right angles to the current than the downstream edge. Hence greater forces act on the upstream than on the downstream half of the blade, and the asymmetrical forces tend to twist the blade more nearly at right angles to the current. If the gorgonin is not fully cross-linked it will flow very slowly in viscous fashion and a constant current (or one which flows backwards and forwards along the same line, like wave motion) will gradually twist it. It is believed that this is what happens.

Quinone-tanned proteins are also found in the perisarcs that stiffen and protect hydrozoan colonies. These consist of chitin and tanned protein, the same combination of materials as is found in insect cuticle. Chitin is a polysaccharide built up from acetylglucosamine units: that is, from glucose units in which one of the OH groups has been replaced by $NHCOCH_3$. Its presence can be demonstrated by histochemical tests. The evidence that the protein is quinone-tanned is indirect. Histochemical tests on *Campanularia* show that some of the epidermal cells under the perisarc contain the enzyme phenol oxidase and also a substance (dopamine) which it can oxidize to an *ortho* quinone.

Gorgonian and hydroid colonies avoid the large forces that waves and water currents might exert on them, by bending. The drag forces that water currents exert on rigid structures are about proportional to the square of the current speed (provided the Reynolds number is not too low; see section 2.8). Colonies of the hydroid *Abietenaria* were attached to a strain gauge force transducer and held in their natural posture in a flow tank. The drag force on them, measured by the transducer, was proportional to the speed of the water current (not to its square). However, the force on colonies that had been stiffened artificially, with wire and plastic, was proportional to the square of the speed. This is a big difference: doubling current speed only doubles the drag force on the flexible hydroid colonies, but increases the force on rigid structures by a factor of four.

Stony corals have rigid skeletons, and cannot avoid drag forces in this way. Those that are exposed to fast currents have to be massively built. None of them can have delicate branching structures because their skeletons are made of calcium carbonate with very little organic matter and so are brittle, like pottery. The physical basis of brittleness is explained in secion 6.2, in an account of mollusc shell.

From these rigid skeletons we return to a much more flexible skeleton, the mesogloea of medusae. Fig. 3.15(*c*) shows an experiment that demonstrated the elastic properties of *Polyorchis*. The edge of the medusa is stuck to a glass plate by

Superglue. When the pipette is lowered, the pressure inside the medusa's bell is reduced and the bell contracts, much as it does in swimming. The movement of the meniscus in the pipette shows how much its volume has changed. When normal pressure is restored by raising the pipette the medusa recoils elastically, springing back to its original shape. In swimming, muscles make the bell contract, but it expands by elastic recoil.

The mesogloea of medusae has a second function, as a buoyancy organ. It is less dense than sea water, which at first seems surprising: even a very dilute gel, made up in sea water, would be denser than sea water. The reason is that it has the same total osmotic concentration as sea water, but with less sulphate and more of other ions (sulphate solutions are denser than isotonic chloride solutions). The effect can only be slight, because sea water contains less than 0.3% sulphate. *Aurelia* mesogloea is about 0.1% less dense than sea water.

The mesogloea is less dense than sea water but the cells of marine medusae are denser than sea water. Some, such as *Pelagia* (Scyphozoa), have overall approximately the same density as sea water. Others such as *Polyorchis* (Hydrozoa) sink quite rapidly whenever they stop swimming.

Swimming involves alternate contraction and expansion of the bell (Fig. 3.15). Contraction drives water downwards out of the bell and propels the medusa upwards. Expansion draws water up into the bell and must tend to pull the medusa down again, but a swimming medusa can make progress because the contraction is more rapid than the expansion.

Polyorchis swims upwards by contracting a few times and then allows itself to sink before swimming up again. As it sinks its flexible tentacles tend to spread out around it (Fig. 3.15a). Food objects in the water which passes between them can be caught. It may be an advantage to *Polyorchis* to be considerably denser than the water. If it sank more slowly it would not have to use so much energy for swimming, but it would have less opportunity to catch food. However, it would be disadvantageous to sink too fast.

The buoyant mesogloea of *Polyorchis* must tend to keep it the right way up as it sinks. If the edge of the bell (with the tentacles and velum) is cut off, the remainder floats. Thus the denser parts of the animal are concentrated around the edge of the bell.

3.6.   Reef-building corals

Three groups of Cnidaria contribute to coral reefs. The most important is a group of stony corals which are members of the Zoantharia. The blue coral (*Heliopora*) and the organ pipe coral (*Tubipora*) are Alcyonaria and can be recognized as such by the eight pennate tentacles on each polyp. Finally the milleporine corals are not Anthozoa at all, but members of the class Hydrozoa.

Reef-building corals have unicellular algae in their cells. These are generally found only in the gastrodermis, but they may contain as much as 50% of the total protein content of the coral. They are oval cells about 10 μm long with chloroplasts and thick cell walls. They can be separated from the host tissues by grinding and centrifuging and grown in culture in suitable solutions. When this is done some of them grow flagella and take the shape of a typical dinoflagellate. The alga usually (and perhaps universally) found in reef-building corals is the dinoflagellate *Gymnodinium microadriaticum*, another species of the genus illustrated in Fig. 2.1(*c*).

The algae grow satisfactorily in solutions of inorganic salts (plus traces of two vitamins), getting their energy by photosynthesis. The coral can also live without the algae. If a coral is kept in darkness for long enough the algae, unable to photosynthesize, degenerate and are expelled by the host cells. In such experiments, the coral must be fed to keep it alive. However, corals kept in the light have survived without food for over a year, which suggests that they can benefit from the photosynthesis of the algae. This has been confirmed in experiments in which corals were kept in the light in the presence of $^{14}CO_2$. Polyps were subsequently sectioned and examined by autoradiography to find the ^{14}C. Radioactivity was found in the tissues of the coral, as well as in the algae. It was found even in the epidermis where there were no algae. It could only have been captured by photosynthesis so it must have been captured by the algae and passed to the coral.

This could happen if the coral periodically digested some of the algae, but there is no good evidence that this happens. Another possibility is suggested by the observation that when the algae are kept in culture, glycerol and a few other organic compounds diffuse out into the culture medium. If these compounds also diffuse out when the alga is in a coral cell, they will become available to the coral. Evidence that they do has been obtained by rather a subtle experiment. Pieces were cut off corals (using wire cutters) and put in sea water to which $NaH^{14}CO_3$ had been added. They were kept in this solution in bright shade out of doors so that the algae took up ^{14}C by photosynthesis. After a while the water was analysed. Acid was added to it to drive off the $^{14}CO_2$ from the bicarbonate, so that any remaining radioactivity could only be due to organic compounds produced by photosynthesis. Very little radioactive organic matter was found in the water, presumably because any organic compounds which diffused out of the algae were taken up by the coral cells before they could escape to the water. The experiment was repeated, adding (non-radioactive) glycerol as well as $NaH^{14}CO_3$ to the initial solution. When this was done a radioactive organic compound escaped into the water. It was shown by chromatography that this compound was [^{14}C]glycerol. In this experiment, so much glycerol was diffusing into the coral cells that they could not use it all, and some of the radioactive glycerol diffused right through them into the water. Similarly in the presence of glucose and of the amino acid alanine, [^{14}C]glucose and [^{14}C]alanine escape into the medium. Fructose, glutamic acid and several other compounds which were tested had no effect: they did not cause release of radioactivity into the water. It is concluded that glycerol, glucose and alanine leak out of the algae into the coral cells. The quantities of these compounds are quite large. In typical experiments with glycerol, the glycerol in the water at the end of the experiment contained 10% of the ^{14}C fixed by photosynthesis (the rest was in the coral). The glycerol and glucose which the coral gets from its algae presumably serve as energy sources. The alanine may be used in part to synthesize other amino acids and build proteins, providing for growth as well as metabolism.

Though the algae lose so much of the products of their photosynthesis, the coral cells are a very favourable habitat for them. This is because the metabolism of the coral releases nitrogenous wastes, and nitrates are apt to be scarce in the sea, especially in the tropics. The association between coral and alga seems beneficial to both. Such associations are called symbioses (singular, symbiosis). The association between cattle and rumen ciliates is another example (section 2.3).

Further Reading

Sponges

Bergquist, P.R. (1978). *Sponges*. Hutchinson, London.
Mackie, G.O. *et al.* (1983). Studies on hexactinellid sponges, I to III. *Phil. Trans. R. Soc. Lond.* B**301**, 365–428.
Vogel, S. (1974). Current-induced flow through the sponge *Halichondria*. *Biol. Bull.* **147**, 443–456.

Cnidarians

Barham, E.G. (1966). Deep-scattering layer migration and composition: observations from a diving saucer. *Science* **151**, 1399–1403.
Diamond, J.M. (1986). Clones within a coral reef. *Nature* **323**, 109 only.
Kamplan, S.W. (1983). Intrasexual aggression in *Metridium senile*. *Biol. Bull.* **165**, 416–418.
Mackie, G.O. (1986). From aggregates to integrates: physiological aspects of modularity in colonial animals. *Phil. Trans. R. Soc. Lond.* B**313**, 175–196.
Muscatine, L. & Lenhoff, M.M. (1974). *Coelenterate biology. Reviews and new perspectives.* Academic Press, New York.
Picken, L.E.R. & Skaer, R.J. (1966). A review of researches on nematocysts. *Symp. zool. Soc. Lond.* **16**, 19–50.

Multicellular structure

Curtis, A.S.G. (1973). Cell adhesion. *Prog. Biophys. molec. Biol.* **27**, 317–386.
David, C.N. & Gierer, A. (1974). Cell cycle kinetics and development of *Hydra attenuata*. III. Nerve and nematocyte differentation. *J. Cell Sci.* **16**, 359–375.
Gierer, A. (1974). *Hydra* as a model for the development of biological form. *Scient. Am.* **231**(6), 44–54.
Hicklin, J., Hornbruch, A., Wolpert, L. & Clarke, M. (1973). Positional information and pattern regulation in *Hydra*: the formation of boundary regions following grafts. *J. Embryol. exp. Morph.* **30**, 701–25.
Wolpert, L., Hicklin, J. & Hornbruch, A. (1971). Positional information and pattern regulation in regeneration of *Hydra*. *Symp. Soc. exp. Biol.* **25**, 391–415.

Muscles and nerves

Lawn, I.D., Mackie, G.O. & Silver, G. (1981). Conduction system in a sponge. *Science* **211**, 1169–1171.
McFarlane, I.D., Graff, D. & Grimmelikhuijzen, C.J.P. (1987). Excitatory actions of Antho-RFamide, an anthozoan neuropeptide, on muscles and conducting systems in the sea anemone *Calliactis parasitica*. *J. exp. Biol.* **133**, 157–168.

Reimer, A.A. (1971). Chemical control of feeding behaviour in *Palythoa* (Zoanthidea, Coelenterata). *Comp. Biochem. Physiol.* **40A**, 19–38.

Shelton, G.A.B. (ed.) (1982). *Electrical conduction and behaviour in 'simple' invertebrates.* Clarendon, Oxford.

Spencer, A.N. & Arkett, S.A. (1984). Radial symmetry and the organization of the central neurones in a hydrozoan jellyfish. *J. exp. Biol.* **110**, 69–90.

Skeletons

DeMont, E. (1988). Mechanics of jet propulsion in a hydromedusean jelly fish *Polyorchis penicillatus*, I to III. *J. exp. Biol.* **134**, 313–361.

Goldberg, W.M. (1974). Evidence of a sclerotized collagen from the skeleton of a gorgonian coral. *Comp. Biochem. Physiol.* **49B**, 525–529.

Harvell, C.D. & LaBarbera, M. (1985). Flexibility: a mechanism for control of local velocities in hydroid colonies. *Biol. Bull.* **168**, 312–320.

Koehl, M.A.R. (1982). Mechanical design of spicule-reinforced connective tissue: stiffness. *J. exp. Biol.* **98**, 239–267.

Reef-building corals

Lewis, D.H. & Smith, D.C. (1971). The autotrophic nutrition of symbiotic marine coelenterates with special reference to hermatypic corals. I. Movement of photosynthetic products between symbionts. *Proc. R. Soc. Lond.* **B178**, 111–129.

Lewis, J.B. (1977). Processes of organic production on coral reefs. *Biol. Rev.* **52**, 305–347.

Muscatine, L., Falkowski, P.G., Porter, J.W. & Dubinsky, Z. (1984). Fate of photosynthetic fixed carbon in light- and shade-adapted colonies of the symbiotic coral *Stylophora pistillata*. *Proc. R. Soc. Lond.* **B222**, 181–202.

Taylor, D.S. (1973). The cellular interactions of algal–invertebrate symbiosis. *Adv. mar. Biol.* **11**, 1–56.

4 *Flatworms*

Phylum Platyhelminthes
    Class Turbellaria   (free-living flatworms)
    Class Monogenea   (ectoparasitic flukes)
    Class Digenea   (endoparasitic flukes)
    Class Cestoda   (tapeworms)
Phylum Nemertea

4.1.   Free-living flatworms

The turbellarians are primitive worms which crawl along on the ventral surface of the body or in some cases swim. One end of the body, the anterior, normally takes the lead. This is a very different way of life from that of a sessile polyp or of a medusa which alternately swims up and sinks down but undertakes little directed horizontal movement. There is a related difference in the symmetry of the body.

A circular cake is radially symmetrical if it can be divided into slices (cut as sectors in the usual way) which are identical with each other. A vertical line through the centre of the cake is its axis of symmetry. A loaf is bilaterally symmetrical if the left-hand half of every slice (cut transversely, as bread is sliced) is a mirror image of the right-hand half. It would not matter if one end of the loaf were different from the other: the loaf would still be bilaterally symmetrical. A bilaterally symmetrical object has no axis of symmetry but only a plane of symmetry. This divides it into two equal halves which are mirror images of each other.

Fig. 4.1(*a*) shows that an individual polyp is more or less radially symmetrical. This seems appropriate to its way of life. If it has a mouth at one end (the oral end) and attaches the other (aboral) end to the bottom there is no obvious reason why it should not be more or less radially symmetrical about an oral–aboral axis. Anthozoan polyps are not perfectly radially symmetrical (their symmetry is spoilt by the siphonoglyphs, for instance) but the departures from radial symmetry are quite small. Medusae are similarly more or less radially symmetrical (Fig. 4.1*b*).

A well-adapted flatworm requires differences of structure between the anterior end, which leads as it crawls, and the posterior end, which follows behind. (For instance, its principal sense organs will be most useful at the anterior end.) It also requires differences between the ventral surface, which rests on the ground, and the dorsal one, which does not. These requirements cannot both be met by a radially symmetrical design, and flatworms are in fact bilaterally symmetrical (Fig. 4.1*c*). The plane of symmetry is known as the median plane. Fig. 4.1(*d*) shows the meanings of terms used to describe sections through bilaterally symmetrical animals.

Most of the animals described in the remainder of this book are more or less

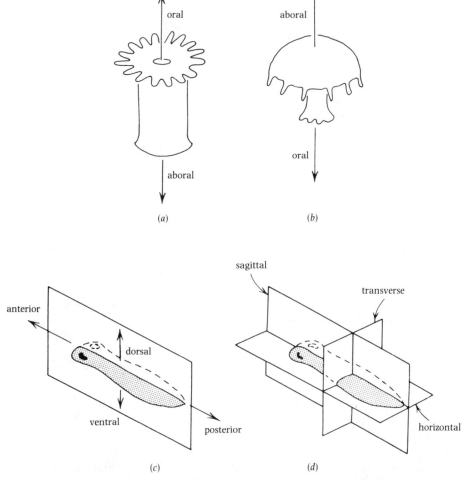

Fig. 4.1.(*a*), (*b*) Diagrams of a polyp and a medusa showing their axes of radial symmetry (arrows). (*c*) A diagram of a turbellarian showing its plane of bilateral symmetry. (*d*) A diagram showing terms used to describe sections through bilaterally symmetrical animals. Each term can be applied to any section parallel to the plane indicated.

bilaterally symmetrical. There are, however, animals which are far from symmetrical, either radially or bilaterally. *Euplotes* (Fig. 2.4*c*) is one example; a snail is another.

The body walls of Cnidaria consist of an outer epidermis and an inner gastrodermis, each a single layer of cells, with a more or less cell-free mesogloea between. Those of sponges are similar, but with more cells in the mesogloea (Fig. 3.1*b*). The Platyhelminthes have no mesogloea. The epidermis is a single layer of cells on the outside of the animal and the gastrodermis is a single layer lining the gut, but the space between them is packed with cells (Fig. 4.2).

The epidermal cells of the ventral surface bear cilia but those of the dorsal surface often do not. Most of the epidermal cells contain rod-shaped bodies called rhabdites which are discharged when the worm is injured and swell up to form a gelatinous

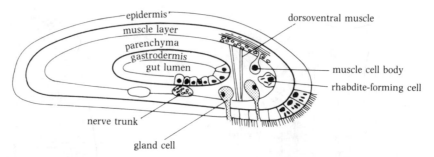

Fig. 4.2. **A diagrammatic transverse section through a turbellarian.**

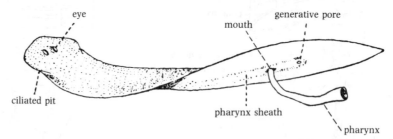

Fig. 4.3. *Dugesia*, a typical triclad 1–2 cm long. From A.E. Shipley & E.W. McBride (1901). *Zoology.* Cambridge University Press.

covering for the body. There is usually a thin layer of fibrous material (the basement membrane) underlying the epidermis and another underlying the gastrodermis.

Immediately under the epidermis is muscle, typically an outer layer of circular muscle fibres and an inner layer of longitudinal ones. Musculoepithelial cells have been found in a few turbellarians but in most species the muscle cell bodies are under the epidermis. Between the muscle layers and the gastrodermis is a tissue called parenchyma, which contains various types of cell. There are gland cells with ducts running through the muscle layers and epidermis to openings in the body surface. There are cells with rhabdites forming in them which seem to be developing epidermal cells. There are dorsoventral muscle fibres and cell bodies of muscle cells which have their fibres in the muscle layers. There are also cells of no obvious function.

Dugesia (Figs. 4.3 and 4.4) will serve as a typical example of the Turbellaria. It is a member of the order Tricladida. It is a flattened worm which growns to lengths around 2 cm and is common in lakes, ponds and slow streams where it crawls over the surface of the mud. It crawls at speeds up to about 1.5 mm s⁻¹ simply by beating its ventral cilia. It can travel a little faster by muscular action, using essentially the same technique as crawling snails (see section 6.4). It has a pair of eyes at the anterior end of the body. The light-sensitive cells of each eye are enclosed in a cup of pigment cells so that the left eye receives light only from the left and the right eye only from the right. On either side of the head are ciliated pits containing sensory cells which respond to chemical stimuli. The mouth is not at the anterior end of the body but on the ventral surface. A muscular pharynx can be protruded from it for feeding. The gut has three main branches: one runs anteriorly and two posteriorly from the base of the pharynx. They give rise to subsidiary branches, which are shown complete only for

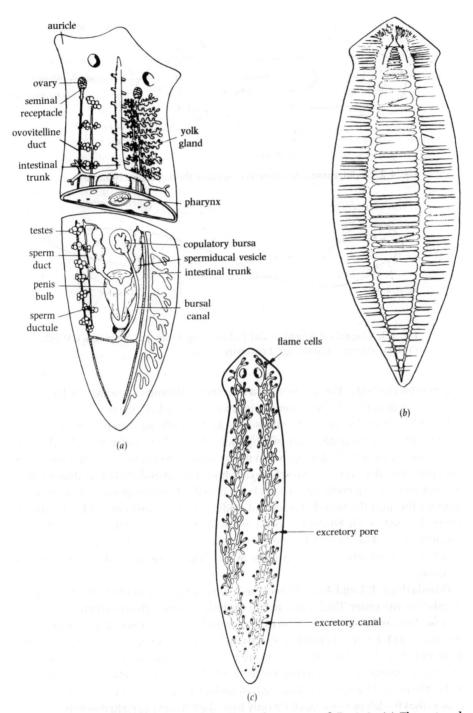

Fig. 4.4. Diagrams showing the internal structure of *Dugesia*. (*a*) The gut and reproductive system. From P.A. Meglitsch (1972). *Invertebrate zoology*, 2nd edn. Oxford University Press. (*b*) The nervous system. From J. Ude (1908). *Z. wiss. Zool.* 89, 308–370. (*c*) The protonephridial system, with the bore of the tubules exaggerated. From K. Schmidt-Nielsen (1975). *Animal physiology: adaptation and environment.* Cambridge University Press.

the right posterior main branch in Fig. 4.4(*a*). Feeding and digestion are discussed in section 4.4.

The nervous system (Fig. 4.4*b*) includes a pair of connected ganglia at the anterior end of the body and a pair of main ventral nerve trunks which run posteriorly from them along the length of the body. There are branches from the main trunks and connections between them. The trunks and their main branches lie immediately internal to the muscle layers of the body wall. The protonephridial system (Fig. 4.4*c*) consists of flame cells (described later in this chapter) and branching tubules which lead from them to pores at the surface of the body. The system probably functions as a kidney, getting rid of excess water and waste products.

Dugesia and nearly all other Turbellaria are hermaphrodite: that is to say, each individual has both male and female gonads. Fig. 4.4(*a*) shows the arrangement of the very complicated reproductive system.

Sperm from the many testes pass along the sperm ducts and are stored in the spermiducal vesicles, which connect to the muscular penis. This normally lies in a cavity just within the genital pore, but can be protruded like the pharynx. The ovovitelline ducts connect the ovaries and yolk glands to the genital pore. In most animals, the store of food required for development is incorporated as yolk in the ovum but in many platyhelminths it is kept in several yolk cells.

When they copulate, two animals place their genital openings in contact with each other and remain thus for a few minutes. The details of copulation cannot be seen by examining intact animals but are revealed by microscope sections of pairs killed in the act. Each partner protrudes its penis and inserts it through the genital pore of the other, into the duct of the copulatory bursa.

The sperm are inactive in the spermiducal vesicle but become active on entering the copulatory bursa. They travel from it up the ovovitelline ducts where they are joined by ripe ova, which they fertilize. The fertilized ova travel down the ducts, mingling with yolk cells. Several ova and hundreds of yolk cells are enclosed in a capsule formed from droplets enclosed in the yolk cells. The capsules are given a sticky coating and adhere after they have been laid onto surfaces such as the undersides of stones. The embryos ingest the yolk cells as they develop and minute worms hatch from the capsules after a few weeks.

Turbellarians kept in isolation sometimes lay capsules, but these do not hatch. Self-fertilization may be prevented by the mechanism that keeps the sperm inactive in the spermiducal vesicles.

Dugesia also reproduces asexually, by splitting in two transversely.

The Nemertea are a small phylum of free-living worms that resemble the Turbellaria in many ways. Most of them are long and slender, and dorsoventrally flattened. Some are astonishingly long. The bootlace worm, *Lineus longissimus*, is commonly found as a tangled mass under stones on W. European shores. It is only a few millimetres wide but is often several metres long.

Like turbellarians, nemerteans have a parenchyma and a protonephridial system. However, the reproductive system is quite simple and most species have separate sexes. There is an anus at the posterior end of the body as well as a mouth at the anterior end. There is a simple blood system. Finally, there is a characteristic proboscis which is normally housed in a cavity in the body but can be extended to attack prey.

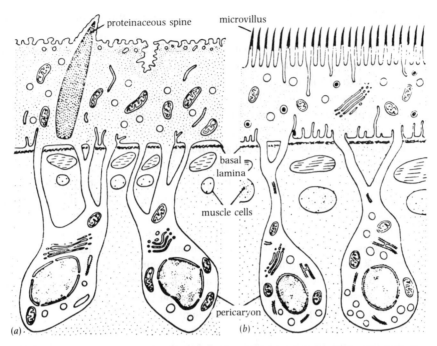

Fig. 4.5. Diagrammatic sections, based on electron micrographs, through the teguments of (*a*) the trematode *Fasciola* and (*b*) the cestode *Abothrium*. From U. Welsch & V. Storch (1976). *Comparative animal cytology and histology*. Sidgwick & Jackson, London.

4.2. Parasitic flatworms

A few of the turbellarians are parasitic, but the worms described in this section belong to the other platyhelminth classes.

Most Monogenea and Digenea are leaf-shaped, like the larger Turbellaria. They are the flukes. The Cestoda have a characteristic ribbon-like (or tape-like) shape, and are called tapeworms. Most of them have constrictions which divide the body into segments called proglottids. All three parasitic classes have parenchyma and nervous, protonephridial and reproductive systems similar to those of Turbellaria. They also have attachment organs of one sort or another, and they have a peculiar epidermis which is known as a tegument (Fig. 4.5).

The tegument is syncytial: that is, it has no cell boundaries although it has many nuclei. Its outer zone is a thin layer of cytoplasm without nuclei, 20 μm thick in the trematode *Fasciola* on which Fig. 4.5(*a*) is based. This layer often includes spines which seem to consist of crystalline protein, and it is perforated by the endings of sensory cells. Immediately under the outer zone is a basement membrane (basal lamina) and under that again are layers of circular and longitudinal muscle fibres. The inner zone of the tegument is under all these. It consists of tegumentary cell bodies (pericarya), each containing a nucleus and each connected to the outer zone by several fine strands of cytoplasm. The cell bodies extend deep into the parenchyma.

The Monogenea consists mainly of ectoparasites which live on the external surfaces of fishes and amphibians. An example is *Entobdella soleae* (Fig. 4.6) which lives on the skin of the common sole, *Solea solea*. It is a flattened hermaphrodite worm with a conspicuous attachment organ, the haptor, at the posterior end. Its internal structure is like that of *Dugesia* (Fig. 4.4) in many ways. There is a mouth with a short protrusible pharynx at the anterior end. The gut has two main branches which run posteriorly from the mouth, giving off subsidiary branches. The gut is shown only on the left side of Fig. 4.5 while the yolk glands (vitelline follicles) are shown only on the right. The yolk glands are scattered widely through the body, as in *Dugesia*. There is only one ovary. The sperm ducts from the testes lead to a protrusible penis. The female reproductive system has an external opening beside the penis and an additional vaginal opening.

The haptor has stiff skeletal elements (sclerites) which seem to be made of a material that resembles keratin, the protein of hair and horn: they give an X-ray diffraction pattern like keratin, and like it incorporate a good deal of the amino acid cysteine. Some of the sclerites have hooks which help to attach the worm to its host, but the haptor seems to attach mainly by suction. If its edge is lifted with a needle it releases immediately (except for the hooks) just as a rubber sucker can be released by raising its edge. The suction is generated by a pair of muscles which lift the centre of the haptor.

Entobdella can also attach itself for a while by the adhesive areas at the anterior end (Fig. 4.6). These seem not to attach by suction, but by the adhesive action of a secretion from their gland cells. They can attach quite firmly; *Entobdella* removed from soles will attach to the glass surface of a dish and cannot easily be dislodged by a jet of water from a pipette, even when attached only by the adhesive areas. They move about in the dish, attaching themselves by the adhesive areas while moving the haptor to a new site.

Large *Entobdella* are normally found only on the lower surface of the sole, so that they are not easily observed in an ordinary aquarium. Infected sole have been kept in glass-bottomed tanks and examined from below every few hours. It was found that the flukes moved about over their skin.

Feeding is most easily observed by removing flukes from a sole, starving them for a day and replacing them on the skin of a freshly killed sole. In these circumstances they generally feed soon. While attached by the haptor they protrude the pharynx and hold it for 5 minutes or so against a single spot on the skin of the host. During this time, peristaltic movements of the pharynx draw fluid into the mouth.

Sole skin consists of a delicate superficial epidermis, typically about six cells thick, and a tougher underlying dermis which contains a feltwork of collagen fibres. The scales are embedded in the dermis. If the skin where an *Entobdella* has been feeding is examined afterwards a circular area is found to be damaged. Microscope sections through this area show that the epidermis has vanished but the dermis below it is intact. The epidermis has apparently been digested off.

After copulation and fertilization of eggs, the zygotes go to the ootype where each is enclosed with a group of yolk cells in a quinone-tanned shell, which is tetrahedral in shape. The worms release the eggs, which sink. Ciliated larvae about 250 μm long hatch from them. They swim around at about 5 mm s^{-1} and can remain active, if they find no sole, for about a day.

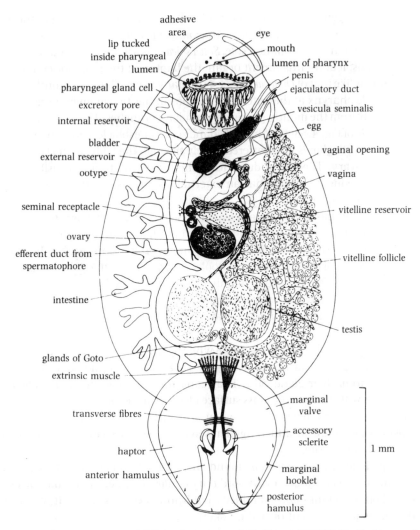

Fig. 4.6. An adult *Entobdella soleae.* From G.C. Kearn (1971). In *Ecology and physiology of parasites* (ed. A.M. Fallis), pp. 161–187. Hilger, London.

Most Monogenea are strictly specific to a single host species, and *Entobdella* is no exception. Divers report that sole live in the sea in close proximity to other flatfish such as plaice (*Pleuronectes*) and dab (*Limanda*), but *Entobdella soleae* is found only on sole.

Do *Entobdella soleae* distinguish between sole and other fish and attach only to sole, or do specimens which settle on other species quickly die? Simple experiments have been performed to try to find out. Larvae were released in the centre of dishes of sea water, with scales of various species of fish arranged around the edge. Nearly all of them settled on sole scales. For instance, in an experiment in which they were given the choice between sole and dab scales, 49 settled on sole scales and only 2 on dab. In other experiments the larvae showed a clear preference for Common sole (*Solea solea*) over Thickback sole (*S. variegata*).

It was suspected that the larvae might be distinguishing between the species by

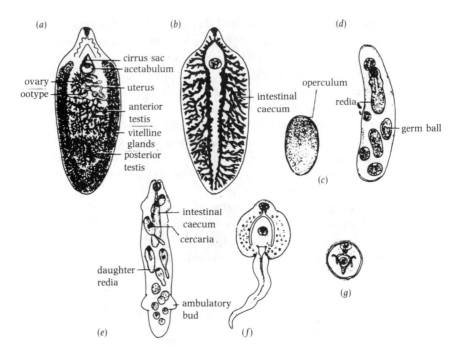

Fig. 4.7. **Stages in the life cycle of the liver fluke,** *Fasciola hepatica.* **(***a***), (***b***) Adult specimens showing the reproductive system and the gut, respectively; (***c***) egg; (***d***) sporocyst; (***e***) redia; (***f***) cercaria; (***g***) metacercaria in cyst. From T.C. Cheng (1973).** *General parasitology.* **Academic Press, New York.**

chemical means, and a further experiment seemed to confirm this. Sheets of agar jelly were strapped to a sole and to a flounder (*Platichthys*) and left in place for half an hour so that any substance diffusing out of the skin would diffuse into the jelly. The jelly was then removed and put in dishes with *Entobdella* larvae, with the face which had not been against the fish uppermost. Larvae attached themselves readily to jelly from the sole but very few attached to jelly from the flounder or to clean jelly.

Most of the Monogenea are ectoparasites of fish but the Digenea are, as adults, endoparasites of all classes of vertebrates including mammals and birds. They also have larval stages parasitic in molluscs (usually gastropods) and often have in addition other larval stages which are parasitic in other hosts. There are even species which pass through four host species in the course of their life cycle.

Several digenean flukes cause important diseases of humans and domestic animals. Three species of *Schistosoma* infect people and the disease they cause, schistosomiasis, is one of the most serious human diseases. It is widespread in the tropics and subtropics and it has been estimated recently that 200 million people suffer from it. Two species of *Fasciola*, the liver flukes, infect sheep and cattle all over the world. They also infect some wild mammals and occasionally people. They cause a disease called liver rot which kills many sheep.

Fasciola hepatica is *not* a typical member of the Digenea, for it is unusually large, but it will serve as an example. The adult is found in the bile ducts of its mammal hosts and grows to a maximum length of about 5 cm. Its structure is shown in Fig. 4.7(*a*) (which shows the reproductive system) and (*b*) (which shows the gut). There is no haptor at the posterior end of the body but there is an attachment organ called the acetabulum on the ventral surface and another round the mouth, which is at the anterior end of the body. The pharynx is not protrusible and has no gland cells. Otherwise the structure is much like that of *Entobdella* (Fig. 4.6). The body is leaf-shaped. The gut has a main branch down each side of the body, and side branches. There is a pair of diffuse yolk glands (vitelline glands), an ovary, two testes and a cirrus. (A cirrus is an organ with the function of a penis, but whereas a platyhelminth penis is simply lengthened by muscle action to make it protrude from the body, a cirrus is protruded by being turned inside-out.) *Fasciola* feeds mainly on blood, but liver cells are also found in its gut.

The eggs are enclosed in quinone-tanned shells. They pass down the bile duct and eventually leave the host's gut in the faeces. While they remain in the faeces, little or no development occurs. If they are washed clean of faeces and remain moist, hatching occurs in a few weeks (depending on the temperature). The operculum (Fig. 4.7*c*) opens like a lid and a ciliated larva hatches out. This larva, the miracidium, can only survive a few hours unless it encounters and enters an appropriate snail of the genus *Lymnaea*. Different species of *Lymnaea* serve as hosts in different countries. In Britain the host is *L. truncatula*, which flourishes in the muddy areas around gateways in damp fields, where there are puddles and deep hoofprints. Wet conditions like this particularly favour the spread of the disease.

If the miracidium encounters a suitable snail it bores its way in through the body wall. The damage to the snail tissues, which is shown by microscope sections of infected snails, indicates that the miracidia secrete enzymes and digest their way through. As the larva enters the snail it loses its cilia and enters the stage in the life cycle known as the sporocyst. This has no mouth and presumably absorbs soluble foodstuffs through its body wall. It is initially only about 70 μm long but it grows to a length of 500–700 μm; as it grows, several larvae of the next stage (rediae) develop inside it (Fig. 4.7*d*). These rediae have a mouth and a simple gut. They break through the wall of the sporocyst and escape from it. They travel to the snail's digestive gland where they feed on its tissues, fragments of which can be found in their guts. They grow and eventually reach a length of about 2 mm. Another generation of asexually produced larvae develop within the rediae: they may be the tailed larvae known as cercariae (Fig. 4.7*f*) or they may be a second generation of rediae. Cercariae and daughter rediae may even develop within the same redia as shown in Fig. 4.7(*e*). Laboratory experiments have shown that cercariae develop at normal room temperatures but that low temperatures promote development of rediae.

Cercariae have a two-branched gut, a sucker round the mouth and an acetabulum. They escape from the rediae in which they develop and congregate in a boil-like structure near the anus of the snail. They are squirted out from this when the snail is in water. They swim until they reach submerged vegetation where they attach, lose their tail to become a metacercaria and form a cyst protected by an outer layer of tanned protein (and several inner layers). The cyst can survive for several months. It need not remain submerged, provided the humidity is high. If it is swallowed by a grazing mammal it releases the metacercaria. Microscope sections of recently infected guinea pigs show the metacercariae boring through the wall of the gut. The

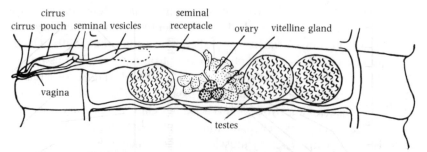

Fig. 4.8. A mature proglottid of the tapeworm *Hymenolepis diminuta*. From A.C. Chandler & C.P. Read (1961). *Introduction to parasitology*, 10th edn. Wiley, New York.

metacercariae reach the liver and tunnel through it, eventually reaching the bile ducts. By this time they are adult flukes.

The tapeworms live as adults in the small intestine of vertebrates. *Taeniarhynchus saginatus*, which infects people, is a slender worm up to 12 m long. This is considerably longer than the human small intestine, but since the worm is rather tangled the intestine can contain it. It has a larval stage which lives in cattle; the adult infects over 10% of the human population in parts of Africa where infected beef is eaten without adequate cooking. Though so large it usually does little harm, and may cause no symptoms other than pieces of worm in the faeces. In other cases it may cause pain, nausea, weakness and loss of weight. Most other species likewise have little obvious effect on the host.

Hymenolepis species have been studied more thoroughly than other tapeworms because the adults live in rodents and are easy to keep in laboratory rats and mice. *H. nana* grows only to about 10 cm but *H. diminuta* grows to a maximum length of about a metre, about the length of a rat small intestine. They consist of a scolex or attachment region and several hundred segments called proglottids, one of which is shown in Fig. 4.8. The proglottids are flattened so the worm is tape-like, but it is a tapering tape because the proglottids near the scolex (the most recently formed proglottids) are narrower than the rest. The scolex has four 'suckers' that grip the host's gut wall, usually near the anterior end of the small intestine, and the proglottids extend posteriorly down the intestine.

There is no gut, but in other respects the internal structure of the proglottids is much like the internal structure of other platyhelminths. Nerve trunks and excretory canals run unbroken along the body but each proglottid has a separate hermaphrodite reproductive system very much like those of *Dugesia* (Fig. 4.4), *Entobdella* (Fig. 4.6) and *Fasciola* (Fig. 4.7).

Mature eggs are found in the oldest proglottids, the ones furthest from the scolex. These proglottids burst and the eggs pass out in the host's faeces. The eggs infect insects that happen to eat them, developing in the insect into a form called the cysticercoid larva. If a rodent eats an infected insect (and rats and mice do eat a very wide range of foods) the cysticeroid develops in it into an adult tapeworm.

Nearly all tapeworms require at least two host species to complete their life cycle. The host for the adult is nearly always a vertebrate but the larvae may inhabit

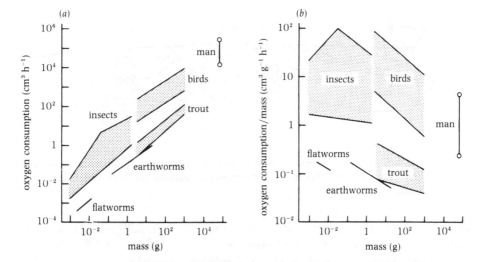

Fig. 4.9. Graphs showing rates of oxygen consumption (*a*) and oxygen consumption per unit body mass (*b*) plotted against body mass. Note that logarithmic co-ordinates have been used. Where a range of rates of oxygen consumption is indicated for animals of the same type and mass the bottom of the range is the rate at rest and the top is the maximum rate recorded in fast swimming, running or flight. From R. McN. Alexander (1971). *Size and shape*. Arnold, London.

vertebrates or invertebrates: *Taeniarhynchus* has its larval stage in cattle, but *Hymenolepis* normally use insects.

4.3. Diffusion of gases and nutrients

The Turbellaria cover a wide range of sizes, with the smallest no larger than many ciliate protozoans and the largest many centimetres long. Large Turbellaria are generally more flattened than small ones and their guts branch more. This section aims to explain why.

The rates at which turbellarians use oxygen have been measured by putting specimens in a jar of water and analysing samples of the water from time to time to determine the concentration of dissolved oxygen in it. The results of such experiments are shown in Fig. 4.9, together with data obtained by other methods for other animals. Within groups of similar animals oxygen consumption increases with body mass, but not in direct proportion to body mass. It is generally about proportional to (body mass)$^{0.75}$, so that graphs on logarithmic co-ordinates of oxygen consumption against body mass tend to be straight lines of gradient 0.75 (Fig. 4.9*a*). Oxygen consumption per unit body mass therefore tends to be smaller for large animals than for small ones, but it lies between 0.1 and 0.2 cm^3 g^{-1} h^{-1} for all the flatworms represented in Fig. 4.9.

Flatworms have no gills or other special respiratory organs, and they have no blood system to distribute oxygen round the body. The oxygen they use must

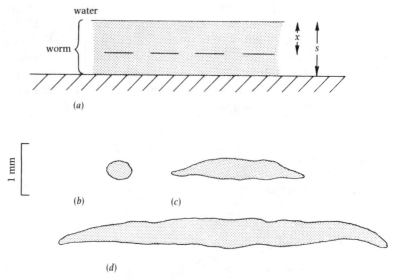

Fig. 4.10.(*a*) A diagram illustrating the account of diffusion through flatworms. (*b*), (*c*), (*d*) Transverse sections drawn to the same scale through three turbellarians of different sizes. From R. McN. Alexander (1971). *Size and shape.* Arnold, London.

presumably reach the tissues by diffusion from the surface of the body. No part of the body can be too far from the surface, or oxygen would not diffuse to it fast enough. This is probably why large flatworms are flattened, as the following discussion will show.

Gases diffuse from regions where their partial pressure is high to regions where it is lower. The partial pressure of a gas in a mixture of gases is the pressure it would exert if it occupied the whole volume of the mixture. Air contains 21% oxygen by volume so the partial pressure of oxygen in air at atmospheric pressure is 0.21 atm. The partial pressure of a gas in solution is its partial pressure in the mixture of gases which would be in equilibrium with the solution. Hence the partial pressure of oxygen in well-aerated water is 0.21 atm.

The rate of diffusion of a gas is proportional to the gradient of partial pressure (that is to the gradient of a graph of partial pressure against distance). Consider a gas diffusing at a rate J (volume per unit time) across a surface of area A. Then if the gradient of partial pressure is dP/dx

$$J = -AD.dP/dx. \tag{4.1}$$

The negative sign indicates that diffusion occurs down the gradient. D is a constant, for a given gas diffusing through a given medium, and is known as the diffusion constant. If J is expressed in $mm^3 s^{-1}$, A in mm^2 and dP/dx in $atm\ mm^{-1}$, D has units $mm^2\ atm^{-1}\ s^{-1}$. The diffusion constant for oxygen diffusing through water is $6 \times 10^{-5}\ mm^2\ atm^{-1} s^{-1}$ and for oxygen diffusing through frog muscle and connective tissue about $2 \times 10^{-5}\ mm^2\ atm^{-1}\ s^{-1}$.

Consider a flatworm of thickness s and area A (Fig. 4.10*a*). Oxygen diffuses in from the dorsal surface but not from the ventral surface because it is resting on a rock or some other impermeable base. It will be convenient to assume that the length and breadth of the flatworm are large compared to s, so that diffusion through the vertical surfaces at the edges of the animal can be ignored. All parts of the worm use oxygen at a rate m per unit volume of tissue. Consider the horizontal plane at a distance x below the dorsal surface of the worm. The tissue below it has volume $A(s-x)$ and uses oxygen at a rate $Am(s-x)$ so oxygen must diffuse through the plane at this rate. Using this rate as the value for J in the diffusion equation we find

$$Am(s-x) = -AD \cdot dP/dx,$$
$$m(s-x)\,dx = -D \cdot dP.$$

If the partial pressure of dissolved oxygen is P_0 at the dorsal surface (where $x=0$) and P_s at the ventral surface (where $x=s$)

$$m\int_0^s (s-x)\,dx = -D(P_s - P_0),$$
$$\tfrac{1}{2}ms^2 = D(P_s - P_0),$$

and since P_0 cannot be negative

$$s \leqslant (2DP_s/m)^{\frac{1}{2}} \tag{4.2}$$

Fig. 4.9 shows that m for flatworms is 0.1 cm^3 oxygen g^{-1} h^{-1}, or a little more. Since the density of flatworm tissue is about 1 g cm^{-3} this is about 10^{-1} mm^{-3} oxygen mm^{-3} h^{-1} or 3×10^{-5} s^{-1}. If the flatworm is in well-aerated water $P_s = 0.21$ atm. Assume that the diffusion constant is about the same as for frog tissues. Then

$$s \leqslant (2 \times 2 \times 10^{-5} \times 0.21/3 \times 10^{-5})^{\frac{1}{2}}$$
$$\leqslant 0.5 \text{ mm}.$$

This calculation indicates that the maximum possible thickness for a flatworm using oxygen at the observed rate is about 0.5 mm if oxygen diffuses in only from the dorsal surface, or 1.0 mm if it diffuses in equally from the ventral surface (as it well might if the worm lived on the surface of sand or mud). A similar calculation for a cylindrical turbellarian indicates that the maximum possible diameter would be 1.5 mm.

Fig. 4.10(b), (c), (d) are transverse sections through three turbellarians. Though very different in size all are about 0.5 mm thick. Oxygen would probably not diffuse into them fast enough for their requirements, if they were much thicker than this. This is probably why large flatworms are flat.

Some sea anemones and jellyfish are of course much larger than flatworms, but the diffusion distances for the oxygen they use in metabolism are small because their cells form single layers of epidermis and gastrodermis. A *Cyanea* medusa 30 cm in diameter may be 3 cm thick but most of its thickness is occupied by mesogloea and almost all the cells are very close to the outside surface of the animal or to the gastric cavity and canals. Water is circulated through the canals by the cilia of the gastrodermis. Even the folded swimming muscles of the *Cyanea* would be less than 0.5 mm thick.

A similar argument applies to diffusion of sugars, amino acids, etc., the final

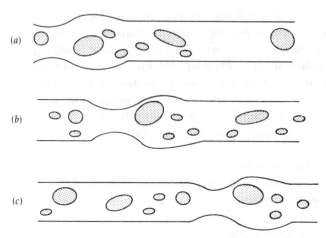

Fig. 4.11. A diagram showing the principle of peristalsis. Constrictions travel along the tube towards the right, moving the contents of the tube in the same direction.

products of digestion, which must diffuse from the gut to the tissues that need them, throughout the body. There is no blood circulation to carry them. The smallest turbellarians have simple guts but larger ones have branching guts arranged in such a way that no part of the body is far from a branch (Fig. 4.4*a*). Monogeneans and trematodes have similar branching guts (Figs. 4.6 and 4.7*b*). The products of digestion need not diffuse far, provided the food is distributed reasonably uniformly to the branches of the gut before digestion.

4.4. Feeding and digestion in turbellarians

Most turbellarians are predators. Some of the small ones swallow prey whole but many have a muscular pharynx like that of *Dugesia* (Fig. 4.3) which they use to suck in semi-liquid food. This kind of pharynx is housed in a cavity in the body but can be protruded for feeding: presumably its circular muscles contract and make it elongate. It is pushed into prey (using enzymes to digest its way in, as we shall see) and sucks up their body contents by the process of peristalsis, which is illustrated in Fig. 4.11.

It is usually possible to identify the natural foods of animals by capturing a sample and examining their gut contents, but this method is of little use for turbellarians that suck their prey in a partly digested state. This difficulty was overcome in a study of some British lake-dwelling turbellarians by using an immunological method. Preparations of proteins were made from various animals that were common in the lakes and seemed likely to be eaten by the turbellarians. These preparations were injected into rabbits, which produced antibodies against them. Serum from a rabbit that had been immunized against proteins from the crustacean *Asellus*, for example, reacted with preparations made from turbellarians that had recently fed on *Asellus* to produce a visible precipitate. Tests on recently captured turbellarians showed that *Dugesia polychroa* fed mainly on water snails, *Dendrocoelum lacteum* fed mainly on *Asellus* and two species of *Polycelis* fed mainly on oligochaete worms. The immunological technique did not distinguish between different oligochaetes, but these worms are among the few foods that can be recognized visually in the gut because some of their chaetae (bristles) get swallowed. Different families of oligochaetes have different-shaped bristles, and it was found that the two species of *Polycelis* ate different proportions of the various families. The four turbellarian species were often found in the same part of the same lake, even on the same stone, but each had a principal food that was eaten less often by the other three. This helps to explain how they can coexist without any one of them outcompeting and eliminating the others.

Fragments of food can often be seen in microscope sections of flatworms, enclosed in vacuoles in gastrodermal cells. Sections of *Polycelis* which had been fed starch or clotted blood showed starch grains or blood corpuscles in vacuoles. Presumably digestion occurs in vacuoles, as in protozoans. Digestion also occurs in the gut cavity: this is particularly obvious in species that swallow their prey whole. What are the roles of the two types of digestion? Histochemical tests have been made on turbellarians, to find out.

Amino acids have the general formula $NH_2CHRCOOH$, where R is one of many alternative groups. The large molecules of proteins and the smaller ones of peptides consist of amino acids linked together, the carboxyl (COOH) group of one joining the amino (NH_2) group of the next:

$$NH_2CHRCOOH + NH_2CHR'COOH = NH_2CHRCO.NHCHR'COOH + H_2O.$$

Digestion involves breaking proteins down to their constituent amino acids, and several types of enzyme are involved. Aminopeptidases remove the terminal amino acid from the end of the chain which has the free amino group, carboxypeptidases remove the terminal amino acid from the other end and endopeptidases break the chain further from its ends. Arrows in the diagram below show the points of attack of the enzymes.

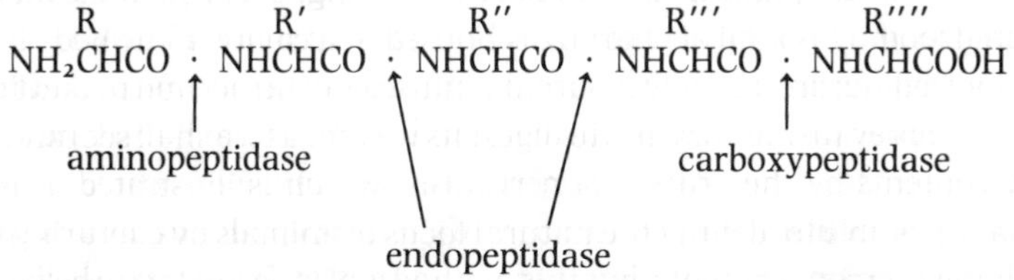

As well as breaking amino acids off the ends of peptides, aminopeptidases can split compounds in which amino acids are combined with amines. For instance, they can split leucyl-β-napthylamide into the amino acid leucine and β-napthylamine:

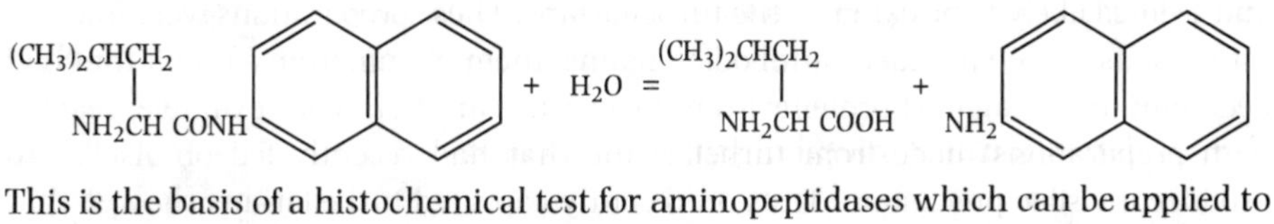

This is the basis of a histochemical test for aminopeptidases which can be applied to microscope sections provided they have been prepared by a method which does not inactivate the enzymes. The sections are left for a few hours in a solution containing leucyl-β-napthylamide and garnet GBC, a diazonium salt. They are then rinsed and examined. Wherever aminopeptidases are present they will have broken down the leucyl-β-napthylamide, releasing β-napthylamine which combines immediately with garnet GBC to give a bright red precipitate (an azo dye). The precipitate shows precisely where in the section the aminopeptidase is. The test is not wholly reliable: there are probably aminopeptidases which will not attack leucyl-β-napthylamide and there may well be enzymes which break down leucyl-β-napthylamide but have no effect on peptides. Nevertheless the test has been found a very useful guide to the distribution of aminopeptidases. Other tests designed on similar principles are used to locate endopeptidases and many other enzymes.

These tests have been applied to sections of turbellarians (mainly *Polycelis*). The pharynx has gland cells which open to its outer surface, especially near its tip. The tests showed that many of these cells contained endopeptidase, and were shrunken after feeding. Presumably they secrete endopeptidases which attack proteins in the prey, enabling the pharynx to penetrate it and disrupting the tissues so that they can be sucked into the gut.

There are two types of cell in the gastrodermis, columnar cells which engulf food particles and gland cells which do not. The gland cells contain numerous small spheres which react strongly to the histochemical test for endopeptidases. There are traces of endopeptidase in the gut immediately after feeding (perhaps due to endopeptidase secreted by the pharynx and taken in with the food), but the concentration builds up to a maximum in the next 4 hours. During the same period the gland cells lose a lot of their spheres. It seems that the gland cells produce endopeptidase, and that they secrete it into the gut lumen after a meal. No other enzymes have been found in the gut lumen. Once the gland cells have released their spheres of enzyme they need 1 or 2 days to make a new supply.

Endopeptidase can also be detected in the cytoplasm of the columnar cells, and it is present in food vacuoles for the first 10 hours or so after a meal. Thereafter it disappears from the vacuoles, and aminopeptidase and lipase appear in them (lipases are enzymes that break down fats). Starch is also digested in the vacuoles at this stage. Digestion may not be complete until 2 days or more after a meal.

Thus digestion starts externally (around the tip of the pharynx) and in the gut lumen. It continues in vacuoles in the columnar cells. Initially the food is attacked only by endopeptidases but later other enzymes are involved. The endopeptidases reduce protein molecules to larger numbers of shorter peptide chains and so increase the number of chain ends which can be attacked by aminopeptidases (and carboxypeptidases if they are present).

The final products of digestion in the vacuoles are presumably amino acids from proteins, fatty acids and glycerol from fats, and sugars from polysaccharides. These must diffuse out from the gut to the tissues that need them.

4.5.   Nutrition and metabolism of tapeworms

Carbohydrates seem to be particularly important for the nutrition of tapeworms. In one series of experiments rats infected with about 10 *Hymenolepis diminuta* each were fed equal rations (in terms of energy content) of various diets. After a week they were killed and the worms were dissected out and weighed. It was found that worms in rats fed on a protein-free, starch-rich diet had grown larger than those in rats fed a normal diet, or a protein-free fat-rich diet. Omission of proteins from the diet does not imply absence of peptides and amino acids from the intestine because the digestive enzymes of the host are proteins and because amino acids diffuse out of the gut wall into the lumen when their concentrations in the lumen are low.

Glycogen seems to be the principal food reserve of tapeworms and is often present in remarkably large quantities. For instance, it was found in one investigation that glycogen made up 46% of the dry weight of *Hymenolepis diminuta*.

Tapeworms have no guts but absorb through their tegument foodstuffs that have been digested in the host's gut. Each of these will tend to diffuse into a tapeworm if its concentration C_1 in the gut lumen is greater than its concentration C_w in the worm. The rate of diffusion will be proportional to $(C_1 - C_w)$. Products of digestion might also be taken up by active transport, a process which uses energy and can operate against a concentration gradient. This would presumably involve an enzyme so its rate should depend on substrate concentration in the manner predicted by the Michaelis–Menten equation, which applies to most enzyme-catalysed reactions:

$$V = V_{max} \, C_1/(K_m + C_1) \tag{4.3}$$

where C_1 is the concentration of the substrate on which the enzyme is acting (in this case the digestion product in the gut lumen), V is the rate of the reaction (in this case the rate of uptake) and K_m is the Michaelis constant. If C_1 is much smaller than K_m, V is more or less proportional to C_1. If, however, C_1 is much larger than K_m, V is nearly equal to V_{max}, the maximum rate: there is plenty of substrate and the reaction proceeds as fast as the enzyme can operate. Further explanation of the equation can be found in textbooks of biochemistry. The equation can also be written

$$1/V = [(K_m/C_1) + 1]/V_{max}, \tag{4.3a}$$

so for reactions which proceed according to the equation, a graph of $1/V$ against $1/C_1$ is a straight line.

Fig. 4.12 shows the results of an experiment on glucose uptake by *Hymenolepis*. The tapeworms were removed from their rat hosts and put in a solution of radioactive glucose for 1 minute. They were removed and washed, and their radioactivity was measured. This indicated the amount of glucose which had been absorbed. The experiment was repeated with different concentrations of glucose and the graph of $1/V$ against $1/C_1$ is more or less a straight line. The result would not be obtained if glucose entered the worms mainly by diffusion but it would if glucose were taken up mainly by active transport. Further experiments showed that the worms could take up glucose until the concentration in the body was greater than the concentration in the external solution. Plainly, active transport must occur. It has also been shown that amino acids are taken up by active transport.

All these experiments used tapeworms removed from their hosts. More elaborate experiments on tapeworms *in situ* in the intestines of rats showed that glucose was taken up mainly by active transport but partly by the solvent drag effect: when the worm absorbs water, dissolved glucose tends to enter as well. Diffusion was relatively

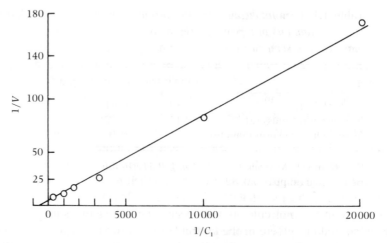

Fig. 4.12. A graph of the reciprocal of the rate of uptake of glucose (in arbitrary units) by *Hymenolepis*, against the reciprocal of the concentration of the glucose solution (in mol l^{-1}). From K. Phifer (1960). *J. Parasitol.* 46, 51–62.

unimportant. Similarly the cells of the wall of the small intestine take up glucose mainly by active transport.

The small intestine is a habitat in which foodstuffs are often plentiful, the concentration of carbon dioxide is high and oxygen is often scarce. Dissolved oxygen is not completely absent for it diffuses in from the blood vessels of the wall of the intestine and is present in the secretions of the digestive glands. Tapeworms can use oxygen or survive without it, depending on its availability. *Hymenolepis nana* has been grown from cysticercoid to maturity in solutions from which oxygen was excluded.

When oxygen is absent, animals cannot get energy by oxidizing foodstuffs, but they can get it by other processes that do not require oxygen. One of these 'anaerobic' processes, that is used by tapeworms and many other animals, involves converting glucose to lactic acid:

$$C_6H_{12}O_6 = 2CH_3.CHOH.COOH + 2 \sim,$$
$$\text{glucose} \qquad \text{lactic acid}$$

where $\sim$ represents an energy-rich phosphate bond (i.e. conversion of ADP to ATP). This yields only a small fraction of the energy that could be got by oxidizing the glucose

$$C_6H_{12}O_6 + 6O_2 = 6CO_2 + 6H_2O + 38 \sim,$$

but that may not matter if food is plentiful and oxygen scarce.

There is a third major process, whereby tapeworms get energy from glucose. It uses carbon dioxide (which is present in high concentrations in the small intestine) and produces succinate:

$$C_6H_{12}O_6 + CO_2 = (CH_2.COOH)_2 + CH_3.CO.COOH + H_2O + 3 \sim.$$
$$\text{succinic acid} \qquad \text{pyruvic acid}$$

The details of the process were worked out by keeping tapeworms in anaerobic conditions in solutions containing $NaH^{14}CO_3$, or glucose incorporating ^{14}C at one

Table 4.1. *Food intake and growth (during 24 h)*
*of 150 g rats and of tapeworms (*Hymenolepis
diminuta*) in some of them*

	Uninfected	Infected
Food intake (g)	11.54	10.91
Increase in rat mass (g)	2.17	1.77
Increase in tapeworm mass (g)	—	0.10

Data from D.F. Mettrick (1973). *Can. J. Public Health*
64 (monogr. suppl.), 70–82.

position only in the molecule, and analysing the radioactive succinate that they produced to find out where in the molecule the ^{14}C was.

Lactate and succinate are excreted by tapeworms, often in similar quantities. It is not clear what happens to the pyruvate from the succinate-producing process.

The excretory canals of *Hymenolepis* are up to 0.5 mm in diameter, so it is possible to insert micropipettes and draw out fluid for analysis. It has been found in this way that the principal cation in the urine is sodium and the principal anions are chloride, lactate and succinate.

Rats infected with *Hymenolepis* have been anaesthetized and opened so that the worms could be observed. Peristaltic waves travelled along the worms, squeezing the contents of the excretory canals towards the free end of the worm and so, eventually, out of the body. Colloidal graphite was injected into the canals and its subsequent movement observed, to find out how fast the fluid was being passed. It was found that the rate was sufficient to account for most or all of the succinate and lactate excretion.

These experiments demonstrate the role of the protonephridial system in tapeworms, in excreting metabolic waste products. It does not seem to be capable of regulating the osmotic concentration of the body fluids in the way the protonephridia of rotifers do (section 4.8).

To survive in the intestine, tapeworms need some protection against digestive enzymes. Since their outer surfaces consist of protein and lipid, protection against peptidases and lipases is necessary. There is experimental evidence of such protection. It has been shown that solutions of trypsin (one of the peptidases of mammal intestines) digest protein less fast if a *Hymenolepis diminuta* is left in them for a while, and then removed, before the protein is added. The effect is probably not simply due to absorption of the enzyme on the worm, as the same worm put successively into a series of trypsin solutions has the same effect on each. Similar results have been obtained in experiments with α- and β-chymotrypsin (which are other intestinal peptidases).

Most tapeworms produce no obvious symptoms of disease but they must deprive their host of nutrients. An experiment was carried out with rats and *Hymenolepis diminuta*, to find out how severe the deprivation was (Table 4.1). One group of rats was infected with *Hymenolepis* which were allowed to grow to maturity, and another group was kept free from infection. Both groups were allowed as much food as they wanted and the amount they took was recorded. It was about the same in both cases, and analysis of the faeces showed that both groups excreted the same fraction of the energy intake. The infected rats grew 0.4 g per day less than the uninfected ones but their worms grew only 0.1 g per day. Some of the lost growth must have been due to energy used in the metabolism of the tapeworms. Also, tapeworms alter the pH and chemical composition of the gut contents and may increase the energy the host has to use in absorbing products of digestion from the lumen.

4.6.   Strategies for reproduction

Parasites such as flukes and tapeworms produce enormous numbers of eggs, devoting to reproduction a very large proportion of their resources of energy and materials. For example, many tapeworms produce thousands of eggs daily while typical free-living flatworms produce only about one egg cocoon per week (during the summer months only), each cocoon containing 2–10 eggs. That comparison is not a fair one because the tapeworms in question are bigger than the free-living flatworms, and lay smaller eggs. However, a fairer comparison gives the same impression: the tapeworm *Hymenolepis diminuta* devotes 35–40% of its food energy intake to egg production, but the free-living flatworm *Polycelis tenuis* devotes only 20%. (The animals being compared here both survive to breed for several seasons.)

Why should parasites produce huge numbers of eggs? The conventional explanation is that they have to, to survive at all. For example, a *Hymenolepis* egg will not survive to maturity unless it has the good luck to be eaten by a suitable insect which in turn gets eaten by a rodent. Each individual egg has only a tiny chance of survival, but there is a reasonable chance that a few will survive from among the millions that a worm may produce in its lifetime.

That argument may explain why parasites have to produce enormous numbers of eggs, but if fails to explain why free-living animals produce fewer. Suppose that, in a free-living population, a mutant appeared that laid an exceptionally large number of eggs. If they survived as well as the eggs of other members of the species, there would be increasing proportions of the mutant in successive generations. Free-living species, as well as parasitic ones, should evolve to produce huge numbers of eggs.

The reason why they do not seems to be that reproduction entails a cost. The more of its resources an animal uses for egg production, the less likely it is to survive to reproduce again. This has been demonstrated for the rotifer *Asplanchna*, in a laboratory experiment: females that produced fewer offspring per day had longer average life spans. The optimum number of eggs depends on the balance between the advantage and the cost of producing more.

Natural selection favours the fittest genes and groups of genes, the ones that make their possessors breed faster or live longer, enabling them to produce more offspring. Formal definition of fitness requires an awkward equation, but simpler, more manageable equations apply in special circumstances. Consider an animal that takes a year to grow from birth to maturity and then breeds at intervals of a year for the rest of its life. Few parasites have life histories like that, but the simple equation for this special case will help us to grasp some basic principles. The simple equation gives the fitness F of the animal's set of genes in terms of juvenile survival S_j (that means the probability of surviving to breeding age); adult survival S_a (the probability of survival

from one breeding season to the next); and fecundity n (the number of offspring produced by an individual in each breeding season).
The equation is

$$F = \log_e(\tfrac{1}{2}S_j n + S_a).\qquad(4.4)$$

The formal derivation of this equation can be found in more specialized books. Here it seems sufficient to notice that it seems sensible. $S_j n$ is the number of offspring from each season's breeding that are likely to reach maturity. Any increase in it or in adult survival (without reducing the other) will obviously increase fitness.

However, these quantities do affect each other: in particular any increase in n can be expected to reduce S_a, as the experiment on rotifers showed. The curve in Fig. 4.13(a) shows how this trade-off seems likely to work. If fecundity is zero (i.e. if the animal does not breed) adult survival has its maximum value $S_{a,max}$. Moderate fecundity reduces this only a little but high fecundity has more than a proportionate effect, because increasing fecundity makes it necessary to divert resources from increasingly important bodily functions. If an animal uses all its resources for reproduction it may achieve the maximum possible fecundity this season (n_{max}), but it will certainly die ($S_a = 0$).

The thin, straight lines in the same diagram are contours of equal fitness: ($\tfrac{1}{2}S_j n + S_a$) (see equation 4.4) has the same value for all points on any particular one of these lines. The further up and to the right a contour lies, the higher the fitness that it represents. The highest possible fitness occurs at the dot, where the trade-off curve reaches furthest up the sequence of equal-fitness lines, so n_{opt} is the optimal fecundity.

Fig. 4.13(b) shows the same trade-off curve, but the lines of equal fitness have been drawn for a smaller value of S_j, which makes them less steep. The optimum fecundity is lower than in (a).

This analysis indicates that animals with poor juvenile survival should generally have relatively low fecundities. This should enable them to survive well as adults so that, though they produce only a few offspring each season, those that reach maturity survive to breed for many seasons. There is some evidence that this happens in free-living flatworms. In laboratory tests in which juvenile flatworms were starved, *Polycelis tenuis* lost weight much faster than *Dendrocoelum lacteum*, and survived less well. (This seemed to be because *Polycelis* becomes more active when starved, searching harder for food, while *Dendrocoelum* becomes less active, conserving its energy until more food appears.) *Polycelis* adults devote only 20% of their food energy intake to egg production, and generally survive for several breeding seasons. *Dendrocoelum*, however, devote 49% of their food energy intake to egg production during their first breeding season, and generally die at the end of the season. *Polycelis* puts about three eggs in each egg cocoon, and *Dendroceolum* about nine.

The theory seems to explain the different breeding habits of these (and other) free-living flatworms, but it predicts precisely the opposite of what is observed for parasites. It predicts low fecundity when juvenile survivorship is low, but parasites such as *Hymenolepis* have high fecundities. The most likely explanation is that endoparasites generally have a superabundance of food, which probably increases both $S_{a,max}$ and n_{max} (Fig. 4.13c): if they did not breed they would survive better than free-living worms because they would always have enough food, and if they used all

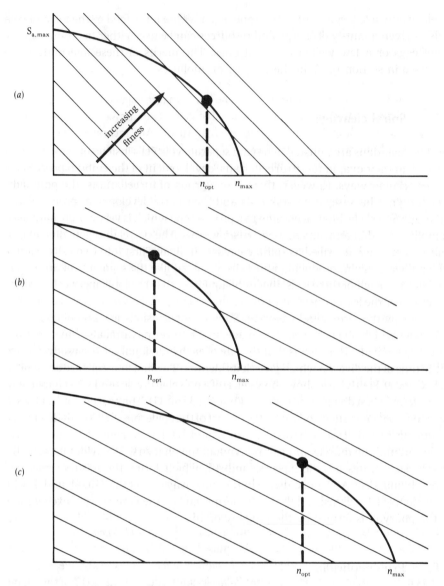

Fig. 4.13. Schematic graphs of adult survival against fecundity, used in the discussion of optimum fecundity.

their food intake for breeding they could produce an enormous quantity of eggs. Fig. 4.13(c) shows how they could have high values of n_{opt} even if juvenile survivorship were low. (The lines of equal fitness have been drawn with the same shallow slope as in Fig. 4.13(b)). Tapeworms devote more of their food energy to reproduction than do free-living flatworms, because they have more energy to spare. This conclusion may seem simple and obvious, but it would have been misleading to have stated it without showing some of the subtlety of the theory that leads to it.

Tapeworms produce large numbers of small eggs. Typical free-living flatworms produce only a few eggs, but they supply each with a lot of food in the form of yolk cells. A given quantity of energy and resources can be used either to produce many small eggs or a few well-provisioned ones. The merits of these alternatives are discussed in section 6.12, in the chapter on molluscs.

4.7.   Spiral cleavage

Most turbellarians are unusual in having separate yolk cells instead of incorporating the yolk in their ova. In these turbellarians, development of the embryo proceeds in rather peculiar ways. However, there is one group of turbellarians (the polyclads) which do not have separate yolk cells and their early development proceeds in a strikingly orderly fashion, according to a programme which is very closely paralleled in molluscs and in annelid and nemertean worms. The polyclad zygote divides into cells in a manner described as spiral cleavage. It divides first into two cells, each of which divides again, producing four cells arranged like the segments of an orange (Fig. 4.14*a*). Each of these cells divides obliquely so that the four upper cells are not directly over the lower ones but in the grooves between. This obliquity of division is described by the term spiral cleavage. Subsequent divisions are also oblique.

Zygotes which divide by spiral cleavage, in whatever phylum, also have determinate development. This means that the fate of each individual cell is already decided at the time of the division which gives it its identity. The first four cells are designated A, B, C and D. Fig. 4.14(*b*) shows how cell D of a polyclad divides and what happens to its progeny. It first divides into cells designated 1d and 1D. The progeny of 1d all go to form the nervous system and the anterior part of the epidermis. Cell 1D divides again, producing 2d and 2D, and 2D divides into 3d and 3D. The progeny of 2d and 3d form epidermis and the muscle and parenchyma of the pharynx. 3D divides to form 4D, which degenerates and disappears, and 4d, which forms the gastrodermis and parenchyma. Cells A, B and C divide in the same way except that 4a, 4b and 4c (the equivalents of 4d) vanish: all the gastrodermis and all the parenchyma except some in the pharynx is formed by the progeny of 4d.

4.8.   Protonephridia

The protonephridial system of tapeworms has an important role in excreting metabolic waste products (section 4.5). This section is about a different role of protonephridia, in salt and water balance.

Fig. 4.15(*a*) is a diagram of part of the protonephridial system of *Dugesia*, based on electron microscope sections. Flame cells close the blind ends of the branching ducts. Each contains a bundle of 35–90 flagella, which are packed close together and beat in unison with a frequency of about $1.5 \, \mathrm{s}^{-1}$. The bundle is visible by light microscopy in living animals and is called the flame because of its flickering appearance.

Fig. 4.15(*b*) shows the flame cell in more detail. It is perforated by groups of slits which connect the intercellular space outside the cell to the lumen within. The slits are about 35 nm wide, and seem to be crossed by fine, closely spaced filaments.

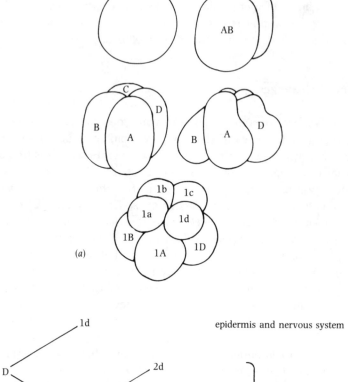

(a)

(b)

Fig. 4.14.(a) **Successive stages of cleavage of the zygote of the polyclad *Hoploplana*.**
(b) **A diagram showing the fate of the progeny of cell D of *Hoploplana*.**

Projections from the flame cell hold adjacent cells at a distance so that they do not block the slits.

Waves of bending travel along the flame from its base to its tip, tending to drive fluid down the tubule and reduce the pressure at the base of the flame. Fluid is probably drawn from the intercellular spaces of the parenchyma through the slits and so into the lumen. Ultrafiltration probably occurs: the fine filaments probably allow water and salts to pass through but stop larger molecules such as proteins. (Protein molecules would pass easily through the slits themselves, for even the huge molecules of snail haemocyanin have diameters of only about 7 nm.)

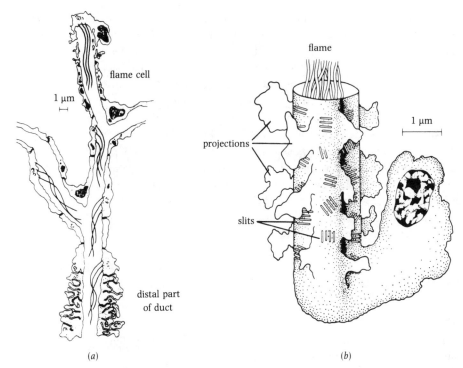

flame cell

1 μm

flame

projections

slits

1 μm

distal part
of duct

(a)   (b)

Fig. 4.15. Protonephridia of *Dugesia*. (*a*) A diagrammatic section. The number of cells has been reduced so that the duct is shown much shorter than it actually is, and the number of flagella has also been reduced. (*b*) A diagram of the flame cell. From J.A. McKanna (1968). *Z. Zellforsch.* 92, 524–535 and 509–523, respectively.

There are bundles of flagella at intervals along the protonephridial ducts as well as in the flame cells at the ends of the ducts. The cells which line the distal parts of the duct have many mitochondria and the cell membrane of their outer surface is deeply infolded. Cells which look like this in various other animals are known to secrete salts actively against a concentration gradient across the cell membrane (Fig. 9.10). The ducts presumably lead to external openings but a zoologist who examined several thousand sections by electron microscopy failed to find a single opening.

Proteins and other large molecules contribute to the osmotic pressure of fluids containing them: their contribution is known as the colloid osmotic pressure. Where ultrafiltration occurs, the colloid osmotic pressure tends to draw fluid back, and a pressure difference is needed to overcome it. Blood pressure serves this purpose in kidneys, and the pressure difference set up by the flame may suffice in protonephridia.

There is little direct evidence about the functions of flatworm nephridia. More is known about the rather similar protonephridia of rotifers. The rest of this section describes experiments on rotifer protonephridia because for the present we can only assume that flatworm protonephridia work in the same way.

The rotifer *Asplanchna priodonta* is only 1 mm long but its protonephridia open into a single duct and it is possible to collect tiny samples of fluid from this duct by drawing the fluid into a micropipette. We will call this fluid urine, because the protonephridia seem to serve as kidneys. The osmotic concentration of the urine has been measured, by determining its freezing point.

Like other rotifers, *Asplanchna* has a fluid-filled space in its body. Fluid from this was found to have an osmotic concentration of 80 mmol l^{-1}. The osmotic concentration of the urine was 42 mmol l^{-1}, when the rotifer was kept in a medium of concentration 18 mmol l^{-1}. When the rotifer was put into distilled water the osmotic concentration of the body fluid was little changed but that of the urine fell to 15 mmol l^{-1}, and urine was produced a little faster.

These results seem to show that at least in the rotifer, protonephridia are involved in osmotic and ionic regulation. A rotifer living in fresh water must lose salts by diffusion and take up water by osmosis. To compensate for this it must have means of taking up salts and getting rid of water. *Asplanchna* eats protozoans and other small animals. If it excretes a urine which has a lower concentration of salts than this food it can get rid of excess water without at the same time losing all the salts from the food. In distilled water it will tend to gain water and lose salts faster so it must excrete a larger volume of more dilute urine.

Further reading

Free-living flatworms

Ball, I.R. & Reynoldson, T.B. (1981). *British planarians*. Cambridge University Press.
Gibson, R. (1972). *Nemerteans*. Hutchinson, London.
Schockaert, E.R. & Ball, I.R. (eds) (1981). The biology of the Turbellaria. *Hydrobiologia* **84**, 1–300.

Parasitic flatworms

Arme, C. & Pappas, P.W. (eds) (1984). *Biology of the Eucestoda*. Academic Press, London.
Kearn, G.C. (1971). The physiology and behaviour of the monogenean skin parasite *Entobdella soleae* in relation to its host (*Solea solea*). In *Ecology and physiology of parasites* (ed. A.M. Fallis), pp. 161–187. Hilger, London.
Smith, J.D. & Halton, D.W. (1983). *The physiology of trematodes*, 2nd edn. Cambridge University Press.

Feeding and digestion in turbellarians

Jennings, J.B. (1974). Digestive physiology of the Turbellaria. In *Biology of the Turbellaria* (ed. N.W. Riser and M.P. Morse), pp. 173–197. McGraw-Hill, New York.
Jennings, J.B. (1962). Further studies on feeding and digestion in triclad Turbellaria. *Biol. Bull.* **123**, 571–581.
Reynoldson, T.B. & Davies, R.W. (1970). Food niche and co-existence in lake-dwelling triclads. *J. Anim. Ecol.* **39**, 599–617.

Nutrition and metabolism of tapeworms

Mettrick, D.F. (1973). Competition for ingested nutrients between the tapeworm *Hymenolepis diminuta* and the rat host. *Can. J. Public Health* **64** (monograph supplement), 70–82.
Mettrick, D.F. & Podesta, R.B. (1974). Ecological and physiological aspects of helminth–host interactions in the mammalian gastrointestinal canal. *Adv. Parasitol.* **12**, 183–278.

Saz, H.J. (1981). Energy metabolism of parasitic helminths. *A. Rev. Physiol.* **43**, 323–341.

Webster, L.A. (1971). The flow of fluid in the protonephridial canals of *Hymenolepis diminuta. Comp Biochem. Physiol.* **39**A, 785–793.

Webster, L.A. (1972). Succinic and lactic acids present in the protonephridial canal fluid of *Hymenolepis diminuta. J. Parasitol.* **58**, 410–411.

Strategies for reproduction

Calow, P. (1979). The cost of reproduction – a physiological approach. *Biol. Rev.* **54**, 23–40.

Calow, P. (1983). Pattern and paradox in parasite reproduction. *Parasitology* **86**, 197–207.

Sibly, R.M. & Calow, P. (1986). *Physiological ecology of animals.* Blackwell, Oxford.

Protonephridia

Braun, G., Kummel, G. & Mangos, J.A. (1966). Studies on the ultrastructure and function of a primitive excretory organ, the protonephridia of the rotifer *Asplanchna priodonta. Pflügers Arch. ges Physiol.* **289**, 141–154.

McKanna, J.A. (1968). Fine structure of the protonephridial system in planaria. I and II. *Z. Zellforsch.* **92**, 509–23 and 524–535.

Wilson, R.A. & Webster, L.A. (1974). Protonephridia. *Biol. Rev.* **49**, 127–160.

5 Rotifers and roundworms

Phylum Rotifera (rotifers)
Phylum Nematoda (roundworms)

5.1. Rotifers

This chapter is about two phyla which are very different in many ways, but which also show similarities that have led some zoologists to place them together (with some other groups) in a single phylum Aschelminthes. I will describe the rotifers in this section and point to the similarities only in the next one where I will describe the nematodes.

The rotifers are among the smallest of the multicellular animals. Most are between 100 and 500 μm long, similar in size to ciliate protozoans. Most of them live in fresh water, and they are very numerous. Almost anyone who has looked at pond water under a microscope will have seen rotifers.

Epiphanes (Fig. 5.1) is a fairly typical rotifer, and is particularly common in pools contaminated with manure. It feeds mainly on green flagellates, which are abundant in the same pools. Like many other rotifers it is superficially rather like *Stentor* (Fig. 2.4b). It has a ciliated corona at its anterior end and tapers to a narrow foot at the posterior end. The cilia of the corona are arranged more or less in two rings, with the mouth in the gap between them. The outer ring (the circumapical band) consists of single cilia but the inner one (the pseudotroch) consists of membranelles which, like the membranelles of *Stentor*, are clumps of closely-packed cilia. The cilia and membranelles beat outwards, setting up currents in the surrounding water just as in *Stentor* (Fig. 2.4b). They beat metachronally and the metachronal waves travel round the corona, making it look as though it were rotating. Because of this, rotifers used to be called 'wheel animalcules'. The rotifer can attach its foot to solid surfaces, probably by means of the secretion of the pedal glands. When it is attached the water currents driven by the corona serve simply as feeding currents bringing potential food near the mouth. When it is detached the currents propel it through the water.

Stentor has only one micronucleus and one macronucleus. *Epiphanes*, which is smaller, has almost 1000 nuclei. The discrepancy is less remarkable than it might at first appear, for the large macronucleus of *Stentor* contains a great many duplicate sets of genes. Some of the tissues of *Epiphanes* are divided up into separate cells but others are syncytial. The number of nuclei in each tissue is constant and the position of each nucleus is more or less identical in different individuals. Thus there are 280 nuclei in the epidermis, 157 in the gut, 246 in the nervous system, 104 in the muscles and the rest elsewhere. This indicates a remarkably inflexible pattern of development. The nuclei do not divide after the rotifer has hatched from its egg and

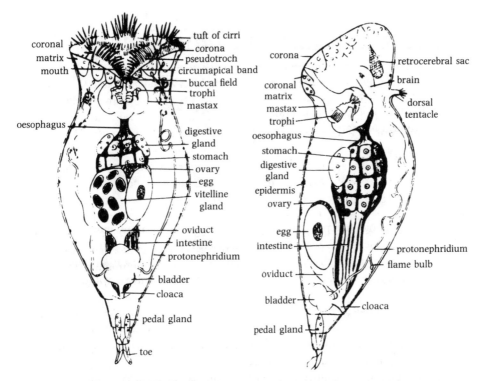

Fig. 5.1. Ventral and lateral views of *Epiphanes senta* (length about 500 μm). From P.A. Meglitsch (1972). *Invertebrate zoology*, 2nd edn. Oxford University Press.

any that are lost if the animal is damaged cannot be replaced. This is a striking difference from *Hydra* and triclad flatworms, which readily regenerate lost parts.

The epidermis is syncytial, and is covered by a thin cuticle which seems to consist of protein. The bands of circular and longitudinal muscle are attached to it at their ends. There is no parenchyma such as flatworms have. Instead, there is a fluid-filled space called a pseudocoel between the epidermis and the gut. This is not a coelom like the body cavities of many of the animals described in later chapters: to be called a coelom, a cavity must be lined by a sheet of epithelium.

The most conspicuous parts of the gut are the pharynx or mastax and the stomach. The mastax contains complicated jaws (trophi) which are used to grind small food and can also be protruded from the mouth to seize large protozoans. Digestion occurs in the stomach, which is enclosed in a network of muscles. These constrict it from time to time, stirring its contents. The posterior end of the gut opens into a cloaca, a combined external opening of the gut and of the excretory and reproductive systems.

The excretory system consists of a pair of syncytial protonephridia with flame bulbs very like the flame cells of platyhelminths. The protonephridia are connected to a bladder which discharges periodically through the cloaca. Four muscle cells around the bladder constrict it and expel its contents. Analyses of urine samples taken from the bladder of a rotifer have already been described (section 4.8).

The nervous system consists of a principal ganglion or brain with nerves radiating to all parts of the body. There are a few ciliated sensory cells on the corona and on three short tentacles, one of which is shown in Fig. 5.1(*b*). The eyes of *Epiphanes* are rudimentary but other rotifers have better-developed eyes.

Some rotifers live in lakes where their food supply fluctuates as the plankton blooms and dwindles. Others live in pools and puddles which grow and shrink as the weather changes and may even dry up altogether. Yet others live among damp moss which may dry out. All these environments are unstable. There are times when rotifers can flourish and others when they must dwindle to very small numbers. When a pool dries out it often leaves a crust of dry algae. If this crust is put in water, rotifers will generally emerge from it within an hour. In the same way laboratory cultures of certain rotifers can be dried out and most of the dried animals subsequently revived by adding water. Dry rotifers do not form a protective cyst, but lose most of their water and shrivel up. Survival in this state is called cryptobiosis. Many seeds such as peas are capable of it, as well as rotifers and some other animals.

It is difficult to investigate the internal structure of dried rotifers because normal methods of preparation of specimens for sectioning involve treatment with aqueous solutions. These solutions tend to make the specimens swell up. However, it has been possible to prepare electron microscope sections of rotifers very nearly in the fully shrunken state, by using a quick-acting fixative. The sections showed that the cell membranes were shrivelled, as well as the outer cuticle. The cilia of the gut were intact, but packed closely side-by-side. The mitochondria were shrivelled but their cristae could still be distinguished. It seems that a great deal of the structure of the cells survives in dried rotifers although the cells are severely distorted and greatly reduced in volume.

Not only can rotifers survive unfavourable conditions such as drought: they also have a remarkable capacity for multiplying when conditions are favourable. Natural populations have been known to multiply tenfold in a week. This is partly due to parthenogenesis, a form of asexual reproduction: the advantage of asexual reproduction for fast multiplication was explained in section 2.6. In parthenogenesis, eggs are produced but develop without being fertilized. One group of rotifers reproduces only parthenogenetically, and males are unknown. Most other rotifers reproduce parthenogenetically for much of the time, but occasionally males appear and sexual reproduction occurs. Eggs for parthenogenesis are produced without meiosis, and so are diploid. The rapidity with which rotifers can multiply is also partly due to their very short generation times. Parthenogenetic *Asplanchna* in laboratory cultures produce about 10 offspring each in a life of only 3–4 days. The first daughter may be born only 30 hours after the birth of the mother. This is not as fast as the reproduction of ciliate Protozoa which, at 25°C, may divide every 12 hours and so produce 4 offspring in a day or 16 in 2 days. It is, however, remarkably fast for an animal with about 1000 nuclei. Only one mitosis is needed when a protozoan divides but the production of 1000 nuclei in a developing rotifer requires at least 10 rounds of mitosis ($2^{10} = 1024$). These 10 rounds of mitosis occur in *Asplanchna* in only 6 hours, an average of only 36 minutes per mitosis.

DNA must be synthesized between nuclear divisions, to replicate the chromosomes. In addition in normal growing tissues RNA must be synthesized, so that proteins can in turn be made. DNA and RNA synthesis each occupy the chromosomes for substantial times. Division is followed by RNA synthesis which is followed in turn by DNA synthesis before division occurs again.

Synthesis of nucleic acids and protein in rotifers has been investigated by experiments with radioactive precursors. Thymidine is a constituent of DNA (but not of RNA) and if rotifers are kept in a solution containing [³H]thymidine while they are synthesizing DNA they incorporate some of this thymidine in their new DNA and become radioactive. Their radioactivity can be detected afterwards by autoradiography. Similarly [³H]uridine can be used to detect RNA synthesis and [³H]leucine (an amino acid) to detect protein synthesis. It has been shown in this way that developing rotifers synthesize DNA and protein, but not RNA.

The cells of the vitelline gland supply the egg with a large stock of food materials and it seems that they also supply all the RNA needed for development. Though RNA is not synthesized by the developing oocyte it is synthesized rapidly in the nuclei of the vitelline gland. The cytoplasm which streams from the gland into the maturing oocyte contains polyribosomes, which presumably contain the RNA needed for development. Since this RNA is supplied ready-made there is no need for the embryo to make its own. The time which would otherwise be needed for RNA synthesis between divisions is saved.

There are three main types of adaptation which help species to succeed in competition against others. First, there are adaptations which increase the rate at which they are capable of reproducing, by increasing fecundity or reducing generation time. Secondly, there are adaptations which enable them to make more economical use of food, space or other resources. Thirdly, there are adaptations which make them less susceptible to competition, for instance by enabling them to use foods not available to the competitor. It can be shown mathematically that the first type of adaptation is particularly effective in unstable environments in which disasters (such as a pool drying up) periodically destroy a large proportion of the population in an unselective way. This type of adaptation is relatively ineffective in stable environments, in which the second type of adaptation is particularly effective.

5.2. Roundworms

The nematodes or roundworms are generally slender worms, circular in cross-section. They are remarkably uniform in structure but vary a lot in size and in way of life. Some live as parasites in other animals; some are parasites of plants and some live free in soil, marine muds and decaying organic matter.

Some of the parasitic nematodes cause serious diseases, of man or of domestic animals. Elephantiasis is a particularly nasty human disease, caused by *Wuchereria bancrofti*. This nematode is no more than about 8 cm long and 0.3 mm in diameter but it blocks lymph ducts and causes swellings quite disproportionate to its size. Infected legs may swell to a circumference of 75 cm or the scrotum may swell till its mass is 25 kg. The hookworms *Ancylostoma* and *Necator* are more widespread, and more insidious. It has been estimated that 450 million people suffer from hookworm disease, most of them in underdeveloped countries in the tropics and subtropics. The worms live in the intestine, abrading it and feeding on blood. They cause anaemia, indigestion and debility. Patients are apt to be weak and apathetic, with slow mental processes. The 'poor white trash' of the southern United States owed their sad condition to hookworm disease. Eggs are passed in the faeces of the host, and larvae enter new hosts through the skin, so a great deal can be done to combat the disease by improving sanitation and providing shoes. *Trichinella* is another nematode which causes human disease. Adults live in the intestines of various mammals including people, pigs and rats, and cause digestive disturbances. Their larvae bore through the wall of the gut and travel to muscles where they form cysts and cause pain, swelling and even death. The disease is transmitted when infected flesh is eaten. Pigs catch it by eating infected rats or contaminated pig swill and people catch it by eating undercooked pork or raw sausage. It is common in eastern Europe but rare in Britain where raw sausage is less popular and pig swill is usually well cooked.

Nematode parasites of plants cause heavy losses in agriculture. *Globodera rostochiensis* is an important pest of potato crops, almost everywhere they are grown. It damages the roots and the plants become stunted. *Ditylenchus dipsaci* is a pest of oats, rye, onions, sugar beet and many other crops and also of tulips and narcissus. It attacks the stem, so that the plants become stunted and twisted with swollen stem bases. Strains of oats and of some other crops have been bred which are resistant to it. *Meloidogyne* causes serious damage to various crops, including tobacco. As well as causing damage directly, nematodes transmit some virus diseases of plants.

Small nematodes are generally plentiful in the top few centimetres of soils and marine sediments. They belong to the interstitial fauna, which consists of animals small enough to fit into the spaces between one sand grain and the next. When they crawl they do not have to force the grains apart, but wend their way through the network of existing spaces. They are particularly abundant in agricultural soils where there may be as many as 10 million individuals, with a total mass exceeding 10 g, per square metre. Some of the nematodes found in soil are juveniles of species which in later life become parasites of animals or plants. Others spend their whole life in the soil.

Soil nematodes and nematode parasites of plants are generally small. The smallest species are only about 200 μm long and since they are slender they are much smaller in volume than many ciliate protozoans. In contrast, some nematode parasites of animals are quite large. *Ascaris lumbricoides*, which lives in the intestines of pigs and people, grows to a length of 40 cm and a diameter of 6 mm.

Nematodes have pseudocoels. The number and arrangement of cells in the nervous system is constant, in normal members of the same species. In these respects they resemble rotifers but in many others they are very different.

Fig. 5.2 shows the structure of typical small nematodes, which have been drawn unrealistically stout for clarity. The body is circular in cross-section, enclosed in a cuticle which is discussed in the next section. The figure shows the arrangement of the gut and sex organs: most species have separate sexes. There are longitudinal muscles, which have the obliquely striated structure also found in many of the muscles of molluscs and annelid worms (Fig. 6.12). There are no circular muscles, and I will explain in the next section why they are unnecessary.

There is a nerve ring around the pharynx, from which nerves run posteriorly along the body. The largest of these nerves are a median dorsal one and a median ventral

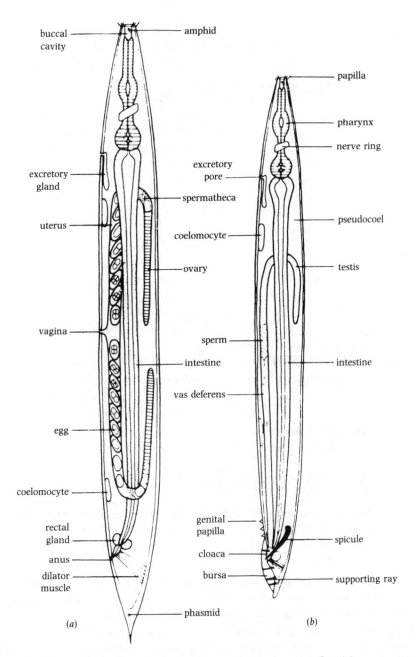

Fig. 5.2. Diagrams showing the structure of typical nematodes; (*a*) represents a female and (*b*) a male. From D.L. Lee & H.J. Atkinson (1976). *The physiology of nematodes*, 2nd edn. Macmillan, London.

one. There are no motor nerves to the muscles: instead the muscle cells have slender processes which run to the nerve ring or to the dorsal or ventral nerve and synapse with neurons there.

There are sensory papillae and a pair of sense organs called amphids on the head,

(a)

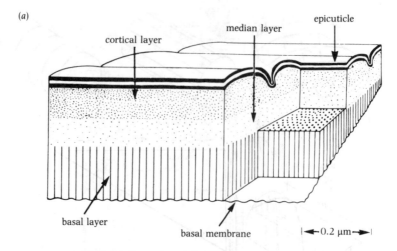

(b)

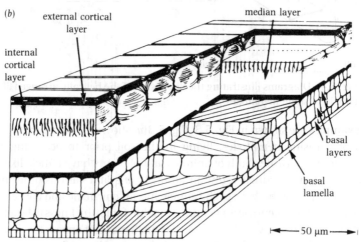

Fig. 5.3. Diagrams showing the structure of the cuticle of (*a*) a typical nematode larva and (*b*) adult *Ascaris*. These diagrams are based on electron microscope sections cut in various directions. From A.F. Bird (1971). *The structure of nematodes.* Academic Press, New York.

and often a pair of posterior organs called phasmids (Fig. 5.2). The amphids are pits containing sensory cells with modified cilia and the phasmids are rather similar. There is evidence that the papillae are sensitive to touch and the amphids to chemical stimuli, but the function of the phasmids remains uncertain.

Many nematodes have one or two cells called excretory glands, with a duct leading to the exterior. Many have excretory tubes which run along the sides of the body and open through the same duct. There is some evidence that these structures function in osmotic regulation.

5.3.   Cuticle

Whatever the size of the nematode the thickness of the cuticle is generally about 0.07 times the radius of the body. The structure of the cuticle varies a lot between species

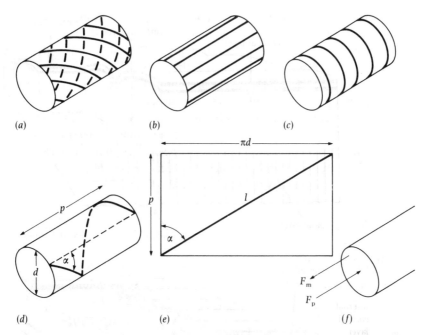

Fig. 5.4. Diagrams illustrating the discussion of the helically arranged fibres in the cuticle of some nematodes.

but two examples are shown in Fig. 5.3. Fig. 5.3(*a*) shows the structure typical of the juveniles of parasitic nematodes, living in the soil prior to becoming parasitic. Juveniles of many parasites both of animals and of plants have cuticle like this. Fig. 5.3(*b*) shows a type of cuticle found in adult *Ascaris* and various other nematode parasites of animals. In both cases the cuticle has an epicuticle and cortical, median and basal layers. The epicuticle is very thin (too thin for its thickness to be apparent in Fig. 5.3*b*). It stains strongly with osmic acid (which makes it appear black in electron microscope sections) so it presumably consists largely of lipid. The other layers of the cuticle are mainly protein. *Ascaris* cuticle, and presumably the cuticle of other nematodes, contains a protein similar to vertebrate collagen. It contains a similar mix of amino acids and gives similar X-ray diffraction patterns. Histochemical tests on several nematodes have indicated that the internal cortical layer is quinone-tanned but that the deeper layers are not.

The principal difference between the two types of cuticle illustrated is in the basal layer. In the larval cuticle it contains radially arranged rods about 20 nm apart. In *Ascaris*, however, it consists of three layers of stout fibres running helically at about 75° to the long axis of the body. In successive layers the fibres form left- and right-handed helices.

Fig. 5.4(*a*) shows diagrammatically how these fibres are wrapped round the body. If their properties are like those of vertebrate tendon collagen (as they presumably are) they can be stretched less than 10% before breaking and they have a fairly high Young's modulus, so that large stresses are needed to stretch them by even a few percent. What are the consequences for the worm of this wrapping of relatively inextensible fibres?

The effects of the fibres depend on the angle they make with the worm's long axis.

Think first about their effect on changes of length. Without the fibres a worm provided with suitable muscles could make itself long and thin or short and fat, its volume remaining constant as its length changed. If the fibres were longitudinal (Fig. 5.4b) they would restrict lengthening. If they ran circularly round the body (Fig. 5.4c) they would restrict shortening by preventing the body from getting fatter. More thought is needed to work out the effect of helical fibres.

Let the fibres run at an angle α to the worm's long axis. Think of a segment of the body of length p, just long enough to include one turn of a helical fibre (Fig. 5.4d. The length p is the pitch of the helix). Cut the cuticle of this segment lengthwise and lay it flat to form a rectangle (Fig. 5.4e), and consider the fibre of length l that forms the diagonal of the rectangle. In the intact worm this fibre made one complete turn of the helix. Fig. 5.4(e) shows that

$$\pi d = l \sin \alpha,$$
$$d = (l/\pi) \sin \alpha \qquad\qquad (5.1)$$
$$\text{and} \quad p = l \cos \alpha. \qquad\qquad (5.2)$$

The segment is a cylinder of radius $d/2$, length p, so its volume is

$$V = \pi d^2 p/4. \qquad\qquad (5.3)$$

By substituting (5.1) and (5.2) in (5.3),

$$V = \pi \, (l/\pi)^2 \, \sin^2\alpha . l \, \cos \alpha /4$$
$$= (l^3 \sin^2\alpha \, \cos\alpha)/4\pi;$$
$$l^3 = 4\pi V/\sin^2\alpha \, \cos\alpha. \qquad\qquad (5.4)$$

V is constant, so l and α are the only variables in this equation. If you calculate $\sin^2\alpha \, \cos\alpha$ for different angles α you will find that it is greatest when $\alpha = 55°$. Thus the length l is *smallest* when α has this value. If α is less than $55°$, the fibres resist lengthening of the worm (which would make α still smaller and so make l larger). If α is greater than $55°$, the fibres resist shortening of the worm (which would make both α and l larger). The fibres in the cuticle of *Ascaris* (Fig. 5.3b) run at about $75°$, so they resist shortening. When the animal's longitudinal muscles contract, they must stretch the fibres and raise the pressure inside the worm.

The principal movements made by nematode worms are bending. Fig. 5.4(f) shows how the longitudinal muscles can be used to bend them. This diagram shows forces acting on one end of a segment of a worm. The dorsal longitudinal muscles are active, exerting the force F_m and increasing the pressure in the worm above that in the water outside. The increased pressure exerts a force F_p which is actually distributed over the whole cross-section but can be considered to act at the centre. Forces F_m and F_p act parallel to each other but in opposite directions. They constitute a couple that tends to bend the worm. If the worm had no cuticle, circular muscles would be needed to build up the pressure needed for forceful bending, but the helical fibres in the cuticle make circular muscles unnecessary.

The pressure inside *Ascaris* has been measured by sticking a hypodermic needle connected to a pressure gauge through the cuticle. In a typical record the pressure fluctuated rhythmically, with an average value of about 9 kN m^{-2} (90 cm water). This is much more than the pressures of about 1 kN m^{-2} that have been measured in sea anemones (section 3.4).

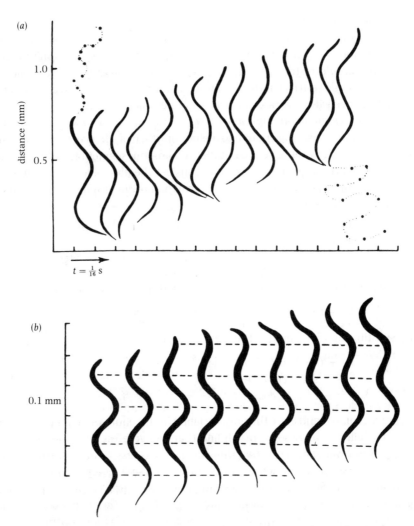

Fig. 5.5. Silhouettes traced from films of nematodes: (*a*) shows the position at intervals of 1/16 s of *Turbatrix* swimming, and (*b*) shows the position at intervals of about 1/3 s of *Haemonchus* crawling on the surface of an agar gel. In each case successive silhouettes have been displaced laterally so as not to overlap their predecessors. From J. Gray & H.W. Lissmann (1964). *J. exp. Biol.* 41, 135–154.

The effect of the helical fibres on bending is more complicated than has been admitted so far, because bending alters the lengths of the fibres. It makes the body wall shorter on the inner side of the bend and longer on the outer side. The overall effect (which has been worked out mathematically) is to allow fibres that make complete turns round the body to shorten slightly: for example, bends like those shown in Fig. 5.5(*b*) would allow 75° helices to shorten by about 1%. If the cuticular fibres of *Ascaris* are stretched by contraction of the longitudinal muscles (as the pressure records show they generally are), bending will tend to slacken them. That means that a straight nematode with its cuticular fibres taut is in unstable equilibrium and will tend to bend one way or the other. (Similarly, a pencil balanced on its point is in unstable equilibrium and tends to fall one way or the other.) That may be why living nematodes nearly always adopt bent positions.

However, not all nematodes have helical fibres in their cuticle: some have cuticle of the kind shown in Fig. 5.3(*a*). It seems likely that the radially arranged fibres in the basal layer of this kind of cuticle may have the same sort of effect as the helical fibres of *Ascaris*, preventing shortening of the worm and making circular muscles unnecessary. When the worm shortens its surface area decreases so the cuticle must get thicker, and the fibres would resist that.

Fig. 5.5 shows how nematodes use the bending movements that the interaction of muscles and cuticle makes possible. They swim and crawl by passing waves of bending posteriorly along the body. Eels and snakes swim and crawl by bending from side to side but nematodes bend dorsally and ventrally (a nematode crawling on a surface is lying on its side). The nematode *Turbatrix*, which lives in vinegar, swims quite well at an average speed of about 0.7 mm s^{-1} (Fig. 5.5a). The waves of bending travel backwards relative to the liquid, as in a swimming flagellate. Most other nematodes are ineffectual swimmers which thrash about in water but cannot keep themselves off the bottom. However, they can crawl over solid surfaces (Fig. 5.5b). In crawling as in swimming waves of bending travel posteriorly along the body, but they are stationary relative to the ground (compare Fig. 5.5b with 5.5a). The same is true of snakes crawling. It can be shown by oblique lighting that a nematode crawling on agar jelly leaves behind it a shallow sinuous groove which marks its path. When waves of bending move posteriorly along the worm, the worm must either slide sideways out of the groove or it must slide forwards, extending the groove forwards. The latter requires less energy so it is what happens.

Nematodes grow a new cuticle and shed the old one several times in the course of their life history. This is called moulting, and occurs four times in most species. It is not clear why it occurs, since the cuticle can grow: nematodes grow between moults and after the final moult. In the extreme case of *Ascaris* the body may be as little as 6 mm long after the final moult but grows to 20 cm or more without further change of cuticle. Nematodes do not become sexually mature until after the final moult and all stages prior to this moult are regarded as juveniles.

Further reading

Rotifers

Birky, C.W. & Gilbert, J.J. (1971). Parthenogenesis in rotifers: the control of sexual and asexual reproduction. *Am. Zool.* **11**, 245–266.
Dickson, M.R. & Mercer, E.H. (1967). Fine structural changes accompanying desiccation in *Philodina roseola* (Rotifera). *J. Microsc.* **6**, 331–348.
Shorrocks, B. & Begon, M. (1975). A model of competition. *Oecologia, Berl.* **20**, 363–367.

Roundworms

Lee, D.L. & Atkinson, H.J. (1976). *The physiology of nematodes*, edn. 2. Macmillan, London.
Maggenti, A. (1981). *General nematology*. Springer, New York.
Poinar, G.O. (1983). *The natural history of nematodes*. Prentice Hall, Englewood Cliffs, New Jersey.
Wharton, D.A. (1986). *A functional biology of nematodes*. Croom Helm, London.

Cuticle

Alexander, R.McN. (1987). Bending of cylindrical animals with helical fibres in their skin or cuticle. *J. theor. Biol.* **124**, 97–110.

Gray, J. & Lissmann, H.W. (1964). The locomotion of nematodes. *J. exp. Biol.* **41**, 135–154.

Harris, J.E. & Crofton, H.D. (1957). Structure and function in nematodes: internal pressure and cuticular structure in *Ascaris*. *J. exp. Biol.* **34**, 116–130.

Wright, K.A. (1987). The nematode's cuticle – its surface and the epidermis. *J. Parasitol.* **73**, 1077–1083.

6   Molluscs

Phylum Mollusca
   Class Polyplacophora   (chitons)
   Class Gastropoda,
      Subclass Prosobranchia   (winkles etc.)
      Subclass Opisthobranchia   (sea slugs etc.)
      Subclass Pulmonata   (snails and slugs)
   Class Bivalvia,
      Subclass Protobranchia   (*Nucula* etc.)
      Subclass Lamellibranchia   (most clams)
   Class Cephalopoda,
      Subclass Nautiloidea   (pearly nautilus)
      Subclass Ammonoidea   (ammonites, extinct)
      Subclass Coleoidea   (octopus, squid, cuttlefish)
   and other classes

6.1.   Introduction

The molluscs form a large phylum, including animals as different as snails and squids. Despite the superficial differences, they show a basic similarity of plan which makes it appropriate to include them all in a single phylum.

The molluscs are the first group of animals described in this book to have the type of body cavity known as a coelom. The distinctive features of coeloms are that they are lined with epithelium (that is, by a continuous sheet of cells) and that they usually house the gonads and excretory organs and have an opening to the exterior. The animals described in the remaining chapters of this book have coeloms of some sort, though they may be small. In annelid worms, echinoderms and vertebrates the main body cavities are coeloms. In most molluscs the coeloms are small and the main body cavity is a blood-filled haemocoel. This cavity, like the pseudocoel of nematode worms, has no epithelial lining but consists simply of spaces in other tissues. The coelom consists merely of the cavity of the gonads, the kidneys and the pericardial cavity (a space round the heart which is part of the excretory apparatus).

The chitons (Polyplacophora) have peculiar shells. In other respects they are perhaps more like the ancestors of the molluscs than are any of the other common groups. They will serve to introduce the phylum (Fig. 6.1). They are rather flat oval animals, up to 0.3 m long but usually much smaller. They live in the sea where they crawl on the surface of rocks. They have a shell of eight overlapping plates which do not completely cover the body: the girdle which runs round the edge is flexible tissue with spicules of calcium carbonate embedded in it. When attached to rocks they are

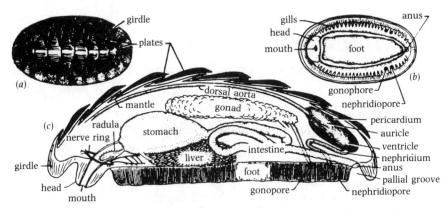

Fig. 6.1. Dorsal (*a*) and ventral (*b*) views of a chiton, and a diagram of its internal structure (*c*). From T.I. Storer & R.L. Usinger (1957). *General zoology*, 3rd edn. McGraw-Hill, New York.

well protected by the shell and girdle. If dislodged, they roll up like woodlice so that the undersurface is still protected. This undersurface is occupied by the muscular foot. Chitons feed on filamentous algae which they scrape off the surfaces of rocks using a rasp-like organ called the radula. The mouth is at the anterior end of the body and the anus at the posterior end. The digestive gland (labelled 'liver' in Fig. 6.1) is a diverticulum of the gut where digestive enzymes are secreted and much digestion occurs. There is a ring of nervous tissues round the pharynx from which two main nerves (the pedal cords) run to the foot and two (the pleural cords) to the viscera. Fig. 6.1 shows only the anterior part of the pedal cords. Note the cross-connections, like the rungs of a ladder. There is a line of up to 70 gills on each side of the foot, under the girdle. They are the type of gills called ctenidia, which are described later in this chapter. The heart has three chambers, two auricles and a ventricle. Each auricle receives blood from the gills of its side of the body and passes it to the ventricle, which pumps it through the aorta and other arteries to the tissues. It returns to the heart through the haemocoel, the main body cavity. There is a pair of excretory organs which discharge through openings called nephridiopores on either side of the foot. There is a single gonad.

Chitons have very large numbers of ctenidia, housed in grooves on either side of the foot. Most other molluscs have just two ctenidia and have them housed in a well-defined mantle cavity, whose wall is called the mantle. The variations on the basic mollusc plan shown by the different classes involve different arrangements of foot, ctenidia and mantle (Fig. 6.2).

Gastropods crawl on a large, flat foot, like chitons, but the shell (if any) is in one piece. Fig. 6.3 shows a few gastropods with variously shaped shells.

The ancestral gastropods probably had bilaterally symmetrical bodies with the mantle cavity opening posteriorly (Fig. 6.2*b*). The most primitive modern gastropods, the prosobranchs, are also symmetrical in their early larval stages but their symmetry is destroyed by the process of torsion, a sudden twisting (completed, in some species, in a few minutes) that brings the mantle cavity opening to an anterior position over the head (Fig. 6.2*c*).

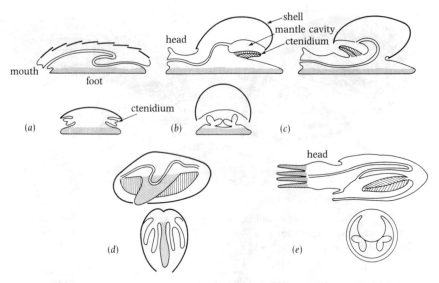

Fig. 6.2. Diagrams showing the arrangement of the principal parts of the body in (*a*) a chiton; (*b*) a hypothetical ancestral gastropod; (*c*) a prosobranch gastropod; (*d*) a bivalve and (*e*) a cephalopod.

Haliotis (Fig. 6.3*a*) and *Buccinum* (Fig. 6.3*b*) are both marine prosobranchs. *Haliotis* species live below low tide level in most parts of the world and feed on red algae. They attach themselves so firmly to rocks that a lever is needed to dislodge them. The significance of the holes in the shell will be explained in section 6.6. The whelk *Buccinum undatum* (Fig. 6.3*b*) is a North Atlantic species, found at and below low tide level. It is a carnivore and scavenger, feeding on crabs, worms and bivalve molluscs and on dead animals. There is a groove (like the spout of a jug) in the shell opening. When the whelk is extended a long tube (the siphon) protrudes from this groove. The respiratory current enters the mantle cavity through the siphon. There is a sense organ in the mantle cavity, and the siphon seems to be used for sniffing prey. The foot bears a hard plate (operculum) which closes the mouth of the shell when the animal retracts into the shell (as in Fig. 6.3*b*).

In gastropods of the subclass Opisthobranchia, torsion has been reversed. The mantle cavity opens to the side or posteriorly or is lost, as in the example shown in Fig. 6.3(*c*). This is *Aeolidia*, one of the sea slugs, which lives on European shores at and below low tide level. It has no shell and no ctenidia. It feeds on sea anemones, and the greyish-brown tentacle-like projections on its back make it inconspicuous on certain species of anemone. These projections serve as gills, and they house extensions of the digestive gland. Nematocysts from anemone prey are somehow transported undischarged to the tips of the projections, where they accumulate. It is believed that they protect *Aeolidia* against predators, by discharging when the sea slug is attacked.

Nearly all the Pulmonata live on land or in fresh water. They include the familiar terrestrial snails and slugs, and also water snails such as *Planorbis* (Fig. 6.3*d*). They have no ctenidia, but the mantle has a rich blood supply and the mantle cavity functions as a lung. Even the aquatic pulmonates breathe air, visiting the surface to breathe. Those that live in well-aerated water get some of their oxygen by diffusion

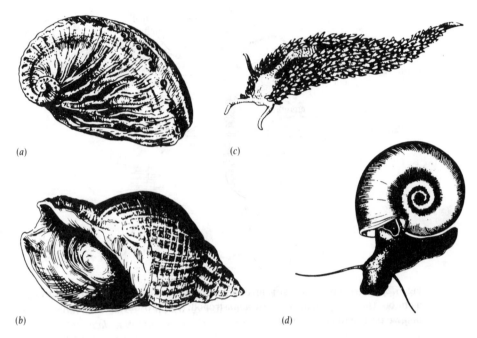

Fig. 6.3. A selection of gastropods. (*a*) An abalone, *Haliotis lamellosa* (Prosobranchia, length up to 7 cm); (*b*) a whelk, *Buccinum undatum* (Prosobranchia, up to 12 cm); (*c*) a sea slug, *Aeolidia papillosa* (Opisthobranchia, up to 8 cm); (*d*) a pond snail, *Planorbis corneus* (Pulmonata, diameter of shell 3 cm). Parts (*a*)–(*c*) are from W. de Haas & F. Knorr (1966). *The young specialist looks at marine life, Burke, London.* Part (*d*) is from W. Engehardt (1964). *The young specialist looks at pond life.* Burke, London.

from the water, through the body surface, but visits to the surface need not be very frequent even in water that contains little dissolved oxygen. For example, the mantle cavity of *Biomphalaria*, which lives in swamps in Uganda, holds enough air to keep the animal supplied with oxygen for 25 minutes.

Most bivalve molluscs live in the sea but a few live in fresh water. They are a large group which have great economic importance, partly because some of them are collected or cultivated as human food and partly because they make up a large part of the diet of some commercially important fishes. For instance, plaice (*Pleuronectes platessa*) in the Irish Sea feed largely on the bivalves *Cultellus*, *Ensis* and *Abra*.

The most characteristic feature of the Bivalvia is the bivalve shell, consisting of two 'valves', one on each side of the body. Bivalves cannot crawl like snails, but many of them use the foot for burrowing in sand or mud. They have no distinct head (Fig. 6.2*d*).

Nucula (Fig. 6.4, right) is a primitive bivalve, one of the Protobranchia. The foot has a flat sole, though it is not used for crawling, and the ctenidia resemble those of gastropods. All the other bivalves in Fig. 6.4 belong to the other subclass, the Lamellibranchia, which have greatly enlarged ctenidia. These are specialized filter feeding organs. The current of water which serves both for respiration and for filter feeding enters the mantle cavity anteriorly in *Nucula* and leave posteriorly. In

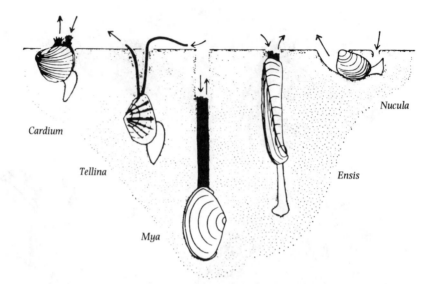

Fig. 6.4. Some burrowing bivalves in the positions they normally adopt. Arrows indicate water currents. *Cardium edule* has valves up to about 5 cm long; *Tellina tenuis* 3 cm; *Mya arenaria* 12 cm; *Ensis ensis* 16 cm and *Nucula nucleus* 1.4 cm. (These are probably the species which the artist intended to represent.) From J.E. Morton (1967). *Molluscs*, 4th edn. Hutchinson, London.

Lamellibranchia it both enters and leaves posteriorly, often through tubes called siphons which are extensions of the mantle (Fig. 6.4). *Tellina* uses its long inhalant siphon as a vacuum cleaner to suck up detritus (fragments of dead organic matter) from the surface of the mud. *Nucula* has a pair of long tentacles which emerge from the shell and collect food particles. Both *Tellina* and *Nucula* probably also practise filter feeding, which is how the other genera in Fig. 6.4 feed.

All the bivalves in this figure burrow, but some other bivalves do not. Mussels (*Mytilus*) anchor themselves to rocks by threads of quinone-tanned collagen, known as byssus. Juvenile scallops (*Pecten*) anchor themselves with byssus but adults live unattached and can swim by flapping the shell open and closed. *Mytilus* is cultivated as food in Europe and *Pecten* is fished commercially by dredging.

The Cephalopoda, the squids and their relatives, all live in the sea. The mantle is muscular, and they use its muscle both to pump water through the mantle cavity for respiration, and to squirt water out forcibly to propel themselves by jet propulsion. The nerve ganglia in the head are much larger than in other molluscs, and form a very effective brain. There is no distinct foot, but the ring of muscular arms that surrounds the mouth may be homologous with the foot of other molluscs.

Nautilus (Fig. 6.5*a*) is the most primitive modern cephalopod. It lives in coastal waters in the S.W. Pacific where it can be caught in traps like lobster pots, set on the bottom at depths around 100 m. Its beautiful shell contains gas-filled chambers which give it buoyancy, as discussed in section 6.5. It has numerous small tentacles and no ctenidia. Squids (Fig. 6.5*b*) have no externally visible shell, only a chitinous rudiment hidden in the body wall. They have eight short arms and two long tentacles with suckers (that help to grip things) both on the arms and on the 'clubs' at the ends

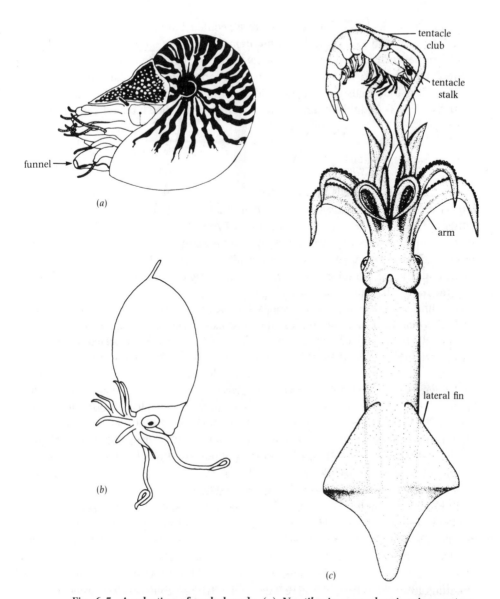

Fig. 6.5. A selection of cephalopods. (*a*) *Nautilus* in normal swimming posture (diameter of shell about 15 cm). (*b*) *Helicocranchia* in the position it adopts when not swimming actively (length, excluding tentacles, 5 cm). (*c*) *Loligo* capturing a shrimp, seen from above, traced from a film (length of mantle 19 cm). (*a*) From M.J. Wells (1962) *Brain and behaviour in cephalopods.* Heinemann, London. (*b*) Drawn from a photograph in E.J. Denton, J.B. Gilpin-Brown and T.I. Shaw (1969). *Proc. R. Soc. Lond.* B174, 271–279. (*c*) From W.M. Kier (1982). *J. Morph.* 172, 179–192.

of the tentacles. They, and all other members of the Coleoidea, have only two ctenidia. Squids of the genus *Loligo* (Fig. 6.5*c*) are abundant off European and N. American coasts. In aquaria they swim perpetually, never resting on the bottom. *Helicocranchia* (Fig. 6.5*b*) is a small oceanic squid caught at depths between 100 and

400 m. It owes its bloated appearance to a peculiar buoyancy mechanism that is described in section 6.5. The giant squid *Architeuthis* reaches overall lengths of 15 m (with the tentacles extended). *Octopus* is a coleoid with eight long arms but no tentacles. It can swim but spends most of its time on the bottom, feeding on crabs and lobsters. It is found offshore on both sides of the Atlantic, and in the Pacific. The cuttlefish *Sepia* is a squid-like animal with a chambered shell hidden inside its body.

6.2.   Mollusc shells

Most molluscs have shells into which they can withdraw, so as to be completely enclosed. The main constituent of the shell is calcium carbonate, which may be in either of two crystalline forms, calcite and aragonite. The remainder is an organic matrix which consists largely of a protein known as conchiolin and usually makes up less than 5% of the mass of the shell. Bone also consists of salt crystals in an organic matrix but the salt is mainly calcium phosphate and the matrix is mainly collagen and makes up 30% of the mass.

The fine structure of mollusc shells has been studied by various techniques including scanning electron microscopy of broken surfaces. Fig. 6.6 shows some of the arrangements which are found. In each of them blocks or strips of calcium carbonate are separated by thin layers of conchiolin. Much the most common structure is crossed-lamellar (Fig. 6.6c). The shell consists of long strips of aragonite laid down in groups. The members of each group are parallel, but different groups lie in different directions. Fig. 6.6(a) shows nacre, or mother of pearl. It consists of tiny blocks of aragonite arranged in layers. Many shells have an inner layer of nacre, though they have a different structure through most of their thickness. A few, including the pearly nautilus (*Nautilus*) and the pearl oyster (*Pinctada*), consist mainly or entirely of nacre. Pearls are balls of nacre formed around sand grains and other particles which get into the mantle cavity.

To protect the mollusc, shells must be rather strong. Strips have been cut from the shells of various molluscs and tested in engineers' testing machines. In most cases the tensile strength lay between 30 and 100 MN m^{-2}, and Young's modulus (which indicates stiffness) was about 50 GN m^{-2}. Thus mollusc shell is weaker and stiffer than compact bone, which has a tensile strength of 150–200 MN m^{-2} and Young's modulus around 18 GN m^{-2}. Nacre is generally stronger than crossed-lamellar shell. It is not surprising that shell is stiffer than bone, since it contains a higher proportion of inorganic crystals. It is less obvious why it should be weaker and why nacre should be stronger than crossed-lamellar shell, which contains a smaller proportion of organic matter. The next few paragraphs point to a possible reason for they show that the organic matter strengthens shell and bone.

How does the conchiolin affect the strength of mollusc shell? Breaking a material involves separating two layers of atoms, and the stress required can be calculated. However, the measured strengths of materials are generally far less than theoretical strengths obtained in this way. For instance, the theoretical strength of glass is about 10 GN m^{-2} but ordinary glass breaks at about 200 MN m^{-2}. Fibreglass, which consists of fine glass fibres embedded in a plastic resin, can have tensile strengths up

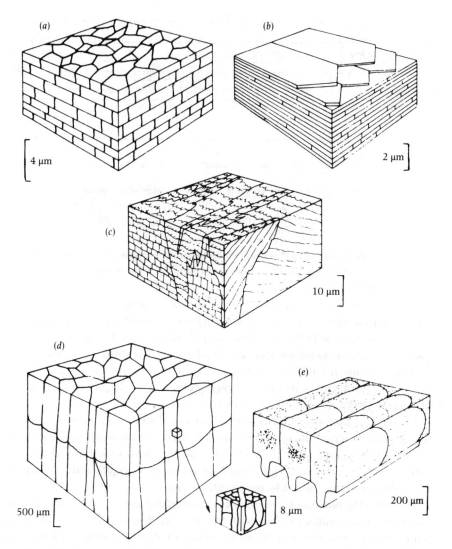

Fig. 6.6. Diagrams of mollusc shell structure. (*a*) Nacre; (*b*) foliated; (*c*) crossed-lamellar; (*d*) simple prisms; (*e*) composite prisms. From S.A. Wainwright, W.D. Biggs, J.D. Currey & J.M. Gosline (1976). *Mechanical design in organisms.* Arnold, London.

to about 1 GN m^{-2}. Mollusc shell and bone are composite materials like fibreglass: they consist of tiny crystals embedded in an organic matrix. They may owe much of their strength to effects like those which make fibreglass stronger than bulk glass.

Materials tend to break at stresses far below their theoretical strength because stress concentrations develop near irregularities in them. This is illustrated by Fig. 6.7(*a*), which represents a notched bar in tension. The thin lines show the direction of tensile stress and where they are evenly spaced the stress is uniform across the bar. They are diverted round the notch and are close together at the tip of the notch: this indicates that the stress there is high. As the force on the bar is increased the stress at

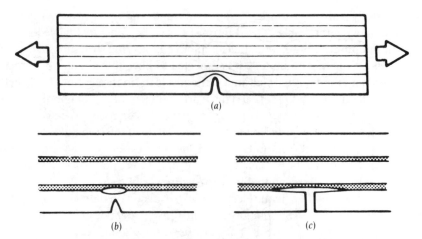

Fig. 6.7. Diagrams of (*a*) a notched bar in tension and (*b,c*) a crack in a composite material. From R. McN. Alexander (1975). *Biomechanics.* Chapman & Hall, London.

the tip of the notch may pass the tensile stress and a crack may form there, while the stress in the bulk of the bar is still far lower. Once a crack has formed it will tend to spread across the bar because a crack is in effect a very sharp notch: the sharper the notch, the greater the stress concentration. There are always unintentional irregularities in objects made of glass and other brittle materials: they cannot be made so perfect as not to be in danger from stress concentrations.

Cracks may form in fibreglass but cannot spread through it easily. Fig. 6.7(*b*) and (*c*) show why. They represent a bar consisting of strong fibres (shown white) separated by weaker glue (stippled). As a crack approaches, the glue splits so that when the crack reaches the glue it no longer has the sharp tip needed to enable it to continue to the next fibre. A bar in tension has tensile stress acting along it but where the force is diverted round a notch or crack the stress has a transverse component. It is the transverse component near the tip of the crack in Fig. 6.7(*b*) which makes the glue split. The layers of conchiolin in shells are believed to stop cracks in this way.

The most primitive molluscs have nacre, but crossed-lamellar shell, which is a little less strong, is much the commonest type. Why should natural selection have preferred the weaker shell type? Two suggestions that have been made are that crossed-lamellar shell may be cheaper to make (cheaper in terms of metabolic energy) and that it may be more resistant to attack by predators such as the dog whelk *Nucella*, a gastropod which bores through the shells of other molluscs to feed on the flesh inside. There is some experimental evidence for both suggestions.

A shell could always be made stronger by making it thicker, but a very thick shell would be expensive to build and unwieldy to carry around. It might be argued that shells should be made just strong enough not to get broken, but how strong is strong enough? The maximum force that a shell will have to bear cannot generally be predicted. For example, many molluscs are attacked by crabs which break their shells to get at them, but crabs vary in strength.

Many shells fail when crabs attack them, and the mollusc gets eaten. Others successfully withstand attack. 359 shells of winkles (*Littorina saxatilis*) had 129 scars where damage (probably caused in most cases by crabs) had healed.

The optimum strength for a shell, or for an engineering structure such as a bridge, depends on the maximum load it is expected to have to bear, on how much doubt there is about the expected load and on the cost of the materials (including, in the case of the shell, the energy cost of carrying it around). Engineers commonly build bridges to a safety factor of about two: that is to say, they design them to be twice as strong as is expected to be necessary. However, a lower factor of safety might be preferred if the weight of traffic over the bridge was to be very carefully controlled (so that the maximum load could be predicted precisely) or if it were necessary for some reason to build the bridge of exceptionally expensive material.

Littorina saxatilis lives in the Fleet, a brackish lagoon on the south coast of England, and on nearby shores. Specimens living just inside the mouth of the lagoon, in full-strength sea water, had thick shells but nevertheless had many healed scars that seemed to be signs of crab damage. Specimens living further up the lagoon, where the salinity and calcium concentration of the water were only half as great, had much thinner shells, although crabs were plentiful where they were living. The low calcium concentration apparently made shell material more expensive, changing the optimum strength. Specimens living on an exposed shore just outside the lagoon also had thin shells but that seemed to be because the exposed shore was not inhabited by crabs.

It is usually difficult to estimate safety factors for animal skeletons, but it seems possible in the case of the chambered shells of cephalopods, which contain gas under reduced pressure (section 6.5). These shells would collapse under the increased outside pressure, if the molluscs swam too deep. Shells of the cuttlefish *Sepia* collapsed in a pressure chamber when the pressure was raised to an average of 20 atmospheres. The greatest depth at which *Sepia* are caught is 150 metres and the pressure there is 16 atmospheres, so the safety factor is 20/16 = 1.3. Shells of *Spirula*, however, did not collapse until the pressure reached 170 atmospheres. *Spirula* lives at depths down to 1200 metres, where the pressure is 121 atmospheres, so its safety factor is 1.4. Low safety factors may be optimal in these cases because the pressure at any particular depth is constant and highly predictable.

Now that we have discussed the strength of mollusc shells we must examine their shape. They are not living material. They can be added to as the mollusc grows but existing parts cannot be altered. The shell of a young snail remains in the adult shell, as the small whorls at the apex.

Most mollusc shells keep the same shape as they grow: the young shell is in effect a scale model of the adult one. This is only possible for a limited range of shapes. If the shell were cylindrical and grew by adding material to its end it would become more slender in its proportions as it grew. If it were a hollow hemisphere it could not be added to without a change of shape. Two shells which can be added to without change of shape are shown in Fig. 6.8. One is a hollow cone like the shells of limpets (*Patella*). The other resembles a *Planorbis* shell (Fig. 6.3d) and is in effect a cone coiled into a spiral. The spiral line is a logarithmic spiral: that is, a spiral which increases its radius by the same factor in every turn. The geometry of spiral shells was discussed by Sir D'Arcy Thompson in his classic *Growth and Form*, which was published in 1917.

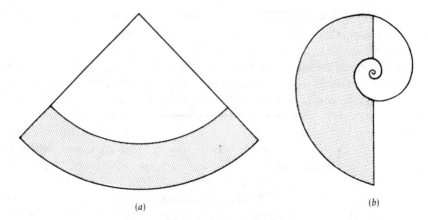

Fig. 6.8. Diagrams of a conical and a spiral shell. Addition of the stippled region to the original untinted shell leaves the shape unchanged. Fom R. McN. Alexander (1971). *Size and shape.* Arnold, London.

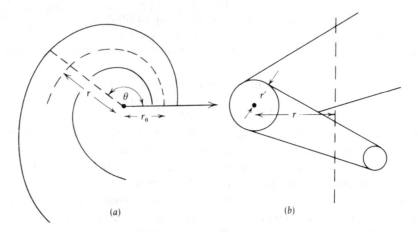

Fig. 6.9. Diagrams of a helical logarithmic spiral shell viewed (*a*) along its axis and (*b*) at right angles to its axis.

New light has been thrown on it much more recently by Dr David Raup.

Fig. 6.9(*a*) represents a spiral shell. Its centre line is a spiral, with radius r_0 at a point which we will take as a starting point. If we move an angle θ round the spiral we find that the radius r there is given by the equation

$$r = r_0 W^\theta, \tag{6.1}$$

where W is a constant. This is the equation of a logarithmic spiral. We will assume that the spiral does not lie in a plane but rises helically, like a snail shell (Fig. 6.9*b*). As we move along the spiral through an angle θ we find that the spiral moves a distance y along its axis. If growth is not to change the shape of the shell

$$y = T(r - r_0), \tag{6.2}$$

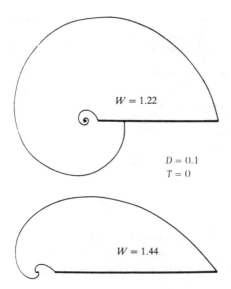

Fig. 6.10. Diagrams of two shells which differ only in the value of the parameter *W*. From D.M. Raup (1966). *J. Palaeontol.* 40, 1178–1190.

where *T* is another constant. For simplicity, the whorls are shown circular in section. At the point where the radius of the spiral is *r* the radius of the section is *r'*. To keep the shape of the shell, *r'* grows in proportion to *r*:

$$r' = r(1 - D)/(1 + D), \tag{6.3}$$

where *D* is a third constant. Many very different-looking shells can be described by varying the parameters *W*, *T* and *D*. *W* indicates the rate at which the shell expands: the radius increases by a factor *W* for every radian of revolution. Fig. 6.10 shows that small values of *W* give tightly coiled shells like snail shells and large values give shells more like the shells of bivalve molluscs: the lower shell resembles a cockle (*Cardium*). *T* indicates the extent to which the shell rises in a spire. *Buccinum* (Fig. 6.3*b*) has a large value of *T* and *Planorbis* (Fig. 6.3*d*) a very small one. *D* shows how far the inner edge of the shell is from the axis of the spiral. If *D* = 0, *r'* = *r* and the inner edges of the whorls reach the axis, as in *Buccinum*, but not in *Planorbis*.

Careful measurements have been made on many mollusc shells and most of them have been found to be described rather accurately by equations 6.1–6.3. In some cases, however, *W* and *T* change in the course of growth and the adult shell is not the same shape as the young one.

Fig. 6.11 illustrates the range of shape which can be obtained by varying *W*, *T* and *D* and shows the ranges actually found in various groups of animals. Most gastropods have low values of *W*, but *T* and *D* vary over wide ranges. Bivalve molluscs have high values of *W* and fairly low values of *T* and *D*. *Nautilus* and most of the extinct ammonoids (which were also cephalopod molluscs) have plane spiral shells (*T* = 0) with low values of *W*. Brachiopods have bivalve shells and look superficially like bivalve molluscs, but are very different inside (see chapter 10). They have high values of *W* and fairly low values of *D* but unlike bivalve molluscs most of them have *T* = 0.

The gastropod and bivalve types of shell are alternative containers which give good

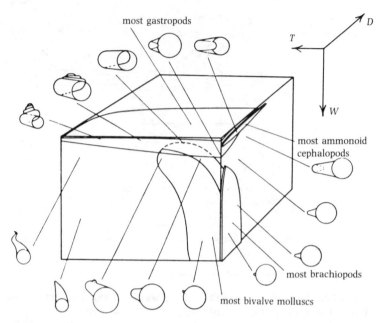

Fig. 6.11. A three-dimensional block showing the range of possible shell forms indicated by equations 6.1 to 6.3 and the ranges actually found in some groups of animals. Some examples of shell shape are shown: they were generated from the equations by means of a computer with a graphic display. From J.D. Currey (1970). *Animal skeletons.* Arnold, London. (After D.M. Raup.)

protection. A typical gastropod shell has a small opening so that when the snail retracts little of its body is exposed. In some cases the opening can be closed by an operculum. A pair of bivalve shells fit together to enclose the mollusc completely. Some gastropod shells such as those of limpets (*Patella*) and of *Haliotis* have large openings, but these molluscs live on rocks and can protect themselves by holding the shell opening firmly against the surface of the rock.

6.3.   Muscle

There are two main kinds of muscle, smooth and striated. They are present in cnidarians and other simple multicellular animals, but discussion of their properties has been deferred to this chapter because the properties of mollusc muscles have been investigated much more thoroughly than those of muscles from more primitive animals.

Smooth and striated muscle seem to work in essentially the same way, by interaction of the proteins actin and myosin, but striated muscle has the protein filaments arranged in very regular ways, and smooth muscle does not. The arrangement of filaments can be seen only by electron microscopy, but it makes striated muscle fibres look striped under the light microscope. Cross-striated muscle (Fig. 6.12*a*) has the stripes running transversely across the fibres and obliquely striated muscle (Fig. 6.12*b*) has them running helically. The muscles of sea anemones are

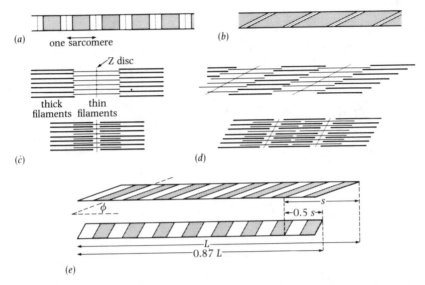

Fig. 6.12.(*a*), (*b*) Sketches of (*a*) cross-striated and (*b*) obliquely striated muscle fibres, as seen by light microscopy. (*c*), (*d*) Diagrams showing the arrangement of protein filaments in (*c*) cross-striated and (*d*) obliquely striated muscle fibres, as seen by electron microscopy. The diagrams also show how the filaments slide past each other when the fibres shorten. (*e*) A diagram of an obliquely striated muscle fibre, showing how an increase in the angle affects its length.

smooth, or irregularly cross-striated, but the circular muscles of the medusa *Polyorchis* have regular cross-striations. Many of the muscles of molluscs, and of nematode and annelid worms, are obliquely striated, but some mollusc muscles are cross-striated, as we will see.

The thin filaments in striated muscles consist largely of actin, and their diameter is about 6 nm. The thick filaments of cross-striated muscle consist of myosin and have a diameter of about 11 nm, but those of obliquely striated muscle are thicker and have the myosin arranged around a core of another protein, paramyosin. The ends of the myosin molecules project from the thick filaments and can attach to the thin filaments, forming cross-bridges. Partitions called Z-discs divide the fibres into segments called sarcomeres, which are the units of the pattern of stripes that can be seen under the light microscope.

Electron micrographs of extended and contracted muscles show that the thick and thin filaments remain constant in length but slide between each other as the muscle lengthens and shortens (Fig. 6.12*c,d*). The forces exerted by active muscles are due to the cross-bridges pulling on the thin filaments. When a muscle shortens, the cross-bridges apparently 'walk' along the thin filaments.

Muscles or parts of muscles dissected from freshly killed animals can be made to contract by electrical stimulation. Transducers have been used to measure the forces that such preparations exert. Fig. 6.13 shows results of experiments in which the forces were measured, while the muscles were held at various lengths. Lengths are given relative to the length at which the largest force can be exerted. Forces are represented by stresses (force per unit cross-sectional area).

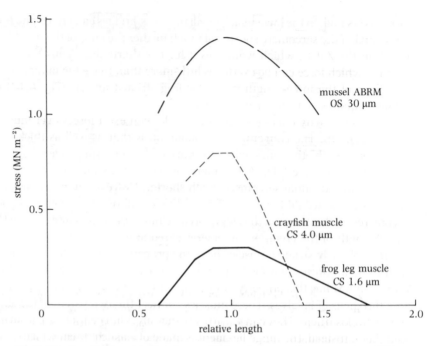

Fig. 6.13. Graphs showing the stresses that various muscles can exert, at different lengths. The length of the thick filaments of each muscle is shown, and also whether it is cross-striated (CS) or obliquely striated (OS). The data are from A. Csapo (1960). In *The structure and function of muscle*, vol. 1 (ed. G.H. Bourne). Academic Press, New York; and from J. Zachar & D. Zacharova (1966), *Experientia* 22, 451–452.

Fig. 6.13 shows that different muscles exert different maximum forces. The extremes are represented by a frog leg muscle (which is typical of vertebrate striated muscles generally) and the anterior byssus retractor muscle (ABRM) of the bivalve mollusc *Mytilus*. The vertebrate muscle is cross-striated and the ABRM obliquely striated but the important difference between them, from the point of view of stress development, is that the ABRM has much longer filaments. The thick filaments are about 20 times as long in the ABRM as in frog muscle, so an actin filament lying alongside one can be attached to it by 20 times as many cross-bridges. The force that can be exerted on each actin filament should be about 20 times as great as in the frog muscle, but there are only about a quarter as many actin filaments as in the frog muscle, per unit cross-sectional area. This is partly due to the thick filaments being much thicker in the ABRM (110 nm) than in the frog muscle (11 nm), and so occupying more of the cross-sectional area. Therefore, the ABRM should be able to exert $20/4 = 5$ times as much stress as the frog muscle. Fig. 6.13 shows that this is approximately true. The crustacean muscle has filaments of intermediate length and exerts an intermediate stress.

Fig. 6.13 also shows that the obliquely striated muscles can exert forces over much wider ranges of length than can the cross-striated ones. Fig. 6.12(c) shows a cross-striated sarcomere extended (above) and much shorter (below). In the upper diagram the thick and thin filaments overlap only a little, so few cross-bridges can attach and the stress is small. In the lower diagram all the cross-bridges can attach and the stress is maximal. If the sarcomere shortened much further the ends of the thick filaments would hit the Z-disc, which would resist further shortening. Thus the maximum length at which force can be exerted is little more than twice the minimum length. The shape of the force–length graph for frog striated muscle (Fig. 6.13) can be explained quite satisfactorily in this way.

It is not clear why obliquely striated muscles can exert forces over much larger ranges of length. The conventional explanation is that as well as thick filaments sliding between the thin ones, when the muscle shortens, they slide relative to each other, so that the angle ϕ (Fig. 6.12e) increases. Fig. 6.12e shows that an increase in ϕ can make an individual sarcomere much shorter. However, it also shows that the change of angle would have less effect on the length of a fibre consisting of many sarcomeres. In the diagram, the length of each individual sarcomere is halved but the overall length of the chain of sarcomeres is reduced by only 13%.

Some obliquely striated muscles have the property called 'catch' which enables them to maintain tension for long periods with very little expenditure of energy. *Mytilus* ABRM has this property, and so also do parts of the adductor muscles that close the shells of bivalves. These muscles connect the two valves of the shell, and shorten to close them. They can pull to close the valves but cannot push from inside to open them. Instead, the hinge ligament is made of elastic protein which is strained when the valves close, and recoils elastically to re-open the valves when the adductor muscle relaxes. The disadvantage of this arrangement is that muscle must remain taut for as long as the shell is kept closed, which may be for a long time. Shore-living bivalves such as *Mytilus* may have to keep their valves closed (preventing evaporation of water) for several hours each day, while they are exposed by the tide. Keeping a muscle taut uses metabolic energy.

The adductor muscles of most bivalves have two parts, an opaque part which has the 'catch' property and a translucent part which does not. Fig. 6.14 shows an experiment with the opaque part of the adductor muscle of the oyster *Crassostrea*. A small piece of the muscle was dissected out, together with pieces of shell at either end which were used to attach it to the apparatus. It was fixed between a rigid support and a force transducer and stimulated electrically. Pulses of alternating current or trains of very short DC pulses (for instance, 10 pulses per second) made the muscle develop tension quickly. The tension declined rapidly when the stimulus ended. The time constant for the decay of tension (i.e. the time in which the tension fell to $1/e = 0.37$ of its initial value) was on average about 9 s. If the stimulus was a long one the muscle remained contracted until it ended, and then relaxed rapidly. Single pulses of direct current lasting several seconds made the muscle develop tension just as quickly, but the tension decayed very slowly when the stimulus ended. The average time constant was 5 minutes, 30 times as long as after an AC stimulus. Thus AC stimulated only a normal contraction but DC put the muscle into the catch state. Further, an AC stimulus given to a muscle in a state of catch made it relax quickly. The muscle could also be put into catch or released from catch by drugs. Addition of a little acetylcholine to the solution containing the muscle stimulated catch, which persisted after the drug had been washed away but could be released by addition of 5-hydroxytryptamine (5-HT). It is presumed that separate nerves stimulate catch and

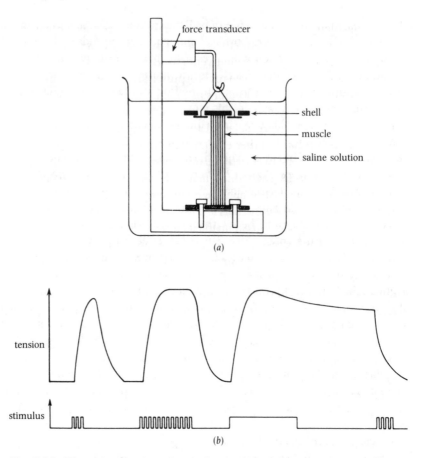

Fig. 6.14. Diagrams showing an experiment with *Crassostrea* opaque adductor muscle, and the results of the experiment.

release and it is suspected that the transmitter substances at the nerve–muscle junctions are acetylcholine and 5-HT, respectively. The mechanism of catch is unknown.

Experiments have been performed on *Mytilus* ABRM, to find out how economical catch is. Oxygen electrodes were used to measure the quantities of oxygen taken up by the muscle from the surrounding solution, during and immediately after periods of normal and catch contraction. Since metabolism using 1 cm^3 oxygen released 20 J, the power requirements could be calculated. The results are conveniently expressed as the power required to maintain unit stress in unit mass of muscle: if power is expressed in W, stress in N m^{-2} and mass in kg this quantity has units W N^{-1} m^2 kg^{-1} or m^3 kg^{-1} s^{-1}. It was found to be 3×10^{-6} m^3 kg^{-1} s^{-1} for ABRM in normal contraction, but only 5×10^{-7} m^3 kg^{-1} s^{-1} in catch. Even the normal value is very low compared with values for vertebrate striated muscle. For instance, frog sartorius muscle uses 8×10^{-4} m^3 kg^{-1} s^{-1}.

The economy of catch can be expressed in another way. When the ABRM is in catch, maintaining a rather high stress of 0.5 MN m^{-2}, it uses oxygen only 1.5 times as fast as when it is resting.

The long filaments of the ABRM help to make it economical: because they are so long, there are only one twentieth as many sarcomeres in it as in an equal length of frog striated muscle, so only one twentieth as many cross-bridges need attach to maintain a given stress in a given mass of muscle. However, this cannot explain why ABRM uses only about 1/300 times as much power, even in normal (non-catch) contraction. The mechanism of catch is not understood but is suspected to involve direct connections between adjacent thick filaments.

Filament length has another effect, on the rates at which muscles can shorten. Since there are only one twentieth the number of sarcomeres in ABRM as in an equal length of frog muscle, each thick filament must move the neighbouring actin filaments 20 times as fast to achieve contraction at the same speed. It is actually found to be about one twentieth as fast as frog sartorius: it can contract at up to 0.3 lengths s^{-1} and frog sartorius at up to 6 lengths s^{-1}, at 16–18°C. It is fortuitous that the ratio of speeds agrees so closely with the ratio of filament lengths: vertebrate striated muscles with similar filament lengths have quite a wide range of speeds.

The translucent (non-catch) part of the adductor muscle of bivalves has thick filaments which are shorter than in the opaque (catch) part. Consequently it can shorten faster, but cannot exert as much stress. It is presumably used to close the valves rapidly, for example to protect the animal from a predator.

The scallop *Pecten* is exceptional among bivalve molluscs, in being able to swim by flapping its valves open and closed, 2–3 times per second. It has a third part in its adductor which unlike the others is cross-striated, not obliquely striated. The sarcomeres of this cross-striated part are only about 3 μm long and it can contract at up to 3 lengths per second (at 14°C). It is the fastest part of the muscle, and the part used in swimming.

When it swims, *Pecten* opens and closes its shell through about 20°. Each time it closes, the swimming muscle shortens by about 0.25 of its length in about 0.4 s, shortening at about 0.6 lengths s^{-1}. This is only 0.2 of the maximum rate of contraction, determined in experiments with isolated muscles.

A muscle can shorten at its maximum rate only when there is no force opposing shortening. The greater the opposing force the more slowly can it shorten; if the force is too great the muscle will stretch instead of shortening. Fig. 6.15(a) shows an experiment to investigate the relationship between force and rate of shortening. When the muscle is stimulated electrically it contracts, lifting the weight attached to the opposite side of the lever. A vane attached to the lever partly interrupts a beam of light aimed at a photoelectric cell. As the lever moves, the illuminated area on the cell changes, so the output of the cell at any instant indicates the position of the lever. This device is used to determine the rate of shortening, when different weights are being lifted.

The results of an experiment with the fast part of the adductor of *Pecten* is shown in Fig. 6.15(b). The piece of muscle tested in this particular experiment could exert 0.15 N when contracting isometrically (i.e. when the shortening rate was zero), and could shorten at 130 mm s^{-1} when there was no resisting force. It was estimated that the rate of shortening in swimming is about 0.2 of the maximum rate. At this rate of shortening the muscle can exert about half the force exerted in isometric contraction.

Squids use the muscles of the mantle to pump water in and out of the mantle cavity,

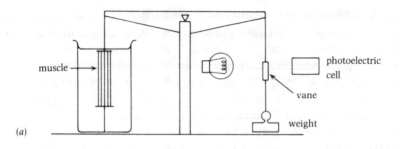

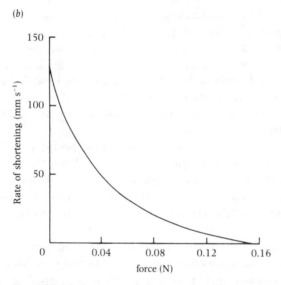

Fig. 6.15.(*a*) Apparatus for measuring the rates at which a muscle can shorten against different forces. (*b*) A graph of rate of shortening against force for a portion of the fast part of the adductor muscle of *Pecten*. From J. Hanson & J. Lowy (1960). In *The structure and function of muscle*, vol. 1 (ed. G.H. Bourne). Academic Press, New York.

both for breathing (section 6.6) and for jet propulsion (section 6.5). The muscles are obliquely striated but their filaments are remarkably short: the thick filaments of *Lolliguncula* are only 1.0–1.5 μm long. Short filaments are to be expected in fast muscles, but this is shorter even than in vertebrate striated muscle.

The mantle muscle consists mainly of circular fibres which when they contract reduce the volume of the cavity, driving water out. There is a thick layer of white circular fibres sandwiched between two thin layers of yellowish ones. The two types of fibre contain very different proportions of mitochondria, the organelles that contain the enzymes of the Krebs cycle: the white fibres contain only 6% by volume of mitochondria, but the yellowish ones contain 47%. This suggests that the yellowish fibres are much better adapted than the white ones for aerobic respiration, in which foods are oxidized completely to carbon dioxide and water. The impression is confirmed by histochemical tests, which show that the activities of mitochondrial

Table 6.1. *Changes in Sepia mantle muscle during exercise*

Glycogen concentrations are represented by the equivalent molar concentrations of glucose

	Concentrations (mmol kg^{-1}) in muscles of	
	rested animals	exhausted animals
ATP	9	2
Arginine phosphate	34	4
Arginine	30	45
Glycogen	24	5
Octopine	0	9

Data from K.B. Storey & J.M. Storey (1979). *J. comp. Physiol.* **131**B, 311–319.

enzymes are much higher in the yellowish muscle than in the white, but that other metabolic enzymes are about equally active in both. Further, the yellowish muscle is much better supplied with blood vessels than the white: aerobic muscle needs a good blood supply to bring oxygen to it as fast as it is needed.

These observations suggest that squids may use aerobic and anaerobic metabolism in much the same way as humans and other vertebrates do. We power prolonged activities by aerobic metabolism, oxidizing foodstuffs to carbon dioxide and water, but we power short bursts of violent activity (for example, sprints) by anaerobic activity, converting glucose to lactic acid as explained in section 4.5. We accumulate lactic acid in our bodies, building up an 'oxygen debt' which is repaid later, when we oxidize some of the lactic acid to obtain the energy needed to re-convert the rest to glucose. It seems likely that the yellowish, aerobic muscle is used for breathing and for prolonged swimming, but that the white, anaerobic muscle is brought into use for bursts of fast swimming. The advantage of using anaerobic metabolism in fast swimming is that speed is not then limited by the rate at which the gills and blood system can supply oxygen to the muscles. The disadvantage is that the burst of swimming has to stop when the oxygen debt rises too high: the animal becomes exhausted.

ATP is the immediate energy source for muscles in vertebrates, molluscs and (apparently) all animals, but it is never stored in large quantities. Larger energy stores are kept as creatine phosphate (in vertebrates) or arginine phosphate (in molluscs and many other invertebrates), which can be used to re-convert ADP to ATP according to the equations

$$\text{creatine phosphate} + \text{ADP} \rightleftarrows \text{creatine} + \text{ATP};$$
$$\text{arginine phosphate} + \text{ADP} \rightleftarrows \text{arginine} + \text{ATP}.$$

A vertebrate, or a squid, becomes exhausted in a bout of violent activity, when it has used up its stocks of ATP and creatine or arginine phosphate and has also built up the maximum tolerable oxygen debt by anaerobic metabolism.

Table 6.1 shows the results of chemical analysis of the mantle muscles of small cuttlefish (*Sepia*), which were killed either after a period of rest in well-aerated water, or after they had been chased around in their tank until they were exhausted. The exhausted animals had used up most of the ATP, arginine phosphate and glycogen in their muscles, as might have been expected. However, the increase in the arginine content of the muscles was less than the decrease in arginine phosphate, and octopine accumulated. It seems that glucose is metabolized to pyruvic acid in the same way as in vertebrates but that this then combines with arginine to form octopine. The overall reaction is

$$\text{glucose} + 2 \text{ arginine} = 2 \text{ octopine} + 2 \sim.$$

It yields the same number of high-energy bonds, per molecule of glucose, as if lactic acid had been produced (section 4.5).

Thus molluscs have a wide range of types of muscle. They have muscles with long filaments that can exert large stresses and ones with shorter filaments that exert less stress but can contract faster. They have cross-striated and obliquely striated muscle, and some of the latter is capable of catch. They also have muscles specialized for aerobic and anaerobic metabolism.

The remainder of this section illustrates how muscles can be used to produce complex movements of structures that have no skeleton. The feet of gastropod and bivalve molluscs and the arms and tentacles of cephalopods consist largely of muscle, with no skeleton, like the human tongue. As in the tongue, muscle fibres running in many directions make elaborate movements possible.

Fig. 6.5(c) shows how squid use their tentacles for catching prey. They can extend them rapidly, to about 1.7 times their resting length. They have to be able to bend them to aim them in the right direction, and they may have to twist them about their long axes to place the suckers of the tentacle club against the prey.

Fig. 6.16 shows how the muscles are arranged in the tentacle stalk. Transverse muscle fibres run across the stalk, in two directions at right angles, and circular muscles run circumferentially round it. When they contract they make the tentacle thinner and therefore longer (because its volume is fixed). The bundles of longitudinal muscle fibres shorten the tentacle or (if they contract asymmetrically) bend it. The two layers of helical muscles (wound in opposite directions) can twist the tentacle clockwise or anticlockwise.

Suppose that the circular and transverse fibres shorten by 20%, thinning the tentacle to 0.8 of its initial diameter and $0.8^2 = 0.64$ of its initial cross-sectional area. It must extend to $1/0.64 = 1.56$ times its initial length. Thus a small contraction of the circular and transverse muscles causes a large extension of the tentacle. These muscles are cross-striated with very short filaments, which presumably enables them to contract fast to shoot out the tentacle, but the longitudinal muscles are obliquely striated and presumably slower.

6.4.   Crawling and burrowing

Gastropods crawl on their large, muscular foot. Their speeds are low; no more than about 2.5 mm s^{-1} for snails (*Helix*) and 1 mm s^{-1} for limpets (*Patella*). A hint as to how they do it can be got by watching them from below, while they are crawling on

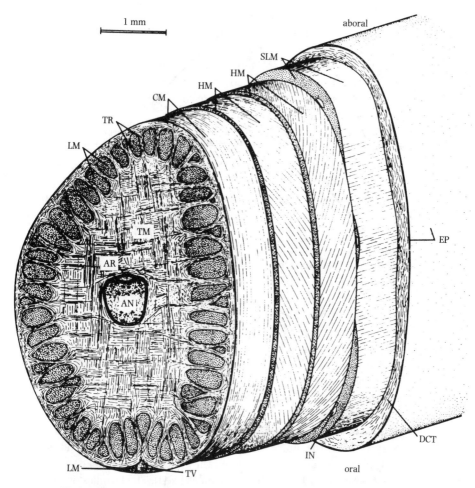

Fig. 6.16. A cut-away diagram of the tentacle stalk of a squid, showing the arrangement of muscle fibres. AN, axial nerve; AR, artery; CM, circular muscle; DCT, connective tissue; EP, epithelium; HM, helical muscle; IN, nerve; LM, SLM, longitudinal muscle; TR, TM, transverse muscle; TV, vein. From W.M. Kier (1982). *J. Morph.* 172, 179–192.

glass. Light and dark transverse bands can be seen on the foot and these move along the foot as the animal crawls. It seems that the animals crawl by passing waves of muscular activity along the foot. However, the waves are retrograde (travel backwards) in some molluscs and direct (travel forwards) in others. There must be two variants of the basic mechanism.

Marks have been made on the soles of gastropod feet, before placing them on glass. It has been shown in this way that each point remains stationary at one stage of the passage of a wave of muscular activity, and moves forward at another. The diagrams of molluscs crawling (Fig. 6.17) are based on these observations. In these diagrams, the hatched parts of the sole of the foot are stationary and the others are moving forward. The foot is divided into numbered segments to make it easier to show which parts are shortening and which are elongating.

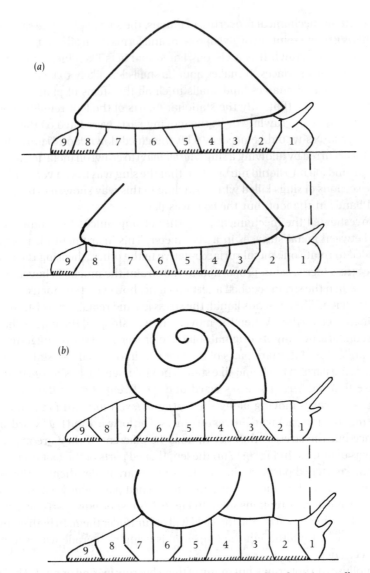

Fig. 6.17. Diagrams of successive stages of crawling by (*a*) a mollusc such as the limpet (*Patella*) that uses retrograde pedal waves and (*b*) one such as the snail (*Helix*) that uses direct ones. The stationary parts of the foot are indicaed by hatching. From R. McN. Alexander (1982). *Locomotion of animals*. Blackie, Glasgow.

In Fig. 6.17(*a*), segment 6 is shortening (pulling segment 7 forward) and segment 8 is lengthening (pushing 7 forward). These changes make the wave move backwards and part of the foot move forwards. A wave travelling the length of the body moves the whole mollusc forwards. In Fig. 6.17(*b*) also, segment 6 is shortening and 8 is lengthening, but in this case the segments that are free to move are the shortened ones, and the wave as well as the snail is travelling forwards.

It remains to explain how the hatched parts of the sole of the foot are held stationary, while the others are free to move. Dr Mark Denny showed that this was due to the peculiar mechanical properties of mucus, the slimy substance secreted by the foot. This is a dilute solution of glycoprotein, and forms the trails that snails leave behind them as they crawl. It is a viscous liquid and acts like glue, attaching the animal firmly to solid surfaces. It makes aquatic snails less likely to be dislodged by water movements, and enables land snails to climb the stems of plants.

It used to be thought that only the stationary parts of the foot remained on the surface the mollusc was crawling over: the moving parts were lifted off the surface. However, large forces would be needed to lift the glued-down foot. Denny showed that no lifting occurred by allowing a slug (*Agriolimax*) to crawl on metal foil and then dropping slug and foil into liquid nitrogen, so that the slug was frozen very suddenly. Microscope sections of slugs killed while crawling in this way showed extended and contracted bands in the foot, but the sole was flat.

Denny investigated the mechanical properties of slug mucus by sandwiching a thin layer between metal surfaces in a viscometer. This instrument measured the torque needed to rotate one metal surface relative to the other, shearing the mucus. At first, the stress increased in proportion to the strain: the mucus behaved like an elastic solid. When the strain reached a certain value, however, the mucus suddenly yielded and behaved like a viscous liquid: the stress fell and remained constant while the strain increased further. When the movement was stopped, the mucus 'healed' within a second, recovering its original solid-like properties. Denny suggested that the mucus under the stationary parts of the foot was in the solid-like state and that under the moving parts in the liquid-like state. The viscometer tests showed that they could be like that if different stresses acted under different parts of the foot.

The muscular waves moving along the foot must exert forward forces on some parts of the mucus, and backward forces on others. The forward and backward forces must balance but the stresses they set up may be different, if they are applied to different areas of mucus. In Fig. 6.17(a) the lengthened parts of the foot occupy less area than the shortened parts, so the greater stresses are under them. If the forces become large enough the mucus under the lengthened parts will yield, while that under the shortened parts remains solid. In Fig. 6.17(b) shortened parts occupy less area than the lengthened parts, and the yield will occur under them. In both cases the mollusc moves forward although the limpet (a) is using retrograde waves and the snail (b) direct ones.

Bivalve molluscs do not crawl, but many of them burrow in sand or mud. The West Indian surf clam *Donax denticulatus* (which is about 2 cm long) can bury itself completely in about 4 s but most bivalves burrow much more slowly. For instance, a British species of *Donax* takes about 1 min to bury itself. If bivalves simply spent their lives buried in one place there would be no need to burrow fast, but at least some move about. The most remarkable movements are made by surf clams, which migrate up and down the shore with the tide. They emerge from the sand from time to time and allow the waves to move them up and down the shore before burying themselves again. Other bivalves which burrow in sandy shores are liable to be exposed by storms, and to have to burrow again. Bivalves such as *Ensis* climb near the surface at high tide so that their siphons project above the surface, but dig deeper when they are exposed at low tide.

Burrowing can be watched if a glass aquarium is partly filled with sand and a

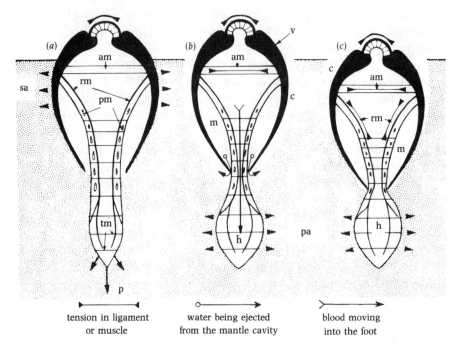

Fig. 6.18. Diagrams showing the sequence of events when a bivalve mollusc burrows. Labels indicate: am, adductor muscle; c, sand loosened by water ejected from the mantle cavity; h, haemocoel; m, mantle cavity; p, probing movements of foot; pa, swollen foot anchored in sand; pm, protractor muscle of foot; rm, retractor muscle of foot; sa, open shell anchored in sand; tm, transverse muscles of foot; v, valve of shell. From E.R. Trueman (1968). *Symp. zool. Soc. Lond.* 22, 167–186.

bivalve induced to burrow close to the glass. The sequence of events is shown in Fig. 6.18. So that as many structures as possible can be shown the diagram has been drawn as if the hinge were horizontal; in fact, bivalves normally burrow with the hinge vertical (Fig. 6.4).

Burrowing works like this. The shell is anchored firmly in the sand by allowing the valves to open, while the foot is pushed down deeper (Fig. 6.18a). The foot is then anchored by making it swell and the shell is pulled down towards it (Fig. 6.18c). Neither shell nor foot can be anchored in this way until the animal has penetrated some distances into the sand, so burrowing starts with the shell lying on the surface and the foot probing to obtain a purchase.

The foot is not a solid lump of muscle like a squid tentacle, but has a cavity (the pedal haemocoel) that can be inflated with blood or allowed to empty. It is connected to the main body cavity (the visceral haemocoel) by a valve (Keber's valve) that can be closed or opened. In Fig. 6.18(a) the adductor muscles are relaxed so that the elasticity of the hinge ligament keeps the valves jammed tightly against the sand. The transverse muscles of the foot are contracting, forcing the foot to elongate and so pushing it down into the sand. Keber's valve is presumably closed, so that the transverse muscles do not merely squeeze blood out of the foot. In (b) the adductor muscles contract, reducing the volume enclosed by the shell. This drives water out of the mantle cavity round the edge of the shell, and also drives blood from the visceral to the pedal haemocoel. The transverse muscles near the distal end of the foot are relaxed so the foot becomes bulbous, and is anchored firmly in the sand. However, the shell is free to move because the adductor muscles have narrowed it and because the water squirted out of the mantle cavity has loosened the surrounding sand. In (c) the retractor muscles contract. These muscles run from the shell to the distal end of the foot so their contraction pulls the shell down into the sand.

Pressures have been measured in burrowing *Ensis*. A hypodermic needle attached to a pressure transducer was inserted between the valves, into the pedal haemocoel. This made it possible to get records of blood pressure during the early stages of burrowing, until the base of the needle came into contact with the surface of the sand and prevented the animal from burrowing deeper. Small fluctuations of pressure up to about 1 kN m^{-2} occurred while the foot was probing the sand (Fig. 6.18a). Very much larger pressures, up to about 10 kN m^{-2}, occurred briefly while the valves were being adducted (Fig. 6.18b). It was calculated that the stress in the adductor muscles would have to reach 0.5 MN m^{-2} if they alone produced such pressures. *Ensis* has other muscles between the valves which may co-operate with the adductors, in which case the stress required may be considerably lower.

The retractor muscles can exert large forces. This was demonstrated in an experiment in which *Ensis* were allowed to bury the foot in a beaker of sand. While the shell was still clear of the sand it was clamped to a retort stand and the mollusc, trying to pull it down into the sand, lifted the beaker and its contents, which weighed 0.4 kg. In further experiments with the foot held by a cord instead of buried, a 13 cm *Ensis* lifted up to 1.1 kg. It was calculated that stresses up to 0.2 MN m^{-2} must have acted in the retractor muscles.

Some bivalves burrow in substrates which are far less easy to penetrate than sand and mud. Piddocks such as *Zirphaea* bore in rock and the shipworm *Teredo* bores in timber. They use their shells for boring but do not depend on them for protection, and their shells are too small to enclose the whole body.

6.5. Jet propulsion and buoyancy

Cephalopods can swim by jet propulsion, by squirting water from the funnel. When the funnel is pointing forwards as in Fig. 6.2(e), squirting drives the mollusc backwards, but it can bend the funnel round to squirt backwards and drive itself forwards. *Octopus* swim relatively slowly but squid can swim fast, at speeds comparable to the top speeds of teleost fish. A 20 cm (100 g) *Loligo* has been filmed in an aquarium accelerating from rest to 2.1 m s^{-1} (backwards) by a single contraction of its mantle. The oceanic 'flying squid' *Onycoteuthis* is presumably fast, for it has been known to leap from the water and land on the decks of ships. The muscles of the mantle wall, that power jet propulsion, make up 35% of the mass of the body in *Loligo* but much less in *Octopus*.

We will compare squid swimming with the swimming of fishes, asking first, how much energy does each use? At low speeds, at which the muscles work aerobically, the rate of oxygen consumption can be used to calculate the rate of energy consumption. It would in principle be possible to have squid swimming around in a tank from

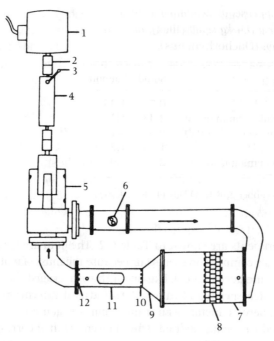

Fig. 6.19. A water tunnel of a type used for investigations of squid and fish swimming. The labels are explained in the text. From G.J. Farmer & F.W.H. Beamish (1969). *J. Fish. Res. Bd Can.* 26, 2807–2821.

which air was excluded, and to measure the rate at which the dissolved oxygen content of the water fell. However, a big tank would be needed to give the animals adequate space for swimming, so it would take a long time for the animals to reduce the oxygen concentration by a measurable amount. Also, it would be difficult to control the speed of swimming.

These difficulties were avoided by using a water tunnel like the one shown in Fig. 6.19, in which a limited volume (92 litres) of water was circulated around a loop of pipes. It was driven clockwise around the loop by a centrifugal pump (labelled 5) powered by an electric motor (1). The squid was confined in the test section (11) by nylon mesh at (10) and (12). It was trained to swim against the current, at just such a speed as neither to gain nor lose ground. Its situation was then the same as if the water were stationary and it were swimming at the same relative speed. For this to be true, the water in the test section had to flow smoothly, and at the same velocity at all points in the cross-section: the grids (8) and reduction cone (9) were designed to achieve this as nearly as possible. An oxygen electrode at (6) was used to measure the oxygen content of the water. The animals would swim steadily for several hours, long enough to measure rates of oxygen consumption at several different speeds. However, if the speed was raised above the critical speed at which the maximum possible rate of aerobic metabolism was needed to power swimming, they soon became exhausted. The dissolved oxygen concentration in the water fell only a little during the experiments, too little to affect the rate at which the animal would take up oxygen.

Table 6.2. *Energy costs of swimming at their critical speeds for a 0.4 kg squid (Illex) and a 0.5 kg sockeye salmon (Onchorhynchus)*

	Squid	Salmon
Critical speed (m s^{-1})	0.76	1.35
Metabolic rate while swimming (W)	2.33	1.33
Metabolic rate, not swimming (W)	0.69	0.11
Difference (W)	1.64	1.22
Energy cost of swimming (J m^{-1})	2.2	0.9

Data from D.M. Webber & R.K. O'Dor (1986). *J. exp. Biol.* **126**, 205–224.

Results from the experiments are shown in Table 6.2. The data are for a 0.4 kg squid and a 0.5 kg fish: this comparison seems fair because the mass of water in the squid's mantle cavity (not included in the 0.4 kg) would have averaged about 0.1 kg. When swimming at its critical speed (0.76 m s^{-1}) the squid used 419 cm^3 oxygen per hour. As a general rule, aerobic metabolism using 1 cm^3 oxygen releases 20 J of energy, whatever the foodstuff being oxidized. Thus 419 cm^3 O_2 h^{-1} corresponds to 8380 J h^{-1} or 8380/3600 = 2.33 W. When the squid were not swimming they used 125 cm^3 O_2 h^{-1}, corresponding to 0.69 W. Thus the energy cost of swimming was 2.33 − 0.69 = 1.64 W. The speed was 0.76 m s^{-1}, so the energy cost of swimming a metre was 1.64/0.76 = 2.2 J. This performance is poor in comparison with the fish, which had nearly twice the critical speed and less than half the energy cost per metre.

It may seem surprising that the squid does so badly, in this comparison, because it seems well shaped for efficient swimming. It was 0.42 m long, so when it swam at 0.76 m s^{-1} its Reynolds number (see section 2.8) was 3×10^5. This is well up in the range of Reynolds numbers in which bodies leave wakes: the pattern of flow around it must have been as shown in Fig. 2.15(*b*). If it were unstreamlined it would leave a broad wake, and the drag would be high, but streamlining can keep the wake narrow and the drag low. The best shape is like a torpedo, rounded in front and tapering gently behind. Squid swim with their tentacles retracted and their arms bunched together, and in that posture seem well streamlined, as fish also are.

There seems to be a different reason for the high energy cost of squid swimming. Consider a squid swimming steadily. Each jet accelerates it and it decelerates between jets, but its mean velocity is u and the mean drag on its body is D. It emits jets with frequency n, taking in each time a mass M of water and ejecting it with velocity U. In unit time it accelerates a mass nM of water to velocity U, giving momentum nMU to water. By Newton's second law of motion, force equals rate of change of momentum:

$$D = nMU. \tag{6.4}$$

The power required to drive the squid through the water is Du, which, from this equation, equals $nMUu$. In addition, power is used accelerating the water: in unit time a mass nM of water is given kinetic energy $1/2\ nMU^2$, so this much power is needed.

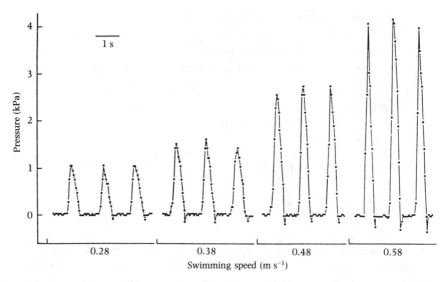

Fig. 6.20. Records of pressure in the mantle cavity of a squid (*Illex*) swimming in a water tunnel. From D.M. Webber & R.K. O'Dor (1986). *J. exp. Biol.* 126, 205–224.

$$\text{Efficiency} = \text{useful power/total power}$$
$$= nMUu/(nMUu + \tfrac{1}{2}nMU^2)$$
$$= 2u/(2u + U). \qquad (6.5)$$

This efficiency will be low if U (the jet velocity) is high compared with u (the swimming velocity). Equation 6.4 shows that U will have to be high if nM is low; that is, if the mass of water accelerated in unit time is low. Squid squirt out quite a small mass of water in each jet but the tails of fish accelerate large masses of water in unit time. This may be why squid need much more energy than salmon to swim at the same speed.

The jet velocity has not been measured directly, but can be calculated from records of pressure in the mantle cavity. These have been obtained by means of a thin plastic tube pushed through the mantle and connected at the other end to a pressure transducer. The tube was long enough to allow the squid to swim in the water tunnel, with the transducer outside the tunnel. Some pressure records are shown in Fig. 6.20. Higher pressures are needed at higher speeds. Even higher pressures, up to 20 kPa, have been recorded in squid accelerating from rest.

In unit time, a squid squirts out mass nM of water, with volume nM/ρ (ρ is the density of the water). The work it has to do in unit time can be calculated by multiplying this volume by the pressure drop P: it is nMP/ρ.

This must equal the total power already calculated:

$$nMP/\rho = nMUu + \tfrac{1}{2}nMU^2$$
$$U^2 + 2(uU - P/\rho) = 0. \qquad (6.6)$$

This quadratic equation can be solved to obtain U, if the other quantities are known. Fig. 6.20 shows a peak pressure of 4 kPa (4000 N m^{-2}) at a swimming speed of 0.58 m s^{-1}; we will use these values for P and u. The density ρ of seawater is 1026 kg m^{-3}.

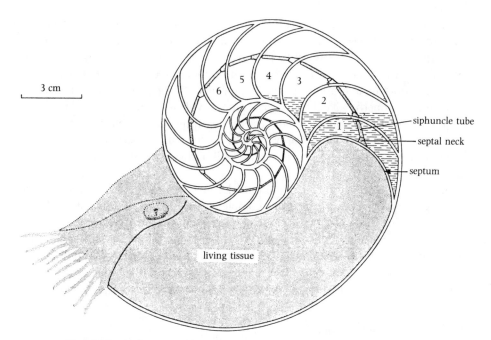

Fig. 6.21. A diagrammatic sagittal section through *Nautilus*. A strand of living tissue, the siphuncle, extends along a tube through all the chambers of the shell. Some of the chambers contain liquid. From E.J. Denton (1974). *Proc. R. Soc. Lond.* B185, 273–299.

With these values, equation 6.6 gives $U = 2.3 \text{ m s}^{-1}$: the jet speed is *much* larger than the swimming speed. By putting this value in equation 6.5 we get an efficiency of 0.34 but this may be an underestimate because some of the water must have been squirted out at lower speeds while the pressure difference was much less than the peak value used in the calculation.

Squids are denser than seawater and have to swim to avoid sinking. Octopus are also denser than seawater and spend most of the time resting on the bottom. Some other cephalopods are about the same density as seawater and can float almost motionless in mid-water. Sir Eric Denton and his colleagues have investigated their buoyancy mechanisms.

They measured the densities of cephalopods by weighing them in air and water. Consider an animal of weight (in air) W and density ρ_a. If it is immersed in water of density ρ_w an upthrust $W\rho_w/\rho_a$ acts on it, by Archimedes' principle, so its weight in water W_w is given by

$$W_w = W - (W\rho_w/\rho_a)$$
and
$$\rho_a = W\rho_w/(W - W_w). \tag{6.7}$$

If W and W_w are measured, ρ_a can be calculated. It has been shown in this way that the density of the squid *Loligo* is about 1070 kg m^{-3}. In contrast, *Nautilus* and the cuttlefish *Sepia* have densities very close to that of seawater, 1026 kg m^{-3}.

Fig. 6.21 is a section through *Nautilus*. The animal occupies the quarter-turn of the shell nearest the opening, and the rest of the shell is divided into compartments.

Compartment 1 is the most recently formed and is still full of fluid but the other compartments are filled mainly or entirely with gas. This makes the animal float in the attitude shown. The only contact between the gas spaces and the living tissue is through the wall of a thin tube which runs through the centres of the compartments. This tube contains a strand of living tissue called the siphuncle. The walls of the chambers are impermeable shell material but the wall of the tube is porous, as can be shown by a simple experiment. A small piece of blotting paper is rolled up and pushed into the siphuncle tube of an empty shell. The adjacent compartment is filled with coloured water, which seeps through the wall of the tube and stains the blotting paper.

The walls of *Nautilus* shells are quite thick; they are made of material of density $2700 \, \text{kg m}^{-3}$. When the chambers are completely full of gas the density of the shell is about $910 \, \text{kg m}^{-3}$. This is not very much less than the density of sea water so quite a large shell is needed to reduce the density of the animal only a little. Consider an animal made up of a volume V_t of tissues of density ρ_t and a volume V_s of shell (or other buoyancy organ) of density ρ_s. The masses of tissue and shell are $\rho_t V_t$ and $\rho_s V_s$ respectively, so the density ρ_a of the whole animal is given by

$$\rho_a = (\rho_t V_t + \rho_s V_s)/(V_t + V_s).$$

Rearrangement of this equation gives

$$V_s/V_t = (\rho_t - \rho_a)/(\rho_a - \rho_s). \tag{6.8}$$

The soft parts of *Nautilus*, removed from the shell, have density about $1060 \, \text{kg m}^{-3}$. Take this as the value of ρ_t, together with $\rho_s = 910 \, \text{kg m}^{-3}$ and $\rho_a = 1026 \, \text{kg m}^{-3}$. This gives $V_s/V_t = 0.29$. To give a *Nautilus* the same density as sea water, the volume of the shell must be at least 29% of the volume of the tissues. This is the volume required if the spaces in the shell are completely filled with gas. In fact, an average of 5% of the space in the chambers is filled with liquid, which increases the density of the shell considerably. The volume of the shell is about 60% of the volume of the soft parts. The shell of *Sepia* is much less dense than the shell of *Nautilus* and its volume is only about 10% of the volume of the soft parts.

It might be expected that the pressure in the chambers of a *Nautilus*, caught deep in the sea, might be high. It is not. If a freshly caught *Nautilus* is held under water while a hole is bored into a chamber no gas bubbles out: rather, water is sucked in. The pressure in the chambers is less than 1 atm and can be calculated from the volume of water sucked in, which can be determined by weighing. In one particular (fairly typical) *Nautilus* the newest chamber was full of liquid, the second chamber contained gas at 0.37 atm and successive older chambers contained gas at successively higher pressures. The sixth chamber (the oldest investigated) contained gas at 0.82 atm. The gas was analysed by mass spectrometry and found to be mainly nitrogen with a little oxygen, argon and water vapour. In the sixth chamber the partial pressures of nitrogen and argon were 0.74 and 0.01 atm.

The partial pressures of nitrogen, oxygen and argon in the atmosphere are 0.78, 0.21 and 0.01 atm. The surface water of the sea is in equilibrium with the atmosphere so these are the partial pressures of the gases dissolved in it. Deeper in the sea the partial pressure of oxygen is less because organisms use it in respiration, but the partial pressures of nitrogen and argon are 0.78 and 0.01 atm at all depths, irrespective of the hydrostatic pressure. They are dissolved in the blood and other tissues of animals at the same partial pressures.

The observations on *Nautilus* indicate that the chambers are not filled by pumping gas into them under pressure; rather, their original liquid contents are drawn out, leaving a vacuum into which gases diffuse until they are in equilibrium with the dissolved gases of the tissues. At equilibrium, the partial pressures of nitrogen and argon would be 0.78 and 0.01 atm. It is only in the older chambers that equilibrium is approached.

Water is probably removed from the chambers by an osmotic mechanism. Samples have been taken of the liquid left in the chambers and their freezing points determined, so that their osmotic concentrations could be calculated. They were all less concentrated than sea water, some as little as 0.4 times as concentrated as sea water. The blood of *Nautilus* has about the same osmotic concentration as sea water. Hence water must tend to move from the chambers to the blood in the siphuncle, through the porous siphuncle tube, until the pressure differences between the blood and the contents of the chamber equals the difference of osmotic pressure.

The osmotic concentration of sea water is about $1 \text{ mol } 1^{-1}$. Hence by equation 2.2 its osmotic pressure is 24 atm. This is also the osmotic pressure of *Nautilus* blood. Even if the liquid in the chambers were pure water, the difference in osmotic pressure between it and the blood would not exceed 24 atm. Water could not be withdrawn, leaving gas at less than 1 atm, when the animal was at pressures greater than 25 atm. Since the pressure is 1 atm at the surface of the sea and increases by 1 atm for every 10 m descent, the blood should not be able to withdraw water by osmotic means at depths greater than about 240 m. There have been a few records of *Nautilus* being caught at greater depths, but it is not certain whether they are reliable. However, there is another cephalopod with gas-filled spaces in its shell which is known to live at depths down to 1200 m. This is *Spirula*. It must extract the liquid from its shell by something other than simple osmotic means. It seems possible that the osmotic concentration of the blood in the siphuncle can be raised well above the osmotic concentration of the rest of the blood. A possible mechanism has been suggested, and probably operates in *Nautilus* as well as in *Spirula*.

Helicocranchia (Fig. 6.5b) has about the same density as sea water, but has no gas-filled shell. It is given buoyancy by the fluid in its coelom, which has about the same osmotic concentration as sea water but contains far less sodium and an extraordinarily high concentration of ammonium ions. Its composition is almost the same as that of a mixture of one part of sea water with four parts of an ammonium chloride solution of the same osmotic concentration. The density of this fluid is much less than that of sea water, about 1010 kg m^{-3}. The density of the remainder of the body is about 1050 kg m^{-3}. Hence by equation 6.8 the volume of the coelom must be 1.5 times the volume of the rest of the body, to give the whole animal the density of sea water (1026 kg m^{-3}). Its volume is about that, which is why the animal looks bloated.

Helicocranchia is not a unique oddity. Many other species of squid have low-density ammoniacal fluids, either in the coelom or in vacuoles in other tissues. They are seldom collected for they live in mid-ocean, many of them at depths of several hundred metres, but they seem to be common. Sperm whales (*Physeter*) feed on squid

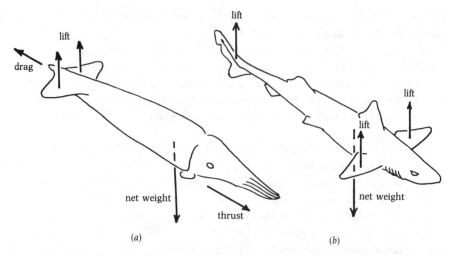

Fig. 6.22. Sketches of (*a*) a squid and (*b*) a shark, both of them denser than the water they live in, showing forces that act on them during swimming. From R. McN. Alexander (1982). *Locomotion of animals*. Blackie, Glasgow.

and it seems from their stomach contents that ammoniacal squid make up 50–80% of their diet. It has been estimated that there are 1.25 million sperm whales of total mass 11 million tonnes and that they must eat over 100 million tonnes of ammoniacal squid each year. The mass of fish taken annually by the fishing fleets of the world is only 60 million tonnes. Plainly, ammoniacal squid are not rare.

Fig. 6.22 shows the forces that enable squid such as *Loligo* to avoid sinking, although they are denser than the water. The animal is swimming forwards, with the drag on its body matched by the forward thrust from the jet. The net weight (W_w in equation 6.7) is probably matched by the lift on the fins at the posterior end of the body, functioning like aeroplane wings. The lift acts far behind the centre of mass and would tend to tilt the animal tail over head, were it not that the thrust acts through the funnel, below the centre of mass, and exerts a balancing moment.

The maximum lift that a fin, or an aeroplane wing, can give is proportional to its area and to the square of the speed. *Loligo* would need very big fins to enable it to swim at very low speeds. In contrast, *Nautilus* and *Helicocranchia* can remain almost stationary in the water but their bulky buoyancy organs would be a hindrance for fast swimming. Buoyancy organs like theirs are probably advantageous for slow-swimming animals but not for fast swimmers like *Loligo*.

6.6.   Respiration and blood circulation

Platyhelminths have no circulatory system. Oxygen has to diffuse to the tissues from the external surface of the body and foodstuffs have to diffuse to them from the gut (section 4.3). In contrast, molluscs have blood circulatory systems that transport oxygen, foodstuffs etc. around the body. They also have respiratory organs where the blood is brought very close to the water or air outside the body, and takes up oxygen.

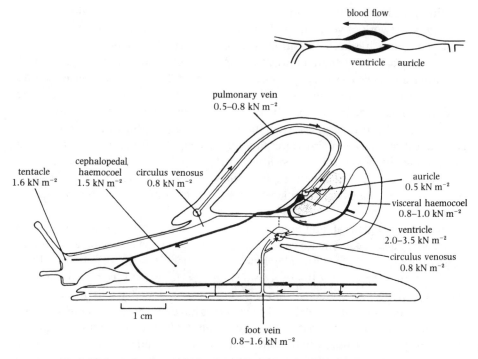

blood flow

ventricle   auricle

pulmonary vein
0.5–0.8 kN m^{-2}

tentacle
1.6 kN m^{-2}

cephalopedal
haemocoel
1.5 kN m^{-2}

circulus venosus
0.8 kN m^{-2}

auricle
0.5 kN m^{-2}

visceral haemocoel
0.8–1.0 kN m^{-2}

ventricle
2.0–3.5 kN m^{-2}

circulus venosus
0.8 kN m^{-2}

1 cm

foot vein
0.8–1.6 kN m^{-2}

Fig. 6.23. A diagram of blood circulation in *Helix pomatia*. The ventricle and principal arteries are shown black and the kidney is stippled. The arrows show directions of blood flow. The numbers show blood pressure relative to the atmosphere. The structure of the heart is shown in an inset. Modified from B. Dale (1973). *J. exp. Biol.* 59, 477–490.

The blood circulation of the snail *Helix* has been studied more thoroughly than those of most other molluscs, and is shown in Fig. 6.23. The finer details can be seen most clearly in dissection, if coloured latex is injected into the vessels. The heart has two chambers, a thin-walled auricle and a more muscular ventricle. It can be watched beating in the living animal if a transparent window is made in the shell over it, by dropping on concentrated hydrochloric acid. It beats with frequencies up to 1 s^{-1}. It will also beat after removal from the animal provided it is kept in a suitable saline solution. Its contraction is muscular and its re-expansion is presumably an elastic recoil. The blood can only flow through it in one direction (from auricle to ventricle) because the valves (Fig. 6.23, inset) stop reverse flow. These valves are flaps of flexible tissue which open for blood flowing in one direction but close as soon as it starts to move in the opposite direction.

The blood leaving the heart flows either along the anterior aorta to the head and foot or along the posterior aorta to the digestive gland and other viscera (Fig. 6.23). These arteries divide up into fine branches but there are no capillaries like those of vertebrates. Instead, the fine branches of the arteries open into spaces in the tissues which connect with the foot veins or directly with one or other of the two main divisions of the body cavity. These divisions are the cephalopedal haemocoel in the head and foot and the visceral haemocoel in the shell. From there the blood enters the circulus venosus, a ring of veins around the edge of the mantle cavity. It flows through a network of veins in the lung, and so back to the auricle.

Blood pressure has been measured in various parts of the circulatory system. The snail was held by its shell in a burette clamp. A hypodermic needle connected to a pressure transducer was held in a micromanipulator. The needle was inserted into the heart through a hole in the shell, or into one of the haemocoels, so that the transducer registered the pressure there. The pressure in the ventricle rose to 2000–3500 N m^{-2} above atmospheric each time it contracted and fell to about 500 N m^{-2} each time it relaxed again. Pressures are shown in Fig. 6.23. The pressure falls (as it must do) as the blood travels from the ventricle, round the body and back to the auricle again. The pressure in the cephalopedal haemocoel is high enough to keep the tentacles stiffly inflated and to support the shell on the stalk of tissue that connects it to the foot.

Gastropods in general, and bivalves, have blood circulatory systems like that of *Helix*, with haemocoels and no capillaries. In contrast, cephalopods have systems in which the blood travels its whole circuit within blood vessels. A network of fine capillaries runs through the tissues, connecting the arteries to the veins. They are not mere gaps in the tissue, but have thin walls formed by endothelial cells. In the muscle of the arms of *Octopus*, for example, the capillaries have a mean diameter of 8 μm and are spaced about 170 μm apart. Thus oxygen need diffuse no more than 85 μm, to reach any part of the tissue. In the aerobic muscle of squid mantle they are about 60 μm apart, giving a maximum diffusion distance of 30 μm. These are much shorter than the diffusion distances of up to 500 μm found in flatworms (section 4.3). Short diffusion distances allow high metabolic rates.

Now we will examine the oxygen-carrying capacity of the blood. Water equilibrated with atmospheric air dissolves about 8 cm^3 oxygen l^{-1} and sea water dissolves 6 cm^3 l^{-1}. If mollusc blood were simply a saline solution which carried oxygen in physical solution, it would have to be pumped very fast around the body to supply oxygen fast enough.

The blood is not simply a saline solution, but contains protein. In many molluscs more than 90% of this protein is haemocyanin, which contains copper and can combine with oxygen. *Helix* haemocyanin is a huge molecule, of relative molecular mass 9 million. It consists of about 180 units of relative molecular mass about 50 000, each containing two copper atoms and capable of combining with one molecule of oxygen. Molluscs and other animals which have haemocyanin in their blood generally have enough to carry about 20 cm^3 oxygen l^{-1}, or $2\frac{1}{2}$–3 times as much as will dissolve in physical solution. *Octopus* blood carried 40 cm^3 l^{-1}.

The haemocyanin adds to the colloid osmotic pressure of the blood. Since 1 mole of oxygen occupies 22.4 l, 20 cm^3 oxygen l^{-1} is about 1 mmol l^{-1}. If the haemocyanin molecules were relatively small, each taking up only one molecule of oxygen, their concentration would have to be 1 mmol l^{-1}. Hence, by equation 2.2, the osmotic pressure of the haemocyanin would be about 2500 N m^{-2}. This is similar to the pressure in the contracting ventricle of *Helix* (Fig. 6.23). Since the colloid osmotic pressure of the blood has to be overcome by blood pressure in the kidney, as will be explained in the next section of this chapter, too high a value must be avoided. Aggregation of the haemocyanin into giant molecules containing 180 of the basic units will reduce its contribution to the colloid osmotic pressure by a factor of 180.

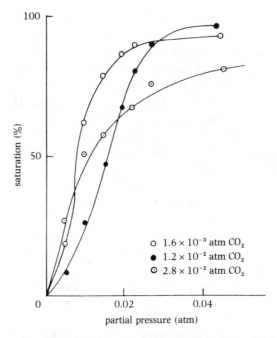

Fig. 6.24. Properties of the haemocyanin of *Helix pomatia* blood. Percentage satura-tion is plotted against partial pressure of oxygen, for three different partial pressures of carbon dioxide. The measurements were made at 15°C. Re-drawn from G.L. Spoek, H. Bakker & H.P. Wolvekamp (1964). *Comp. Biochem. Physiol.* 12, 209–221.

The total colloid osmotic pressure of *Helix* blood, due to haemocyanin and other proteins, is only 100–300 N m^{-2}.

Haemocyanin is blue when it is oxygenated and colourless when it is not. Blood from the heart of *Helix* is blue and blood from the veins is colourless. It appears that the haemocyanin takes up oxygen in the lung and gives it up in the tissues. It does this because it takes up oxygen where the partial pressure of oxygen is high and releases it where the partial pressure is low. Fig. 6.24 shows that it is about 50% saturated (i.e. it carries about 50% of the maximum amount of oxygen) when the partial pressure of oxygen is about 0.01 atm. At higher partial pressures it becomes more nearly saturated and below about 0.003 atm it gives up nearly all its oxygen. The partial pressure of oxygen in air is 0.21 atm so the haemocyanin can be expected to become more or less saturated in the lung, but the partial pressure in the tissue must be remarkably low if it loses most of its oxygen there.

Fig. 6.24 shows that the amount of oxygen taken up by the blood depends on the partial pressure of carbon dioxide, as well as oxygen. An increase in the partial pressure of carbon dioxide from 0.0016 to 0.012 atm increases the partial pressure of oxygen needed to saturate the haemocyanin. This is the Bohr effect, which is also shown by mammal blood. Further increases in the partial pressure of carbon dioxide to 0.028 atm has a peculiar effect on snail blood: it reduces the partial pressure of oxygen needed for 50% saturation. High partial pressures of carbon dioxide do not affect mammal blood in this way.

The Bohr effect is important in mammals because it aids the release of oxygen from

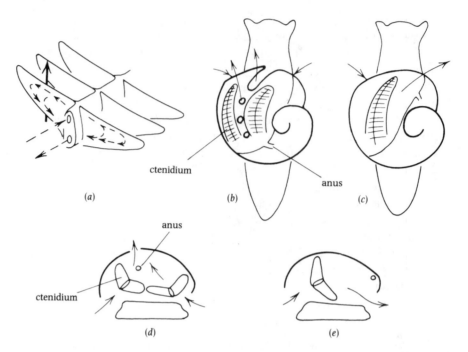

(a)   (b)   (c)

ctenidium

anus

anus

ctenidium

(d)   (e)

Fig. 6.25.(*a*) A diagram of part of a ctenidium. (*b*), (*d*) Diagrammatic dorsal view and transverse section of a primitive gastropod. In (*b*) the shell is represented as transparent. (*c*), (*e*) Similar diagrams of a more advanced gastropod. Broken arrows show the direction of movement of blood and continuous arrows show water movement.

haemoglobin in the tissues, where partial pressures of carbon dioxide tend to be high. It has not been shown to be important in snails.

Most molluscs have ctenidia (gills) in the mantle cavity. The structure of a gastropod ctenidium is shown in Fig. 6.25(*a*). It has a central axis with a row of gill filaments attached on either side. There are two blood vessels in the axis, one carrying blood to the filaments and other carrying it from them to the heart. The blood travels from one vessel to the other through blood spaces in the gill filaments as indicated in the figure by broken lines. Cilia on the flat faces of the filaments drive water through the spaces between the filaments so that the water and the blood cross the filament in opposite directions. This makes it possible for more oxygen to be transferred to the blood than if the water and blood flowed in the same direction, because the blood is exposed to the highest concentrations of oxygen in the water just before it leaves the ctenidia.

The more primitive prosobranchs such as *Haliotis* (Fig. 6.3*a*) have two ctenidia which lie more or less horizontally (Fig. 6.25*b*, *d*). Water enters the mantle cavity below them and passes between them to leave through a slit or one or more holes in the shell. These water currents can be shown by releasing drops of a suspension of carmine particles into the water near the mollusc. The anus opens into the dorsal part of the mantle cavity near where the water leaves so faeces are carried away in the current. The kidney opens near it so the urine is carried away without passing over the ctenidia where waste products might diffuse back into the blood. In more advanced prosobranchs such as *Buccinum* (Fig. 6.3*b*) there is only one ctenidium, and water passes through it from left to right (Fig. 6.25*c, e*). No special slits or holes are needed in the shell. The anus and kidney open near where the water leaves the mantle cavity.

The ctenidium of a *Buccinum* of mass 20 g (excluding the shell) has about 250 filaments of total area 160 cm². The external surface area of the animal is very roughly 50 cm². Much of this area is normally in contact with a rock or covered by a shell so the whelk could not use it for respiration, but the surface area of the gills is in any case much larger than the external surface area. The same is true of other prosobranchs.

Oxygen diffuses from the water to the blood according to equation 4.1. For fast diffusion the area of the gills should be large and the gradient of partial pressure should be steep (i.e. $-dP/dx$ should be large). The cilia maintain a flow of water through the ctenidia so that the water in contact with the ctenidia is constantly being changed. Only a thin layer of tissue separates the blood in the ctenidia from the water. Thus the distance the oxygen has to diffuse is small and the gradient of partial pressure is correspondingly large.

Ctenidia generally do not work well out of water. Surface tension makes the gill filaments clump together, reducing the effective surface area. Pulmonate gastropods have lost their ctenidia but the mantle cavity has become a lung. Its wall has evolved a rich blood supply, and has become ridged so that its surface area is enlarged. The area of the respiratory surface in a large snail (*Helix pomatia*) is about the same as in a *Buccinum* of the same mass (excluding the shell). Only a thin layer of tissues separates the blood from the air. The anus and kidneys do not open into the lung, but on the external surface of the body.

Gastropods which extract oxygen from water need ciliary currents to bring fresh supplies of aerated water to the gills. Simple diffusion suffices in the lungs of pulmonates because the diffusion constant for oxygen diffusing through air is 3×10^5 times the constant for oxygen diffusing through water. The oxygen has to diffuse into the lung through the small pneumostome, across the lung and then through a thin layer of tissue into the blood.

Bivalve molluscs have very large ctenidia, which are used for filter feeding as well as for respiration, as described in section 6.10.

Cephalopods also have ctenidia, but their respiratory water currents are driven by the muscles of the mantle, not by cilia. The movements involved resemble those of jet propulsion but are much more gentle; the peak pressures in the mantle cavity, measured in the same way as jet-propulsion pressures, are only 0.2–0.5 kPa.

Fig. 6.26 shows how the respiratory water currents flow. While the mantle is expanding (Fig. 6.26*a*) water enters through a slit at the base of the funnel. It is prevented from entering through the funnel by a one-way valve. While the mantle is contracting (Fig. 6.26*b*), the pressure in the mantle closes the slit at the base of the funnel, and water is driven out of the funnel itself.

The ctenidia divide the mantle cavity into an upper inhalant chamber, into which the slit enters, and a lower exhalant chamber which leads to the funnel. Thus the water current over the ctenidia always flows in the same direction, from inhalant to exhalant. As in gastropods, blood flows through the gill filaments in the opposite direction to the water flow.

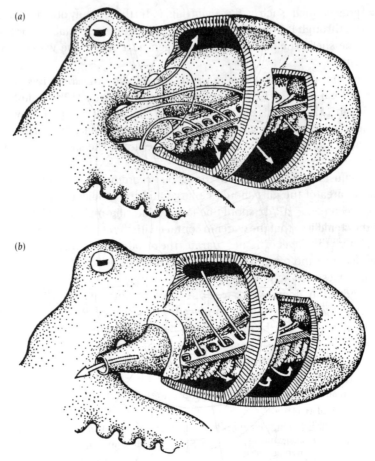

Fig. 6.26. Diagrams showing water flow through the mantle of *Octopus* during (*a*) inspiration and (*b*) expiration. From M.J. Wells & P.J.S. Smith (1985). *J. exp. Biol.* **116**, 375–383.

During inspiration (Fig. 6.26*a*) the inhalant and exhalant chambers are both expanding. Water enters the inhalant chamber via the slit, and some of it travels through the ctenidia to the exhalant chamber. During expiration, both chambers are contracting. Water is squeezed from the inhalant chamber to the exhalant one, and out through the funnel. Thus water flows over the ctenidia both during inspiration and during expiration. This was confirmed by experiments in which pressure transducers were used to record the pressures in both chambers in *Octopus*. During inspiration, the pressures in both chambers were lower than in the water outside and during expiration they were higher than outside, but the pressure in the inhalant chamber was slightly above that in the exhalant one, almost throughout the cycle.

6.7.   Excretion

Fig. 6.27 shows the structure of the excretory organs of *Nerita fulgurans*, a prosobranch gastropod. This particular species has one ctenidium and one kidney:

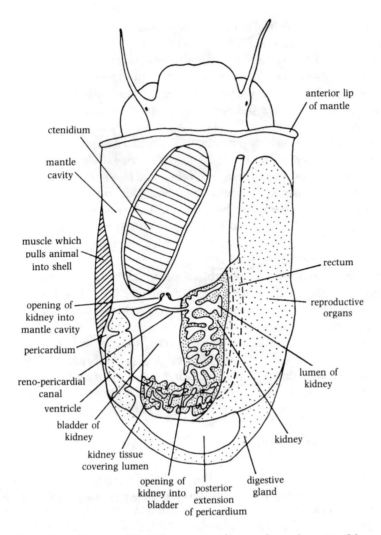

Fig. 6.27. A diagram of the kidney, pericardium and mantle cavity of the gastropod *Nerita fulgurans.* From C. Little (1972). *J. exp. Biol.* 56, 249–261.

more primitive prosobranchs have two of each. The heart is enclosed in a fluid-filled pericardium, which is connected by the renopericardial canal to the kidney, which leads in turn to the mantle cavity. The pericardium is lined by epithelium which covers the whole of its surface including the outer surface of the heart. The epithelium covering the ventricle has not been examined in *Nerita* but in *Viviparus* (another prosobranch) its cells have interlocking processes with slits between them, which probably serve as ultrafilters like the slits in the flame cells of turbellarians (Fig. 4.15). The muscular wall of the ventricle is spongy and has no inner epithelium, so the peculiar epithelium that covers the ventricle is the only effective barrier between the blood in the ventricle and the pericardial fluid. When the ventricle contracts the pressure in it rises far above the pressure in the pericardium (Fig. 6.23) and it is believed that ultrafiltration occurs through this epithelium in most molluscs (but it occurs elsewhere in pulmonates). The colloid osmotic pressure of the blood may be too high for ultrafiltration by flame cells to be feasible but it is only a small fraction of the blood pressure (see section 6.6).

Samples of the blood, pericardial fluid and urine of several prosobranchs have been analysed. One set of experiments was done on the queen conch, *Strombus gigas*, from the Florida Keys. This species was found convenient because it is so big, up to 30 cm long. Part of the shell, over the pericardium and kidney, was cut away with a miniature circular saw. It was replaced by a sheet of wax, which could be lifted whenever samples were to be taken or injections made. Fine glass pipettes were used to take samples of blood and pericardial fluid, which were analysed. It was found that these fluids were almost identical to each other, and to sea water, in ionic composition. However, the haemolymph was bright blue (due to haemocyanin) and contained 3.4% organic matter while the pericardial fluid was clear and contained only 0.8% organic matter. This suggests that the pericardial fluid is an ultrafiltrate of the blood. The small quantity of organic matter in it presumably consists of molecules small enough to pass through the filtering epithelium. A further experiment confirmed that fluid was moving from the blood into the pericardium. Radioactive inulin was injected into the blood. Shortly afterwards samples of blood, pericardial fluid and urine (from the kidney) were taken and their radioactivity was measured to find out the concentrations of radioactive inulin. The same concentration was found in all three. Inulin is a polysaccharide which is often used in experiments of kidney physiology because its molecule is small enough to pass through kidney filters and because animals seem unable to metabolize it, to secrete it or to absorb it. The rate at which inulin escaped into the sea water indicated that ultrafiltration was occurring and urine being passed at a rate of 3 cm^3 (kg soft tissue)$^{-1}$ h^{-1} (or 12% of blood volume per day).

Urine from the kidney is brown and contains 2.4% organic matter, three times as much as the pericardial fluid. This suggests that waste products are secreted into it, and experiments with *Haliotis* confirm this. Various substances were injected into the blood and samples were later taken and analysed. In one experiment, phenol-sulphonphthalein was injected. Later it was found in much higher concentration in the urine than in either the blood or the pericardial fluid, so it is presumably secreted by the kidney cells into the urine. In other experiments glucose was injected and found in lower concentration in the urine than in the blood or pericardial fluid. Its molecule is small enough to pass through the wall of the ventricle and the animal avoids losing it in the urine by reabsorbing it in the kidney.

The animal may have to get rid of insoluble particles as well as soluble wastes. This has been investigated by injecting a suspension of thorium dioxide into yet another marine prosobranch. Thorium has a high atomic number so it is very opaque to X-rays. X-rays of the mollusc were taken periodically after the injection. They showed that the thorium dioxide spread round the body within 10 minutes but that after a week most of it had accumulated in the heart and pericardium and was moving into the kidneys. It took 4–6 weeks to disappear. Its fate was checked by making microscope sections of tissues of the injected snails. The particles were easily seen in the sections and since thorium is radioactive it was possible to check their identity by autoradiography. Within a few days all the particles had been engulfed by amoebocytes, blood cells which resemble the macrophages of vertebrates (see section 2.4). At this stage the amoebocytes are all in the blood, most of them attached to the

Table 6.3. *Concentrations of ions* (mmol l^{-1}) *in
the blood, muscle cells and urine of a freshwater
gastropod* (Viviparus viviparus) *and in the water
in which it lives*

	Na$^+$	K$^+$	Cl$^-$
Blood	34.0	1.2	31.0
Muscle cells	13.6	14.6	10.0
Urine	9.0	—	10.0
Water	2.5	0.2	8.0

Data from C. Little (1965). *J. exp. Biol.* **43**, 23–37
and 39–54.

walls of blood vessels. Later they are found in solid tissues, especially the wall of the heart and in the pericardium. It seems that they travel to the heart, through its wall into the pericardium and from there to the kidney and out in the urine. Their passage from pericardium to kidney may be helped by the cilia of the wall of the renopericardial canal, which beat towards the kidney. It was also found in this investigation that the total number of amoebocytes increased greatly after the suspension had been injected.

For marine gastropods, the water they live in has almost exactly the same ionic composition as the blood. For freshwater ones, there is a marked difference. This is illustrated by Table 6.3, which gives data for *Viviparus*, a prosobranch found in slow rivers and ditches. The blood contains much higher concentrations of sodium and chloride than either the external water or the cell contents. The sodium is necessary, if action potentials are to be conducted in the nerves and muscles. It has been shown that *Viviparus* nerves do not conduct action potentials in sodium-free solutions: the reason should be apparent from the explanation of action potentials in section 3.4.

The kidneys have an important role in regulating the ionic composition of the blood. Since the blood has a higher osmotic concentration than fresh water, water will diffuse into the snail and must be excreted if the snail is not to swell. Also, salts will diffuse out and must be replaced. By eating food which contains water and salts and excreting urine containing lower concentrations of salts, the snail can keep both its water content and its salt content constant.

The samples of urine (Table 6.3) were taken by means of a cannula slipped into the kidney opening. Samples of fluid were also taken from the pericardium. The pericardial fluid had the same ionic composition as the blood but the urine was much more dilute. This indicates *either* that water is added to the urine in the kidney *or* that salts are extracted from it. To find out which, radioactive inulin was injected into *Viviparus* and samples were taken after an interval. It was found that the concentrations of inulin in the blood and the pericardial fluid were about the same, but that the concentration in the urine was more than twice as high. Since animals seem unable to secrete or absorb inulin, water must have been absorbed from the urine in the kidney. If some of the water was absorbed from it and it was nevertheless getting more dilute, salts must also have been absorbed.

Like other animals, molluscs have to get rid of the nitrogenous waste products formed by metabolism of proteins. In aquatic gastropods most of this waste is produced as ammonia and no special processes are needed to get rid of it: it simply diffuses out of the blood into the water. It must diffuse most rapidly from the ctenidia where the blood comes closest to the water.

Ammonia is removed very effectively by the water passing over the ctenidia, because it is extremely soluble. To illustrate this, imagine a snail metabolizing only protein. For simplicity, assume that the protein is composed entirely of alanine (an amino acid of moderate molecular mass) and is oxidized according to the equation

$$-NH.CH(CH_3)CO- + 3O_2 = NH_3 + H_2O + 3CO_2.$$

One molecule of ammonia will be produced for every three molecules of oxygen which are used. One volume of ammonia will be produced for every three volumes of oxygen. The solubility of ammonia in water is about 140 times the solubility of oxygen. If the water passing over the gills gives up enough oxygen for the partial pressure of the oxygen dissolved in it to fall by n atm, it will receive enough ammonia to build up a partial pressure of only $n/(3 \times 140) = 0.0024\, n$ atm.

Terrestrial pulmonates cannot get rid of ammonia so easily. Ammonia will of course diffuse out of the body, particularly at the lung. The diffusion constants of gases diffusing through gas are inversely proportional to the square roots of their relative molecular masses so ammonia (molecular mass 17) diffuses $(32/17)^{\frac{1}{2}} = 1.4$ times as fast as oxygen (molecular mass 32). Suppose again that ammonia has to be lost one third as fast as oxygen is used. Suppose further that gases move between the lung and the outside air entirely by diffusion. Then if the partial pressure of oxygen in the lung is m atm less than in the outside air, the partial pressure of ammonia in the mantle cavity must be $m/(3 \times 1.4) = 0.24\, m$ atm. The partial pressures of oxygen in air and in well-aerated water are equal, but the partial pressure difference m between air and lungs may not be the same as the partial pressure change n at the gills. None the less, it seems certain that $0.0024\, n$ will be less than $0.24\, m$. The partial pressure of ammonia in the blood of the aquatic snail would have to be greater than $0.0024\, n$ atm, or the ammonia would not diffuse out at the required rate. The partial pressure in the blood of the terrestrial one would have to be higher, over $0.24\, m$ atm. The difference might well be critical, for ammonia is toxic. I have no data on its toxicity to snails but it kills alderfly larvae (*Sialis*) when its partial pressure in their blood reaches 3×10^{-4} atm. Diffusion of ammonia from the respiratory organs is an adequate means of getting rid of nitrogenous waste for aquatic gastropods, but not for terrestrial ones.

Terrestrial pulmonates get rid of most of their nitrogenous waste as purines, compounds containing rings of carbon and nitrogen atoms. These insoluble compounds accumulate in the kidney and are voided only occasionally. Kidneys dissected from *Otala* of tissue mass 5 g commonly contain as much as 0.1 g purines, mainly uric acid. They contain very little ammonia.

The rates at which *Otala* excretes nitrogen as purines, and the water snail *Lymnaea* excretes nitrogen as ammonia, have both been measured. In each case the rate was found to be about 3 mg (kg tissue)$^{-1}$ h^{-1}. Both snails probably used oxygen at rates around 0.1 cm^{-3} (g tissue)$^{-1}$ h^{-1}. If this had been used entirely to oxidize protein, about 20 mg nitrogen (kg tissue)$^{-1}$ h^{-1} would have had to be excreted. (This follows from the equation on p. 190, if the nitrogen is excreted as ammonia. The rate would not be very different for purine excretion.) Since this is much more than the observed rate of excretion, protein was only being used for a small proportion of the snail's metabolism. It would probably account for a much larger proportion of the metabolism in carnivorous species.

In the calculation about ammonia excretion by aquatic and terrestrial gastropods it was assumed that all the metabolism was of protein. This assumption was unrealistic but the conclusion is still valid: if an aquatic and a terrestrial snail use equal proportions of protein in their metabolism and excrete all the nitrogen as ammonia, the terrestrial one is more likely to accumulate a harmful concentration of ammonia in its body.

The kidneys of other molluscs seem to work in essentially the same way as those of gastropods.

6.8.   Eyes

Medusae and free-living flatworms have very simple eyes, pigment-lined cups that can distinguish the general direction from which light comes but cannot form an image (section 3.4). *Nautilus* has a rather better eye, which works on the principle of the pinhole camera. It is simply a cavity of about 9 mm diameter with an opening to the exterior of about 1 mm diameter (Fig. 6.28*a*). Light can reach each point on the retina only from a narrow range of directions, so it forms an image of a sort. However, the range is not very narrow: the diagram shows that it is about 1/9 radians or 6°. Consequently, *Nautilus* cannot distinguish fine detail. It turns in response to a moving pattern of stripes if each stripe subtends an angle of 11° at the eye, but not if it subtends only 5.5°. It presumably cannot distinguish the narrower stripes. It could distinguish finer detail if the aperture were even smaller, but a smaller aperture would admit even less light and would work well only in very bright conditions.

An eye with a lens can admit light through a large aperture, and nevertheless form a detailed image. Gastropod and cephalopod molluscs have eyes with lenses. Fig. 6.28(*b*) represents a squid eye. The aperture at the front of the eye is plugged by a lens, which focusses light onto the retina, which is a densely packed layer of light-sensitive (retinula) cells.

Specimens of the squid *Illex*, of about 30 cm overall length, had almost spherical lenses of diameter 8 mm. Measurements with a refractometer showed that the fluid in the eye had a refractive index of 1.34 but slices cut from the lens had higher refractive indices: 1.44 near the surface and 1.49 at the centre of the lens. (For comparison, the refractive index of water is 1.33 and most kinds of glass have refractive indices between 1.50 and 1.65.)

The gradient of refractive index within the lens has important effects. A sphere of uniform refractive index 1.49 in sea water of refractive index 1.33 would have a focal length about five times its radius (i.e. it would bring parallel light to a focus about five radii from its centre). The image it formed would be poor, because of spherical aberration: light passing through the edges of the lens would not focus in the same plane as light passing through the centre (Fig. 6.28*c*). The gradient of refractive index

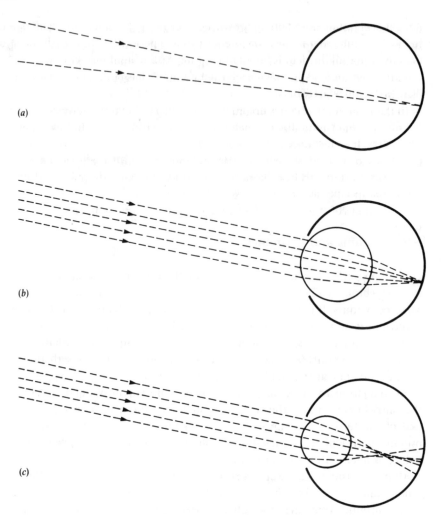

Fig. 6.28. Diagrammatic sections through eyes of (*a*) *Nautilus*, (*b*) a squid, and (*c*) a hypothetical eye with a lens of uniform refractive index. Broken lines represent light rays.

shortens the focal length and tends to correct the spherical aberration. Experiments in which narrow laser beams were passed through a squid lens in sea water showed that the focal length is only about 2.4 radii and that the aberration is over-corrected, producing a fault that is the opposite of the one shown in Fig. 6.28*c*.

Cephalopod eyes have muscles that may be used to move the lens, to bring near or distant objects into focus on the retina. Other muscles, on the outside of the eyeball, are used to turn the eye in its socket. There is also an iris which contracts in bright light, making the pupil smaller.

Diffraction makes it impossible for light of wavelength λ passing through an aperture of radius r to form distinct images of objects separated by angles smaller than about $\lambda/2r$ radians. Consider light entering a cephalopod eye with the pupil wide open so that r is the radius of the lens. Since the focal length of the lens is 2.5r the images of objects $\lambda/2r$ radians apart will be 1.3λ apart on the retina. It would be useful to have retinula cells as close together as this in a perfect eye, so that they could take full advantage of the detail shown in the image. Blue light has wavelengths around 0.4 μm in air, or 0.3 μm in water. Hence it might be useful to have retinula cells as little as 0.4 μm apart. This would require extraordinarily small cells: even in the fovea of the human eye, the cones are about 2 μm apart. The retinula cells of *Octopus* are about 5 μm apart, and the eye seems to be less acute even than this suggests. In a series of tests *Octopus* seemed unable to distinguish black and white striped boards from plain grey ones if the width of the strips subtended an angle less than 0.3° at the eye. (This corresponds to a distance of 50 μm or more on the retina.) Humans can distinguish stripes subtending only 0.01°.

6.9. Brains and hormones

Cephalopods have the largest brains among the invertebrates, as well as the largest eyes. The brain of an *Octopus* is about equal in size to the brain of a teleost fish of the same body mass. It contains about 170 million cells. The brain of *Octopus* and the behaviour it makes possible have been studied intensively by Professor J.Z. Young and his associates. *Octopus* is more convenient than squid or cuttlefish for experiments on behaviour, because it is easier to keep in aquaria. The experiments have been carried out at the Naples Zoological Laboratory.

An octopus in an aquarium generally spends most of the day at one end of its tank. If a pile of bricks is put in the tank it will sit among them. If a crab is put at the other end of the tank the octopus will leave its end, and pounce on the crab by jet propulsion. It will also attack other moving objects such as a shape cut from sheet plastic, waved on the end of a wire.

Many experiments have been performed with pairs of shapes, for instance a cross and a square. From time to time the cross or the square is dipped into the tank and moved around. Whenever the octopus attacks the square, for instance, it is given food. Whenever it attacks the cross it is given a mild electric shock. If this is repeated often enough the octopus learns to discriminate between the shapes: it will usually attack the square and usually leave the cross alone. Fifty or so trials over a period of about 5 days is enough to train an octopus to make a simple discrimination reasonably reliably. The octopus may make the discrimination successfully several weeks later even if it has not been shown the shapes in the interval.

This method has been used to investigate the ability of octopus to distinguish one shape from another. It has been found that octopus have some unexpected limitations. They distinguished a narrow vertical rectangle from a narrow horizontal rectangle, making the correct choice in 81% of trials. They apparently could not distinguish a rectangle sloping at 45° to the left from one sloping to the right for they made the correct choice only 50% of the time (i.e. they apparently chose at random). They made frequent errors when required to distinguish a disc from a square.

Similar experiments have been carried out to investigate the ability of octopus to distinguish objects by touch. It was necessary to blind the octopus for these experiments by cutting their optic nerves. An untrained octopus will usually grasp an

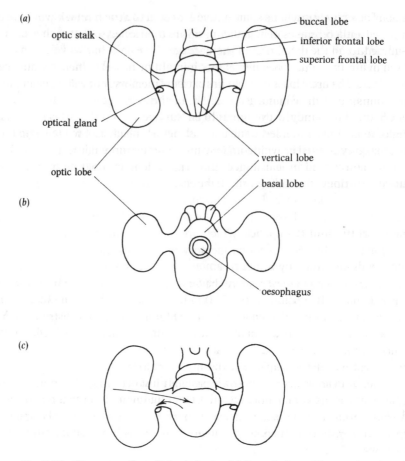

Fig. 6.29. Diagrams of an *Octopus* brain. (*a*) Dorsal view; (*b*) posterior view; (*c*) a dorsal view showing probable inhibitory pathways between the eye and the optic gland.

unfamiliar object and pull it under the body close to the mouth. If the experimental octopus did this with one object of a pair it was rewarded with food but if it did it with the other it was given a mild electric shock. It was found that octopus could discriminate between plastic cylinders with closely spaced grooves and similar cylinders with similar grooves more widely spaced. However, they failed to distinguish a cylinder with lengthwise grooves from one with similarly spaced circumferential grooves.

These unexpected limitations to the ability of octopus to distinguish objects by sight and by touch are apparently due to deficiencies in the analytical powers of the nervous system, rather than in the sense organs. The octopus can see (or feel) the details but fails to perceive some differences of pattern.

Now look at the structure of the brain (Fig. 6.29). It consists of a ring of ganglia encircling the oesophagus, with very large optic lobes on either side. Attempts have been made to discover the functions of its various parts, by cutting them out. The octopus is anaesthetized, part of its brain is cut out and it is allowed to recover. Only an hour or so later it will be sufficiently recovered to attack prey. Operated animals are tested for deficiencies in behaviour and are then killed so that their brains can be examined to check the extent of the damage.

Removing the optic lobes blinds the animal but has no other obvious effect on behaviour. (The optic lobes may be sites of visual memory, but visual memory cannot be demonstrated in a blind octopus.) If all the parts of the brain dorsal to the oesophagus are removed the animal can survive for several weeks but lies in a tangled heap and cannot feed. The buccal lobes are needed for feeding and the basal lobes for co-ordinated crawling and swimming. The more dorsal parts of the brain seem to be involved in learning. If the vertical lobe is removed from an octopus trained to distinguish shapes by sight, the effect of the previous training is lost and the animal is very slow to re-learn. This operation also makes animals trained to discriminate by touch less accurate, but it does not abolish touch memory. Complete removal of the subfrontal lobes (which are hidden below the inferior lobes in Fig. 6.29a) apparently destroys touch memory and leaves the animal with very little ability to discriminate by touch.

These and other experiments give the impression that particular memories are not located at particular points in the brain, in the way that items of information are stored at particular places on floppy discs. Rather, memories are distributed through the brain, and removal of a part of a brain may make a memory less reliable without eliminating it.

The parts of the brain that seem to be important in learning contain large numbers of neurons with short axons. These are probably involved in the learning process, but we have little idea how memory is stored. One clue comes from experiments with the sea slug *Hermissenda* in which it was shown that learning was accompanied by chemical changes in the membranes of certain neurons, which altered their electrical properties.

Evolution of a blood circulation made possible the evolution of hormones carried by the blood. Molluscs have hormones, and I have chosen an octopus hormone as an example. The first evidence of its existence was obtained by accident in the experiments on learning that have just been described. It was noticed that young octopus which had been blinded or had had parts of their brains removed tended to reach sexual maturity precociously. Once this had been noticed it was investigated systematically.

The optic lobes are connected to the main part of the brain by short stalks (optic stalks) which bear glands (optic glands). These glands are much larger in adult octopus than in immature ones. They were found to be enlarged in all experimental animals which had matured precociously. They are believed to secrete a hormone which is released into the blood and stimulates the gonad to develop.

Most of the experiments were performed on immature female *Octopus vulgaris* weighing 0.4 kg or less. At this size their ovaries normally make up less than 0.3% of the mass of the body. Females seldom mature naturally until they weigh at least 1 kg, and mature ovaries may be up to 20% of the body mass. Several operations cause great increases in the masses of the ovaries of immature females, sometimes making them grow within 2 months to 10% of body mass. The most dramatic effects were obtained after removing the subpedunculate lobe (which lies ventral to the vertical lobe) or cutting the proximal end of the optic stalk. Presumably the optic gland normally receives messages from the subpedunculate lobe which inhibit it from enlarging until the appropriate time. Less rapid enlargement follows cutting of the distal end of the optic stalk, or the optic nerves. This suggests that inhibitory messages start from the eyes. Cutting the distal part of the optic stalk and removing the subpedunculate lobe simultaneously is no more effective than the latter operation alone, which suggests that the inhibitory messages from the eye may not go direct to the optic glands but may travel via the subpedunculate lobes (Fig. 6.29c).

Similar operations also make males mature precociously.

Octopus normally live about 2 years, dying after their first and only breeding season. Maturation of the gonads starts in the autumn preceding the spring in which they breed. It may occur in response to shortened day length, which would explain the role of the eyes. *Sepia* kept in aquaria with abnormally short daily periods of light matured precociously, but there seem to have been no similar experiments on *Octopus*.

So far it has been shown that the gonad and the optic glands enlarge in response to the same stimuli, but it has not been shown how the one effect is related to the other. In a series of experiments a blood space behind the eye was opened under anaesthetic and a piece of tissue from another octopus dropped in. Several weeks later the animal was killed and its gonad was examined. In some of the experiments pieces of testis or optic lobe were put into the blood space, and there was no discernible effect on the gonad. In others optic glands were inserted and in about a third of these experiments the gonad enlarged, becoming larger than in any unoperated animals of the same size. This suggests that the optic glands produce a hormone which stimulates enlargement of the gonad. Optic glands from male and female donors were both effective in bringing about maturation both in males and in females, so the hormone is probably the same in both sexes. It made no difference whether the implanted gland was enlarged or not at the time of the operation, presumably because it was free from inhibition in its new site.

It may seem unsatisfactory that only a third of the optic gland implants resulted in enlarged gonads. However, in many cases when no enlargement was observed the implant could not be found at post mortem: it has presumably slipped out of the wound after the operation or been destroyed by the host's amoebocytes. In most of the cases in which enlargement did occur the implanted gland was found at post mortem, attached to the wall of the blood space. In several cases Indian ink was injected into the blood system of the host and entered the gland, showing that it had become connected to the host's blood system.

6.10.   Filter feeding

Most bivalves are filter feeders, filtering unicellular algae and other suspended particles from the water that they pass over their ctenidia. Protobranchia have small ctenidia like gastropods (Fig. 6.25) but Lamellibranchia have greatly enlarged ones (Fig. 6.30).

The process of filter feeding has been studied by watching a ctenidium through a

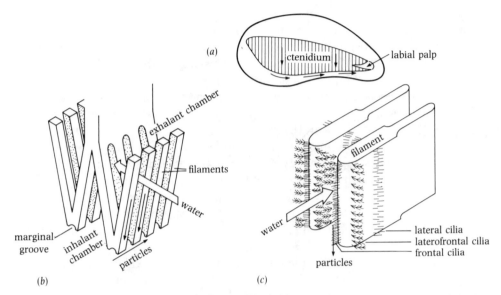

Fig. 6.30. Diagrams of the filter feeding apparatus of *Mytilus*. (*a*) shows the animal in side view with one valve removed; (*b*) shows a few filaments of a ctenidium and (*c*) shows a portion of two adjacent filaments.

microscope, either through a window cut in the shell or after removing one valve completely. Different observers have interpreted their observations in different ways, and some of the statements that follow are controversial. Water is driven through the gaps between the filaments by the lateral cilia (Fig. 6.30*c*). The particles seem to be strained from it by the big laterofrontal cilia, which are actually bundles of (in *Mytilus*) about 50 cilia of varying length, with each cilium bent near its tip to give the bundle a feather-like structure (Fig. 6.31). Films taken through a microscope show that adjacent laterofrontal cilia beat alternately. The two (A and C) shown in Fig. 6.31 are not immediate neighbours: the line (B) between them marks the plane of the intervening cilium, which is out of the picture because it is beating out of phase with them. The beating of the laterofrontal cilia seems to pass captured particles to the frontal cilia, where they get entangled in mucus. The frontal cilia beat parallel to the length of the filament, moving particles and mucus down the filament to the marginal groove. They are moved anteriorly along this groove and passed to ciliated flaps of tissue called labial palps which can be held in either of two positions. In one, they pass the mucus and food particles to the mouth. In the other, they pass them to the mantle to be carried by cilia to the edge of the shell and discarded as 'pseudofaeces'.

Many experiments have been performed to discover the size range of particles that are filtered and to measure the rate of filtration. In some, bivalves were put in a bowl of suspension which gradually became less concentrated as they passed it through their ctenidia. Samples of the suspension were taken from time to time and their concentrations measured by colorimetry or by means of an electronic particle counter. In other experiments the bivalves have been kept in vessels with the suspension flowing slowly through and the concentrations entering and leaving

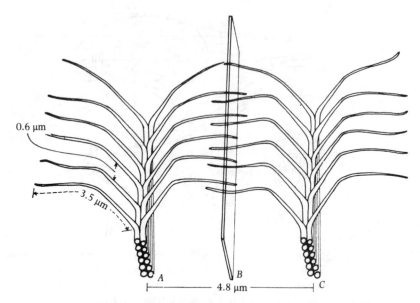

Fig. 6.31. A diagram based on scanning electron micrographs of two laterofrontal cilia on a gill filament of *Mytilus*. From G. Owen (1974). *Proc. R. Soc. Lond.* B187, 83–91.

have been measured. Let the suspension flow through at a rate V (volume per unit time). Let the number of particles per unit volume of suspension be n_{in} in the inflow and n_{out} in the outflow. Then the bivalves must be extracting $(n_{in} - n_{out}) V$ particles in unit time. They are getting these particles from a suspension of mean concentration $\frac{1}{2}(n_{in} + n_{out})$ so they must be filtering in unit time a volume $2V(n_{in} - n_{out})/(n_{in} + n_{out})$ of suspension. This is the rate at which the suspension must be passing through the ctenidia if the filter is perfect, stopping all the particles. If only a fraction are stopped the suspension must be passing through faster.

Experiments with suspensions of latex particles of different sizes have shown that *Mytilus* filters successfully only particles of at least 1 μm diameter. Fig. 6.31 suggests that the laterofrontal cilia ought to be able to stop slightly smaller particles than this (it shows the branches of the cilia spaced 0.6 μm apart).

In experiments with particles well above the minimum size, *Mytilus* has been found to filter water at rates around 250 cm³ (g soft tissue)⁻¹ h⁻¹ or (in some experiments) rather faster. Two hundred and fifty cubic centimetres of well-aerated sea water would contain 1.5 cm³ dissolved oxygen. In some of the experiments the rate of oxygen consumption was measured at the same time as the filtration rate, by using an oxygen electrode to measure the partial pressure of oxygen in the water entering and leaving the experimental chamber. Rates around 0.06 cm³ oxygen (g soft tissue)⁻¹ h⁻¹ were found. Thus only about 4% of the dissolved oxygen was being removed from the water as it was filtered. Water was being driven through the ctenidia far faster than would have been necessary if they served solely for respiration.

Neither the rate of filtration nor the rate of consumption of oxygen changed much when the concentration of the suspension was changed, provided the concentration

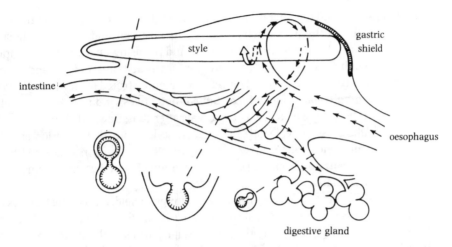

Fig. 6.32. A highly stylized diagram of a bivalve stomach with internal structures revealed by removing the right wall. Three sections cut at right angles to the plane of the paper are also shown. Short and long arrows show movements of particles which do and do not enter the digestive gland, respectively.

was not raised too high above the range which is usual in coastal water. This makes it possible to calculate how much food is needed in the water to enable *Mytilus* to grow. One litre of oxygen oxidizes about 1.2 g polysaccharide or protein or about 0.5 g fat. Hence the 0.06 cm³ oxygen used by 1 g of *Mytilus* tissue in an hour will oxidize 0.03–0.07 mg dry organic matter. If *Mytilus* filters 250 cm³ water each hour and gets this much food from it, the food content of the water must be at least 0.1–0.3 mg dry organic matter l⁻¹. It must actually be rather higher, since analyses of *Mytilus* faeces show that only 70–85% of the food is assimilated.

Mytilus cannot grow unless it assimilates food faster than it uses it in metabolism. The experiment just described shows that it cannot be expected to grow in water containing less than 0.1 mg dry organic matter l⁻¹. Concentrations are generally lower than this in mid-ocean, but higher in coastal water. For instance, the phytoplankton concentration in the English Channel generally varies, according to season, between 0.1 and 1.0 mg dry organic matter l⁻¹. It is mainly in coastal waters that bivalves flourish.

Bivalves in turbid coastal waters must often filter out a mixture of digestible algae and indigestible silt particles, and they seem to swallow both indiscriminately. In one set of experiments, *Mytilus* and other species were put in a mixed suspension of a unicellular alga (diameter about 30 μm) and alumina particles (18 μm), and their faeces and pseudofaeces were collected and analysed. The alumina was not rejected preferentially in the pseudofaeces: pseudofaeces were formed only in concentrated suspensions and when they were formed they contained algae and alumina in the same proportion as the suspension. However, the alumina passed faster through the gut than the algae did: the first faeces to be passed were white, consisting mainly of alumina, but later ones were greener.

Digestible and indigestible particles are separated in the stomach, which has the complicated structure shown in Fig. 6.32. The style is a translucent rod of protein and carbohydrate, bound together as mucoprotein. It is kept revolving about once in five seconds by cilia in the pocket which houses its posterior end. It rubs against the chitinous gastric shield, wearing itself down and releasing enzymes that are incorporated in its substance. It is continuously worn away at its anterior end and replaced by secretion at the posterior end. This has been demonstrated by putting bivalves in water containing trypan blue, which stained their styles, and then back into clean water. When they were examined at intervals after that, newly secreted, colourless material could be seen replacing the stained style, from the posterior end. The enzymes of the style (which digest carbohydrates), and enzymes secreted by the digestive gland, carry out some digestion in the stomach, but digestion occurs largely in the digestive gland.

Parts of the stomach wall are ridged, like a ploughed field. Cilia in the furrows beat along the furrows, towards the intestinal groove in the floor of the stomach. Cilia on the ridges beat at right angles to the ridges. Some particles fall into the furrows and travel direct to the intestinal groove, where there are cilia which transport them to the intestine. Others tend to remain on the ridges and are passed by the cilia from ridge to ridge, finally entering the main ducts of the digestive glands. This sorting process has been watched in the opened stomachs of dissected bivalves.

The ducts of the digestive glands are partly divided into two channels, one without cilia and the other with cilia beating towards the stomach. The cilia probably drive fluid out of the glands, drawing fluid and particles into the non-ciliated channel. Evidence that particles enter by the non-ciliated channel has been obtained by feeding oysters on algal cultures labelled with radioactive carbon. Radioactivity was detectable in the non-ciliated channels of oysters killed after only 10 minutes, but appeared in the ciliated channels only after 90 minutes.

The ducts of the digestive gland branch, leading to blind-ended tubules. Electron microscopy shows that the walls of these tubules contain three kinds of cell. One type is thought to be immature, capable of developing into one or both of the other types. The second type of cell is packed with rough endoplasmic reticulum, like known protein-secreting cells such as the ones that secrete the enzymes in the mammalian pancreas. They presumably secrete enzymes for extracellular digestion. The remaining cells have conspicuous vacuoles containing fragments of food. It seems that in bivalves, as in flatworms (section 4.4), small particles of food are digested intracellularly. Electron micrographs of cockles (*Cardium*) fed pigeon blood showed that the blood corpuscles broke up in the ducts of the digestive gland and that the haemoglobin so released was taken into the digestive cells by phagocytosis.

6.11.   Life on the shore

Many gastropods live on rocks between the tide marks, so that they are covered by water for part of each tidal cycle, and exposed to the air for the rest. Some bivalves such as *Mytilus* live attached to intertidal rocks and others such as *Ensis* and *Cardium* (Fig. 6.4) burrow in sand or mud that is submerged for only part of the tidal cycle.

Shore life involves several problems. First, there is the danger of drying out while exposed. This danger is especially severe for gastropods that have holes in their shells (like *Haliotis*, Fig. 6.3*a*) or no shell at all (like *Aeolidia*, Fig. 6.3*c*). These particular gastropods are confined to the lower part of the shore, where they are not uncovered for long. Some other gastropods are found higher on the shore but mainly in damp places, for example under boulders. However, limpets (*Patella* spp.) are common on bare rock faces between the tide marks, where they may be left dry for several hours at every low tide. The flesh of a limpet contains 86% water and *Patella vulgata* can survive loss of over 50% of this water. It dries out very slowly because it can clamp its shell down firmly on the surfaces of rocks (though not so tightly as to stop respiration). Limpets weighing 7 g allowed to attach to glass and left in a desiccator for 12 h lost only 2.5% of their flesh mass (3% of their water).

Many bivalve molluscs clamp firmly shut, when exposed by the tide. This greatly reduces evaporation, but it also reduces the rate at which they can take up oxygen for respiration. When the mussel *Mytilus edulis* is taken out of water, its rate of oxygen consumption falls to 5% of the rate in water, and succinate and other products of anaerobic metabolism accumulate in the body. Mussels out of water seem to use anaerobic mechanisms like those used by parasitic worms (section 4.5). Even molluscs that do not clamp shut may not be able to maintain their rates of aerobic metabolism at low tide because ctenidia generally do not work well out of water.

Another problem for shore animals is that they are apt to be exposed to extremes of temperatures. The temperature of the sea changes only a little in the course of a day, because of its high heat capacity, but an animal exposed at low tide may be in severe danger of overheating in the sun, or of freezing on a cold night. Their problems of heat balance are the same as for terrestrial animals, which are discussed in section 16.4.

Finally, shore animals may be in danger of being broken or dislodged by waves. Wave forces are much more serious for them than for animals living deeper in the sea, where the water movements associated with waves are smaller. The force exerted on a stationary animal by moving water has two parts. There is the drag force due to the velocity of the water, which is

$$\tfrac{1}{2}\rho A v^2 C_{\mathrm{D}},$$

where ρ is the density of the water, A is the frontal area of the animal (i.e. the area of a view from the direction of the approaching water), v is the speed of the water and C_{D} is the drag coefficient, a quantity that depends on the shape of the body and on the Reynolds number. There is also a second part of the force, due to the acceleration dv/dt of the moving water. It is

$$\rho C_{\mathrm{m}} V . dv/dt,$$

where V is the volume of the body and C_{m} is a quantity called the added mass coefficient, which depends on the shape. It is difficult to measure the maximum wave forces that shore animals must withstand, because severe storms are rare and dangerous, but it seems clear that the acceleration forces on animals often exceed the drag. Other things being equal, the acceleration force will be proportional to the animals' volume.

Gastropods are in danger of being washed off rocks, and the forces they can exert to hold on are proportional to the area of the foot. *Mytilus* is in danger of having its byssus broken, and the strength of the byssus is proportional to its cross-sectional

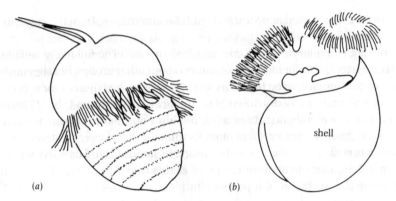

Fig. 6.33.(*a*) An early (trochophore) larva of a chiton (*Katherina*, diameter 40 μm); and (*b*) a later (veliger) larva of a gastropod (*Rostanga*, diameter 300 μm). Based on scanning electron micrographs in F.-S. Chia, J. Buckland-Nicks & C.M. Young (1984). *Can. J. Zool.* 62, 1205–1222.

area. Corals are in danger of having branches broken, and their strength is proportional to cross-sectional area. The acceleration forces are proportional to volume but the forces available to withstand them are proportional to areas. The ratio of volume to area is larger for big animals than for small ones of the same shape. Therefore large animals are in more danger than small ones, of being detached or broken by waves. Consequently, animals living on wave-swept shores generally do not grow as large as similar animals on more sheltered shores, or in deeper water.

6.12. Planktonic larvae

Many marine animals that live on the bottom or attached to rocks as adults, have larvae that swim for a while in the plankton before settling. They include chitons, gastropods and bivalves, and also some flatworms, crabs, etc. Fig. 6.33 shows two mollusc larvae.

The sea is an enormous continuous area of water but only patches of its bottom are generally suitable as habitats for any particular species. If chitons, for instance, produced creeping larvae instead of planktonic ones, each patch of rocky shore would have its own isolated population of chitons. If that population were to die out for any reason, larvae from other populations could not take advantage of the vacant patch. In fact chitons produce trochophore larvae, which are distributed by tidal and other currents. (The currents move faster than they can swim, so swimming serves only to keep them off the bottom). This advantage of planktonic larvae has been explored quantitatively, by computer simulation.

Some planktonic larvae, such as those of the worm *Spirorbis*, do not feed, so they must settle and start their adult life quite soon after being released. Most of them settle within 12 hours, which is long enough to take advantage of the distributing effect of a cycle of tides without too much danger of being washed right away from the coast. Other larvae, such as those of oysters (*Ostrea*), feed and grow, and can survive much longer. Oyster larvae in the laboratory settle at an age of 10–14 days. Larval molluscs which feed get their food like *Stentor* (Fig. 2.4*b*) and rotifers, from the currents set up by the ring of cilia.

How long can a larva which does not feed survive? The following calculation is based on oyster larvae for lack of information about other groups, but the conclusions probably hold for gastropod larvae as well. Newly hatched oyster larvae use oxygen at a rate of about 6 cm^3 (g dry mass)$^{-1}$ h^{-1} or 1 cm^3 (g wet mass)$^{-1}$ h^{-1}. This is much higher than the metabolic rate of adult molluscs but metabolic rate per unit body mass is generally higher for small animals than for their larger relatives (Fig. 4.9*b*). *Paramecium* is about the same size as a young oyster larva and uses oxygen about as fast. How long can metabolism continue at this rate if the larva does not feed?

Consider the oxidation of a polysaccharide

$$(C_6H_{10}O_5)_n + 6nO_2 = 6nCO_2 + 5nH_2O.$$

This equation shows that one mole of polysaccharide (162*n* g) is oxidized by 6*n* moles of oxygen (6*n* × 22.4 = 134*n* l), so 1 l oxygen oxidizes about 1.2 g polysaccharide. Alternatively, it can oxidize about the same mass of protein (as the equation in section 6.7 shows) or about 0.5 g fat. If the larva used 6 cm^3 oxygen (g dry mass)$^{-1}$ h^{-1} and did not feed, it would destroy 8–17% of its dry mass in a day and could probably survive 2 or 3 days, considerably longer than non-feeding larvae usually remain in the plankton. Oyster larvae normally feed but young oyster larvae have been kept without food for 2 days and have survived, losing mass at about the calculated rate.

Planktonic larvae must settle in appropriate places if they are to mature and breed successfully. The method of choosing an appropriate site has been studied particularly thoroughly for barnacles and is discussed in section 8.5.

Though planktonic larvae are common in the sea they are rare in fresh water. The freshwater prosobranch *Viviparus* gives birth to creeping young and pulmonates hatch from eggs as creeping young. Planktonic larvae might have a value in distributing a species between patches of suitable habitat in a lake but if they hatched in one pond they could not colonize another, and in a river they would be washed downstream. The swan mussel *Anodonta* (a bivalve mollusc) releases swimming larvae in ponds and slow rivers but these larvae are exceptional. When fish breathe water containing them they cling to their gills and live there for a while as parasites before dropping off and settling. The fish are as likely to carry them upstream as down.

There are also some shore-living gastropods that produce creeping rather than planktonic young. For example, the winkle *Littorina littorea* lays eggs that hatch as planktonic larvae, but *Littorina saxatilis* (= *rudis*) gives birth to creeping young. *Lacuna vincta* and *Lacuna pallidula* both lay eggs that adhere in clusters to seaweed until they hatch, but the eggs of *L. vincta* hatch after two weeks as planktonic larvae while the much larger eggs of *L. pallidula* hatch after about ten weeks, as miniature snails. Both *Littorina saxatilis* and *Lacuna pallidula* are protected during the larval stages of development, one by the mother's body and the other by being contained in a tough egg mass, but this relative safety is gained at a price: the eggs have to contain large food reserves, because the larvae cannot feed, and much fewer can be produced than in the species that have planktonic larvae. The *Lacuna* species devote approximately equal proportions of their food energy intake to egg production, but *L. pallidula* lays eggs that are 30 times larger than those of *L. vincta* and lays correspondingly fewer. The optimum egg size for a species must depend on how the probability of survival changes with increasing egg size (see section 2.6).

There is another consequence of producing creeping young directly, without a planktonic larval stage. These young will not crawl far, so differences may evolve between populations that are only a short distance apart. Such differences were described in section 6.2 for *Littorina saxatilis*: different populations have shells of different thicknesses, depending on the salinity of the water and the commonness of crabs. Local differences cannot be maintained in species with planktonic larvae, because larvae from a substantial length of shore get mixed together in the sea. There is no apparent difference in shell thickness between *Littorina littorea* populations on sheltered shores where crabs are common, and nearby exposed shores where there are few crabs.

Further Reading

General

Fretter, V. & Peake, J. (1975–9). *Pulmonates*. Academic Press, New York.

Hughes, R.N. (1986). *A functional biology of marine gastropods*. Croom Helm, London.

Morton, J.E. (1967). *Molluscs*, 4th edn. Hutchinson, London.

Nixon, M. & Messenger, J.B. (1977). *The biology of cephalopods* (*Symposium of the Zoological Society no. 38*). Academic Press, New York.

Purchon, R.D. (1977). *The biology of Mollusca*, 2nd edn. Pergamon, Oxford.

Ward, P.D. (1987). *The natural history of Nautilus*. Allen & Unwin, Boston.

Wells, M.J. (1978). *Octopus*. Chapman & Hall, London.

Wilbur, K.M. (1983). *The Molluscs* (6 vols). Academic Press, New York.

Mollusc shells

Alexander, R.McN. (1981). Factors of safety in the structure of animals. *Sci. Progr.* **67**, 109–130.

Brandwood, A. (1985). The effect of environment upon shell construction and strength in the rough periwinkle *Littorina rudis*. *J. Zool., Lond.* A**206**, 551–565.

Currey, J.D. (1980). Mechanical properties of mollusc shell. *Symp. Soc. exp. Biol.* **34**, 75–97.

Gabriel, J.M. (1981). Differing resistance of mollusc shell materials to simulated whelk attack. *J. Zool., Lond.* **194**, 363–369.

Palmer, A.R. (1983). Relative cost of producing skeletal organic matrix versus calcification: evidence from marine gastropods. *Mar. Biol.* **75**, 287–292.

Raup, D.M. (1966). Geometric analysis of shell coiling: general problems. *J. Palaeontol.* **40**, 1178–1190.

Muscle

Baguet, F. & Gillis, J.M. (1968). Energy cost of tonic contraction in a lamellibranch catch muscle. *J. Physiol., Lond.* **198**, 127–143.

Bone, Q., Pulsford, A. & Chubb, A.D. (1981). Squid mantle muscle. *J. mar. biol. Ass. U.K.* **61**, 327–342.

Kier, W.M. (1982). The functional morphology of the musculature of squid (Loliginidae) arms and tentacles. *J. Morph.* **172**, 179–192.

Kier, W.M. & Smith, K.K. (1985). Tongues, tentacles and trunks: the biomechanics of muscular hydrostats. *Zool. J. Linn. Soc.* **83**, 307–324.

Millman, B.M. (1964). Contraction in the opaque part of the adductor muscle of the oyster (*Crassostrea angulata*). *J. Physiol., Lond.* **173**, 238–262.

Pfitzer, G. & Ruegg, J.C. (1982). Molluscan catch muscle: regulation and mechanics in living and skinned anterior byssus retractor muscle of *Mytilus edulis*. *J. comp. Physiol.* **147**, 137–142.

Storey, K.B. & Storey, J.M. (1979). Octopine metabolism in the cuttlefish *Sepia officinalis*. *J. comp. Physiol.* **131**, 311–319.

Crawling and burrowing

Denny, M. (1980). The role of gastropod pedal mucus in locomotion. *Nature* **285**, 160–161.

Denny, M.W. (1981). A quantitative model for the adhesive locomotion of the terrestrial slug *Agriolimax columbianus*. *J. exp. Biol.* **91**, 195–217.

Trueman, E.R. (1975). *The locomotion of soft-bodied animals.* Arnold, London.

Jet propulsion and buoyancy

Denton, E.J. (1974). On buoyancy and the lives of fossil and modern cephalopods. *Proc. R. Soc. Lond.* B**185**, 273–299.

Gosline, J.M., Steeves, J.D., Harman, A.D. & de Mont, M.E. (1983). Patterns of circular and radial mantle muscle activity in respiration and jetting of the squid *Loligo opalescens*. *J. exp. Biol.* **104**, 97–109.

O'Dor, R.K. & Webber, D.M. (1986). The constraints on cephalopods: why squid aren't fish. *Can. J. Zool.* **64**, 1591–1605.

Webber, D.M. & O'Dor, R.K. (1986). Monitoring the metabolic rate and activity of free-swimming squid with telemetered jet pressure. *J. exp. Biol.* **126**, 205–224.

Respiration and blood circulation

Browning, J. (1982). The density and dimensions of exchange vessels in *Octopus pallidus*. *J. Zool., Lond.* **196**, 569–579.

Dale, B. (1973). Blood pressure and its hydraulic function in *Helix pomatia* L. *J. exp. Biol.* **59**, 477–480.

Duval, A. (1983). Heartbeat and blood pressure in terrestrial slugs. *Can. J. Zool.* **61**, 987–992.

Wells, M.J. & Smith, P.J.S. (1985). The ventilation cycle in *Octopus*. *J. exp. Biol.* **116**, 375–383.

Excretion

Boer, H.H. & Sminia, T. (1976). Sieve structure of slit diaphragms of podocytes and pore cells of gastropods molluscs. *Cell Tiss. Res.* **170**, 221–229.

Brown, A.C. & Brown, R.J. (1965). The fate of thorium dioxide injected into the pedal sinus of *Bullia* (Gastropoda: Prosobranchiata). *J. exp. Biol.* **42**, 509–519.

Harrison, F.M. (1962). Some excretory processes in the abalone, *Haliotis rufescens*. *J. exp. Biol.* **39**, 179–192.

Little, C. (1965). Osmotic and ionic regulation in the prosobranch gastropod mollusc, *Viviparus viviparus* Linn. *J. exp. Biol.* **43**, 23–37.

Little, C. (1965). The formation of urine by the prosobranch gastropod mollusc, *Viviparus viviparus* Linn. *J. exp. Biol.* **43**, 39–54.

Little, C. (1967). Ionic regulation in the queen conch, *Strombus gigas* (Gastropoda, Prosobranchia), *J. exp. Biol.* **46**, 459–474.

Sattelle, D.B. (1972). The ionic basis of axonal conduction in the central nervous system of *Viviparus contectus* (Millet) (Gastropoda: Prosobranchia). *J. exp. Biol.* **57**, 41–53.

Speeg, K.V. & Campbell, J.W. (1968). Purine biosynthesis and excretion in *Otala* (= *Helix*) *lactea*: an evaluation of the nitrogen excretory potential. *Comp. Biochem. Physiol.* **26**, 579–595.

Eyes

Land, M.F. (1980). Optics and vision in invertebrates. In *Handbook of sensory physiology* (ed. H. Autrum), vol. 7, part 6B, pp. 471–592. Springer, Berlin.

Muntz, W.R.A. & Raj, U. (1984). On the visual system of *Nautilus pompilius*. *J. exp. Biol.* **109**, 253–263.

Sivak, J.G. (1982). Optical properties of a cephalopod eye. *J. comp. Physiol.* **147**, 323–327.

Brains and hormones

Alkon, D.L. (1983). Learning in a marine snail. *Scient. Am.* **249**(1), 64–74.

Wells, M.J. (1974). A location for learning. In *Essays on the nervous system* (ed. R. Bellaris & E.G. Gray), pp. 407–470. Clarendon, Oxford.

Wells, M.J. (1976). Hormonal control of reproduction in cephalopods. In *Perspectives in experimental biology* (ed. P.S. Davies) vol. 1, pp. 157–166. Pergamon, Oxford.

Young, J.Z. (1983). The distributed tactile memory system of *Octopus*. *Proc. R. Soc. Lond.* **B218**, 135–176.

Filter feeding

Foster-Smith, R.L. (1975). The effect of concentration of suspension and inert material on the assimilation of algae by three bivalves. *J. mar. biol. Ass. U.K.* **44**, 411–418.

Owen, G. (1974). Feeding and digestion in the Bivalvia. *Adv. comp. Physiol. Biochem.* **5**, 1–36.

Silvester, N.R. & Sleigh, M.A. (1984). Hydrodynamic aspects of particle capture by *Mytilus*. *J. mar. biol. Ass. U.K.* **64**, 859–879.

Widdows, J. & Bayne, B.L. (1971). Temperature acclimation of *Mytilus edulis* with reference to its energy budget. *J. mar. Biol. Ass. U.K.* **51**, 827–843.

Life on the shore

Denny, M.W., Daniel, T.L. & Koehl, M.A.R. (1985). Mechanical limits to size in wave-swept organisms. *Ecol. Monogr.* **55**, 69–102.

McMahon, R.F. (1988). Respiratory response to periodic emergence in intertidal molluscs. *Am. Zool.* **28**, 97–114.

Newell, R.C. (1970). *Biology of intertidal animals*. Logos, London.

Planktonic larvae

Crisp, D.J. (1976). The role of the pelagic larva. In *Perspectives in experimental biology* (ed. P.S. Davies), vol. 1, pp. 145–155. Pergamon, Oxford.

Currey, J.D. & Hughes, R.N. (1982). Strength of the dogwhelk *Nucella lapillus* and the winkle *Littorina littorea* from different habitats. *J. Anim. Ecol.* **51**, 47–56.

Grahame, J. (1982). Energy flow and breeding in two species of *Lacuna*: comparative costs of egg production and maintenance. *Int. J. Invert. Repro.* **5**, 91–99.

7 *Segmented animals*

Phylum Annelida,
   Class Polychaeta   (ragworms, lugworms etc.)
   Class Myzostomaria
   Class Oligochaeta   (earthworms etc.)
   Class Hirudinea   (leeches)
Phylum Chelicerata,
   Subphylum Merostomata   (horseshoe crabs)
   Subphylum Arachnida   (mites, spiders, scorpions)
Phylum Crustacea   (crustaceans, chapter 8)
Phylum Uniramia,
   Subphylum Onychophora
   Subphylum Myriapoda   (centipedes, millipedes)
   Subphylum Hexapoda   (insects, chapter 9)
   and other subphyla

7.1.   Segmentation

This chapter is about the annelid (segmented) worms, and also about the arthopods, the huge assemblage that includes the crustaceans, insects, spiders and many other groups. Traditionally, all the arthropods have been put together in a single phylum Arthropoda, but many zoologists now divide them into three phyla, as in the classification above. The crustaceans and insects are particularly important and interesting groups and have chapters to themselves. This chapter concentrates on the other arthropods, and on the annelids.

Annelids and arthropods have their bodies built up from large numbers of more or less similar segments, arranged one behind the other in single file. This is clearly shown by the ragworm *Nereis* (Fig. 7.1a); the whole of its body, apart from the head and the tip of the tail, consists of almost identical segments. Annelids have soft bodies, though some such as *Serpula* (Fig. 7.1b) build protective tubes that are external to the body. Most arthropods have their bodies enclosed by tubes and plates of stiff cuticle, like the jointed armour of a mediaeval knight. The scorpion *Buthus* (Fig. 7.2a) illustrates this well. Notice that the stiff cuticle makes jointed legs possible: the projections from the sides of *Nereis* are not legs, but relatively short, soft structures called parapodia. The composition and properties of arthropod cuticle are discussed in section 8.2.

Annelids have most of the body composed of segments that are nearly uniform, at least in external appearance. The same is true of centipedes (Fig. 7.8), but most other arthropods are less uniformly segmented. For example, scorpions have their bodies

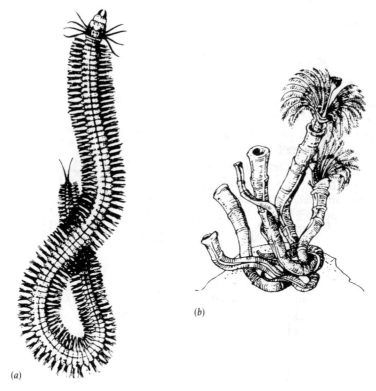

Fig. 7.1. Polychaete worms. (*a*) *Nereis diversicolor* (up to about 10 cm long) and (*b*) *Serpula vermicularis* (5 cm), from W. de Haas & F. Knorr (1966). *The young specialist looks at marine life.* Burke, London.

divided into an anterior prosoma (which bears the legs and the pincer-like pedipalps) and a posterior opisthosoma. The opisthosoma is segmented but its anterior segments are broad and its posterior ones (which form the tail) are narrow. The prosoma is covered by a single plate of cuticle but the limbs suggest an originally segmented structure in which (as in centipedes) each segment bore a pair of limbs. In spiders (Fig. 7.2*b*) even the opisthosoma is no longer divided into segments.

Fig. 7.3 illustrates the modified segmental structure of the interior of an earthworm, as an example of the annelids. The nerve cord runs the length of the body, ventral to the gut, giving off three pairs of nerves in each segment, except at the anterior end of the body. Here it splits to encircle the gut and attach to a dorsally placed pair of nerve ganglia, and gives off numerous nerves that show no obvious segmental arrangement. There are three main longitudinal blood vessels (dorsal, ventral and subneural), which send branches in each segment to the body wall and the gut. The segmental pattern is modified near the anterior end, where five pairs of hearts (one pair in each of five successive segments) connect the dorsal to the ventral blood vessel. The space between the body wall and the gut is a coelom. Earthworms and many other annelids have the coeloms of successive segments separated by partitions called septa. Every segment of *Lumbricus* except the first three and the last has a pair of nephridia. These are slender coiled tubes with one opening into the

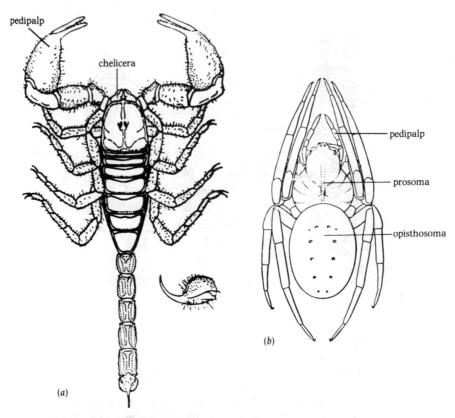

Fig. 7.2. Arachnids. (*a*) A scorpion, *Buthus occitanus*, and a side view of its sting. (*b*) A spider, *Araneus diadematus* (length 12 mm, excluding legs). (*a*) From T. Savory (1964). *Arachnida*. Academic Press, New York. (*b*) From W.S. Bristowe (1971). *The world of spiders*, 2nd edn. Collins, London.

coelom and one to the exterior of the body, that function like the kidneys of molluscs (section 6.7).

The embryology of annelids clarifies the nature of segmentation. They develop from the zygote by spiral cleavage, like polyclad flatworms (Fig. 4.14). However, cells 4a–c and 4A–D survive and form the lining of the gut. Fig. 7.4 shows the course of development of typical polychaetes. Division initially produces a hollow ball of cells, the blastula (Fig. 7.4*a*). The cells at one end of the ball are the products of division of cells 1a–d. Those at the other end come from 4A–D, and have the products of 4a–d next to them. This end invaginates (folds in) to form the next embryonic stage, the gastrula (Fig. 7.4*b*). The products of 4a–c and 4A–D are now in the appropriate position to form the lining of the gut. The products of 4d, however, have slipped into the cavity of the blastula and formed two bands of cells, one on each side of the gut. The gastrula develops to a trochophore, very like the trochophores of molluscs (Fig. 7.4*d*; compare Fig. 6.33). The single opening of the gut of the gastrula becomes the mouth, and an anus forms at this stage.

It is convenient to call the cells derived from 4a–c and 4A–D the endoderm, those derived from 4d the mesoderm, and the rest the ectoderm. In general, the ectoderm

(a)

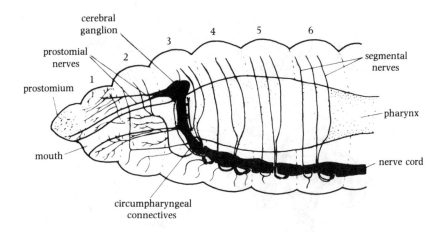

(b)

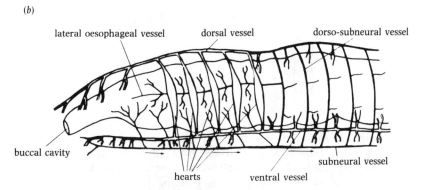

Fig. 7.3. (*a*) A lateral view of the nervous system of the first six segments of *Lumbricus*. (*b*) A lateral view of the anterior blood vessels of *Lumbricus*. From C.A. Edwards & J.R. Lofty (1972). *Biology of earthworms*. Chapman & Hall, London.

covers the external surface of the body, the endoderm lines the gut and the mesoderm lies between. However, there are exceptions. For instance, the anterior part of the gut is formed separately from the rest, by invagination of ectoderm, at the time the mouth forms. It is therefore lined by ectoderm.

The mesoderm bands grow and form cavities which become the coeloms of successive segments (Fig. 7.4*e*). The most anterior segments are formed first and more segments are added at the posterior end so that the most posterior segment is always the newest one. This is quite different from the situation in tapeworms, where the newest proglottid is always the one next to the scolex. The larva sinks and settles on the bottom as it becomes a young worm.

The mesoderm forms the peritoneum and gonads. It also forms the muscles of the body wall and the walls of the principal blood vessels, as indicated in Fig. 7.4(*c*). However, the nerve cord is formed from ectoderm which sinks into the body so as to be enclosed by the body wall. The nephridia also develop from ectoderm.

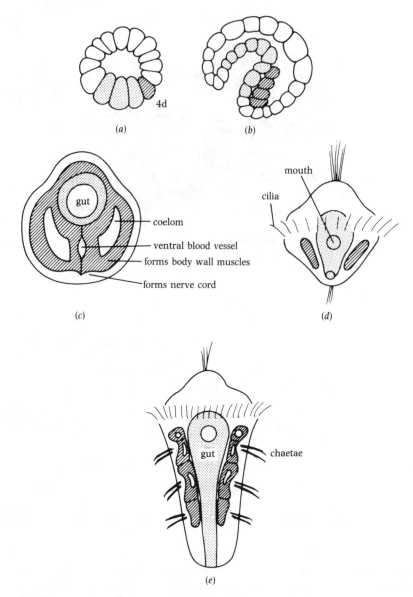

Fig. 7.4. **Diagrams illustrating the development of typical polychaetes.** (*a*), (*b*) and (*c*) Sections of a blastula, a gastrula and a later larva. (*d*) and (*e*) Sketches of a trochophore and a later larva. Mesoderm (derived from cell 4d) is hatched; endoderm (derived from cells 4a–c and 4A–D) is stippled.

By the stage shown in Fig. 7.4*e*, the developing worm has segmentally arranged chaetae as well as coeloms. The chaetae are bristles made partly of chitin borne (in the adult worm) on the ends of the parapodia. The nerves, blood vessels and nephridia develop after the coeloms, matching their segmental arrangement.

Arthropods have hardly any trace of coeloms: their main body cavities are haemocoels, like those of molluscs. Nerves, muscles and blood vessels may be

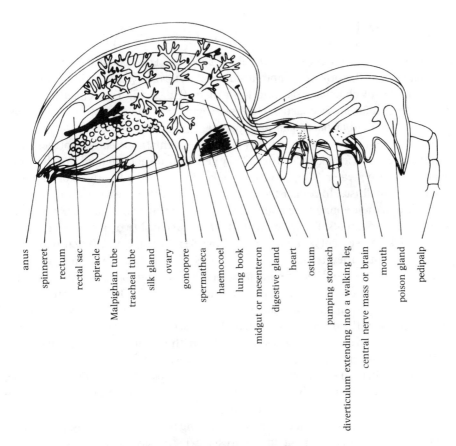

Fig. 7.5. A diagram showing the internal organs of a female spider. From K.R. Snow (1970). *The arachnids: an introduction*. Routledge & Kegan Paul, London.

arranged segmentally but there is little or no sign of segmentation in other internal organs, as Fig. 7.5 illustrates. However, it is generally believed that arthropods evolved from soft-bodied ancestors which had their segments more or less alike along most of the length of the body. These ancestors would have been very like annelid worms.

Hints about how segments have become different from each other in advanced arthropods come from observations on fruit flies (*Drosophila*). Mutants have been discovered that make segments develop more like their neighbours than in normal flies. For example, normal flies have legs on the three thoracic segments and none on the abdomen, but a mutant develops extra legs on the first abdominal segment.

Segments might be thought of as modules repeated along the length of the animal, but there is an important difference between a segmented animal such as an annelid worm and a modular one such as a cnidarian colony (section 3.2). The annelid forms a certain number of segments and then stops, but the colony grows by developing an indefinite number of polyps. Segmented animals are not generally thought of as modular.

7.2.   Annelids

There are three main classes of annelid worm, the Polychaeta (marine worms), Oligochaeta (earthworms, etc.) and Hirudinea (leeches). Polychaetes have parapodia with bundles of chaetae projecting from them (Fig. 7.1a). Oligochaetes have no parapodia and far fewer chaetae (eight per segment in the earthworm *Lumbricus*). Leeches have neither parapodia nor chaetae.

Polychaetes develop gonads in many of their segments, and nearly all of them have separate male and female individuals. They generally do not copulate, but shed their gametes into the water. They have trochophore larvae very like those of gastropod and bivalve molluscs. In contrast, oligochaetes and leeches are hermaphrodite, they copulate and their young hatch out as miniature worms. Oligochaetes have gonads in only a few segments. For example, *Lumbricus* have just two pairs of testes (in the 10th and 11th segments) and one pair of ovaries (in the 13th). When two earthworms copulate, they bring the anterior ends of their bodies together, facing in opposite directions, and secrete mucus which binds them together. Sperm emerge from the male apertures of each worm, which are on the 15th segment (several segments posterior to the testes) and travel along grooves in the outer surface of the body to pores on the other worm's 9th and 10th segments where they enter sperm-storing sacs called spermathecae. After copulation the worms separate. Each secretes a short tube (like a belt around its body), discharges sperm from the spermathecae and eggs into it, then slips it off the anterior end of its body. The ends of the tube close to form a cocoon from which young worms eventually hatch. Several fertile cocoons can be produced after a single copulation.

Fig. 7.1 shows two polychaetes with contrasting ways of life. The ragworm *Nereis diversicolor* is common on muddy shores where it occupies a semi-permanent burrow. Anglers dig it up for use as bait. It is often particularly common in estuaries where sea water and fresh water mix, and may have to withstand large fluctuations of salinity. It leaves its burrow and crawls around, using its parapodia as explained in section 7.4, and feeds on dead animals and pieces of algae. The pharynx can be turned inside out so that it projects as a proboscis in front of the mouth, as in Fig. 7.1(a). It bears a pair of sharp jaws which are hidden inside it when it is retracted but appear when it is protruded.

Serpula (Fig. 7.1b) is a very different polychaete. It does not burrow but secretes a protective tube of calcium carbonate (with a small proportion of organic matter) on stones, shells and seaweeds. It has stiff feathery tentacles with cilia which drive a current of water upwards between them. This serves for filter feeding. The tentacles can be withdrawn into the tube, and a modified tentacle serves as a stopper to close the mouth of the tube. It remains perpetually inside its tube and has only small parapodia.

The earthworms are the most familiar oligochaetes, too familiar to need illustration. They live in burrows in the soil. *Lumbricus terrestris* feeds on fallen leaves which it collects from the surface and pulls into its burrow. Other earthworms eat leaves or other plant materials, and dung. Particles of soil are often swallowed with food and the gut may seem to be filled mainly with soil. The enchytraeids are much smaller members of the class, which also live in soil. Adults of different species are 1–50 mm long. There are also aquatic oligochaetes such as *Tubifex*, which lives in mud in the bottoms of rivers and streams and is sold as a food for aquarium fish. All these oligochaetes are fairly similar to each other, except in size.

Most leeches live in fresh water. They look rather like flattened earthworms but have a sucker at the posterior end of the body and often another round the mouth. Some fed on invertebrates but others attach themselves to fishes and suck their blood.

7.3.   Various arthropods

The classification set out at the beginning of this chapter treats the arthropods as three phyla, the Chelicerata, Crustacea and Uniramia. A major reason for separating them is that their anterior appendages (antennae, mouthparts etc.) are so differently arranged as to suggest that the three phyla have evolved independently from a very primitive ancestor.

That ancestor presumably had one pair of appendages on each segment, all of them more or less alike. It must have been rather like polychaete worms, which have a short head region followed by a long chain of segments, each bearing a pair of parapodia (Fig. 7.1a). Among modern arthropods, the ones most like it are the subphylum Onychophora (*Peripatus* etc.) which live in damp places in the tropics and subtropics. They look rather like worms or caterpillars with numerous pairs of stumpy legs, but even they have three pairs of anterior appendages that are different from the rest: antennae, jaws and knobs called oral papillae.

The Chelicerata have six pairs of appendages: chelicerae, pedipalps and four pairs of legs. In scorpions, the chelicerae are small pincers and the pedipalps are much larger ones (Fig. 7.2a). In spiders, the chelicerae are the fangs used to inject poison into prey and the pedipalps look like small legs (Fig. 7.2b). Notice that there are only two pairs of appendages anterior to the walking legs.

Crustacea have two pairs of antennae (appendages modified for sensory functions), a pair of mandibles (biting jaws) and many more pairs of mouthparts called maxillae and maxillipeds, used for manipulating food. Another difference from other arthropods is that many crustacean appendages are two-branched (for example, the antennae of *Daphnia*) (Fig. 8.3a). It is believed that ancestral crustaceans had two-branched appendages all along their bodies. The Crustacea are described in chapter 8.

Typical Uniramia have on their heads one pair of antennae, a pair of mandibles and two pairs of maxillae (but Onychophora have only one pair of oral papillae instead of the maxillae). The labium of insects (Fig. 9.3) is the second pair of maxillae, fused together. The insects are described in chapter 9.

The remainder of this section consists of brief accounts of some examples of the Arachnida and Myriapoda, important groups that are not described in later chapters.

The scorpion *Buthus* (Fig. 7.2a) is common in Mediterranean countries. It strikes insects and other prey with the sting at the end of its tail, a spine with a pair of poison glands opening near its tip. Poison is squeezed from the glands as it strikes. This sting causes fever in people and considerable pain, and has been known to kill children.

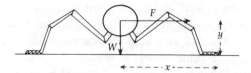

Fig. 7.6. A diagrammatic front view of a standing arthropod, showing forces acting on its body.

Prey are held by the pedipalps while the chelicerae are used to tear through the body wall, so that the scorpion can suck out the juices.

Araneus diadematus (Fig. 7.2*b*) is a British member of a genus of spiders that is also found in N. America. They build orb webs to catch insect prey which they subdue by injecting poison from the chelicerae. Many of the internal structures shown in Fig. 7.5 resemble similar structures in crustaceans and insects, whose working is explained in chapters 8 and 9. The silk glands produce the silk used for web-making, which is extended through the spinnerets to form fine threads. The lung books are a pair of respiratory organs, pockets in the ventral surface of the opisthosoma with leaves like those of a book extending into them. These leaves are hollow, with very thin cuticle. They are filled with blood which circulates through them, exchanging gases by diffusion with the air. In addition, most spiders have a second respiratory system: they have air-filled tubes (tracheae) ramifying through the body and connected to the outside air by a pore (the spiracle) on the ventral surface of the opisthosoma. Only the ends of the tracheae are shown in Fig. 7.5. The similar tracheal system of insects is explained in section 9.3.

The mites (Acari) are another important group of arachnids. They look rather like tiny spiders and are extremely common in soil.

Centipedes such as *Scolopendra* and *Scutigera* (Fig. 7.8) are predators that feed on worms, slugs and arthropods, generally hunting by night and hiding by day under leaf litter or in crevices. They have antennae and three pairs of mouthparts on their heads, and 15 to 177 pairs of legs. The first trunk segment bears a pair of pincers with venom glands opening at their points, that are used to grasp prey and to immobilize it by injecting venom.

Millipedes look rather like centipedes but generally have more cylindrical bodies and relatively shorter legs. Most of their segments seem to bear two pairs of legs but that is an illusion due to rings of cuticle enclosing pairs of segments. Millipedes burrow in soil and feed as earthworms do on dead and decaying vegetation.

Most adult arthropods have long legs and stand with their feet well out on either side of the body (Fig. 7.6). This makes them very stable. Suppose a wind or water current acts towards the right of the diagram, exerting a force F on the body. It will tend to bowl the arthropod over, but the weight W will tend to prevent this. The animal will be bowled over when the clockwise moment exerted by F about the downwind feet exceeds the anticlockwise moment exerted by W, i.e. when

$$Fy > Wx. \tag{7.1}$$

The force needed to bowl the animal over is Wx/y, so the greater the value of x/y the more stable the animal will be.

A stance with the legs well apart is particularly important for Crustacea which live on shores. The density of a typical crab is about 1150 kg m^{-3} so when it is submerged in sea water (density 1026 kg m^{-3}) the effective weight W available to stabilize it is only 11% of its weight in air (see equation 6.7). Waves and tides are liable to set up quite rapid water movements which may exert large forces tending to bowl the animal over.

Terrestrial arthropods such as insects stand with their feet wide apart but most mammals do not. For instance x/y is typically about 2 for a blowfly and 0.2 for a dog. However, F is proportional to the area of the animal; for animals of the same shape but different size it will be proportional to (body weight)$^{0.67}$. At the same wind speed F/W will thus be larger for a small animal than a large one. Small animals are more liable to be blown over than large ones unless they stand as arthropods do with their legs well apart. (A wind is slower close to the ground, but not sufficiently so to invalidate the argument.)

7.4. Metachronal rhythms

Metachronal rhythms have already been described for cilia: successive rows of cilia beat slightly out of phase with each other so that waves seem to travel over a ciliated surface (Fig. 2.19). Metachronal rhythms are also apparent in the parapodial movements of polychaete worms and in the leg movements of centipedes and millipedes.

Fig. 7.7 shows *Nereis* crawling, with the same two parapodia marked on each outline. Consider the one nearer the left of the page. In (*a*) it has just been set down on the bottom of the aquarium, and is reaching forward (the worm is crawling to the left). In (*b*) and (*c*) it is being swung backwards, pushing the worm forwards, and in (*d*) it is off the bottom, moving forward for the next step. Each leg moves slightly after the one behind, so waves of leg movement travel forward along the body. *Nereis* can crawl slowly simply by this stepping movement of the parapodia but to go fast (as in Fig. 7.7) it also undulates the body, making waves of bending travel forward along the body at the same rate as the waves of parapodial movement. This increases the angle through which the parapodia swing, increasing the distance that the worm advances in each step.

Fig. 7.8 shows that the running movements of centipedes are very like the crawling of *Nereis*. To run slowly, *Scolopendra* moves only its legs and keeps its body straight but to go faster it undulates the body (Fig. 7.8*a–d*). However, the waves of leg movement and of bending travel backwards along the body. Consequently, when two successive feet are on the ground the more anterior one is at a later stage of its step, and the two legs converge. In contrast, waves travel forward along *Nereis* and successive parapodia diverge while on the ground.

At the low speed shown in Fig. 7.8(*a*), each leg of *Scolopendra* is 1/5 cycle out of phase with the next so that the first walking leg moves in phase with leg 6, leg 2 with leg 7, and so on. At the high speed shown in Fig. 7.8(*d*) each leg is 1/13 cycle out of phase with the next so that leg 2, for example, moves in phase with leg 15. The number of feet on the ground at any instant also changes with speed. At low speed,

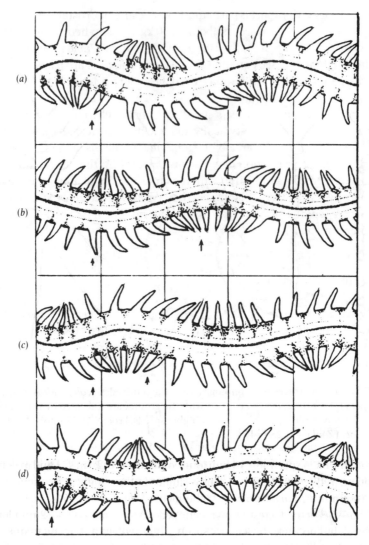

Fig. 7.7. *Nereis diversicolor* crawling fast towards the left, drawn from a cine film. Arrows show successive positions of two parapodia. The lines on the floor of the aquarium are 1 cm apart. From R.B. Clark (1964). *Dynamics in metazoan evolution.* Clarendon, Oxford (after a film by Sir James Gray).

half the feet are on the ground at any instant (and each foot is on the ground for half the time). At high speed only 0.15 of the feet are on the ground at any instant (and each foot is on the ground for 0.15 of the time).

Suppose a centipede has legs of length l. It will probably swing each leg through about 60° and so advance a distance about equal to l while a particular foot is on the ground. If each foot is on the ground for a fraction β of the time, the centipede will advance l/β in each complete cycle of leg movements. To run at velocity u it must make $u\beta/l$ cycles of leg movements per unit time (i.e. each leg must step with frequency $u\beta/l$). Very high stepping frequencies would be needed at high speeds if β were not reduced as u increased. Even with the reductions of β which occur, very high frequencies are used. For instance, *Cryptops* is a centipede which resembles *Scolop-*

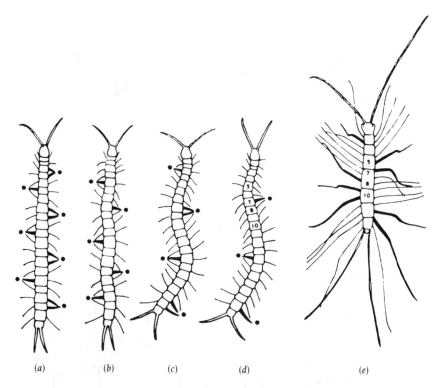

Fig. 7.8. Outlines traced from photographs of the centipedes *Scolopendra* (*a–d*) and *Scutigera* (*e*) running. Legs in contact with the ground are indicated by thick lines; (*a*) to (*d*) show running at progressively higher speeds. From S.M. Manton (1965). *J. Linn. Soc.* (*Zool.*) 45, 251–484.

endra in its proportions. A specimen 38 mm long could run at 0.26 m s^{-1} and in doing so used a stepping frequency of 25 s^{-1}. This is similar to the wingbeat frequencies of large insects (the moth *Manduca*, 28 s^{-1}; the locust *Schistocerca*, 17 s^{-1}).

Scutigera (Fig. 7.8*e*) is the fastest known centipede. A specimen only 22 mm long ran at 0.42 m s^{-1}, but because the legs are so long this involved a stepping frequency of only 13 s^{-1} (when *l* is large, $u\beta/l$ is relatively small for given values of *u* and β).

Adjacent legs of *Scolopendra* converge while they are on the ground. If *Scutigera* ran in the same way its long legs would cross over each other while on the ground and would be apt to get in each other's way. It moves differently, as Fig. 7.8(*e*) shows. The metachronal waves run forwards along the body so that adjacent legs diverge while on the ground. A foot which is off the ground may have to be lifted over one which is on the ground but legs are not crossed while both are on the ground.

The crawling of earthworms also involves a metachronal rhythm. Each segment is made alternately short and fat (by contraction of its longitudinal muscles) and long and slender (by contraction of its circular muscles). Peristaltic waves travel backwards along the body. Fat segments take most of the weight of the worm and tend to remain stationary, so the worm moves forwards. The principle is the same as in the crawling of limpets, which move forward by passing waves of contraction backwards along the foot (section 6.4). The chaetae of earthworms are protruded while the segment is fat, helping to anchor it, and retracted while it is thin. They are particularly effective because they slope posteriorly, as can be felt by running a finger

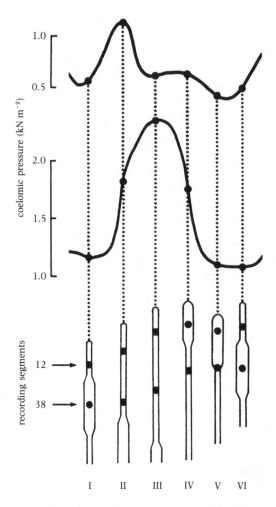

Fig. 7.9. Simultaneous records of pressure from segments 12 (upper record) and 38 (lower record) of *Lumbricus* crawling over damp earth. Positions of the worm are shown below. The worm advanced 2–3 cm between positions I and VI. This movement probably took about 5 s. From M.K. Seymour (1969). *J. exp. Biol.* 51, 47–58.

forwards and back along a worm. Hence they allow their segments to slide forward freely but tend to prevent them sliding back.

The crawling action also serves for burrowing, as can be seen when an earthworm burrows alongside the wall of a glass tank filled with soil. The fat segments are anchored by being jammed against the wall of the burrow.

Pressures have been recorded from the coelom of *Lumbricus*, through fine steel tubes slipped into the coelom and connected by flexible tubing to pressure transducers. Fig. 7.9 shows records made simultaneously from two segments of a crawling worm. The pressures are quite small, and are greatest in each segment when it is slender. The pressure is not the same all along the worm. At position III, for instance, the pressure is only 0.5 kN m^{-2} in segment 12 but 2.5 kN m^{-1} in segment 38. Since the septa are thin they cannot maintain much pressure difference between successive segments, but quite substantial pressure differences can build up along a series of segments.

Records have also been obtained of pressures in the coeloms of worms burrowing in loose earth. They show that in burrowing, in contrast to crawling, the pressure is greatest when the segment is fat. While the anterior segments are slender and moving forward, they probably slip into existing crevices in the soil. This requires little force so the pressure in the coelom can be quite low. When the longitudinal muscles contract the segments get fatter, enlarging the crevice, and it is for this that high pressures are required. It is difficult to get representative records of pressures during burrowing because the ground gets in the way of the tube connecting the worm to the transducer. Only records of burrowing in loose earth have been obtained and the pressures they show are no more than about 2.5 kN m^{-2}. Pressures up to 7.5 kN m^{-2} have been recorded from *Lumbricus* squirming in air and it is believed that much higher pressures may act in burrowing.

The longitudinal muscles of *Lumbricus* occupy 28% of the cross-sectional area of the body, so the stress required in them to produce a pressure of 7.5 kN m^{-2} is only 7.5/0.28 = 27 kN m^{-2}. Data for other muscles (section 6.3) suggest that they can probably exert much larger stresses than that. The pressure that can be developed by a worm squirming in air is probably limited by the relative weakness of the circular muscles. When the worm is in air with excess pressure in its coeloms both the longitudinal and the circular muscles must be active: the longitudinal muscles to prevent the worm from lengthening and the circular muscles to prevent it from swelling. When the worm is in a burrow contracting its longitudinal muscles there is no need for the circular muscles to be active: indeed, it is better for them to be slack so that the force on the wall of the burrow is as large as possible. High pressures can be developed in the swollen parts of a burrowing worm while other parts are at much lower pressure, because the septa make pressure differences possible.

7.5.   Life in the soil

Among the animals mentioned in this chapter, oligochaetes, mites, centipedes and millipedes are all important members of the soil fauna. This section is mainly about oligochaetes.

Pastures, woods and orchards in Europe often have 50–120 g earthworms per square metre. An average pasture will support cattle at a density of about 0.5 t ha^{-1} or 50 g m^{-2}, so many pastures support a greater mass of earthworms than of cattle. Enchytraeid worms are also common in soil, particularly in moorland soils where up to 50 g m^{-2} have been found.

Animals in woodland soil get most of their energy from the leaves that fall from the trees. Earthworms feed directly on fallen leaves. Enchytraeids feed largely on fungi and bacteria which have themselves fed on dead leaves. A deciduous wood in England had a mean earthworm population of 120 g m^{-2}. The mean mass of individual worms was 2.7 g and it was estimated from laboratory measurements of metabolic rate at various temperatures that their mean metabolic rate, over a whole year, would have been 0.02 cm^3 oxygen g^{-1} h^{-1}. Expressed as oxygen consumption per unit area of wood, this amounts to 20 l oxygen m^{-2} a^{-1}. A coniferous wood in Wales had a mean enchytraeid population of 11 g m^{-2}. The mean body mass of these worms was only 0.1 mg so their metabolic rate could be expected to be high (Fig. 4.9). It was calculated from laboratory measurements that their mean metabolic rate over a year would have been 0.3 cm³ oxygen g⁻¹ h⁻¹. This amounts to 30 l oxygen m⁻² a⁻¹. It was estimated that the rate of fall of leaves was about 300 g dry weight m⁻² a⁻¹ in each wood. Complete metabolism of the leaves would use about 300 l oxygen m⁻² a⁻¹, so earthworm metabolism accounted for about 7% of the litter in the deciduous wood and enchytraeid metabolism for about 10% of the litter in the coniferous one. There must have been some of the other group of worms in each wood so oligochaete respiration probably accounted for 10–15% of the energy in the litter in each case. Much of the rest must have been uses by bacteria, fungi and nematodes.

Though earthworms metabolized less than 10% of the energy content of the litter in the deciduous wood they probably broke a much larger proportion of the leaves into fragments and passed them through their guts. The importance of earthworms in breaking up litter was illustrated when 2.5 cm diameter discs of oak and beech leaves were buried out of doors, in nylon bags of various mesh. They were examined periodically and the area lost from each disc was recorded. Discs in 7 mm mesh bags (big enough to admit earthworms) disappeared 2–3 times as fast as discs in 0.5 mm mesh bags (which excluded earthworms).

Earthworms in pastures feed largely on dung. Though cattle are much larger than earthworms they have a metabolic rate (per unit body mass) around 10 times as high. About 30% of the energy content of the grass they eat remains in the dung. Hence a given mass of cattle probably produces ample dung to support an equal mass of earthworms.

Earthworms in damp soil face the same osmotic problems as animals living in fresh water. Professor Arthur Ramsay took tiny samples of fluid from nephridia of *Lumbricus* and determined their freezing points. He showed that worms with coelomic fluid of concentration 170 mmol l⁻¹ produced urine of only about 30 mmol l⁻¹. Fluid from near the coelomic opening had the same osmotic concentration as the coelomic fluid, but the concentration fell as the fluid travelled along the nephridium. Thus the nephridia seem to work in the same way as the kidney of the gastropod *Viviparus* (section 6.7).

Polychaete worms that burrow in muddy or sandy shores are exposed to sea water, so their osmotic problems are different from those of earthworms. Less obviously, their respiratory problems are also different. Earthworms generally live in soil with plenty of air space, through which oxygen can diffuse rapidly. Burrowing polychaetes live in waterlogged sand or mud. Oxygen diffuses much faster through air than through water (its diffusion in air is 300 000 times as high as in water). Earthworms are adequately supplied with oxygen by diffusion through the soil but polychaetes must pump water through their burrows to get enough oxygen. Many do this by peristalsis, but intertidal species cannot do this at low tide. The lugworm *Arenicola* is an intertidal burrowing polychaete of about the size and shape of a typical earthworm (*Lumbricus*), but it has gills and earthworms do not. These gills are 13 pairs of delicate tufts projecting from its body. Samples of water taken from *Arenicola* burrows at low tide contained oxygen at partial pressures less than 0.02 atm (less than 10% of the partial pressures in well-aerated water). This may be why *Arenicola* and many other polychaetes have gills.

A simple calculation will help us to understand how earthworms survive without gills in well-aerated soil. Consider a cylindrical worm of length *l* and diameter *d* which uses oxygen at a rate m per unit volume of tissue. Its volume is $\pi ld^2/4$ so it uses oxygen at a rate $\pi ld^2m/4$. Its surface area is about πld. Blood is brought by capillaries to a distance s below the surface, all over the body, so diffusion must occur through a layer of tissue of thickness s. The partial pressure of oxygen is P_0 outside the worm and P_b in the blood. By equation 4.1,

$$\pi ld^2m/4 = -\pi ldD(P_b - P_0)/s;$$
$$P_0 - P_b = sdm/4D. \qquad (7.2)$$

D is the diffusion constant for oxygen diffusing through tissue and is probably about 2×10^{-5} mm^2 atm^{-1} s^{-1}.

The cuticle and epidermis of a typical earthworm are together about 50 μm thick but the most superficial blood vessels form loops within the epidermis and 30 μm (0.03 mm) is probably a realistic estimate of s. This is much larger than the distance from blood to air in a snail's lung (about 10 μm) or from blood to water in most gills, but most specialized respiratory organs are delicate structures in protected positions whereas a worm's epidermis must be able to withstand abrasion. The diameter d of a typical earthworm might be about 5 mm, and if it lived at tropical temperatures (like the ones that supplied the data for Fig. 4.9) it might have used about 0.06 cm^3 oxygen g^{-1} h^{-1}. Since the densities of worms are about 1 g cm^{-3}, this rate of oxygen consumption makes $m = 0.06$ cm^3 oxygen cm^{-3} h^{-1} or 1.7×10^{-5} s^{-1}. These values in equation 7.2 give

$$P_0 - P_b = 0.03 \times 5 \times 1.7 \times 10^{-5}/4 \times 2 \times 10^{-5}$$
$$= 0.03 \text{ atm.}$$

In well-aerated soil, P_0 would be much larger than this, about 0.2 atm, so diffusion through the general body surface is well able to supply the animal's oxygen requirements.

7.6.   Life in estuaries

For animals living in estuaries, the salt concentration of the water around them may fluctuate widely. It will fall as the tide falls and rise again as the tide rises. Its value at a given state of the tide may fluctuate seasonally. *Nereis diversicolor* is common in estuaries and is sometimes most abundant where the salt concentration is most variable. It can survive in fresh water, provided the calcium concentration is not too low. *N. limnicola* is a very similar worm which is found in estuaries and even in fresh water on the Pacific coast of N. America.

Many experiments have been carried out on the ability of *Nereis* to survive changes of salt concentration. In one, *N. diversicolor* were kept in mixtures of different proportions of sea water and distilled water. After 6 or more days samples of their coelomic fluid were taken, for measurements of osmotic concentration. Fig. 7.10(a) shows that worms kept in undiluted sea water (1 mol l^{-1}) or 50% sea water (0.5 mol l^{-1}) had coelomic fluid only slightly more concentrated than the water. Those kept in more dilute mixtures (5–300 mmol l^{-1} had coelomic fluid much more concentrated than the water. If *Nereis* are wiped dry, droplets of urine can be watched forming at the nephridial apertures. These droplets have been collected in fine pipettes, and their

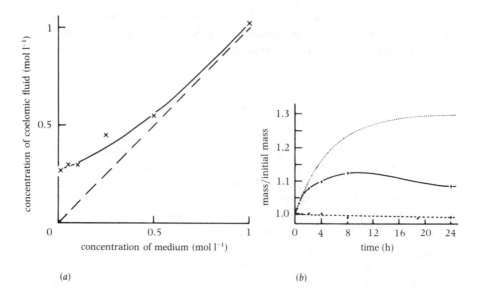

(a)   (b)

Fig. 7.10. Osmotic regulation of *Nereis diversicolor*. (*a*) A graph of the osmotic concentration of the coelomic fluid against that of the water in which the worms had been kept. Data from C.R. Fletcher (1974). *Comp. Biochem. Physiol.* A47, 1199–1214. (*b*) A graph of body mass (expressed as a multiple of initial mass) against time after transfer from 70% sea water to (continuous line) 50% sea water or (broken line) another dish of 70% sea water. The dotted line is explained in the text. After C.R. Fletcher (1974). *Comp. Biochem. Physiol.* A47, 1221–1234.

osmotic concentration determined by measuring their freezing point. It was found that urine from worms kept in very dilute sea water is only 50–60% as concentrated as the coelomic fluid. The worms were apparently maintaining the osmotic concentration of their body fluids in the same way as *Lumbricus*, by secreting a dilute urine.

At concentrations between 50 and 100% sea water, *Nereis* coelomic fluid is only slightly more concentrated than the water and the worms do not face any severe osmotic problem. However, a problem arises when the osmotic concentration of the medium is changed. Suppose for instance that a worm is moved from 70% sea water to 50% sea water. Immediately after transfer the worm's body fluids are quite a lot more concentrated than the water; water tends to diffuse in, diluting the body fluids and making the worm swell. If the surface of the worm were a semipermeable membrane no solutes would diffuse out and equilibrium would not be attained until the quantity of water in the worm was $70/50 = 1.4$ times its initial value. Since the worm would have contained initially about 80% water, its final volume would be about 1.32 times its initial volume. It would swell gradually to this volume as indicated by the dotted line in Fig. 7.10(*b*). What happened when the experiment was tried is shown by the continuous line. The worms stopped swelling after 8 h and after 24 h were only 1.09 times their initial volume.

Some of the worms were dissolved in potassium hydroxide solution and analysed for chloride. It was found that the total quantity of chloride in the body fell by about 25% in the 24 hours following transfer to 50% sea water. It was concluded from this and other observations that the worms must have limited their swelling by passing urine of about the same osmotic concentration as the coelomic fluid. This got rid of some of the water which diffused in and also got rid of some salt, reducing the amount of swelling needed to restore equilibrium.

Further reading

Segmentation

Anderson, D.T. (1973). *Embryology and phylogeny in annelids and arthropods*. Pergamon, Oxford.
Lewis, E.B. (1978). A gene complex controlling segmentation in *Drosophila*. *Nature* **276**, 565–570.

Annelids

Dales, R.P. (1967). *Annelids*, 2nd edn. Hutchinson, London.
Edwards, C.A. & Lofty, J.R. (1972). *Biology of earthworms*. Chapman & Hall, London.

Various arthropods

Barthel, K.W. (1974). *Limulus*, a living fossil. *Naturwissenschaften* **61**, 428–433.
Blower, J.G. (ed.) (1974). *Myriapoda* (*Symposium of the Zoological Society of London* no. 32). Academic Press, New York.
Bristowe, W.S. (1971). *The world of spiders*, 2nd edn. Collins, London.
Lewis, J.G.E. (1980). *The biology of centipedes*. Cambridge University Press.
Manton, S.M. (1977). *The Arthropoda: habits, functional morphology and evolution*. Oxford University Press.
Savory, T. (1964). *Arachnida*. Academic Press, New York.

Metachronal rhythms

Clark, R.B. & Tritton, D.J. (1970). Swimming mechanisms in nereidiform polychaetes. *J. Zool., Lond.* **161**, 257–271.
Mettam, C. (1967). Segmental musculature and parapodial movement of *Nereis diversicolor* and *Nephthys hombergi* (Annelida: Polychaeta). *J. Zool., Lond.* **153**, 245–275.
Seymour, M.K. (1969). Locomotion and coelomic pressure in *Lumbricus terrestris* L. *J. exp. Biol.* **51**, 47–58.

Life in the soil

Burges, A. & Raw, F. (1967). *Soil biology*. Academic Press, New York.
Ramsay, J.A. (1949). The osmotic relations of the earthworm. *J. exp. Biol.* **26**, 46–56.
Ramsay, J.A. (1949). The site of formation of hypotonic urine in the nephridium of *Lumbricus*. *J. exp. Biol.* **26**, 65–75.

Life in estuaries

Fletcher, C.R. (1974). Volume regulation in *Nereis diversicolor* [3 papers]. *Comp. Biochem. Physiol.* A**47**, 1199–1234.
Smith, R.I. (1970). Hypo-osmotic urine in *Nereis diversicolor*. *J. exp. Biol.* **53**, 101–108.

8 Crustaceans

Phylum Crustacea,
    Class Branchiopoda   (water fleas)
    Class Maxillopoda, Subclass Cirripedia   (barnacles)
                      Subclass Copepoda   (copepods) and other subclasses
    Class Ostracoda
    Class Malacostraca   (crabs, shrimps, woodlice, etc.)

8.1. Introduction

It has long been traditional to use the crayfish *Astacus fluviatilis* (Fig. 8.1) as a typical crustacean, to introduce the phylum. It is a member of the class Malacostraca. It is big enough to be easily dissected and it is more typical of the phylum than the equally large crabs. It is a European species found in streams, especially in limestone districts, spending much of the day under stones or in burrows in the bank but emerging at night to feed on water snails, insect larvae, etc.

The body of *Astacus* has two main parts, an anterior cephalothorax which is roofed by a rigid piece of exoskeleton (the carapace) and a posterior flexible abdomen. The most conspicuous limbs are the pair that bear the chelae (pincers) and the four pairs of walking legs, but there are many other appendages. Here is a complete list of them, starting at the anterior end of the body:

(on the cephalothorax)	antennules
	antennae
	mandibles
	2 pairs of maxillae
	3 pairs of maxillipeds
	chelae
	4 pairs of walking legs
(on the abdomen)	5 pairs of swimmerets
	uropods

Each pair of appendages belongs to a different segment of the body. The antennules bear chemosensory organs, spread out along one of their two branches. In addition, there are statocysts (organs of balance) housed in their bases. The antennae are used as organs of touch and are often waved about in front of the animal, but are also often carried trailing posteriorly. The next six pairs of limbs are arranged around the mouth. The mandibles are biting jaws and the maxillipeds are used to hold and manipulate food. The first maxillae have no apparent function but the second maxillae are important in respiration; their movements pump water over the gills

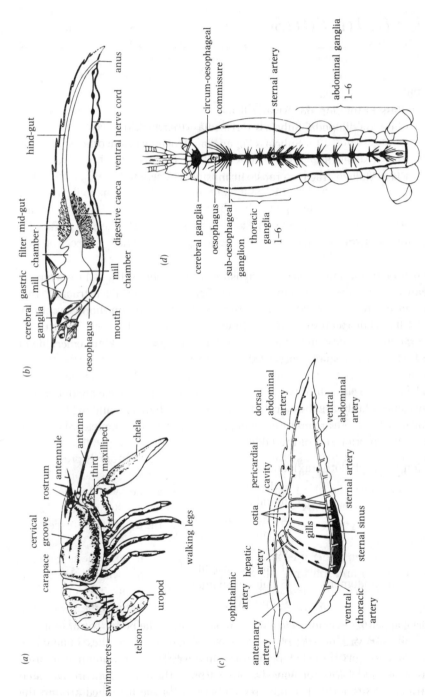

Fig. 8.1. Anatomy of the crayfish *Astacus fluviatilis*. (*a*) A lateral view of the intact animal. (*b–d*) Diagrams of dissections showing the gut, the blood circulation and the nervous system, respectively. Length up to 15 cm. From G. Chapman & W.B. Barker (1966). *Zoology for intermediate students*. Longman, London.

through the paddle-like action of processes called scaphognathites. The chelae are used to seize and crush food and to pass it to the mouth and the walking legs are (of course) used for walking. The swimmerets move forward and back in seemingly futile fashion as the crayfish walks. The pair of uropods form a fan which can be spread or folded. *Astacus* normally moves about by walking on its walking legs but it can swim rapidly backwards by repeatedly bending the abdomen ventrally with the uropods spread.

Fig. 8.1(*b*) shows the arrangement of the gut. Only the mid-gut is endodermal; the mill chamber and filter chamber and the hind-gut are formed from ectoderm which tucks in at the mouth and anus as the embryo develops. These ectodermal parts of the gut are lined with cuticle, similar to the cuticle of the outer surface of the body. Most of the gut cuticle is flexible, like the flexible membranes between the stiff plates of the exoskeleton. However, there are thick stiff plates in the cuticle of the mill chamber. There are muscles to move these plates and they form the gastric mill which breaks up chunks of food into smaller pieces. The food cannot move on to the mid-gut until it is finely ground because it is strained through fringes of setae which project from the cuticle of the filter chamber.

A pair of digestive glands open into the mid-gut. Each consists of a large number of finger-like caeca. They secrete most or all of the digestive enzymes, but enzymes seep forward into the mill chamber and digestion starts there. Digestion seems to be entirely extracellular and the products of digestion are absorbed through the walls of the mid-gut and digestive glands. There are no cilia in the gut; the food is moved along the gut entirely by peristalsis. The gut muscles are striated, not unstriated like the gut muscles of vertebrates.

Fig. 8.1(*b*) and (*d*) show the central nervous system. There is a ventral nerve cord (divided for most of the length of the body into two strands) with ganglia which link the strands together. The cerebral ganglia lie anterior to the mouth and the two strands of the nerve cord pass on either side of the mouth. The central nervous systems of annelids are very like this (Fig. 7.3*a*). Primitive arthropods presumably had a ganglion in each segment, but the crayfish has separate ganglia only for the segments that bear the third maxillipeds and the chelae, walking legs and swimmerets. The cerebral ganglia send nerves to the antennules, antennae and eyes, and the sub-oesophageal ganglion provides the nerves for the mouthparts (except the third maxillipeds).

The gills are hidden under the carapace (Fig. 8.2). They are extensions from the bases of the second and third maxillipeds, the chelae and the walking legs. Each consists of numerous fine filaments, projecting from a central stalk. The total area of the gills of a 40 g crab (*Carcinus maenas*) is about 300 cm^2, about the same as would be expected for a gastropod of the same mass (section 6.6). The gills are covered by a very thin layer of cuticle. The path of water over the gills has been investigated by watching the movements of dyes or milk, released into the water near the animal. Water enters the gill cavity through the gaps between successive pairs of limbs and leaves by a pair of exhalant openings on either side of the mouth. The movement of dyed water within the gill chamber has been observed in crabs which had had part of the carapace replaced by a window of transparent plastic. Fig. 8.2 shows the direction of flow. The water is apparently propelled by the movements of the

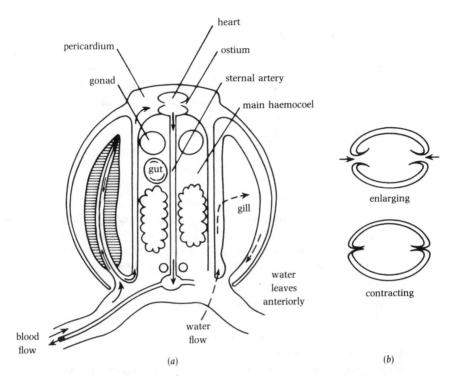

Fig. 8.2.(*a*) A diagrammatic transverse section through the cephalothorax of a crayfish. Short arrows indicate blood flow and long broken arrows indicate the flow of water over the gills. (*b*) Diagrammatic transverse sections of the heart showing how the valves of the ostia open when the heart is enlarging but close when it is contracting.

scaphognathites, which are just inside the exhalant openings. The scaphognathites beat 1–5 times per second, in the crab *Carcinus*, and set up a pressure difference across the gills of 2–6 mm water (20–60 N m^{-2}); this has been measured by means of pressure transducers). The use of scaphognathites instead of cilia to drive water over the gills is a striking difference from gastropods and bivalves.

Figs. 8.1(*c*) and 8.2 show the arrangement of the blood system and the directions of flow of blood. The directions were discovered by watching the movement of injected dyes. There is a dorsal heart enclosed in a blood-filled pericardium. Three pairs of openings, the ostia, admit blood from the pericardium into the heart. They have valves which prevent flow in the reverse direction. When the heart contracts it drives blood out through the arteries. When it expands (presumably by elastic recoil) it draws blood in through the ostia. The frequency of the heart beat is 0.5–1 s^{-1}. Peak pressure in the heart of a lobster (*Homarus*) during contraction was about 1800 N m^{-2} when the lobster was resting and 3600 N m^{-2} when it was active: similar pressures occur in the hearts of snails (section 6.6). There are arteries to all parts of the body, including the legs. The sternal artery pierces the ventral nerve cord (Fig. 8.1*d*). At the tissues the blood escapes from fine branches of the arteries into the haemocoel. It returns to the pericardium by way of the gills. This circuit is rather similar to the circuit of the blood in gastropods (Fig. 6.23).

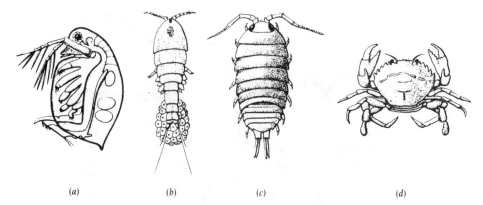

(a)   (b)   (c)   (d)

Fig. 8.3. A selection of Crustacea. (*a*) *Daphnia* (Branchiopoda: 1–4 mm); (*b*) *Tigriopus* (Copepoda: 1 mm); (*c*) *Ligia* (Malacostraca: 25 mm); and (*d*) *Portunus* (Malacostraca: carapace 10 cm wide). From A.P.M. Lockwood (1968). *Aspects of the physiology of Crustacea.* Oliver & Boyd, Edinburgh.

The blood contains a haemocyanin, which enables it to transport (in the crab *Cancer*) 34 cm^3 oxygen l^{-1}. The haemocyanin is present as large polymeric molecules dissolved in the blood, but the molecules of crustacean haemocyanin are not nearly as big as those of snail haemocyanin.

The main body cavity is a haemocoel and there is hardly any trace of coeloms. However, there is a pair of excretory organs (the green glands), each of which consists of a rudimentary coelom and coelomoduct; they can be identified as coeloms because they develop from mesoderm. The green glands lie in the haemocoel anterior to the mouth and their ducts open to the exterior through the pores in the basal segments of the antennae. They are supplied with blood by the antennary artery (Fig. 8.1*c*). Ultrafiltration occurs from the blood into the rudimentary coelom, and urine is excreted through the duct. The urine of freshwater crayfish may be even more dilute than that of the freshwater gastropod *Viviparus* (Table 6.3), though the crayfish has much more concentrated blood. Crayfish with 184 mmol l^{-1} chloride in their blood excreted urine with only 3.4 mmol l^{-1} chloride.

The sexes are separate and the reproductive organs are simple. The gonads lie in the main haemocoel, dorsal to the gut (Fig. 8.2). The female has a pair of ovaries with oviducts which lead to openings near the bases of the second walking legs. The male has a pair of testes with vasa deferentia leading to openings near the bases of the fourth walking legs. Fertilization is external. The eggs adhere to the female's swimmerets and the young which hatch from them at first hold on to her swimmerets by their chelae. The newly hatched young are more or less similar in form to the adults, though much smaller. (This is unusual among crustaceans, most of which have larvae quite unlike their adults.)

Figs. 8.3 and 8.4 show some other crustaceans. *Daphnia* (Fig. 8.3*a*) is one of the water fleas, a member of the class Branchiopoda. It is very common in ponds, where it swims by movements of its large antennae. It uses the paddle-shaped limbs that are enclosed in its carapace for filter feeding: their movements set up water currents and closely spaced setae (bristles) along their edges strain out flagellate protozoans and

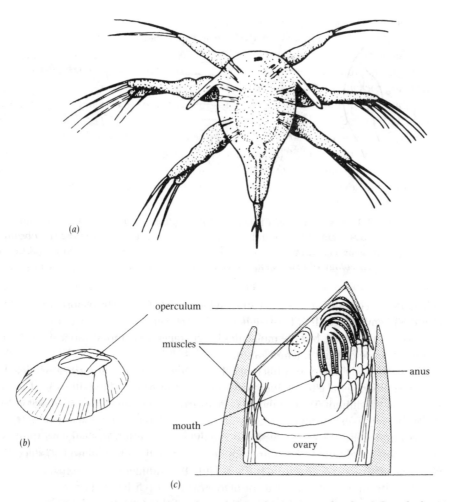

Fig. 8.4. An acorn barnacle, *Balanus balanoides*. (*a*) A nauplius larva (length about 0.3 mm); (*b*) an adult (diameter about 10 mm); and (*c*) a diagram of a barnacle cut in half to show its internal structure.

other small food items. In many species of water flea, as in many species of rotifer (section 5.1), reproduction is usually by parthenogenesis and males occur only occasionally. Rotifers and the smaller water fleas are similar in size and live largely in the same habitats.

Most Branchiopoda live in freshwater but Copepoda are abundant both in fresh water and in the sea. They are among the most plentiful and important constituents of the marine zooplankton. They are the principal food of the herring (*Clupea harengus*), which is one of the most important fish landed commercially in Europe. The copepod illustrated in Fig. 8.3(*b*) is a female, carrying eggs. Many species have a pair of egg sacs but this species has just one. Copepods swim by movements of their antennae and other limbs. They feed mainly by filter feeding, on phytoplankton.

The earliest larval stage of Copepoda, the nauplius, has just three pairs of limbs. Very similar larvae are possessed by Cirripedia (barnacles) (Fig. 8.4*a*) and these two subclasses are grouped together in the class Maxillopoda. However, adult barnacles are very different from adult copepods. They live attached to rocks, floating wood and ships. *Balanus balanoides* lives on intertidal rocks and is the commonest European species. Large numbers are often found packed so closely together that the underlying rock is completely hidden. There are six pairs of two-branched limbs. Fig. 8.4(c) shows them and also shows the positions of the mouth and anus. The animal is attached to the rock by the dorsal surface of its head. It is enclosed in a ring of fixed calcareous plates with a lid (operculum) of movable plates. The lid opens and closes, rather like a double door. The animal can retract into the shell and close the operculum for protection (for instance when it is exposed at low tide). It can open the operculum and extend the limbs for filter feeding. Its foods range from flagellate protozoans to small crustaceans.

There are parasitic copepods and barnacles, as well as free-living ones. Many of them are irregular-shaped bags of tissue, when adult, and it is by no means obvious from their appearance that they are crustaceans. However, they have free-living larvae which resemble the larvae of other copepods and barnacles. The earliest larva is a typical nauplius. *Sacculina*, a barnacle parasite of crabs, is an example of these parasites. It forms a bag of tissue protruding from the crab's abdomen, with root-like processes permeating the whole of the crab's body.

Astacus is one of the Malacostraca, as also are *Ligia* and *Portunus* (Fig. 8.3c, d). All members of this class have the same number of segments. *Ligia* is one of the isopods, a group of depressed (dorsoventrally flattened) Malacostraca which also includes the terrestrial woodlice. *Ligia* is found under stones and weed on the upper part of the shore and feeds on pieces which it bites off seaweeds. *Portunus* is an example of the crabs, which are very similar in structure to the crayfish but have a very broad carapace and a very small abdomen which is folded forward. Crabs eat a wide variety of foods, including molluscs, whose shells they break with their chelae. *Portunus* lives on shores and, unlike most other crabs, can swim.

8.2.   The exoskeleton

Though the arthropods were introduced in chapter 7 it seems convenient to discuss here the exoskeleton (external skeleton) that is characteristic of them all.

This skeleton consists of hardened parts of the cuticle that covers the whole body. It generally has three layers. The outermost is a thin waxy epicuticle. The middle layer, the exocuticle, generally consists of very thin ($c.$ 10 nm) fibrils of the polysaccharide chitin embedded in quinone-tanned protein. (Quinone tanning was explained in section 3.5.) The innermost layer, the endocuticle, consists of chitin and untanned protein. Thus the main thickness of the cuticle (exocuticle plus endocuticle) is composite material, chitin fibres in a protein matrix. The chitin is very strong, with a high Young's modulus, and the protein is weaker with a lower modulus. Many crustaceans have a third constituent, calcium salts (mainly carbonate but with a little phosphate). The highest proportions of calcium salts, up to 90% of the cuticle's dry mass, are found in the hardest parts, for example at the tips of crabs' legs. The advantages of composite materials were explained in sections 3.5 and 6.2.

Table 8.1. *Mechanical properties of some skeletal materials*

	Tensile strength (MN m²)	Young's modulus (GN m⁻²)	Density (kg m⁻³)
Mollusc shells	30–100	50	2700
Crab carapace	32	13	1900
Locust leg cuticle	95	9.5	1200
Locust apodeme	600	19	—
Compact bone	150–200	18	2000

Data from S.A. Wainwright, W.D. Biggs, J.D. Currey & J.M. Gosline (1976). *Mechanical design in organisms*. Arnold, London.

Pieces of crab carapace have been stretched and broken on engineers' testing equipment. They had lower strength than many mollusc shells but they were also less dense so the value of strength/mass was about the same as for typical mollusc shells (Table 8.1). They were less stiff (had a lower Young's modulus) than many mollusc shells, as might be expected since they contained a smaller proportion of calcium salts. Pieces of insect cuticle have also been tested and are generally stronger than crab cuticle. The calcium salts in crustacean cuticle may help to stiffen it (increasing Young's modulus) but apparently contribute little to its strength.

The apodeme in Table 8.1 connects a muscle to the skeleton. The only substantial force that it has to bear is tension, acting along its length, and its chitin fibrils are appropriately arranged, all running in this direction. That is why it is so strong and has so high a Young's modulus (when tested in this direction only).

Most other cuticle structures have to withstand forces in several directions, and have more complex structure. Electron microscope sections of them often show patterns of curving lines formed by the chitin fibrils (Fig. 8.5a). It looks as if the fibrils run in hooped paths but careful study of insect cuticle has shown that this is an illusion. The actual arrangement is like a complex version of plywood. Successive layers in a sheet of plywood have the grain running north–south and east–west. Insect cuticle is laid down in layers with all the chitin fibrils in a given layer running in the same direction, but the directions are not at right angles in successive layers; rather, they differ only by a small angle. Fig. 8.5(b) is a vertical section through a series of layers in which the direction changes from parallel to the page (at the top), to at right angles to the page (half way down), to parallel to the page again (at the bottom). Fig. 8.5(c) shows an oblique section through the same set of layers. At a lower magnification it could easily be mistaken for an array of hooped fibrils. Cuticle is commonly made up of many sets of layers with the fibril direction changing through 180° in each set.

Fig. 8.5(b) and (c) are drawn as though the fibril direction changed $22\frac{1}{2}°$ between each layer and the next. Arthropod cuticle generally has much smaller changes of angle between successive layers: changes as small as 1° have been observed.

Wood splits easily along the grain but is harder to break across the grain. It is also much stiffer along than across the grain: Young's modulus for Douglas fir timber is about 10 times as high for tension along the grain as for tension across the grain.

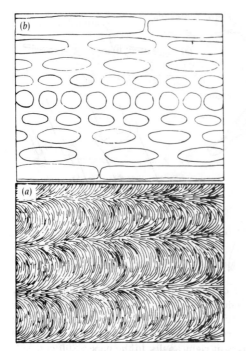
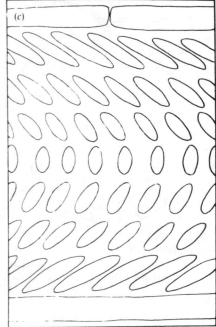

Fig. 8.5.(*a*) An oblique section of a typical arthropod cuticle, as sen by electron microscopy. (*b*) A diagrammatic vertical section and (*c*) a diagrammatic oblique section of the same cuticle, on a larger scale than (*a*). These diagrams are explained in the text. From S.A. Wainwright, W.D. Biggs, J.D. Currey & J.M. Gosline (1976). *Mechanical design in organisms.* Arnold, London.

Plywood with grain running at right angles in successive layers is almost equally strong and stiff for tension in any direction in its own plane. The arrangement of fibres in arthropod cuticle is probably even better because of the smallness of the angle between successive layers makes the layers less likely to come apart under stress.

The cuticle of a crustacean covers the whole external surface of the body. Movement is possible because there are flexible regions between the stiff ones. These arthrodial membranes are continuous with the stiff cuticle and have the same basic structure but most of their protein is not tanned, and they are not impregnated with calcium salts. Fig. 8.6(*a*) shows, very diagrammatically, the exoskeleton of a crayfish. The cephalothorax is covered by an unjointed carapace but the abdomen has a separate dorsal and ventral plate of stiff cuticle for each segment, with strips of arthrodial membrane between. The plates overlap like tiles on a roof (Fig. 8.6*b*) so that the relatively weak arthrodial membranes are not exposed. The antennae are covered by rings of stiff cuticle joined by rings of arthrodial membrane which make them flexible. The legs are covered by tubes of stiff cuticle with arthrodial membrane at the joints. The stiff cuticle of the more distal segment involved in a joint forms a pair of knobs which fit into sockets in the more proximal segment, so that the joint is a hinge and can rotate only about an axis through the two knobs (Fig. 8.6*c, d*: the proximal end of a limb is the end attached to the trunk and the distal end is the free

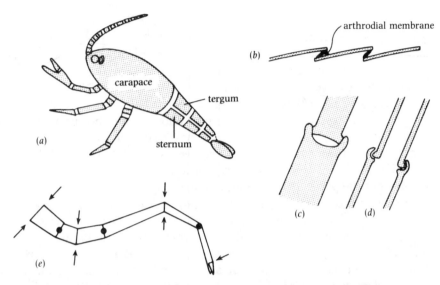

Fig. 8.6. Diagrams of the exoskeleton of a crayfish. (*a*) The complete animal, grossly simplified with the number of parts reduced; (*b*) a longitudinal section through a few of the dorsal plates of the abdomen; (*c*) a sketch of and (*d*) a section through a leg joint; (*e*) a sketch of a leg with arrows representing axes of joints lying in the plane of the paper and dots representing axes at right angles to the paper. Stiff cuticle is stippled in (*a*) to (*d*).

end). Successive joints along the leg have their hinge axes at right angles to each other, making a wide range of leg positions possible.

Vertebrates and arthropods can move in broadly similar ways. They can bend their trunks, walk on their legs, and so on. The exoskeleton of arthropods allows the same sorts of movements as the endoskeleton (internal skeleton) of vertebrates. What are the advantages of each type of skeleton?

Obviously an exoskeleton protects internal organs, but it is also rather easily damaged because there is no soft tissue between it and the body surface to cushion a blow. To understand what this implies we need to consider the mechanics of breaking things.

If a piece of material is stretched until it breaks, a certain amount of work is done stretching it against the elastic restoring force. It cannot be broken unless this amount of work is done. If a cup is dropped on the floor it will not break unless it is dropped from high enough to have enough kinetic energy at the moment of impact to provide the work needed to break it. It is less likely to break if the floor is carpeted because some of the energy is used in deforming the carpet, leaving less to break the cup. Similarly, soft tissues protect endoskeletons from blows. This was demonstrated in a series of experiments with metatarsal bones from the feet of rabbits. In the intact rabbit, the metatarsals are covered by furry skin. They were broken by a blow from a heavy pendulum and it was found that 37% more energy was needed to break them when they were covered with skin than when they were not. Bones deeper in the body are even better protected. The exoskeletons of arthropods do not have this sort of protection.

Imagine a series of animals each with a different mass m. Let them all have exoskeletons of the same material and let the exoskeleton be the same proportion of body mass in each case. Then the energy required to fracture the exoskeleton will be km, where k has the same value for all the animals. Now let each animal collide with a rigid wall while travelling with velocity u. The exoskeleton will break if the kinetic energy $\frac{1}{2}mu^2$ is greater than km, i.e. if $u > \sqrt{(2k)}$. Whatever the size of the animal, fracture can be expected at the same impact velocity. An exoskeleton will thus be dangerously fragile for any animal which moves fast, but there are no really fast arthropods. On the other hand, many vertebrates, which have endoskeletons, can move at great speed. This simple argument will mislead readers if they take it too literally, but the principle it presents is sound.

As well as protecting internal organs, exoskeletons have another advantage. Consider a rod which has to withstand forces tending to bend it. It is neither as strong nor as stiff as a hollow tube of the same length, made from the same mass of the same material. This is why tubes are used for making scaffolding and the frames of bicycles. A stout, thin-walled tube is better than a more slender thick-walled tube of the same length and mass, provided its wall is not so thin that it is apt to kink like a drinking straw. The exoskeletons of crustacean legs are thin-walled tubes, stout enough to accommodate the muscles and other soft tissues within them. Endoskeletons made of the same mass of the same material would be neither as stiff nor as strong in bending.

The skeletons of arthropods, like the shells of molluscs, cannot be re-shaped once they have been formed. Molluscs grow by adding to the existing shell (section 6.2). Arthropods grow by shedding the skeleton and forming a new, larger one. The process is similar to the moulting of nematodes (but nematode cuticle can grow after the final moult). The inner layers of the old cuticle are partly broken down so that the cuticle is detached from the epidermis, and some of the breakdown products are resorbed. New, soft cuticle forms inside the old.

The animal often retires to a relatively safe place before the old cuticle splits: the crab *Carcinus*, for example, hides among seaweed. The new cuticle, formed inside the old, must initially be smaller than the old, but it has to be enlarged to accommodate the next stage of growth. It is stretched while still soft. Observations on lobsters (*Homarus*) show that they increase their body volume by almost 50% in three hours, splitting the old cuticle and then stretching the new. Evidence that they do this by drinking came from experiments in which moulting lobsters were put in a dilute suspension of barium sulphate, an insoluble salt that is opaque to X-rays. Shortly afterwards X-radiographs showed an accumulation of the salt in the mid-gut. It appeared that water was being absorbed there, leaving the barium sulphate behind. Terrestrial insects stretch their new cuticle by inflating themselves with air.

Newly formed crustacean cuticle has no calcium salts in it, which partly explains its extensibility. However, the hardening process also involves changes in the cuticle protein. These changes have been investigated principally in insects, and there is controversy about them. It seems to be generally agreed that the newly formed protein is not cross-linked, so that the molecules can flow past each other like the ones shown in Fig. 3.18(*a*). This means that the chitin fibres can also slide past each other: there is no firm connection between one chitin fibre and the next. The subsequent hardening of the cuticle seems to involve quinone tanning but the number of cross-

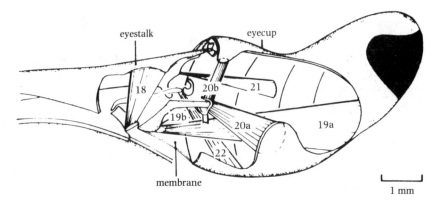

Fig. 8.7. A lateral view of a dissection of the right eyecup of the crab *Carcinus*. Numbers refer to muscles, discussed in the text. From M. Burrows & G.A. Horridge (1968). *J. exp. Biol.* 49, 285–297.

links that this can introduce seems insufficient to explain the very great increase in stiffness that occurs. The water content of the cuticle falls at the same time as (possibly as a result of) the tanning and it has been suggested that the stiffening is due mainly to loss of water, rather than directly to tanning.

8.3.   Muscles

Crabs have movable eyes mounted on stalks. Their eye movements have been studied in great detail by Professor Adrian Horridge and others and provide convenient examples of the ways in which muscles work joints. The eye itself is shown black in Fig. 8.7. It is on the end of a cup-shaped piece of rigid cuticle which fits over the tubular eyestalk. The cup is attached to the stalk by a ring of arthrodial membrane which allows a wide variety of movements: the eyecup can be pointed up or down and to one side or the other, and it can be withdrawn over the stalk like a telescope being shortened. The eyestalk is in a hollow under the carapace so withdrawal pulls the eye into a protected position.

Some of the muscles responsible for these movements are shown in Fig. 8.7. Muscle 18 runs between the eyestalk and the main body skeleton and moves the eyestalk. The others run from the eyestalk to the eyecup and move the cup in various directions.

In the experiments to discover how the muscles are used, small holes were bored in the eyecup and fine wire electrodes were pushed through into the muscles, to record the electrical changes that occur when the muscles are active (section 3.4). The crabs were killed and dissected after the experiments to make sure that the electrodes were lodged in the intended muscles.

When someone watches a train go by their eyes fix on one carriage and follow it for a while, then flick back to fix on another. Crabs watch moving objects in similar fashion. Thus the crabs in the experiments could be made to move their eyes in various directions, by moving patterns of black and white stripes in front of them. It

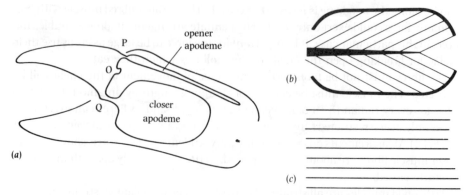

Fig. 8.8.(*a*) The exoskelton and apodemes of a crab's chela. (*b*) A horizontal section through the chela showing the pennate arrangement of the muscle fibres. (*c*) A parallel-fibred muscle of the same dimensions as the pennate muscle shown in (*b*).

was found that muscle 20a, which attaches to the lateral side of the eyecup, was active whenever the eye was turning to that side. Muscles 19b and 21 were active when the eye was turning medially. Muscle 22 was used with 20a when the eye was being turned laterally, but was also used when it was being turned downward. Muscle 20b was used when it was being turned up. Each muscle was used as might be predicted from its position. There are three more eyecup muscles which are not illustrated, so there are more than enough muscles to turn the eye in any desired direction.

Muscle 19a is the biggest of the eyecup muscles but it is not used when the eye is turned from side to side or up and down. It serves to withdraw the eye, but is assisted by all the muscles shown in Fig. 8.7: electrical activity can be recorded from all of them during withdrawal. The eye is presumably extended again, when they relax, by the pressure in the haemocoel.

The muscles shown in Fig. 8.7 are small, with few fibres. Muscle 20a, for instance had about 25 fibres. In contrast, consider the very much larger muscles which work the crab's chelae (Fig. 8.8*a*). The lower claw of each chela is fixed. The upper claw is movable, with knob-and-socket joints on either side at O forming a hinge joint. Muscles attach to it through apodemes, which are plates formed by ingrowth of the cuticle. The plates themselves are stiff but their attachments to the movable claw at P and Q are flexible. Large numbers of muscle fibres run obliquely from the exoskeleton to the apodemes (Fig. 8.8*b*). The muscle which attaches to the upper apodeme opens the claw and the (much larger) muscle of the lower apodeme closes the claw.

The muscles of the crab chela are pennate: that is, their fibres attach obliquely to an apodeme. Many other muscles are parallel-fibred, with all their fibres parallel to each other and to the direction in which the muscle pulls (Fig. 8.8*c*). What are the merits of the two arrangements? Suppose the crab had parallel-fibred claw muscles. If they filled the same space in the chela their fibres would be about twice as long as in the actual crab (compare Fig. 8.8*c* with *b*). However there would only be room for about half as many of these long fibres, so that they could only exert half as much force. Suppose the fibres of the pennate muscle exert a total force F. They run at 30° to the apodeme (this angle changes a little as the claw opens and closes) so the component of force along the apodeme is $F \cos 30° = 0.87 \, F$. The parallel-fibred muscle with half as many fibres could exert only $0.5 \, F$. The pennate arrangement almost doubles the force, in this particular case. It is important for the force to be large as crabs use their chelae for breaking open the chells of the molluscs which they eat.

The fibres must not be too short or the range of movement of the chela will be restricted. The muscle must be capable of changing length by an amount OQ (Fig. 8.8a) for every radian (57°) of movement of the joint, and its fibres can exert large forces only over a restricted range of lengths (see the graph for a crustacean muscle in Fig. 6.13). Measurements on crab chelae by my student Patrick Morris show that the closer muscle fibres are about 25% longer when the chela is fully open, than when it is tightly closed.

Arthropod muscles are all striated. There seem to be no obliquely striated muscles like those of nematodes, molluscs and annelids, nor any unstriated fibres like those of vertebrate guts and blood vessels. Muscle fibres with different properties are often mixed in the same muscle so experiments with intact muscles are apt to give confusing results. Fig. 8.9(a) shows a technique which has been used to study the properties of individual fibres. A crab leg has been cut open to expose a pennate muscle. A small piece of the apodeme has been cut free from the rest and all but one of the attached muscle fibres cut away from it. This piece of apodeme is held in a tiny clamp attached to a force transducer which registers the force when the fibre is stimulated to contract. Two electrodes have been struck into the fibre; current is passed through one to alter the membrane potential of the fibre and the other is used to measure the resulting changes in membrane potential. The fibre only contracts when its membrane potential is increased (i.e. made less negative than its resting value of about -70 mV).

Fig. 8.9(b) shows results from experiments with this apparatus. Records from a series of tests are superimposed: the stimulus was stronger in successive tests. A stimulus which raised the membrane potential only slightly caused no contraction. However, when the membrane potential was raised by more than about 20 mV the fibre developed tension. The further beyond this threshold the membrane potential was taken, the greater the force registered by the transducer.

Some of the muscle fibres developed tension faster than others, at the same membrane potential. A few transmitted action potentials when the membrane potential passed a threshold. However, the responses shown in Fig. 8.9(b) are typical of a great many arthropod muscles. Properties like this make possible a system for controlling muscle force which is quite unlike the system used in mammals.

In mammals each skeletal muscle has a large number of motor axons, each serving a different group of muscle fibres. Each muscle fibre is served by just one axon. If a small force is required, action potentials are sent to the muscle along only a few of the axons and only a few of the muscle fibres are activated. If a large force is required action potentials are sent along all the axons and all the fibres contract.

In arthropods each muscle receives only a few motor axons. Often there is only one motor axon which serves all the muscle fibres, and when there are several they may all serve all the muscle fibres. Fig. 8.10 shows the results of an experiment which indicates how action potentials in a single motor axon can cause different intensities of contraction. The nerve to a crab leg muscle was stimulated by a series of short

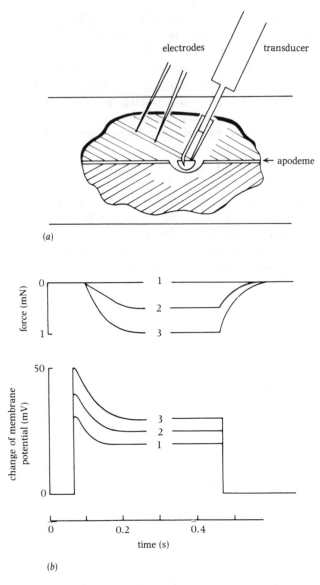

Fig. 8.9.(*a*) Apparatus for measuring the force exerted by a single fibre of a crab leg muscle, in response to changes of membrane potential. (*b*) Records of (above) force and (below) change of membrane potential, from the experiment shown in (*a*). Several records, distinguished by numbers, have been superimposed. After H.L. Atwood, G. Hoyle & T. Smyth (1965). *J. Physiol.* 180, 449–482.

electric shocks, each of which made an action potential travel along the single motor axon. These set up excitatory post-synaptic potentials (EPSPs) in the muscle fibres. When they came at low frequencies each EPSP had died away by the time the next occurred. At higher frequencies each arrived before the previous one had died away and the membrane potential rose by a staircase effect. The higher the frequency of stimulation (within limits) the higher the membrane potential rose and the greater

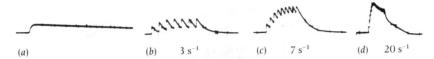

Fig. 8.10. Records of membrane potential in a muscle fibre in a leg muscle of a crab (*Chionoecetes*). The motor axon was stimulated by (*a*) a single shock and (*b–d*) shocks at the frequencies shown. Potential is shown on the same scale in each case but the time scales vary. From H.L. Atwood (1965). *Comp. Biochem. Physiol.* **16**, 409–426.

the force the muscle must have exerted. The crab can control the force by adjusting the frequency of action potentials in the motor axons. In muscles with several motor axons control may be more subtle, with action potentials in different axons producing different sized EPSPs in the same muscle fibre.

As well as motor axons, there are inhibitory axons serving most arthropod muscles. Stimulation of the motor axon alone makes the muscle develop tension. Stimulation of the inhibitory axon alone has no apparent effect. However, if the inhibitory axon is stimulated while the motor axon is being stimulated, the muscle relaxes its tension.

The synapses between motor nerves and muscles work as follows. Action potentials in the axon cause the release of a transmitter substance at the synapse. This substance has an effect on the muscle cell membrane; it increases the permeability of the membrane to sodium ions so the membrane potential rises towards the Nernst potential for sodium (section 2.7). This rise in membrane potential is the EPSP.

Action potentials in an inhibitory axon cause release of a different transmitter substance at the inhibitory synapse. The effect of this substance is to increase the permeability of the muscle cell membrane to chloride, moving the membrane potential towards the Nernst potential for chloride. This is lower (more negative) than the threshold for tension development, so action potentials in the inhibitory axon tend to eliminate tension.

There is evidence that the transmitter substance at crustacean motor nerve–muscle synapses is glutamic acid ($HOOC.CH_2.CH_2.CHNH_2.COOH$) and at inhibitory ones γ-aminobutyric acid ($HOOC.CH_2.CH_2.CH_2NH_2$). The evidence includes the results of experiments in which minute quantities of the substances were injected into synapses.

8.4.   Compound eyes

Most arthropods have compound eyes like the one illustrated in Fig. 8.11. The surface of the eye has a large number of facets, each of which is the end of a separate sensory unit, an ommatidium. The very large eyes of some dragonflies (Odonata) have as many as 28 000 ommatidia.

Crabs and many insects have cells filled with pigment separating each ommatidium from the next. Light travelling down the long axis of an ommatidium reaches the sensory cells at its inner end and stimulates them, but light that enters

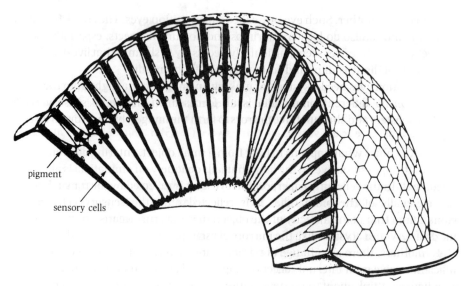

pigment

sensory cells

Fig. 8.11. A cut-away diagram of a compound eye from an insect. From V.B. Wigglesworth (1964). *The life of insects.* Weidenfeld & Nicolson, London.

obliquely strikes the pigment and is absorbed. Each ommatidium points in a slightly different direction and senses light only from that direction. The amount of detail that the animal can see in its surroundings depends on the number of ommatidia.

Bees (*Apis*) have eyes like this. The outer surface of the eye is an incomplete sphere of radius 1.2 mm (1200 μm), and the diameter of the facets is 21 μm, so each ommatidium is set at an angle 21/1200 = 0.018 radians (1.0°) to the next. The eye cannot distinguish objects separated by angles less than this: a bee viewing a page 30 cm away could not distinguish stripes less than 5 mm wide. This makes it much inferior to the octopus eye (section 6.8), which can distinguish stripes only 0.3° wide. Could the bee eye be improved by giving it more ommatidia, of smaller diameter?

The resolution of the eye is limited by diffraction, which makes it impossible for light of wavelength λ passing through an aperture of diameter d to form separate images of objects less than about λ/d radians apart. Blue light of wavelength 0.4 μm entering an ommatidium of diameter 21 μm cannot form separate images of objects less than 0.019 radians apart. This is about the same as the angle between adjacent ommatidia, so the bees' ommatidia are close to the ideal diameter for detailed vision. If the ommatidia were bigger (in the same size of eye) the bee would see less well because the angle between them would be larger: if they were smaller the bee would see less well because the effect of diffraction would be worse. (The wavelength used in the calculation is at the blue end of the visible spectrum for humans but lies well within the visible spectrum for bees, which extends into the ultra-violet.)

The bee's kind of eye, with pigment separating the ommatidia, is called an apposition eye. It works well in bright light but is poor in dim light because a lot of the light that falls on it is absorbed by the pigment between the ommatidia, and wasted. Only light travelling along the axis of an ommatidium reaches the sensory cells.

A different type of compound eye wastes far less of the light. It has no pigment between the ommatidia, and light that enters one ommatidium obliquely goes to the sensory cells of another. Such eyes are called superposition eyes. The crayfish *Astacus* has them, and so also do lobsters, shrimps and nocturnal insects, especially moths and beetles. Sensitive eyes seem particularly necessary to shrimps that live deep in the sea, where the light is dim, and to insects that fly by night.

Superposition eyes need a system for making an image, so that light from the same source entering different ommatidia is all focussed on the same sensory cells. Crayfish, lobsters and shrimps have mirrors to form the images, but nocturnal insects have lenses.

Fig. 8.12 shows what happens to light from a small object, entering different kinds of eye. In the apposition eye (*a*) the light that falls obliquely on ommatidia is absorbed by pigment. In the reflecting superposition eye (*b*) it is reflected by mirror surfaces in the side walls of the ommatidia: it is reflected onto the sensory cells of the ommatidium that points directly at the object. In some crustaceans, the mirrors are stacks of very thin crystals, like the mirrors of fish scales that are described in section 14.4. In others, there is a change of refractive index at the edges of the ommatidia, sufficient to give total internal reflection. Fig. 8.12(*b*) is a two-dimensional diagram but when you think about the system in three dimensions it becomes apparent that it will work only if the facets of the eye are square. Reflecting superposition eyes have square facets but insect eyes generally have hexagonal facets, as shown in Fig. 8.11.

Fig. 8.12(*c*) shows a refracting superposition eye such as nocturnal insects have. Each ommatidium has a lens at its outer end, which bends rays that enter it obliquely. No homogeneous lens would bend light in quite the way that the diagram shows, but these lenses are not homogeneous: observations with interference microscopes show that the refractive index is higher along the axis of the lens than round the edges.

Fig. 8.12(*d*) shows a simple (non-compound) eye, such as squids and vertebrates have. Notice that the object in the upper half of the field of vision forms an image on the lower half of the retina: the image is upside-down. In contrast, the superposition eyes form images the same way up as the object.

8.5. Settlement of barnacle larvae

Many of the Crustacea in the plankton are larvae of species which spend their adult life on the shore and sea floor. If they are to mature and flourish as adults they must settle on a suitable patch of sea floor. Crabs and other crustaceans which walk about on the sea floor as adults have some margin for error. Barnacles which settle on a particular spot and stay there must find a good spot at the outset. The planktonic larvae of molluscs and sedentary polychaetes must similarly find suitable places to settle, but discussion of the problem has been deferred to this chapter because Professor Dennis Crisp has made a particularly enlightening study of settlement by barnacle larvae.

The barnacle *Balanus balanoides* flourishes only on rock and similar surfaces between the tidemarks. A larva may make contact with the bottom in all sorts of unsuitable places, on sand or mud, at extreme high tide level, or offshore. The most reliable sign that a surface is suitable is that there are barnacles living there already.

In their experiments, Professor Crisp and a colleague offered barnacle larvae small

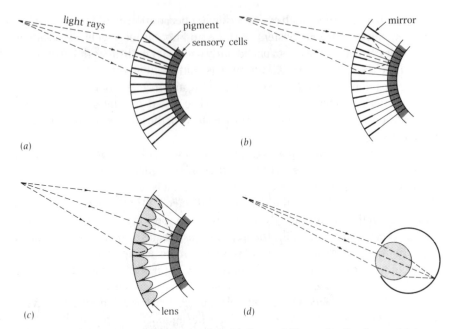

Fig. 8.12. Diagrams showing paths of light in different kinds of eye: (*a*) is an apposition compound eye; (*b*) is a reflecting superposition eye; (*c*) is a refracting superposition eye and (*d*) is a simple (i.e. not compound) eye.

rectangles of slate to settle on. Hollows were ground in the slate to make them more attractive. Slates treated in different ways were laid out round the perimeter of a large trough of sea water. Barnacle larvae were released into the sea water. The trough was rotated slowly so that factors such as uneven lighting should not make the larvae congregate on one side. After a few hours the slates were removed and the numbers of young barnacles which had settled on each were counted.

Before the experiments, all the slates were thoroughly clean. Some were then soaked for a few hours before use in an extract of barnacles, made by chopping and grinding adult barnacles and removing the solid debris by filtration. In a series of experiments larvae were given the choice of equal numbers of clean slates and ones treated with barnacle extract. One hundred settled on the clean slates and 2197 on the treated ones. Some substance present in adult barnacles apparently encourages settlement.

Further experiments showed that extracts of the mollusc *Mytilus* had little or no effect on settlement but that extracts of crabs (*Carcinus*) and cockroaches (*Blaberus*) were very effective. The substance which encourages settlement is apparently present in arthropods in general, not just in barnacles. However, it is not identical in all arthropods: the barnacles *Balanus* and *Elminius* are each more sensitive to extracts of their own species than to extracts of the other.

Extracts of whole crab or of crab carapace favour settlement but extracts of crab viscera, blood or muscle do not. The active substance must be present in the cuticle. It was shown that its molecule was too large to pass through a cellulose filter with pores of diameter 5 nm. It is probably a cuticle protein.

Some other planktonic larvae find suitable sites for settlement by other means. For instance, *Spirorbis borealis* is a polychaete which spends its adult life like *Serpula* (Fig. 7.1*b*) in a permanently fixed tube, which it builds on a frond of the seaweed *Fucus*. The larvae seem to be able to identify *Fucus* by chemical means.

Further reading

Introduction

Bliss, D.E. (ed.) (1982–   ). *The biology of Crustacea*. Academic Press, New York.
Fincham, A.A. & Rainbow, P.S. (eds.) (1988). Aspects of decapod crustacean biology. *Symp. Soc. exp. Biol.* 59. (386 pp.) Cambridge University Press.
Lockwood, A.P.M. (1968). *Aspects of the physiology of crustacea*. Oliver & Boyd, Edinburgh.

The exoskeleton

Currey, J.D. (1967). The failure of exoskeletons and endoskeletons. *J. Morph.* **123**, 1–16.
Dendinger, J.E. & Alterman, A. (1983). Mechanical properties in relation to chemical constituents of postmolt cuticle of the blue crab, *Callinectes sapidus. Comp. Biochem. Physiol.* A**75**, 421–424.
Greenaway, P. (1985). Calcium balance and moulting in the Crustacea. *Biol. Rev.* **60**, 425–454.
Hadley, N.F. (1986). The arthropod cuticle. *Scient. Am.* **255**(1), 98–106.
Mykles, D.L. (1980). The mechanism of fluid absorption at ecdysis in the American lobster, *Homarus americanus. J. exp. Biol.* **84**, 89–101.
Vincent, J.F.V. & Ablett, S. (1987). Hydration and tanning in insect cuticle. *J. Insect Physiol.* **33**, 973–979.

Muscles

Atwood, H.L. (1972). Crustacean muscle. In *The structure and function of muscle*, 2nd edn (ed. G.H. Bourne), vol. 1, pp. 421–489. Academic Press, New York.
Burrows, M. & Horridge, G.A. (1968). The action of the eyecup muscles of the crab *Carcinus* during optokinetic movements. *J. exp. Biol.* **49**, 223–250.
Florey, E. (1975). The integrative capacity of chemical transmission at arthropod neuromuscular synapses. In *Simple nervous systems* (ed. P.N.R. Usherwood & D.R. Newth), pp. 323–341. Arnold, London.
Tse, F.W., Govind, C.K. & Atwood, H.L. (1983). Diverse fibre composition of swimming muscles in the blue crab, *Callinectes sapidus, Can. J. Zool.* **61**, 52–59.

Compound eyes

Land, M.F. (1980). Compound eyes: old and new optical mechanisms. *Nature* **287**, 681–686.

Settlement of barnacle larvae

Crisp, D.J. & Meadows, P.S. (1962). The chemical basis of gregariousness in cirripedes. *Proc. R. Soc. Lond.* B**156**, 500–520.

Crisp, D.J. & Meadows, P.S. (1963). Adsorbed layers: the stimulus to settlement in barnacles. *Proc. R. Soc. Lond.* B1**58**, 364–387.

Yule, A.B. & Crisp, D.J. (1983). Adhesion of cypris larvae of the barnacle, *Balanus balanoides*, to clean and arthropodin treated surfaces. *J. mar. biol. Ass. U.K.* **64**, 429–439.

9 Insects

Phylum Uniramia, subphylum Hexapoda,
   Class Collembola (springtails)
   Class Insecta,
     Subclass Apterygota (silverfish, etc.)
     Subclass Pterygota,
       Order Ephemeroptera (mayflies)
       Order Odonata (dragonflies)
       Order Dictyoptera (cockroaches)
       Order Isoptera (termites)
       Order Orthoptera (grasshoppers)
       Order Hemiptera (bugs)
       Order Diptera (flies)
       Order Siphonaptera (fleas)
       Order Lepidoptera (moths & butterflies)
       Order Coleoptera (beetles)
       Order Hymenoptera (ants, wasps, bees)
and other classes and orders

9.1. Introduction

About three quarters of all known animal species are insects, but they are all fairly similar to each other and it seems possible to give a general account of them in just one chapter.

Fig. 9.1 shows a selection of adult insects. The differences between them are obvious, but so also are some basic similarities. The body is enclosed in an exoskeleton constructed on the same principles as in crustaceans, but not impregnated with calcium salts; it consists simply of chitin and tanned protein with a waxy outer layer. The body has three regions: head, thorax and abdomen. The head bears a pair of antennae, several pairs of mouthparts (which will be described later) and, usually, a pair of compound eyes. It shows no obvious division into segments, but develops from the first six segments of the embryo. The thorax consists of three segments, each of which bears a pair of legs. In addition, the second and third thoracic segments usually each have a pair of wings. The abdomen has up to 11 segments and no limbs, but some insects have two or three tail-like appendages at its posterior end (see Fig. 9.6).

When they walk, most insects move their six legs in groups of three (the first and third legs of one side and the second leg of the other). This ensures that there are always three feet on the ground, forming a stable tripod.

Fig. 9.2 shows the internal arrangement of a cockroach, a reasonably typical

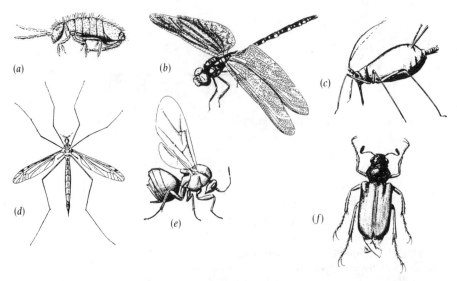

Fig. 9.1. A selection of adult insects: (*a*) a springtail (*Entomobrya laguna*, Collembola, body length 7 mm); (*b*) a dragonfly (*Aeschna multicolor*, Odonata, 70 mm); (*c*) an aphid (*Macrosiphum rosae*, Hemiptera, 3 mm); (*d*) a crane fly (*Tipula* sp., Diptera, 20 mm); (*e*) a gall wasp (*Andricus californicus*, Hymenoptera, 4 mm); and (*f*) a beetle (*Dichelonyx backi*, Coleoptera, about 10 mm). From J.A. Powell & C.L. Hogue (1979). *California insects*. University of California Press, Berkeley.

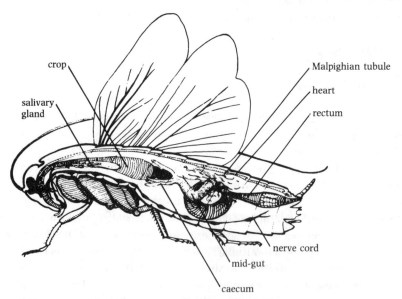

Fig. 9.2. The internal anatomy of a cockroach. After V.B. Wigglesworth (1964). *The life of insects*. Weidenfeld & Nicolson, London.

insect. Notice the similarity to crayfish (Fig. 8.1). The main body cavity is haemocoel. There is a nerve cord ventral to the gut and a heart dorsal to it. The nerve cord has a ganglion in each segment of the thorax and abdomen. In the head it encircles the oesophagus: there is a brain dorsal to the oesophagus and a sub-oesophageal

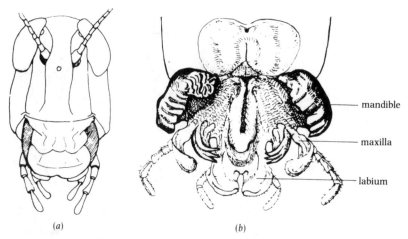

Fig. 9.3. Mouthparts of a grasshopper. (*a*) An anterior view of the head. (*b*) The labrum has been folded dorsally and the mouthparts separated to expose the mouth. After V.B. Wigglesworth (1964). *The life of insects.* Weidenfeld & Nicolson, London.

ganglion ventral to it. Many insects have a short heart like the crayfish but the cockroach has a heart almost as long as the thorax and abdomen, with a pair of ostia in each segment. There is only one artery, which runs forward from the heart and opens into the haemocoel of the head.

We will examine the mouthparts before the gut. Fig. 9.3 shows the mouthparts of a grasshopper, which are very similar to those of cockroaches. The most anterior mouthparts are a pair of biting mandibles. Next are a pair of complicated maxillae, which are used for manipulating food. Finally there is the labium, which is a pair of second maxillae fused together to form a lower lip. There is a pair of large salivary glands (Fig. 9.2) which discharge saliva through an opening in the labium. The saliva contains an enzyme which digests starch, and is spread on the food before it is ingested. Cockroaches feed on crumbs of food and other waste matter left around by people: the natural food is probably dead animals. Grasshoppers, with very similar guts and mouthparts, eat vegetation.

The anterior part of the gut is a large crop where food is stored and some initial digestion takes place. The food in the crop is mixed with saliva and also with other digestive enzymes, which get squeezed forward from the mid-gut. The crop is followed by the gizzard, which is muscular and has its lining of cuticle thickened to form teeth. Food is broken up in the gizzard by the action of the muscle and teeth. The mid-gut is the only part of the gut which is not lined by cuticle, but even here the epithelium is protected from abrasion by a loose chitinous membrane. Enzymes secreted by the mid-gut epithelium diffuse freely through this membrane into the lumen, and products of digestion diffuse out through the membrane to be absorbed by the epithelium. In the rectum, water and inorganic ions are absorbed from the gut contents before they are passed as faeces.

At either end of the mid-gut, blind-ended tubes are attached. The relatively thick caeca at the anterior end are simply extensions of the mid-gut surface: their epithelium secretes enzymes and absorbs products of digestion. The much thinner

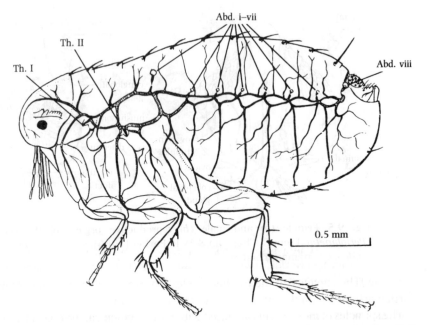

Fig. 9.4. The tracheal system of a flea (*Xenopsylla*). Th I, II, thoracic spiracles; Abd. i–viii, abdominal spiracles. From V.B. Wigglesworth (1935). *Proc. R. Soc. Lond.* B118, 397–419.

Malpighian tubules are excretory organs. Their function and the function of the rectum are discussed later in the chapter, in section 9.4.

Some quite conspicuous organs are omitted from Fig. 9.2. They include the fat body, a diffuse gland dorsal to the gut. In function it parallels the mammalian liver. Globules of fat and granules of glycogen and protein are stored in its cells. It is the major site of intermediary metabolism. For instance, it converts glucose absorbed from the gut to the disaccharide trehalose which is the principal sugar in the blood and one of the principal fuels for flight.

Fig. 9.2 also omits the muscles, the respiratory system and the gonads. Among the muscles, the ones which flap the wings are particularly large. They are described in section 9.2. The respiratory system is a branching system of air-filled tubes, quite different from anything encountered in earlier chapters. The tubes are called tracheae, and open to the atmosphere through holes called spiracles in the sides of the thorax and abdomen. Fig. 9.4 shows the arrangement of the tracheal system of a flea. The tracheae permeate all the tissues of the body. They are lined by cuticle, which is shed when the insect moults. This cuticle is strengthened by thickenings, either rings or helices, which prevent the tracheae from collapsing while leaving them flexible. The finest branches of the tracheae, called tracheoles, have diameters of the order of 1 μm.

Oxygen diffuses from the atmosphere through the spiracles and tracheae to the tracheoles and from them to the tissues. In many insects parts of the tracheae are enlarged to form thin-walled collapsible sacs, and diffusion is supplemented by

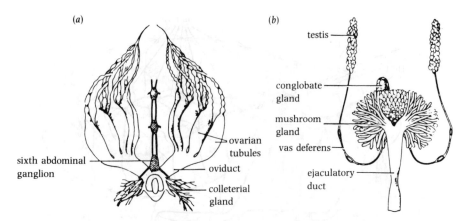

Fig. 9.5. Female (*a*) and male (*b*) reproductive organs of the cockroach (*Periplaneta*). From G. Chapman & W.B. Barker (1966). *Zoology for intermediate students.* Longman, London.

pumping. The capacity of the tracheal system to supply oxygen to the tissues, and particularly to the flight muscles, is discussed in section 9.3.

The spiracles of most insects are fitted with valves which can be opened and closed by muscles as required, to control airflow or restrict water loss.

Male insects have a pair of testes and females a pair of ovaries (Fig. 9.5). They lie in the haemocoel of the abdomen and ducts lead from them to an external opening near the posterior end of the abdomen. The reproductive organs include accessory glands as well as the gonads themselves.

Cockroaches lay their eggs in batches of 12 or more enclosed in an egg capsule. The capsule is soft and white when first formed, but later becomes hard and dark and eventually almost black. It is secreted by the female accessory glands (the colleterial glands, Fig. 9.5). The left gland is the larger. Its contents are white and can be kept indefinitely without change of colour, but they darken when mixed with the (clear) fluid from the right gland, or with compounds such as 3,4-dihydroxybenzoic acid:

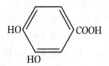

Analysis has shown that the glands contain the following substances:

left	*right*
protein	glucosidase
glucoside of 3,4-dihydroxybenzoic acid	
phenol oxidase	

When the contents of the two glands are mixed the glucosidase (an enzyme) breaks down the glucoside into its components, glucose and 3,4-dihydroxybenzoic acid. The acid is oxidized by the phenol oxidase to a quinone, which tans the protein as explained in the account of gorgonin (section 3.5). This was the first case of quinone

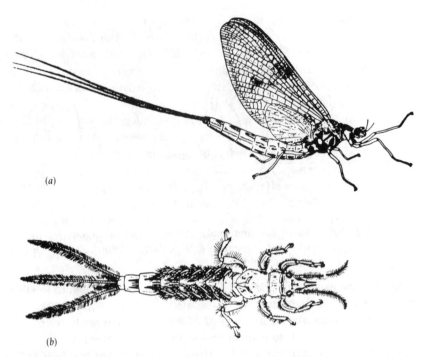

Fig. 9.6. (*a*) Adult and (*b*) nymph of the mayfly *Ephemera vulgata* (order Ephemeroptera). Each is about 2 cm long, excluding 'tails'. From W. Engelhardt & H. Merxmuller (1964). *The young specialist looks at pond life.* Burke, London.

tanning in nature to be recognized and explained. Insect cuticle consists of tanned protein and chitin but cockroach egg capsules consist of tanned protein alone.

Though cockroaches lay their eggs in batches, many other insects lay their eggs singly.

Insects hatch from their eggs as larvae or nymphs. They moult several times, growing between moults. At each moult their structure changes to a greater or lesser extent until they eventually reach the adult stage: thereafter, they do not moult. In some insects the changes from stage to stage are relatively small, in which case the young stages are called nymphs. Each nymphal stage is a little different from its predecessor in shape and only the adult has fully developed wings and reproductive organs. Cockroach and grasshopper nymphs look like tiny adults, apart from differences of body proportions: they have relatively large heads and short abdomens, and rudimentary wings. Mayfly nymphs are more different from their adults (Fig. 9.6) and have to be: they live in water but the adults fly in the air. The feather-like structures along the sides of the abdomen in Fig. 9.6(*b*) are gills. Many other insects have immature stages called larvae that are very unlike the adults and undergo a profound metamorphosis. Adult butterflies are quite unlike their caterpillars and adult flies are quite unlike their maggots. In such extreme cases an inactive non-feeding stage (the pupa) intervenes between the final larval stage and the adult. During the pupal stage the whole insect is re-modelled.

The remainder of this section is a brief review of some of the more important orders of insects.

Collembola (springtails, Fig. 9.1*a*) are probably not very closely related to the other insects. Most of them are much smaller than the one in the illustration, only 1–3 mm long. They are extremely common in soil and leaf litter. Many of them have two long projections on the abdomen, like the prongs of a fork. They are normally kept folded forward (as in Fig. 9.1*a*) but can be swung downwards and backwards very rapidly to make the animal jump.

Ephemeroptera (mayflies, Fig. 9.6) have quite a long life as nymphs in ponds and streams but their adult, aerial life is very short indeed, often less than a day. The nymphs feed on dead plant material and the adults do not feed. Though the nymphs live in water, they have tracheae like terrestrial insects. The gills have very thin cuticle, and have tracheoles looping through them. Oxygen dissolved in the water diffuses through the cuticle into the air-filled tracheoles, and carbon dioxide diffuses in the opposite direction. Diffusion occurs through other parts of the body surface, as well as the gills. The gills flap up and down (faster in water that contains little dissolved oxygen), keeping the water moving over themselves and the rest of the body.

Odonata (dragonflies, Fig. 9.1*b*) also have aquatic larvae and aerial adults, but in their case both are carnivorous. The adults have large eyes and are active by day, feeding on smaller insects which they catch in flight.

Hemiptera (bugs) have sucking mouthparts. Some suck the blood of other animals but the example shown in Fig. 9.1(*c*) is an aphid, a member of a large group of bugs which suck the juices of plants. This particular one feeds on roses and is very troublesome to gardeners. The spike projecting downwards from the head is the bundle of mouthparts. The mandibles and maxillae are long slender stylets, and the labium forms a sheath for them. Only the stylets pierce the plant: the labium folds out of the way as they are driven in. The two maxillary stylets interlock to form a pair of tubes, one to carry saliva into the plant and the other to carry the sap into the mouth. The saliva contains an enzyme (pectinase) which probably attacks the middle lamella between adjacent plant cells, clearing a path for the stylets. Most aphids insert their stylets into phloem which contains sap under pressure, making sucking unnecessary: the pressure drives the sap into the aphids mouth. This method of feeding has made aphids very important vectors of disease. Sixty-five percent of known plant viruses are transmitted from plant to plant by aphids.

The aphid shown in Fig. 9.1(*c*) is a wingless female, but winged offspring, capable of flying to another plant, are produced in crowded conditions. Most aphid populations consist only of females during the summer. They reproduce parthenogenetically, which makes very fast reproduction possible, as in rotifers (section 5.1). It is particularly fast because the females of many species contain embryos at birth and start giving birth themselves only 10 days after being born. (Parthenogenetically produced young are born as nymphs, not laid as eggs.) If the mean generation time is 14 days and each female produces 50 offspring (these are reasonably typical figures) the population is potentially capable of multiplying by a factor of 50 in 14 days or 2500 in 28 days, a very high rate for so advanced an organism. Males may appear in autumn in which case sexual reproduction occurs then, producing eggs which do not hatch until the spring.

The Diptera (flies) have only one pair of wings. The posterior pair is represented by a pair of small club-shaped structures, the halteres, which can be seen in the picture of a crane fly (Fig. 9.1*d*). They look useless, but flies cannot fly without them. Flies with their halteres cut off cannot fly a straight course and are liable to fall on their backs. They can fly again if a thread is attached to the abdomen so as to trail behind in flight. The thread tends to stabilize flight, keeping the fly flying straight, just as feathers at the rear of an arrow stabilize its flight. Thus the halteres seem to have an important function in the control of flight. They vibrate at the same frequency as the wings and act like gyroscopes which keep reversing their direction of rotation. If the insect deviates from a straight path torques must act on the halteres, like the torques that act when you try to turn the axis of a spinning gyroscope to a new direction. The halteres are far too small to have a direct stabilizing effect (like the effect of gyroscopic stabilizers in ships) but they have sense organs in them which detect the torques, as has been shown by electrical recording from their nerves. They serve as rotation detectors, similar in function to the semicircular canals of vertebrate ears (section 12.6) though quite different in mechanism. They control the reflexes that make stable flight possible. Neither haltere can detect rotations about its own axis of vibration but the axes of the left and right halteres are not parallel to each other and the two between them can detect rotation about any axis.

Diptera have mouthparts adapted for drinking liquid food. House flies (*Musca domestica*) have a grooved pad on the end of the labium, with the grooves converging on the food canal that leads to the mouth. Mosquitoes have mouthparts superficially similar to those of aphids and use them for drinking nectar and other sugar solutions. Some feed only on these foods and must draw on reserves accumulated during larval life for the proteins needed for egg production. Others, such as *Aedes aegypti*, need a meal of blood before producing each batch of eggs. They are vectors of animal disease, just as aphids are of plant diseases. *Aedes* transmits the yellow fever virus and other mosquitoes are vectors of the protozoan that causes malaria and the nematodes that cause elephantiasis.

Most Hymenoptera (ants, bees and wasps) have the first segment of the abdomen joined to the thorax in such a way as to seem part of it. The 'waist' in the example shown in Fig. 9.1(*e*) is not between thorax and abdomen, but in the second abdominal segment. This example is a gall wasp: it injects its eggs into oak twigs which swell up as galls ('oak apples'), which house the developing larvae. Wasps and ants have biting mouthparts like those of grasshoppers (Fig. 9.3) but bees have only small mandibles and a long deeply grooved labium used for drinking nectar. Ants are social insects, living in highly organized colonies, as also are many species of bees and wasps. This way of life is discussed in section 9.7.

The Coleoptera (beetles) is the largest order in the animal kingdom. About 0.3 million species have been described, out of 1.0–1.5 million species of animal. The forewings of beetles are stiff elytra (singular: elytron). They are not flapped in flight but protect the delicate hindwings when the insect is resting. The hindwings are generally longer than the elytra but can be folded so as to be completely covered: they are not properly folded, and are projecting slightly, in Fig. 9.1(*f*). Beetles have biting mouthparts both as larvae and as adults. Most of them eat plant material and some are destructive pests, but most ladybirds (Coccinellidae) eat aphids and other insects. A ladybird (*Rodolia cardinalis*) was used in the late nineteenth century in the first successful case of biological control of a pest. The cottony-cushion scale (*Icerya purchasi*, Hemiptera) had been imported accidentally into California from Australia. It had multiplied enormously and had become a serious threat to the citrus fruit industry. It is not a pest in Australia, where it is preyed on by *Rodolia*. *Rodolia* was introduced into California and the epidemic was checked.

Only a few of the major groups of insects have been introduced in this section. The Lepidoptera (butterflies and moths) have been omitted, but appear in section 9.6, where their colours are discussed.

9.2. Flight

Insects are the only flying invertebrates. Their wings develop as outgrowths of the body wall and consist of two cuticle-covered membranes, back to back. Most of the area of the wing is very thin but it is stiffened by a network of tubular veins. The cavities of the veins are connected to the haemocoel and are filled with blood. They also contain tracheae and sometimes nerves. Even the veins are quite thin, too thin to make a flat wing sufficiently stiff. Insect wings are lightly pleated, with pleats running from the base of the wing towards the tip. This stiffens them, just as a sheet of paper can be stiffened by pleating.

Many insects hover like helicopters, keeping themselves stationary in mid air by beating their wings. Moths such as *Manduca* (Fig. 9.7) hover in front of flowers at night, extending their long tongues into them to feed on the nectar. Fig. 9.7(*a*) shows a hovering *Manduca* in side view and Fig. 9.7(*b*) shows one from above. The film (*b*) was taken in a laboratory, of a moth taking sugar solution from an artificial flower. The camera had to be run at a high framing rate because *Manduca* beats its wings 28 times per second.

To understand insect flight, we need to know something about the aerodynamic forces that act on wings. Any body moving through air or water suffers drag, a component of force that acts backwards along the direction of motion. Aerofoils such as insect or aeroplane wings and hydrofoils such as squid fins (Fig. 6.22) are acted on also by lift, a component of force at right angles to the direction of motion. The lift depends on the angle of attack (Fig. 9.7*c*) at which the aerofoil is tilted, relative to its direction of motion. It is much larger than the drag when well-designed aerofoils and hydrofoils are given suitable angles of attack, of 10–20°.

The pictures show that when *Manduca* hovers, it beats its wings forward and back in a horizontal plane. The wings have their dorsal side uppermost in the forward stroke, but turn upside-down so that their ventral surfaces (shown black in (*b*)) are uppermost in the backward stroke. Fig. 9.7(*d*) shows how this keeps the insect airborne. The wings act as aerofoils so lift and drag forces act on them. Both in the forward stroke and in the backward one they have an angle of attack such that the resultant of lift and drag acts upwards. This effect could be achieved without turning the wings upside-down, but turning them over ensures that the anterior edge of the forewing is the leading edge for the backward stroke as well as for the forward one. This edge is stiffer than the remainder of the wings because it has closely spaced veins and rather deep pleating. If the leading edge were not stiff the wing would flutter

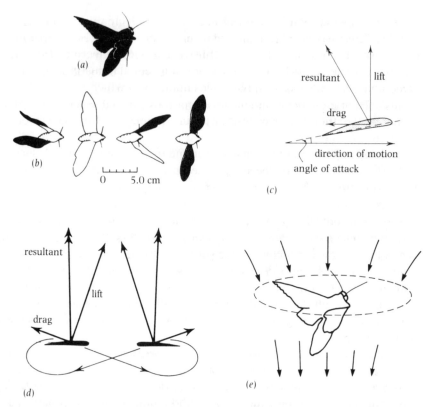

Fig. 9.7. Hovering flight of the Florida tobacco hornworm moth, *Manduca sexta.*
Part (*a*) was drawn from a flash photograph. Part (*b*) was traced from a cine film and
shows the moth from above; successive outlines show it at intervals of 12 ms. The
ventral surfaces of the wings are shown black. Part (*c*) shows the forces on a moving
aerofoil (see in section) and (*d*) shows how the backward and forward strokes of a
hovering moth's wings both produce upward forces. (*e*) shows the downward jet of
air that is driven by the beating wings. Parts (*a*) and (*b*) are from T. Weis-Fogh
(1973). *J. exp. Biol.* 59, 169–230.

uncontrollably. This technique of hovering is similar in principle to the hovering of
helicopters. However, a helicopter rotor keeps turning in the same direction while
insect wings move alternately forward and back.

Insect wings and helicopter rotors produce lift by driving air downwards: they
push down on the air and the air pushes up on them. Fig. 9.7(*e*) shows the downward
jet of air produced by a hovering insect. Newton's second law of motion tells us that,
to keep itself airborne, the insect must give downward momentum to the air at a rate
equal to its own weight (see section 6.5).

When a helicopter is hovering the rotor needs only to produce a vertical force.
When the helicopter flies forward drag acts on the fuselage so the force produced by
the rotor must have a forward component. This is obtained by tilting the rotor
forward. Similarly insects change from hovering to forward flight by altering the
plane of beating of their wings.

Aeroplane wings and helicopter blades have relatively thick, streamlined cross-sections. Insect wings are thin pleated membranes. A magnified insect wing would perform very badly indeed as an aeroplane wing or a helicopter rotor. However, these aerofoils operate at much higher Reynolds numbers than insect wings. Tests with models have shown that at the Reynolds numbers at which insect wings operate, pleated membranes are just as good as smooth streamlined aerofoils. They have the advantage that they can be made light: the importance of lightness will soon become apparent.

Hovering requires a great deal of power and therefore a high metabolic rate. The moth *Manduca* (Fig. 9.7) has been persuaded to hover for periods of several minutes in a large (9 litre) jar. Samples of the air in the jar were analysed beforehand and afterwards, using a paramagnetic oxygen analyser (see section 1.2). It was found in this way that the rate of oxygen consumption during hovering is about 50 cm^3 oxygen (g body mass)$^{-1}$ h^{-1}. This is about a hundred times faster than when the moth is resting. Similar metabolic rates are achieved by flying birds and bats but not, so far as is known, by any other animals (see Fig. 4.9b). The metabolic rates of insects in flight are particularly remarkable because they are almost entirely due to metabolism in the wing muscles, which represent 20–35% of the body mass, and so must be using oxygen at rates around 200 cm^3 g^{-1} h^{-1}. No other tissue in any animal is known to use oxygen as fast.

Metabolism using 1 cm^3 oxygen releases 20 J energy, so *Manduca*'s oxygen consumption represents a power consumption of 1000 J g^{-1} h^{-1} or 280 W kg^{-1}: 0.56 W for a 2 g moth. Some of the power is used to give kinetic energy to the air in the wake, but much of it becomes heat. The moth's thorax warms up until the rate of loss of heat from it equals the rate of heat production by the muscles. Indeed, a cold *Manduca* cannot fly until it has spent two or three minutes contracting its wing muscles, generating heat and warming them up. The temperature of the thorax of *Manduca* has been measured immediately after bouts of hovering, by sticking a thermocouple into it. When the temperature of the laboratory was 25° C, the temperature of the thorax reached 40°C, well up in the temperature range of the 'warm-blooded' birds and mammals. Like other moths, *Manduca* has a fur-like covering of scales on its thorax that serve as heat insulation, making it get hotter than it would if it were nude. The advantages of a high body temperature are discussed in section 16.4, in the chapter on reptiles.

Now we will consider why the power is needed. Mechanical power is used, in insect flight, in two main ways. First there is the aerodynamic power needed to overcome the drag on the wings. The major part of this is needed to give kinetic energy to the air that is driven downwards, and can be calculated by using a standard equation from helicopter theory. The power P_a used in this way to support a helicopter of weight W, with a rotor of radius r, in air of density ρ is

$$P_a = (W^3/2\rho\pi r^2)^{\frac{1}{2}} \tag{9.1}$$

This equation tells us that the bigger the radius of the rotor of a helicopter (or the longer the wings of an insect) the less aerodynamic power is needed. This is because a larger rotor can produce the required upward force by accelerating a larger quantity of air to a lower velocity: we have already met this principle in section 6.5, in the discussion of squid swimming.

Secondly, there is the inertial power needed to give kinetic energy to the wings. Just as the kinetic energy of a body in linear motion is $\frac{1}{2}$(mass) (velocity)^2, so the kinetic energy of a rotating body is $\frac{1}{2}$(moment of inertia) (angular velocity)^2. Let the wings have moment of inertia I (this is the total of the moments of inertia of all the wings about their bases), and let them be accelerated to angular velocity ω in every stroke. They are given kinetic energy $\frac{1}{2}I\omega^2$ in every stroke and, if the wing beat frequency is f, there are $2f$ strokes in unit time. The inertial power P_i is given by

$$P_i = \tfrac{1}{2}I\omega^2 . 2f = I\omega^2 f. \tag{9.2}$$

This is likely to be larger for long wings than for short ones because longer wings will generally have larger moments of inertia. Increasing the length of the wings reduces the aerodynamic power needed for flight but increases the inertial power, and there is probably an optimum wing length that minimizes the power requirement.

The quantities on the right hand sides of equations 9.1 and 9.2 have all been measured for *Manduca sexta*. (The wingbeat frequency was measured by using a stroboscope to illuminate a hovering moth in an otherwise darkened room. The frequency of flashing of the stroboscope was adjusted until flashes were occurring at the same stage of successive wingbeat cycles, making the wings appear stationary.) It was estimated in this way that the aerodynamic power required for hovering by a 2 g *Manduca* was 30 mW and the inertial power 72 mW, a total of 102 mW. These are *mechanical* power requirements, and we have already noted that the *metabolic* power consumption is 560 mW. This is a hundred times higher than the resting rate so nearly all of it is needed for flight. The muscles can supply the mechanical power if they work with efficiency $102/560 = 0.18$. Various vertebrate muscles have been shown to work with efficiencies of this order.

However, it is possible that the moth may not have to supply the inertial power. I have assumed so far that the kinetic energy given to the wings at the beginning of each stroke is lost as heat at the end of the stroke. (Similarly, the kinetic energy lost when a car is braked is lost as heat.) Suppose instead that the lost kinetic energy is stored so that it can be re-used in the next stroke. To see how this could happen, imagine a flexible strip of steel such as a hacksaw blade, clamped in a vice. If you twang the free end of the blade it will vibrate at its natural frequency. In the middle of each vibration it moves fastest and has maximum kinetic energy. At the extremes of the vibration it is momentarily stationary: it has no kinetic energy but it is bent and so has elastic strain energy. As the blade vibrates energy is converted back and forth, from kinetic energy to elastic strain energy and vice versa. A little energy is lost (as heat) at each conversion and the vibration gradually dies down, but the whole kinetic energy does not have to be supplied afresh each time. Similarly insect wings may have elastic structures attached to them which take up kinetic energy as elastic strain energy at the end of each stroke, and restore it in an elastic recoil for the next stroke. If so, and if the insect beats its wings at the natural frequency, the mechanical power which the muscles must supply may be little more than the aerodynamic power.

This would make a tremendous difference. It would reduce the mechanical power needed by our 2 g *Manduca* from 102 to 30 mW, and the efficiency with which the muscles would have to work from 0.18 to $30/560 = 0.05$. This efficiency seems low, but there is evidence from research on the energy cost of mammal and bird running that the muscles of small animals work with lower efficiency than those of large ones. Insects are small compared with most vertebrate animals and their muscles may work with low efficiency.

There are several structures in the thoraxes of insects which store elastic strain energy at the end of one wing stroke, and then return it in an elastic recoil. The cuticle that forms the wall of the thorax has elastic properties. In many insects, small parts of the thorax cuticle are made of a remarkable rubber-like protein called resilin: for example, the wings of locusts are flexibly joined to the thorax wall by small blocks of resilin. Finally, the elastic properties of the wing muscles may be important.

The wing muscles of locusts, dragonflies and some other insects are fairly conventional striated muscles. They can deliver a very high sustained power output and to make this possible contain a lot of mitochondria (30% by volume of locust wing muscle). In other respects they are not very different from the striated muscles of other arthropods and of vertebrates. A burst of action potentials must arrive in the motor axon to initiate every contraction. All insects with this type of muscle beat their wings at fairly low frequencies.

The wing muscles of flies (Diptera), wasps and their relatives (Hymenoptera) and many other insects have different and very remarkable properties. They do not need action potentials to initiate every contraction. In one experiment, action potentials were recorded from the wing muscle of a tethered blowfly. The insect was beating its wings at a frequency of 120 Hz but action potentials were occurring at a frequency of only 3 Hz. Occasional action potentials suffice to keep the muscles oscillating at the much higher frequency. Muscles which have this oscillatory property are known as fibrillar flight muscles.

Professor John Pringle and his colleagues studied the properties of fibrillar flight muscle. They naturally found it more convenient to experiment with reasonably large muscles working at reasonably low frequencies than with smaller muscles working at higher frequencies. They therefore used some of the largest insects, the Indian rhinoceros beetle *Oryctes* and the giant tropical water bug *Lethocerus*. Species of *Lethocerus* grow up to 11 cm long. Though these insects have quite low wingbeat frequencies, their wing muscles are fibrillar.

In the experiments, wing muscles were connected to apparatus that had an adjustable natural frequency of vibration. When the muscle was stimulated electrically it oscillated, lengthening and shortening at the natural frequency of vibration of the apparatus. A muscle from a beetle with a normal wingbeat frequency of 35–40 Hz could be made to oscillate in this way at any frequency from 28 Hz to 72 Hz. The special property of fibrillar muscle seems to be that it will drive any system at the system's natural frequency of vibration. This property can be demonstrated even with isolated fibres, treated with glycerol to break down the cell membrane (see section 2.8) and bathed in a suitable saline solution containing ATP. The oscillation is plainly a property of the muscle alone, not of the nervous system.

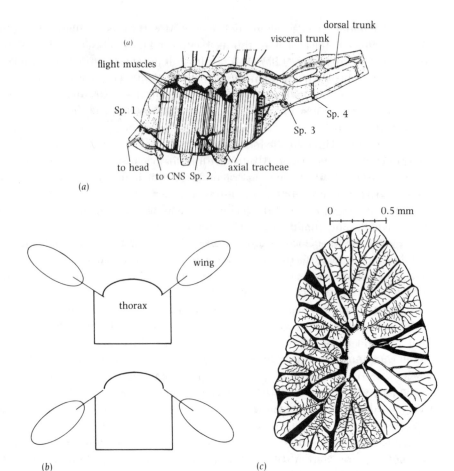

(a)

(b)

(c)

Fig. 9.8. Dragonfly (*Aeschna*) wing muscles and their tracheal supply. (a) A diagram showing the muscles and air sacs. Sp, spiracle. From P.L. Miller (1962). *J. exp. Biol.* 39, 513–535. (b) Diagrammatic transverse sections of the thorax showing how the muscles raise and lower the wings. (c) A horizontal section of one of the muscles. Blood-filled spaces are shown black and tracheae are stippled. From T. Weis-Fogh (1964). *J. exp. Biol.* 41, 229–256.

9.3.   Respiration

Insect wing muscles use oxygen faster than any other known tissue, and need a very efficient system to supply them with oxygen. The supply comes through the tracheae from the spiracles.

The tracheal supply to the wing muscles differs in detail between groups of insects. The arrangement found in dragonflies is shown as an example in Fig. 9.8(a). There are two pairs of spiracles on the thorax and 10 pairs on the abdomen. Each muscle has an axial trachea running through its centre, connected at its ventral end to one or other of the first three spiracles and at its dorsal end to air sacs, which are considerably larger than the diagram indicates. The air sacs in the thorax are also connected to the abdominal tracheae. Flight movements make the thorax enlarge and contract at each wingbeat: the roof of the thorax rises during the downstroke and sinks again during the upstroke (Fig. 9.8*b*). As a result, air is drawn into the air sacs during the downstroke and blown out again in the upstroke. Some of this air may flow from and to the abdomen but much of it must flow through the axial tracheae of the wing muscles. As the thorax expands, air must be drawn from the first three pairs of spiracles through the muscles to the air sacs. As it contracts, air must be blown out by the same route. Thus the air in the axial tracheae is continually changed.

The details have been worked out more fully for locusts, which have a slightly more complicated tracheal system than dragonflies. It is difficult to estimate the sizes of the air sacs in a dissection but a technique has been devised for filling them with coloured jelly and making the rest of the locust transparent. This makes possible measurements which show that the thoracic air sacs occupy $100–150$ mm^3 in a 2 g locust. The roof of the thorax rises and falls through about 0.6 mm, which is enough to enlarge and contract the thorax by about 25 mm^3. This volume may seem small but since the wings beat at a frequency of 17 Hz the total ventilation volume is $17 \times 25 = 425$ mm^3 s^{-1} or 1500 cm^3 h^{-1}.

Some of this airflow bypasses the muscles and so is ineffective in supplying oxygen to them. The effective rate of ventilation has been estimated in another way. Locusts will 'fly' while tethered in a jet of air. Locusts were fitted with a device which took small samples of air from the second thoracic spiracle during tethered flight. The samples were analysed and found to contain, on average, 13% oxygen. Dry air contains 21% oxygen but moist air at the temperature of the locusts' thoraxes contains only 20%, so the oxygen used by the locusts amounted to only 7% of the air. The locusts were using about 28 cm^3 oxygen h^{-1} (rather less than they would probably have used in free flight) and if this is 7% of the volume of the air pumped through the muscles that volume must be 400 cm^3 h^{-1}. This is only about a quarter of the volume calculated from the total volume change of the thorax, but it is still ample.

The pumping action renews the air in the axial tracheae, but oxygen can only move from there to the muscle fibres by diffusion. Sections through the muscles show lobes, each of them a bundle of muscle fibres (Fig. 9.8*c*). Each lobe is served by radial branches from the axial trachea. The branches subdivide into fine tracheoles which pass between the muscle fibres. Though the fibres are only 20 μm in diameter, each has tracheoles between itself and its neighbour. The arrangement allows fast diffusion between the axial trachea and the muscle fibres, since oxygen diffuses much faster in air than in water or tissues (see section 6.6, where this is mentioned in a discussion of snail lungs).

Since movement of oxygen from the axial trachea depends on diffusion there must be a maximum practical radius for insect wing muscles. In the muscle shown in Fig. 9.8(*c*), no muscle fibre is more than 0.8 mm from the axial trachea. How near the limit is this?

The question can be answered in the same way as the question of the maximum feasible thickness for a flatworm (section 4.3). The diffusion constant for oxygen diffusing in air is 20 mm^2 atm^{-1} s^{-1}. A tangential section through a muscle such as the one shown in Fig. 9.8(*c*) (i.e. a section cut at right angles to the direction of diffusion) would show that the air passages made up about 1% of the area. Hence we can estimate that the diffusion constant D for oxygen diffusing through this muscle is 1% of the value for diffusion through air, or 0.2 mm² atm⁻¹ s⁻¹. The metabolic rates m of insect wing muscles in flight are around 0.06 mm³ oxygen (mm³ muscle)⁻¹ s⁻¹. (This is 200 cm³ g⁻¹ h⁻¹: see p. 257.) The partial pressure P_s of oxygen in the axial trachea falls from 0.20 atm when fresh air enters to 0.13 atm when it leaves so we will use a mean value of 0.17 atm. Applying equation 4.2 to determine the maximum diffusion distance, we find

$$s \leqslant (2DP_s/m) \tag{9.3}$$
$$\leqslant (2 \times 0.2 \times 0.17/0.06) = 1.1 \text{ mm}.$$

The diameter of the muscle is $2s$ plus the diameter of its axial trachea, so this calculation indicates that insect wing muscles of diameter 2.5 mm are feasible. The vast majority of insects have wing muscles more slender than this. Species of the giant water bug *Lethocerus* have wing muscles of diameter up to 5 mm, but these muscles have a peculiar arrangement of tracheae which probably ensures that air is pumped through the radial tracheae as well as the axial ones.

Oxygen diffuses radially outwards from the axial tracheae of insect wing muscles, but the fuels for flight have to diffuse in from the blood in the haemocoel. The principal fuels are the sugar trehalose, and fatty acids. The lobed structure of the muscles ensures that the fuels do not have to diffuse far. In the muscles shown in Fig. 9.8, no fibre is more than about 0.2 mm from the nearest blood. When the muscles contract they squeeze blood out of the clefts between their lobes. When they extend they draw fresh blood into the clefts. Nevertheless, high fuel concentrations are needed in the blood. The blood of locusts and bees contains about 2% sugars (mainly trehalose). In contrast, the sugar content of human blood does not normally rise above 0.2%.

Even when they are resting on the ground the larger species of insect pump air through their tracheae. Locusts do this mainly by expanding and contracting the abdomen. They open their anterior and posterior spiracles in turn so that air enters through the anterior ones and leaves through the posterior ones. The rate at which air is pumped through the system is (appropriately) far less than in flight.

9.4.   Water balance

Terrestrial snails and slugs are generally active only at night or in damp weather, when the relative humidity is high. During the day snails retire into their shells and slugs into holes, and the whole of a hot dry season may be passed in a dormant state. Thus snails and slugs avoid losing too much water by evaporation. Many insects, however, are active in very hot dry conditions, even in deserts. They need much more effective protection against water loss than snails.

Evaporation is a particularly severe problem for small terrestrial animals, simply because they are small. Compare two animals of the same shape, one twice as long as the other. The larger one has four times the surface area of the smaller so it is likely to lose water by evaporation about four times as fast. However, it has eight times the volume of the smaller animal and so starts off with eight times as much water. It will take twice as long to dry up.

Water loss from the spiracles is inevitable. Consider an insect in dry air at 40°C (for instance, a desert beetle). In some period of time it pumps a litre of air through its tracheae. This air enters the spiracles dry but it inevitably leaves them saturated with water vapour, containing 60 mg water. The air enters containing 210 cm^3 oxygen and leaves containing considerably less. It is unlikely that more than about a third of the oxygen will be removed for respiration, so that the 60 mg of water is lost for the uptake of 70 cm^3 oxygen. Thus 0.9 mg water must be lost for every cubic centimetre of oxygen used. More would be lost if the spiracles were left perpetually open.

Insects can be made to open their spiracles by mixing carbon dioxide with the air. For instance, the bug *Rhodnius* keeps its spiracles perpetually open in an atmosphere containing 5% carbon dioxide. In an experiment, a *Rhodnius* was kept starved in dry air for 6 days. It was weighed every day. For the first two days and the last three it was kept in normal (though dry) air. On each of these days it lost less than 2 mg, mainly by evaporation of water. On the third day its spiracles were kept open by means of carbon dioxide and it lost 11 mg.

The spiracles are not the only route for water loss, and not necessarily the most important. *Eleodus* is a desert beetle which at temperatures around 40°C uses 0.4 cm^3 oxygen $g^{-1} h^{-1}$. If it opened its spiracles only when necessary to admit or expel air and used one third of the oxygen from this air it would inevitably lose through the spiracles 0.36 mg water $g^{-1} h^{-1}$ or 0.036% of its mass per hour. It actually loses a total of 0.2% of its mass per hour, mostly through the cuticle, which is not perfectly waterproof. It would lose water faster if the cuticle did not have its outer layer of wax.

Water is also lost with faeces and urine. Insects excrete most of their waste nitrogen as uric acid, which can be excreted with far less water than would be needed to get rid of ammonia (section 6.7). Terrestrial snails similarly excrete uric acid and other purines, and gain the same advantage.

The Malphigian tubules and the rectum co-operate in the process of excretion. Many experiments have been done with single Malpighian tubules, removed from the insect. The basic technique was invented by Professor Arthur Ramsay and is illustrated in Fig. 9.9. Everything in the picture is submerged in liquid paraffin. Most of the length of the tubule is in a drop of fluid which represents the blood of the intact insect: this fluid may be an imitation of insect blood or it may be some other solution devised for the experiment. The open end of the tubule (the end which was connected to the gut in the intact insect) is pulled out from this drop of bathing fluid into the paraffin and a cut made in it. The fluid secreted by the tubule emerges from this cut and can be collected for analysis.

The secreted fluid generally has about the same osmotic concentration as the bathing fluid, but contains much less sodium and much more potassium. The rate of secretion is very much reduced if the bathing fluid contains no potassium. These observations have led to the theory that the main active process is secretion of potassium taken from the bathing fluid and secreted into the tubule lumen. The lumen will become positively charged and the potential difference will tend to drive anions from the blood to the lumen; if potassium is moved, anions will follow. The movement of potassium and anions will increase the osmotic concentration of the

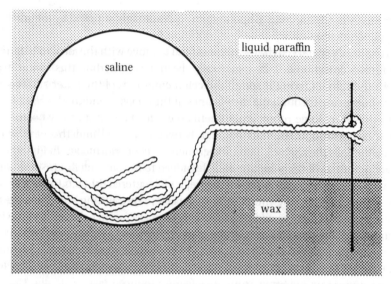

Fig. 9.9. An experiment with a Malpighian tubule. From S.H.P. Maddrell (1971). *Phil. Trans. R. Soc. Lond.* B262, 197–207.

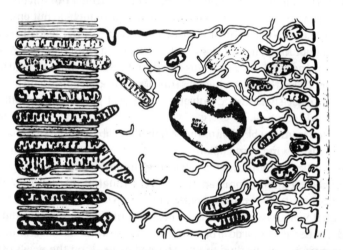

Fig. 9.10. An electron microscope section of a cell in the wall of a cockroach Malpighian tubule. The surface on the right is the outer surface of the tubule and the surface on the left faces the lumen. From J.L. Oschmann & M.J. Berridge (1971). *Fed. Proc.* 30, 49–56.

fluid in the lumen so water will be drawn into the lumen with them.

Measurements with electrodes confirm that the lumen is usually positively charged relative to the bathing fluid: the potential is usually about 30 mV. Thus the potassium moving into the lumen is moving against both a concentration difference and a potential difference. The secretion of potassium must be an active, energy-consuming process. Electron micrographs of sections of Malpighian tubules show many mitochondria, especially at the ends of the cells that face the lumen (Fig. 9.10).

The energy needed to secrete the potassium is presumably supplied by metabolism in these mitochondria.

I have already remarked that water will tend to move with the ions, to balance the osmotic concentrations. It has generally been thought that this water travels through the cells, entering through the cell membranes of the outer surface of the tubule and passing through the membranes at the opposite ends of the cells, into the lumen. However, some recent experiments seem to show that at least some of the water travels through the narrow gaps between one cell and the next without entering the cells themselves. These experiments were performed as in Fig. 9.9, with substances that have fairly large molecules added to the bathing fluid. Some of these were substances such as urea and L-glucose, which entered the cells freely. (This was demonstrated by analysing tubules that had been in the bathing fluid for some time: the concentrations of these additives in the tissue were found to be only a little lower than in the bathing fluid.) Some other added substances, such as sucrose, polyethylene glycol and the polysaccharide inulin, did not enter the cells. If these substances travelled from the bathing fluid to the tubule lumen by diffusion, they would travel at about the same rate, whether urine was being produced fast or slowly. However, most of them travelled much faster when urine was being produced rapidly: they must have been carried along by the flow of water. (This is called the solvent drag effect.) This happened with sucrose and polyethylene glycol, which do not enter the cells, so at least some of the water must have been flowing through gaps between cells. However, it did not happen with large molecules such as inulin, which passed very slowly into the urine, however fast it was being produced. The data seemed to be explained by the hypothesis that the water was travelling through gaps about 1.2 nm wide, wide enough to let sucrose and polyethylene glycol pass through, but too narrow for inulin.

Uric acid is secreted from the blood into the Malphigian tubules. It may remain in solution if the urine is flowing very copiously (for instance in *Rhodnius* just after a meal of blood) but it generally precipitates, making the urine cloudy.

The fluid produced by the Malphigian tubules has roughly the same osmotic concentration as the blood. It is secreted rapidly, at rates up to 50% of the total blood volume per hour. If it were passed as urine the insect would lose water very rapidly indeed. However, water is removed from the urine, and from the faeces, in the rectum. The concentration of the fluid finally excreted depends on the needs of the insect at the time. Cockroaches excrete fluid at osmotic concentrations of up to 1 mol l^{-1} (more than twice the concentration of the blood). Some other insects, which are better adapted to life in dry conditions, excrete solid pellets of faeces and uric acid. They include mealworms (larvae of the beetle *Tenebrio*) which live in grain stores.

Some insects, including mealworms, can take up water vapour from a damp atmosphere. The most remarkable example known is the firebrat *Thermobia*, which lives in bakeries and kitchens. Firebrats kept without food or water gain weight in air of relative humidities down to 45%, presumably by uptake of water. They do not gain weight if the anus is blocked by a ligature. This suggests that uptake is by way of the rectum and that *Thermobia* extracts water vapour from the air by the same mechanism as is used to dry the faeces.

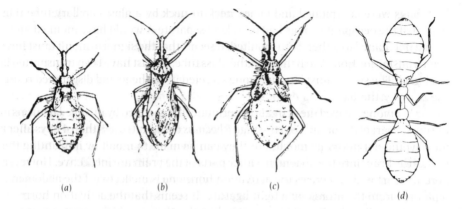

Fig. 9.11. *Rhodnius prolixus.* (*a*) The final (fifth) nymphal stage. (*b*) An adult. (*c*) A sixth nymphal stage which does not occur naturally but was produced in experiments described in the text. (*d*) Two decapitated fourth-stage nymphs joined by a glass tube. From V.B. Wigglesworth (1974). *Insect physiology,* 7th edn. Chapman & Hall, London.

9.5.   Metamorphosis

In the course of its life an insect moults several times, and each time it is transformed. The transformations may all be relatively slight, or there may be dramatic changes from larva to pupa and from pupa to adult. How is moulting initiated, and how are the transformations controlled so that the right change occurs at the right stage in the life history?

The research of Sir Vincent Wigglesworth went a long way towards answering these questions. Its success depended on a wise choice of experimental animal. This animal was *Rhodnius,* a South American blood-sucking bug (Fig. 9.11). One of its advantages is that it survives drastic surgery very well. Another is a peculiarly convenient life history. It passes through five nymphal stages. During each it takes a single gigantic meal of blood, amounting to several times its body mass. Ten to twenty days later (the interval differs between nymphal stages) it moults and is transformed to the next stage. The experimenter can induce moulting at will (by feeding the bug) and once the meal has been given he can predict accurately when moulting will occur.

Many of the experiments used fourth-stage nymphs, which normally moult 14 days after a meal (at 26°C). If one of these nymphs is decapitated more than 4 days after a meal, and the neck is sealed with wax to prevent loss of blood, it will survive and moult at the expected time. It will probably wear the old cuticle like an overcoat instead of shedding it completely, but in other respects the moult is normal. However, if it is decapitated less than 4 days after a meal it fails to moult. It seems that the head gives the signal for moulting about 4 days after a meal and that once the signal has been given the moult will occur even without the head.

A simple experiment revealed the nature of the signal. It used two *Rhodnius* which will be called A and B. A was fed, and a week later B was fed. The day after B's meal both bugs were decapitated and joined neck-to-neck by a glass capillary tube (Fig. 9.11d). If the decapitated bugs had been kept separate A would have moulted and B would not. Joined together, both moulted. It seems that the signal to moult must have been transmitted from A to B through the glass tube: it must have been transmitted in the blood. The signal must be a hormone secreted in the head and distributed round the body by the blood.

The hormone is called the activation hormone. It is secreted by modified neurons in the dorsal part of the brain. Fourth-stage *Rhodnius* decapitated less than 4 days after a meal would not normally moult, but they can be made to moult by implanting this part of the brain into the abdomen. Other parts of the brain are ineffective. However, even the part which secretes the activation hormone is ineffective if the abdomen is separated from the thorax by a tight ligature. It seems that the activation hormone does not act directly. It activates a gland in the thorax which secretes a second hormone, called ecdysone. This has been demonstrated by experiments in which glands taken from the thorax were implanted into isolated abdomens. Ecdysone, like many vertebrate hormones, is a steroid.

The activation hormone and ecdysone give the signal for moulting but a third hormone is needed to decide whether the moult is to produce another immature stage or an adult. This is the juvenile hormone, so called because its presence prevents metamorphosis to the adult form. It is a lipid. Wigglesworth demonstrated its action by two types of experiment. In one he made first-stage nymphs metamorphose to tiny precocious adults. In the other, he made fifth-stage nymphs (which ought to have metamorphosed directly to adults) produce giant sixth- and even seventh-stage nymphs (Fig. 9.11c).

The miniature adults were produced by joining first-stage nymphs, decapitated the day after a meal, to moulting fifth-stage nymphs. The join was made by a tube, as in Fig. 9.11(d), but the head of the larger nymph was left almost intact. The miniature adults produced in this way were not perfect, but they were much more like adults than normal second-stage nymphs.

Giant nymphs were produced by joining fifth-stage nymphs, decapitated the day after a meal, to fourth-stage nymphs fed a week earlier. Alternatively, glands from earlier nymphs were implanted into fifth-stage nymphs. The gland in question is the corpus allatum, which lies immediately posterior to the brain. It secretes juvenile hormone until the fifth nymphal stage is reached, and then ceases.

The chromosomes of insects must carry instructions both for making immature structures and for making adult ones. The concentration of juvenile hormone decides which set of instructions will be used.

9.6. Insect colours

Many insects have beautiful colours; this section concentrates on the Lepidoptera, the butterflies and moths. Their colours are in the small scales that cover their wings and bodies. Many of them are the colours of pigments. For instance, the yellow of the brimstone butterfly (*Gonepteryx rhamni*) and the orange of the orange-tip (*Anthocharis cardamines*) are pterines, and the reds and browns of many other butterflies are probably ommochromes. Some other insect colours do not involve pigments. The coloured parts of the body are built from materials which in bulk would be colourless, and the colours are due to interference or other optical phenomena.

South American butterflies of the genus *Morpho* have extraordinarily vivid blues on their wings. These blues are produced by interference of light from structures which consist, in effect, of alternating layers of cuticle and air. The principle is the same as the one which makes an oil film floating on water look coloured.

Fig. 9.12(*a*) shows a stack of alternating layers of cuticle and air. Some of the light striking the top of the stack is reflected from there, some passes through the top layer of cuticle and is reflected from the lower surface of the layer, and some penetrates deeper before it is reflected. At every cuticle–air interface the refractive index changes and reflection is apt to occur. Consider the various rays which are reflected into the eye. Rays α and γ come from the same source but are reflected from the upper surfaces of successive layers of cuticle. Ray γ travels $2(a + b)$ further than ray α on its way to the eye. Light of wavelength $2(a + b)$ will arrive in phase in the two rays but light of other wavelengths will arrive more or less out of phase and be partly destroyed by interference. Rays β and δ are reflected from the lower surfaces of successive layers of cuticle and light of wavelength $2(a + b)$ will arrive in phase in them. Light of this wavelength in ray β may be expected to arrive out of phase with light of the same wavelength in ray α since the difference in distance for this pair of rays is only $2a$. However, the phase gets reversed on reflection from a material of higher refractive index, but not from one of lower refractive index: α and γ have their phase reversed on reflection, but β and δ do not. Light of a particular wavelength will arrive at the eye in phase in all four rays, if a and b are each one quarter of a wavelength.

It is necessary to explain more precisely what this means. Consider, for instance, blue light of wavelength 400 nm. If it is to be reflected without loss by interference, the air layers should each be 100 nm thick. However, the refractive index of cuticle is about 1.5 so the wavelength of the light in cuticle is $400/1.5 = 267$ nm and the thickness of the cuticle layers should be a quarter of this, 67 nm.

Fig. 9.12(*b*) shows the layered structure which produces the interference colours of *Morpho*. The blue scales on the wings have closely spaced vanes standing vertically on them. Thicker bars run horizontally along the vanes. Since the bars of adjacent vanes are side by side, almost touching, the effect is almost the same as if there were alternate layers of cuticle and air, parallel to the scale surface.

The structure shown in Fig. 9.12(*b*) cannot be seen by light microscopy since the distance between adjacent bars is too small a fraction of the wavelength of light. It was first revealed by electron microscopy, in the very early days of the technique. The first account of it was published in 1942.

For optimum reflection of blue light the bars should all be 67 nm deep and the spaces between them 100 nm deep. In fact the bars vary in depth and the total depth of (bar + space) seemed in the early electron microscope preparations to be only 130 nm. However, it was shown that the structure shrunk badly when the electron beam was turned on.

The colours of Lepidoptera serve various functions. In some cases they are effective as camouflage. The peppered moth (*Biston betularia*), in its original form, is white

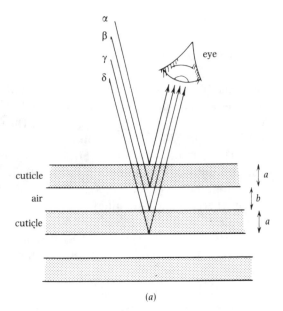

(a)

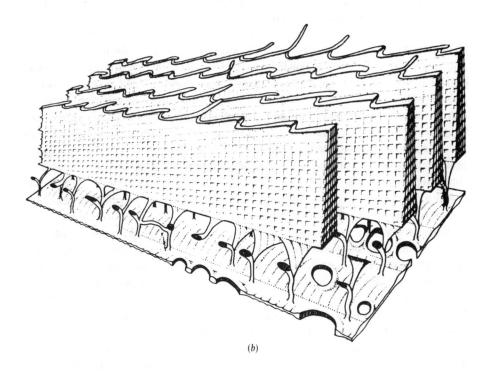

(b)

Fig. 9.12. (*a*) A diagram showing how interference colours are produced by alternating layers of materials of different refractive index. It is explained in the text. (*b*) The surface of a scale of the butterfly *Morpho*, from T.F. Anderson & A.G. Richards (1942). *J. appl. Phys.* 13, 748–58.

Fig. 9.13. A caterpillar of the eyed hawk moth (*Smerinthus ocellatus*) on a twig of willow (*Salix* sp., one of its natural food plants). The caterpillar on the left is in its normal resting position, with the ventral surface uppermost. From H.B. Cott (1975). *Looking at animals: a zoologist in Africa.* Collins, London.

with black speckling on the wings and body. It often rests on tree trunks and is very inconspicuous on a background of lichen. The Industrial Revolution which started in England in the late eighteenth century had a marked effect on trees near industrial areas. Smoke pollution made lichens much less plentiful, leaving bare bark which was blackened by soot. In 1850 a black form of the peppered moth was caught for the first time, in Manchester. It is far less conspicuous than the white form on bare, sooty bark and has become extremely common. It has become the predominant form in the industrial areas of Britain, and to the east of them where pollution is carried by prevailing winds. It remains rare in the rural areas in the west and north of Britain, in Cornwall, N.W. Wales and the Scottish Highlands. It has been shown by crossing the black and white forms that the black form is controlled by a single dominant gene.

Many species of moth, and some of other insects, have evolved black forms since the Industrial Revolution. These forms are called industrial melanics. Their evolution is perhaps more spectacular than any other evolutionary change witnessed by man. (Their only obvious rivals are some cases of the evolution of pesticide resistance.)

The efficacy of the camouflage of the two forms of peppered moth has been tested in field experiments. Equal numbers of marked specimens of the two forms were released in a rural wood in Dorset, where the natural population was entirely or almost entirely white. Flycatchers (*Muscicapa striata*) and other birds were seen feeding on them. Later, traps were set which attracted the moths by strong lights or by chemical attractants. Fourteen per cent of the white moths were recaptured but only 5% of the black ones. Predators had presumably found the black moths more easily than the white ones. The experiment was repeated in a wood near Birmingham where the natural population was predominantly black. There 28% of the black specimens were recaptured, but only 13% of the white ones.

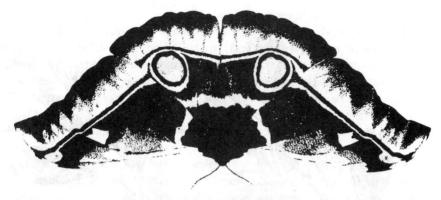

Fig. 9.14. An East African moth, *Bunaea alcinoë*, displaying its eye spots. Wing span about 14 cm. From H.B. Cott (1975). *Looking at animals: a zoologist in Africa.* Collins, London.

There are many examples of camouflage among Lepidoptera (and other animals), some of them involving much more detailed resemblance than the resemblance of the peppered moth to bark. Look at Fig. 9.13. The caterpillar is green with oblique yellow lines, matching the green leaves with their oblique yellow veins. Not only are the colours right but the caterpillar on the left (in its natural position) looks flat like a leaf rather than cylindrical. This is because its uppermost (ventral) surface is dark and its lower (dorsal) one is lighter, counteracting the effect of the shadow thrown by light falling on it from above. When the caterpillar is turned the other way up (on the right) it is far more conspicuous. Many animls are countershaded in this way but since most of them habitually keep the dorsal surface uppermost, most of them are darker on the dorsal than on the ventral surface.

Many moths and butterflies have patterns on their wings which resemble the eyes of mammals. An example from Africa is shown in Fig. 9.14. When the wings are folded the eye spots are hidden by the forewings. When they are spread the moth looks disconcertingly like the face of a predator such as the genet (*Genetta*). When a moth like this is found by a bird which threatens to eat it, it spreads its wings. The bird suddenly finds itself confronted by the appearance of a dangerous predator, and may well take flight. The effect of such displays on birds has been tested. A small horizontal glass screen was arranged so that patterns could be projected onto it from below. It was put in a cage with a bird. A mealworm (*Tenebrio* larva) was placed on the screen and when the bird approached to eat it, a pair of eyes or other patterns were made to appear suddenly on either side of it. It was found that feeding was discouraged to some extent, by each of the patterns that were tried. However, eye-like patterns were more effective than circles, which in turn were more effective than crosses. This was true when the experiments were done with hand-reared birds which had never encountered a vertebrate predator. The birds were chaffinches (*Fringilla coelebs*), yellowhammers (*Emberiza citrinella*) and great tits (*Parus major*).

Warning coloration is another common phenomenon in Lepidoptera and other insects. Many animals are protected against predators by having a sting (like wasps, *Vespula*, etc.), or an unpleasant taste, or by some other device. These devices would not be very effective if predators did not learn to recognize and avoid the protected animals: it is far better to be avoided than to be seized and spat out again, probably mutilated. Protected animals are often conspicuously coloured in bold patterns. For instance, wasps (order Hymenoptera) have black and yellow striped abdomens. Birds stung by wasps learn not to attack them and it has been shown that they remember not to attack them for at least several months. The caterpillars of the cinnabar moth (*Callimorpha jacobaeae*) are striped black and orange. It has been shown that birds find their taste very unpleasant.

Wasps and cinnabar moth caterpillars both have encircling stripes, of black and either yellow or orange. The resemblance may benefit both. It has been shown experimentally that birds which have encountered wasps are less likely to attack cinnabar moth caterpillars, when they first meet them, than birds which have not. Similarities of appearance between noxious species which are believed to give this sort of advantage are described as Mullerian mimicry.

There are also many cases of apparently harmless and palatable insects resembling ones which are protected by stings or unpleasant taste. The moth *Sesia apiformis* has transparent, almost scale-free wings and a black and yellow striped abdomen, like a wasp. Many hoverflies (family Syrphidae, order Diptera) also have black and yellow striped abdomens. The superficial resemblance to wasps probably makes predators avoid these apparently harmless mimics. This type of mimicry is known as Batesian mimicry.

9.7.   Social insects

Termites (Isoptera) are social insects, as also are ants and many species of bees and wasps (Hymenoptera). They live in colonies which may have relatively few members, but may have as many as 2 million in the African termite *Macrotermes natalensis*. However many members there are in a termite colony, only two are sexually mature: a male (the 'king') and a female (the 'queen'). All the other members are the offspring of these two unless the original king or queen has been lost. Most hymenopteran colonies have just one mature female (the queen) and no mature male: the queen is ferilized before founding the colony and stores the sperm she receives. All the other members of the colony are the queen's offspring.

Fig. 9.15 shows the kinds of termite (castes) found in a colony of *Kalotermes flavicollis*. This is one of only two European species of termite (most termites are tropical). It infects dead and diseased vines and trees, eating the wood away. Some tropical termite species cause serious damage to wooden buildings.

A typical *Kalotermes* colony consists of a king and a queen, about 30 soldiers and almost 1000 nymphs. Notice that the queen (Fig. 9.15) has her abdomen so swollen that there are wide strips of arthrodial membrane between the plates of hard cuticle. The king and queen stay together and undertake all the reproduction in the colony.

The soldiers face intruding animals and snap their big mandibles at them, defending the colony. All other work is done by advanced nymphs (including pseudergates, Fig. 9.15). These feed on the wood and dispense food from mouth and anus to king, queen, soldiers and younger nymphs. They secrete no enzyme capable of digesting cellulose but have in their guts flagellate protozoans which digest cellulose

king   queen   pseudergate
                              (juvenile worker)   soldier

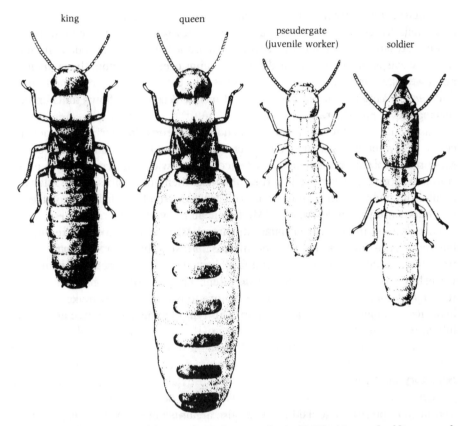

Fig. 9.15. Castes of the termite *Kalotermes flavicollis.* The king and soldier are each about 1 cm long. From M. Chinery (1976). *A field guide to insects of Britain and northern Europe,* 2nd edn. Collins, London.

anaerobically. This produces mainly acetic acid which is absorbed and used by the termite. The termites depend on the flagellates in just the same way as cattle depend on the ciliates and bacteria in the rumen (section 2.3). Young termites become infected with flagellates by feeding from the anus of older termites. The extent to which food passes from individual to individual was demonstrated by an experiment in which termites removed from a colony were fed for 1 or 2 days on filter paper impregnated with radioactive phosphate. These individuals, numbering 9% of the colony, were returned to the colony. Within 20 hours, 70% of the members of the colony were radioactive.

Pseudergates also remove the eggs as the queen lays them and take them to another part of the colony where they are reared.

From time to time some of the nymphs in the colony moult to become winged adults which leave through tunnels bored to the surface of the wood by the pseudergates. They fly off and form pairs which shed their wings and mate. Each successful pair becomes king and queen of a new colony composed entirely of their own offspring. The stumps of the shed wings of the king and queen can be seen in Fig. 9.15.

A pseudergate may moult at intervals without growing or changing its form, remaining a pseudergate until death. Alternatively it may in appropriate circumstances become a winged adult (and eventually a king or queen), or a wingless reproductive adult to replace a dead king or queen, or a soldier. How are the alternatives controlled?

When the king and queen are removed from a colony, even if only for 24 hours, replacements start to develop and reach maturity 4–7 days later. Several are produced but all but two are killed and eaten by the other members of the colony. If a colony is divided by a fine wire gauze screen with the king and queen both on the same side, replacement adults appear on the other side but are promptly eaten. However, a double wire gauze screen allows the orphaned half of the colony to produce its replacement reproductives. It is believed that the information that king and queen are present is conveyed by antennae through the single screen.

If a king or queen is fixed in a partition between two halves of a colony, so that one half has access only to its anterior end and the other to its posterior end, replacement reproductives develop in the anterior half only. This is so even if its abdomen is varnished, but if its anus is blocked replacements appear in the posterior half as well. It seems that some substance released from the anus by the king and queen and eaten by other members of the colony inhibits the development of replacements. The inhibition is passed from member to member of the colony by anal feeding: it can be transmitted from one half of a divided colony to the other by pseudergates fixed in the partition, with their heads towards the king and queen. There must be separate inhibitory substances for the two sexes. They are examples of the class of substances known as pheromones which transmit information between members of a species.

How can the termites and other social insects have evolved? Only a tiny proportion of the members of a species reproduce. There must be genes which prevent the rest from becoming reproductive adults. How could genes which prevented most of their possessors from breeding appear in an increasing proportion of the population from generation to generation and so become established? While the social habit was evolving, termites prevented by the new genes from breeding would have to compete against termites which bred freely.

It is tempting to suggest that division of labour is efficient and good for the species, and therefore evolved; that some individuals gave up the ability to reproduce, for the greater good of the species. This is nonsense. Suppose an insect appeared in a non-social species with a new gene which made it devote its life to the care of its fellows, without reproducing. There might as a result be more insects in the next generation, but none of them would possess the new gene.

An acceptable theory for the origin of social insects has been devised by Professor William Hamilton. A diploid animal (such as a termite) gets half its genes from each parent. A gene in either parent has a 50% chance of appearing in its offspring. A termite which does not become a king or queen has no offspring, but devotes its time to rearing its brothers and sisters. Two siblings each get 50% of their genes from the king and 50% from the queen. They will on average share 50% of their genes (25% from each parent). Hence a termite's siblings are as similar to it as its offspring would be. A termite does as much to propagate its genes by rearing siblings as it would by rearing an equal number of its own offspring. If rearing siblings is more efficient it will be favoured by natural selection. Notice that this argument depends on the non-reproductive members of the colony all being offspring of the same parents.

Hymenopteran colonies differ from termite colonies in two important ways. There is no king in the colony: the queen is fertilized before founding the colony, and stores the sperm she receives. The workers and soldiers, if any, are female. (Only ants have soldiers.)

Hornets (*Vespa* species: large wasps) will serve as an example of social hymenopterans. Only queens survive the winter, but they have been ferilized in the autumn by males whose sperm they store. They spend the winter in a torpid state and each queen establishes a colony in the spring. She builds a small nest, often in a hollow tree, using paper made from chewed wood and saliva. This nest has downward-opening cells, and she lays an egg in each. She feeds the larvae on chewed insects and they develop into adult workers, smaller than herself. Thereafter she continues to lay eggs, but the workers do most of the work of enlarging the nest and feeding the young.

The number of larvae that can be looked after at any time is proportional to the number of workers available to feed them. (I am assuming that food is plentiful.) Therefore the number of workers in the colony increases exponentially: if it doubles in the first month it doubles again in the second, and yet again in the third. This can be expressed mathematically. Let there be N_0 workers initially, and N_t at a time t. The expression dN_t/dt means the rate of increase of N_t, which I have stated to be proportional to N_t:

$$dN_t/dt = kN_t. \tag{9.3}$$

Calculus tells us that the solution to this equation is

$$N_t = N_0\,e^{kt}, \tag{9.4}$$

where e is the base of natural logarithms, approximately 2.72.

The rate of increase was studied in an Israeli species of hornet, which was persuaded to nest in boxes with transparent floors so that they could be counted. The number of workers increased from about 10 on May 15 to about 250 on September 15, approximately according to the exponential equation 9.4.

Late in the season, larger cells are built in the nest, and the hornets reared in them become reproductive adults (queens and males) rather than workers. They are larger than the workers and do not share in the work of the colony. In the Israeli study, no reproductives were produced until October, but all the young produced between early October and mid-November (when reproduction ceased) were reproductives.

If the queen produced only workers, she would have no surviving descendants next year. If she produced only reproductives, she would produce only a few because she would get no help in rearing them. She can produce more reproductives by building up a large team of workers and producing reproductives with their help only at the end of the season. Natural selection has favoured that strategy because genes from queens that produce a lot of reproductives are more plentiful than genes from queens that produce only a few, in subsequent generations.

The Hymenoptera cannot be divided into a primitive non-social group and an advanced social one. Rather, many of the taxonomic groups include both social and non-social species. Zoologists who have studied them closely believe that social habits have evolved at least eleven times within the order. They seem to have evolved only once among other insects, in the termites. This suggests that the Hymenoptera have some special characteristic which makes them particularly apt to evolve social habits.

The characteristic in question seems to be their mechanism of sex determination. The great majority of metazoan animals have both sexes diploid, but one sex has two X chromosomes and the other an X and a Y. In Hymenoptera the males are haploid and the females diploid. Ova are haploid. Ova which are left unfertilized remain haploid and develop into males but ones which are fertilized receive a second haploid set of chromosomes from the spermatozoon and become diploid females.

Since males are haploid, all the spermatozoa produced by an individual male carry the same genes. Two sisters share all the genes they get from their father but on average only half the genes they get from their diploid mother. A gene which is present in a female has a 75% chance of appearing in a sister but only (as in termites) a 50% chance of appearing in a son or daughter. Female Hymenoptera make more of their genes appear in succeeding generations by rearing reproductive sisters than they would by rearing an equal number of their own offspring.

Rearing brothers is less effective. A gene present in a female has only a 25% chance of appearing in her brother, because the only genes he shares with her are half of the ones she got from her mother. A female hymenopteran which devotes its life to rearing brothers and sisters will only do better (from a genetic point of view) than it could by rearing an equal number of sons and daughters, if it devotes more effort to rearing sisters than brothers. Some brothers (or other males) must of course be reared, and the scarcer males are the better the chance each one has of breeding. It can be shown that the best strategy for a worker with the option of rearing sisters carrying 75% of her genes or brothers carrying 25% of them, is to devote three times as much effort to rearing sisters as to rearing brothers. Intact colonies of many species of ant have been dug up, weighed and counted. It has been found that the mass of the young reproductive females being reared in the colony is typically about three times the mass of the young males.

A gene in a male hymenopteran has a 100% chance of appearing in a particular daughter but only a 50% chance of appearing in a brother or sister, so a male gets more genetic advantage from daughters than from rearing an equal number of brothers or sisters (since males develop from unfertilized eggs, males have no sons). He will do better as a father than he would as a worker.

It follows that females are particularly prone to evolve into workers, but males are most unlikely to do so. This is presumably why all the workers are female. The males are all capable of reproduction and play no part in the care of the young.

Further reading

Introduction

Chapman, R.F. (1971). *The insects, structure and function*, 2nd edn. English Universities Press, London.

Imms, A.D. (1951). *A general textbook of entomology*, 8th edn. Methuen, London.

Rockstein, M. (ed.) (1973–4). *Physiology of Insecta*, 2nd edn. Academic Press, New York.

Snodgrass, R.E. (1935). *Principles of insect morphology*. McGraw-Hill, New York.

Wigglesworth, V.B. (1964). *The life of insects*. Weidenfeld & Nicolson, London.

Wigglesworth, V.B. (1974). *Insect physiology*, 7th edn. Chapman & Hall, London.

Flight

Casey, T.M. (1981). A comparison of mechanical and energetic estimates of flight cost for hovering sphinx months. *J. exp. Biol.* **91**, 117–129.

Ellington, C.P. (1984). The aerodynamics of hovering insect flight [six papers]. *Phil. Trans. R. Soc. Lond.* B**305**, 1–181.

Ellington, C.P. (1985). Power and efficiency of insect flight muscle. *J.exp. Biol.* **115**, 293–304.

Heinrich, B. (1971). Temperature regulation of the sphinx moth, *Manduca sexta. J. exp. Biol.* **54**, 141–152.

Usherwood, P.N.R. (ed.) (1975). *Insect muscle*. Academic Press, New York.

Weis-Fogh, T. (1973). Quick estimates of flight fitness in hovering animals, including novel mechanisms for lift production. *J. exp. Biol.* **59**, 169–230.

Respiration

Weis-Fogh, T. (1964). Diffusion in insect wing muscle, the most active tissue known. *J. exp. Biol.* **41**, 229–256.

Weis-Fogh, T. (1967). Respiration and tracheal ventilation in locusts and other flying insects. *J. exp. Biol.* **47**, 561–587.

Water balance

Florey, E. (1982). Excretion in insects: energetics and functional principles. *J. exp. Biol.* **99**, 417–424.

Gupta, B.L., Wall, B.J., Oschman, J.L. & Hall, T.A. (1980). Direct microprobe evidence of local concentration gradients and recycling of electrolytes during fluid absorbtion in the rectal papillae of *Calliphora. J. exp. Biol.* **88**, 21–47.

Maddrell, S.H.P. (1981). The functional significance of the insect excretory system. *J. exp. Biol.* **90**, 1–15.

Noble-Nesbitt, J. (1973). Rectal uptake of water in insects. In *Comparative physiology* (ed. L. Bolis, K. Schmidt-Nielsen & S.H.P. Maddrell), pp. 333–351. North-Holland, Amsterdam.

Whittembury, G., Biondi, A., Paz-Aliaga, A., Linares, H., Parthe, V. & Linares, N. (1986). Transcellular and paracellular flow of water during secretion in the upper segment of the Malphigian tubule of *Rhodnius prolixus*: solvent drag of molecules of graded size. *J. exp. Biol.* **123**, 71–92.

Metamorphosis

Wigglesworth, V.B. (1970). *Insect hormones*. Oliver & Boyd, Edinburgh.

Insect colours

Blest, A.D. (1957). The function of eyespot patterns in the Lepidoptera. *Behaviour* **11**, 209–255.

Guilford, T. (1986). How do 'warning colours' work? Conspicuousness may reduce recognition errors in experienced predators. *Anim. Behav.* **34**, 286–288.

Hinton, H.E. (1976). Recent work on the physical colours of insect cuticle. In *The insect integument* (ed. H.R. Hepburn), pp. 475–496. Elsevier, Amsterdam.

Kettlewell, B. (1973). *The evolution of melanism: the study of a recurring necessity.* Clarendon, Oxford.

Sheppard, P.M. (1975). *Natural selection and heredity,* 4th edn. Hutchinson, London.

Social insects

Free, J.B. (1977). *The social organization of honey bees.* Arnold, London.

Hamilton, W.D. (1964). The genetical evolution of social behaviour [2 papers]. *J. theor. Biol.* **7**, 1–52.

Howse, P.E. (1970). *Termites: a study in social behaviour.* Hutchinson, London.

Macevitz, S. & Oster, G.F. (1976). Modelling social insect populations. II. Optimal reproductive strategies in annual eusocial insect colonies. *Behav. Ecol. Sociobiol.* **1**, 265–282.

Oster, G.F. & Wilson, E.O. (1978). *Caste and ecology in the social insects.* Princeton University Press.

Trivers, R.L. & Hare, H. (1976). Haplodiploidy and the evolution of the social insects. *Science* **191**, 249–263.

10 *Bryozoans and brachiopods*

Phylum Bryozoa
Phylum Brachiopoda

10.1. Bryozoans

This chapter is about two phyla which resemble each other in various ways, most notably in having a characteristic crown of tentacles called a lophophore. The Bryozoa (sometimes called Ectoprocta) is a moderately large phylum of about 4000 species, which has until recently attracted surprisingly few research workers. Most Bryozoa live in the sea but our first example lives in fresh water.

Cristatella (Fig. 10.1) lives in lakes and clear ponds, often on the undersides of waterlily (*Nymphaea*) leaves. A colony looks rather like a feathery slug. Like slugs, and unlike most Bryozoa, it can crawl. Its speed is unimpressive (about 10–15 mm per day) but it is interesting to find a colony which can behave in an integrated way as if it were a single individual. There are of course other examples, notably the swimming colonies of siphonophores (section 3.2).

Each individual member of the colony (zooid) has a lophophore with tentacles. These are shown projecting from the colony in Fig. 10.1(*a*) but are withdrawn into the colony in response to touch or vibration: the muscles involved are shown in Fig. 10.1(*b*). The tentacles are not arranged in a simple ring, but along the perimeter of a crescent. (Fig. 10.1(*b*), showing half a zooid, includes just one arm of the crescent.) The tentacles have cilia on them, which set up filter feeding currents.

Each zooid has a U-shaped gut, with the mouth opening inside the lophophore. It has two coelomic cavities, one in the lophophore and the other interconnecting with the main coeloms of all the other zooids in the colony. It has a nerve ganglion but no circulatory system and no excretory organs.

Cristatella is hermaphrodite with simple gonads in the coelom. Embryos develop in the coelom and are released as young colonies. Asexual reproduction also occurs. The colony may split in two. It also forms statoblasts, which are asexually-formed buds enclosed in protective chitinous shells. The statoblasts are generally not released until the beginning of winter when the colony dies and disintegrates. They remain dormant through the winter and develop into new colonies in the spring.

Fig. 10.2 shows *Flustra*, a bryozoan that lives attached to stones off European shores. Its rather stiff colonies could be mistaken for seaweed but consist of large numbers of tiny exoskeletal boxes, each containing a zooid with a circular (not crescentic) lophophore. The boxes are made of calcium carbonate, protein and a polysaccharide related to chitin.

Attached to the *Flustra* in Fig. 10.2 are colonies of some other bryozoans that look

279

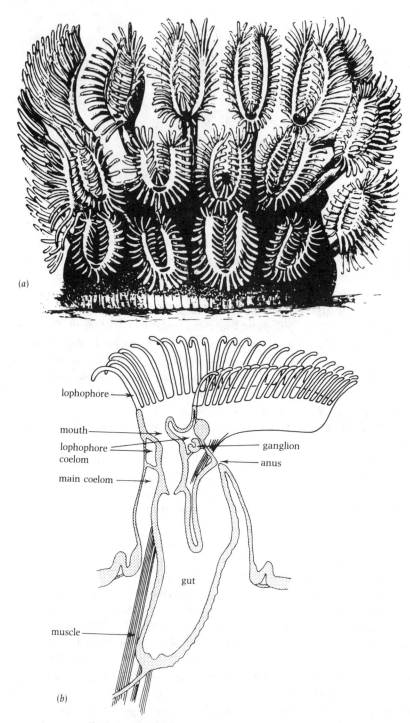

Fig. 10.1. *Cristatella mucedo.* (*a*) A colony. Length about 3 mm. From S.F. Harmer (1896). *Polyzoa.* In *The Cambridge natural history*, ed. S.F. Harmer & A.E. Shipley. Macmillan, London. (*b*) A section through a zooid. After L.H. Hyman (1959). *The invertebrates*, vol. 5. McGraw-Hill, New York.

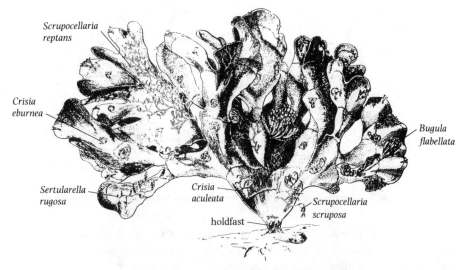

Scrupocellaria
reptans

Crisia
eburnea

Bugula
flabellata

Sertularella
rugosa

Crisia
aculeata

Scrupocellaria
scruposa

holdfast

Fig. 10.2. A colony of *Flustra foliacea* (height 7 cm). Colonies of the bryozoans *Bugula flabellata*, *Crisia aculeata* and *eburnea* and *Scrupocellaria reptans* and *scruposa* are attached to it, as also are colonies of the hydrozoan *Sertularella rugosa*. From A.R.D. Stebbing (1971). *J. mar. Biol. Ass. U.K.* 51, 283–300.

very much like colonial hydrozoans (phylum Cnidaria, section 3.2). Indeed, one of the attached colonies (*Sertularella*) is a hydrozoan and is not very obviously different from the rest.

Though some bryozoans look like hydrozoans, the resemblance is superficial. Hydrozoans are built of just two layers of cells, with mesogloea between, and have no coeloms. Bryozoans have mesoderm as well as ectoderm and endoderm (see section 7.1), and have coeloms. Bryozoan zooids may look rather like hydrozoan polyps, and form similar-shaped colonies, but they are built to an utterly different plan. It is quite clear that they belong to a different phylum.

10.2.   Brachiopods

Living brachiopods are quite scarce (there are only about 300 species) but brachiopod fossils are extremely abundant in marine sedimentary rocks of the Palaeozoic era. They seem to have been the dominant bottom-living animals in many Palaeozoic marine communities. All brachiopods live in the sea.

Brachiopods live in two-valved shells and most laymen would identify them as bivalve molluscs, but there is a marked difference from the molluscs in the symmetry of the shell and of the animal inside (Fig. 10.3). In bivalve molluscs, each valve is typically asymmetrical but one is a mirror image of the other (the scallop, *Pecten*, is an exception). In brachiopods the valves are not mirror images but each is symmetrical about a plane perpendicular to the axis of the hinge. The soft parts are also more or less symmetrical about this plane. Bivalve molluscs have a left valve and a right one but brachiopods have a dorsal valve and a ventral one. Confusingly, brachiopods often rest with the valve which is conventionally called ventral uppermost.

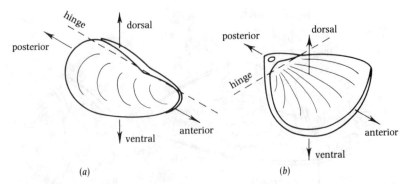

Fig. 10.3. Diagrams of (*a*) a bivalve mullusc shell and (*b*) a brachiopod shell.

Brachiopods, like bivalve molluscs, have logarithmic spiral shells with high values of W (section 6.2; Figs. 6.10 and 6.11). Unlike most bivalve molluscs they have $T = 0$ (and must do, to achieve their symmetry).

Typical brachiopods have shells made of calcium carbonate with a very small proportion of protein. The two valves hinge together but there is no elastic hinge ligament such as bivalve molluscs have (section 6.3) to make the valves spring open when the adductor muscles relax. Instead there are diductor muscles which run posterior to the axis of the hinge, so that when they shorten the valves open.

Most brachiopods live attached by a short stalk to rock, a loose stone or a shell. The stalk is called the pedicle and consists of connective tissue covered by epithelium. The animal has muscles which enable it to swivel on its stalk but the stalk itself contains no muscles.

Fig. 10.4 shows how the lophophore and viscera are arranged in a typical brachiopod. Much of the space between the valves is occupied by a cavity comparable to the mantle cavity of molluscs. This contains the lophophore, which is a simple ring of tentacles encircling the mouth in young brachiopods but extends along coiled supports in most adults. Different shapes of coil are formed in different groups of brachiopods: in the example which is illustrated the lophophore is broken so that it is no longer a complete loop. The tentacles divide the mantle cavity into two lateral inhalant chambers and a median exhalant one. Their cilia draw water in between the valves on either side and drive it out anteriorly. The mouth is on the inhalant side of the row of tentacles so the direction of flow is the same as in bryozoans. Diatoms and other small planktonic organisms are filtered from the water and transported by cilia to the mouth.

Most brachiopods have no anus. Diverticula from the gut form a digestive gland where most of the digestion occurs, as in molluscs. There are one or two pairs of nephridia. There is a blood circulatory system which has not been studied in detail. There is a simple nervous system radiating from a ganglion near the mouth. The coelom is undivided.

The sexes are separate in most species. There are two pairs of gonads in the coelom, and their products escape through the nephridia. Ciliated larvae develop from the fertilized eggs.

One of the unusual features of brachiopods is that they have a second defence

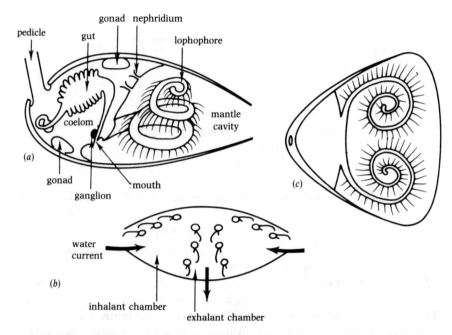

Fig. 10.4. **The structure of a typical brachiopod, based on members of the order Rhynchonellida. (*a*) represents a specimen cut in half sagitally; (*b*) a specimen with the dorsal valve removed; and (*c*) a transverse section.**

against predators, as well as a shell: they taste horrid. They seem to be unpalatable to other animals as well as to people. Carnivorous gastropod molluscs, crabs, starfish and fish all strongly prefer mollusc to brachiopod flesh, when given a choice, and many reject brachiopod flesh even when no other food is offered.

Further reading

Bryozoans

Ryland, J.S. (1970). *Bryozoans*. Hutchinson, London.
Ryland, J.S. (1976). Physiology and ecology of marine bryozoans. *Adv. mar. Biol.* **14**, 285–443.
Woollacott, R.M. & Zimmer, R.L. (eds.) (1977). *Biology of bryozoans*. Academic Press, New York.

Brachiopods

Richardson, J.R. (1986). Brachiopods. *Scient. Am.* **255**(3), 96–102.
Rudwick, M.J.S. (1970). *Living and fossil brachiopods*. Hutchinson, London.

I I *Starfish and sea urchins*

Phylum Echinodermata
    Class Crinoidea   (sea lilies and feather stars)
    Class Asteroidea   (starfish)
    Class Ophiuroidea   (brittle stars)
    Class Echinoidea   (sea urchins)
    Class Holothuroidea   (sea cucumbers)
    (and several extinct classes)

11.1. Introduction

The echinoderms all live in the sea. A few live in dilute sea water, for instance in parts of the Baltic Sea which contain 0.8% salts instead of the usual 3.5%, but none lives in fresh water.

The starfish *Asterias* (Fig. 11.1*b*) will be used to introduce the phylum. The British species *A. rubens* and the American *A. forbesi* are very similar. They live on the bottom of the sea at depths down to about 200 m.

Asterias has a mouth in the centre of its lower surface and an anus near the centre of the upper one. It has five arms, and almost perfect five-fold symmetry. The symmetry is, however, spoilt by a single porous plate, the madreporite, and some internal structures connected to it. The madreporite is on the upper surface, with the anus, and they cannot both be placed centrally. The madreporite is conspicuous and is shown in Fig. 11.1(*b*) but the anus is inconspicuous.

The lower surface bears about 1200 tube feet on which the animal crawls. (Some of them are visible in Fig. 11.1*b*). A starfish may crawl with any one of its five arms leading and the probability that a particular arm will be leading on a particular occasion is not very different from the probability that any other arm will be leading. Starfish crawl less often with two arms leading, side by side.

Since there is no strongly preferred direction of crawling no arm can be called anterior. The terms anterior and posterior cannot be used appropriately in describing adult starfish. Similarly the terms dorsal, ventral, left and right, which are designed for describing bilaterally symmetrical animals, are inappropriate. The surface which bears the mouth is called the oral surface and the opposite one is called the aboral surface, as in describing Cnidaria (Fig. 4.1). 'Oral' and 'aboral' are preferrable to 'lower' and 'upper' because not all echinoderms keep the aboral surface uppermost.

Scattered over the surface of the body are small spines and pedicellariae (Fig. 11.2*d*). The latter resemble tiny crab pincers. They grab, kill and discard small animals (for instance, small polychaetes) which touch them. Their main function

284

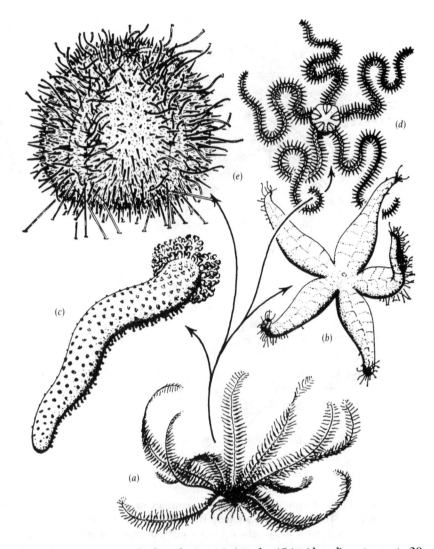

Fig. 11.1. A selection of echinoderms. (*a*) *Antedon* (Crinoidea: diameter up to 20 cm); (*b*) *Asterias* (Asteroidea: 30 cm); (*c*) *Holothuria* (Holothuroidea: length 20 cm); (*d*) *Ophiothrix* (Ophiuroidea: diameter 15 cm); (*e*) *Echinus* (Echinoidea: up to 16 cm). From D. Nichols (1969). *Echinoderms*, 4th edn. Hutchinson, London.

may be to prevent the larvae of sessile animals such as barnacles and bryozoans from settling on the starfish.

Fig. 11.2(*a*) shows how the main internal organs are arranged. The gut has a central stomach and five pairs of caeca, one pair in each arm. There are ten gonads, two in each arm, with separate openings on the aboral surface. There is a water vascular system, a peculiarity of the phylum. Its main vessels are a circumoral water ring encircling the mouth, a radial water canal along each arm and a stone canal (so called because it has hard spicules in its wall) leading to the madreporite.

The radial water canals have branches to the tube feet, which lie on either side of

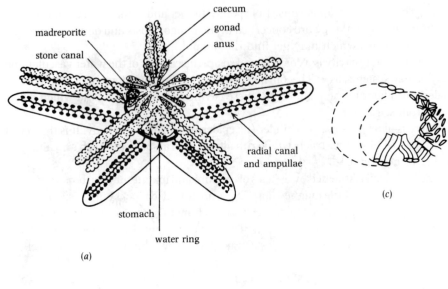

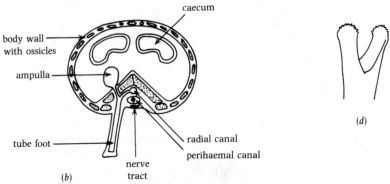

Fig. 11.2. Diagrams illustrating the structure of *Asterias rubens*. (*a*) The internal organs. The gut is lightly stippled, the gonads more heavily stippled and the water vascular system shown black. (*b*) A section through an arm; (*c*) the skeleton of a short piece of arm; (*d*) a pedicellaria.

them (Fig. 11.2*b*). Each tube foot has its own ampulla, a thin-walled sac which lies in the main body cavity. The tube feet are hollow, with muscular walls, and with suckers at their ends. They can lengthen and shorten, or point in any direction. Lengthening is caused by contraction of muscle in the wall of the ampulla, which drives water out of the ampulla into the tube foot.

There are perihaemal canals and haemal strands running largely parallel to the water vascular canals. They are not shown in Fig. 11.2(*a*), but Fig. 11.2(*b*) shows a perihaemal canal running below (oral to) the radial water canal, enclosing a haemal strand. There is a haemal ring oral to the water ring, a haemal strand parallel to the stone canal and other haemal strands to the caeca and gonads. Most of the haemal strands are enclosed in perihaemal canals. The stone canal and the haemal strand which runs parallel to it are both enclosed by a perihaemal canal known as the axial sinus. The haemal strands consist of spongy tissue, and some observations on sea urchins suggest that they are concerned in the phagocytosis and destruction of any micro-organisms which may get into the animal.

The main body cavity is a coelom, and so are the cavities of the water vascular and perihaemal systems, as will become apparent when the development of echinoderms is described later in the chapter. All these cavities are filled with fluid which differs little from sea water.

There is a net of nerve cells under the epidermis, all over the body. It is thickened into more definite nerve tracts which run along the oral surface of each arm and form a ring round the mouth.

The body wall is stiffened by ossicles of calcium and magnesium carbonate (mainly calcium, with just a little magnesium). These are not solid blocks but are spongy, with soft tissue filling the pores. Fig. 11.2(c) shows how the ossicles are arranged in the arms. Adjacent ossicles are connected together flexibly, by collagen fibres laced through their pores, so the muscles can bend the arms orally, aborally or to either side.

Each ossicle seems to be a single crystal (as the spicules of calcareous sponges also are). Ossicles have been broken and examined by scanning electron microscopy. Micrographs of the broken surfaces look just like micrographs of broken surfaces of calcite crystals. They look quite unlike the much rougher surfaces which are obtained by breaking molluscs shells, which consist of tiny crystals separated by protein.

The strength of mollusc shell depends on its composite structure (section 6.2). Cracks which would spread right through a single solid crystal are halted by the layers of protein between the crystals. The same result may perhaps be achieved in echinoderm ossicles by the network structure: the pores may help to stop cracks from spreading. Starfish skeletons are made of very large numbers of tiny ossicles, but some other echinoderms have skeletons built of quite large plates. For them, cracking might be a serious problem.

Asterias feed on molluscs and crabs. It attacks bivalve molluscs by holding them with its tube feet and pulling the shell slightly open. It turns its stomach inside out and slips it through the opening, and starts digesting the tissues of the mollusc. The fluid products of digestion travel into the starfish along ciliated channels in the wall of its stomach.

Asterias is a member of the class Asteroidea. The feather star *Antedon* (Fig. 11.1a) is one of the Crinoidea. It lives offshore, resting on the bottom or swimming by beating its arms. It spreads its arms as a net to catch such small food items as are carried past by currents. Each arm is divided into two branches which are fringed by pinnules, so the circle of arms makes quite an effective net.

Ophiothrix (Fig. 11.1d) is one of the brittle stars, the members of the class Ophiuroidea. It has no anus, and there are no gut caeca in its slender arms. It lives on gravel off European shores, often in extraordinarily dense patches of over 1000 animals per square metre. Scuba divers find that in gentle currents it rests on two or three arms, extending the others vertically to catch food. In stronger currents, in rough weather or when the tide is flowing fastest, it links arms with its neighbours. Individuals detached from their neighbours are apt to be swept along by the current

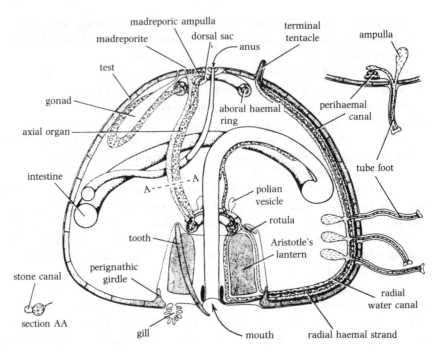

Fig. 11.3. A diagram representing *Echinus* sliced in half. A section across an ambulacrum is shown on the right and one through the stone canal and associated haemal strand (the axial organ) on the left. The water vascular, haemal and perihaemal systems are distinguished by distinctive textures. From D. Nichols (1969). *Echinoderms*, 4th edn. Hutchinson, London.

but the patch as a whole stays put. Experiments with radioactive phytoplankton have shown that *Ophiothrix* is capable of filtering out unicellular algae.

Echinus (Fig. 11.1*e*) and the other sea urchins belong to the class Echinoidea. *Echinus* is almost spherical, and is enclosed in a rigid test made of closely fitting ossicles. The test is not external like the exoskeletons of arthropods and the shells of molluscs, for the ossicles are embedded in the body wall and are covered externally by epidermis. There are much longer spines than *Asterias* has. They are attached to the outer surface of the test by ball-and-socket joints and have muscles which can point them in any direction. There are no arms but there are five double rows of tube feet running from the mouth at the centre of the lower surface to the anus at the centre of the upper one. Each double row of tube feet, with its associated structures, is called an ambulacrum. The water canals are inside the test and are connected to the tube feet through holes in the test.

Fig. 11.3 shows the internal organs of *Echinus*. Most of them are arranged as in *Asterias*, with distortions due to the difference between the star shape and the sphere. However, the gut is quite long and coiled, without caeca. There is also a remarkable structure around the mouth, called Aristotle's lantern. It is a framework of ossicles supporting five long teeth and has muscles to move it. There is a ring of so-called gills round the mouth but most of the oxygen which the animal uses probably diffuses in through the tube feet.

Echinus lives just offshore on gravel or rocks but is sometimes left exposed at low tide. It eats the 'fur' of attached algae which coats the rocks, scraping it off with the five teeth.

The sea cucumber *Holothuria* (Fig. 11.1*c*) is an example of the Holothuroidea. Instead of resting with its mouth pointing downwards and its anus upwards, like a starfish or sea urchin, it crawls with its mouth in front and its anus behind. Accordingly, it has become bilaterally symmetrical. It crawls on three ambulacra which have tube feet with suckers. The other two ambulacra run along its dorsal surface and have irregularly arranged tube feet without suckers.

There is a circle of 20 branched tentacles around the mouth which can be withdrawn into the body when the animal is disturbed. They are highly modified tube feet. Each has its own ampulla and is connected to one of the radial water canals. There is no external madreporite but the water vascular system opens into the main body cavity through pores in a madreporic body. There is little trace of a perihaemal system. The body wall is tough but flexible, for instead of large ossicles it has only small, scattered spicules embedded in it.

There is only one gonad, and there is a pair of branched organs in the body cavity called respiratory trees. These are departures from five-fold symmetry but they fit into the scheme of bilateral symmetry. Movements of the cloaca pump water in and out of the respiratory trees. It has been shown by experiment that the animal gets 60% of its oxygen from this water.

Each respiratory tree has a bundle of Cuvierian organs associated with it. These are long tubules filled with collagen and polysaccharide. If the animal is molested the contents of the Cuvierian organs swell, splitting the tubules and bursting through the wall of the cloaca to escape as a mass of long sticky threads which entangle the molester.

Holothuria crawls along on the bottom of the sea, using its tentacles to sweep up the particles of detritus which it eats.

11.2.   Development

The eggs and larvae of sea urchins have played an extremely important part in cell biology. They have proved convenient for research for various reasons. It is reasonably easy to achieve artificial fertilization in the laboratory by mixing ripe eggs and sperm. Development is straightforward with none of the complications which occur in the development of large yolky eggs. It is also rapid: the zygote becomes a mature larva in only 48 h. The larvae are transparent and have only 1000–2000 cells. *Paracentrotus lividus* has been found particularly convenient for study because one end of the egg is distinguished from the other by a band of orange pigment (Fig. 11.4*a*).

The zygote is enclosed in two layers of membrane, an outer fertilization membrane and an inner hyaline membrane. They are both present at all the stages shown in Fig. 11.4 but are only shown in Fig. 11.4(*a*). The first two divisions of the zygote occur symmetrically with respect to the orange band so the four cells they produce look identical (Fig. 11.4*c*). They seem in fact to be identical. They can be separated by

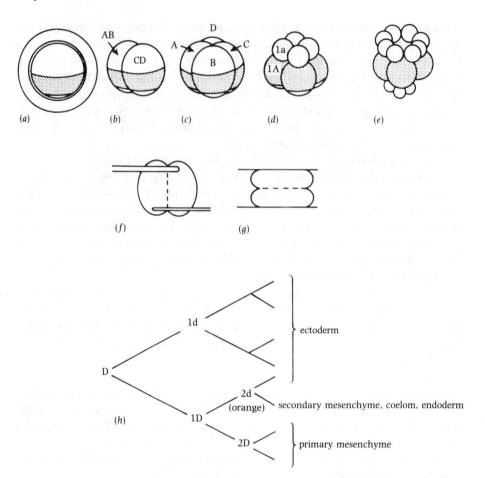

Fig. 11.4.(*a*) A zygote of *Paracentrotus lividus* and (*b*)–(*e*) early stages in its development. (*f*), (*g*) Experiments which are described in the text. (*h*) A diagram showing the fate of the products of division of one of the four cells shown in (*c*).

dissection with fine glass needles and each will develop into a small but otherwise normal larva.

The next division separates four pale cells at one end of the embryo from four mainly orange ones at the other end (Fig. 11.4*d*). Neither the top four cells nor the bottom four are capable, without the others, of producing a normal embryo. Top halves of embryos fail to develop a gut and bottom halves fail to develop the usual complement of cilia. Fig. 11.4(*e*) shows the next stage of division and Fig. 11.4(*h*) shows the eventual fate of the products of the next few divisions. This has been discovered partly by studying the imperfect larvae which develop from early embryos after some of the cells have been removed, and partly by experiments in which parts of an egg or embryo were coloured with dyes and the position of the colour observed at a later stage in development.

Compare Fig. 11.4 with Fig. 4.14, which shows early stages in the development of a polyclad flatworm. Notice first a difference at the eight-cell stage. The sea urchin embryo has each upper cell immediately on top of the corresponding lower cell. The manner of division which gives this effect is called radial cleavage. In the polyclad the top four cells alternate with the bottom four as a result of spiral cleavage. Now look at Fig. 11.4(*h*). It applies equally to any of the four cells A, B, C and D. Any one of the four has parts capable of producing all the tissues of the larva (and can indeed develop into a complete larva). In contrast, Fig. 4.14(*b*) applies only to cell D, for cell 4d gives rise to the gastrodermis and parenchyma while 4a, 4b and 4c vanish. The early development of molluscs and of polychaete worms proceeds in essentially the same way as in polyclads. (For a difference see section 7.1.)

The first division of the sea urchin zygote has often been studied as an example of cell division, in the hope of gaining insight into the process of cell division in general. It is of course easier to observe and interpret than the division of a cell within a tissue. The undivided cell is kept spherical by tension in its membrane, just as surface tension tends to keep a raindrop spherical. When the cell divides, additional tension is needed round the 'waist' to constrict it. Electron microscope sections show a girdle of microfilaments here, just under the cell membrane. This presumably provides the constricting force. The force has been measured by sticking fine glass needles through a dividing zygote and observing how much they are bent by the constriction (Fig. 11.4*f*). It has also been calculated from observations of the shapes which dividing zygotes adopt when they are squeezed (Fig. 11.4*g*). It is found that the force is about 6×10^{-8}N, corresponding to a stress in the girdle of microfilaments of 0.1 MN m^{-2}. This is comparable to the stresses of about 0.3 MN m^{-2} which most muscles can develop.

Later stages in development have been studied by time-lapse cinematography of embryos trapped in the meshes of a nylon net. When the film is projected at normal speed it shows the process of development greatly speeded up. The net is needed to keep the same embryo constantly in the field of the microscope.

The most informative observations have been made on the sea urchin *Psammechinus miliaris*. The zygote divides about 10 times at intervals of about an hour producing about $2^{10} = 1024$ cells. The cells adhere to the hyaline membrane and their planes of division are at right angles to the membrane, so the divisions produce a single layer of cells adhering to the inside of the membrane, which becomes greatly stretched. Thus a hollow blastula is produced (Fig. 11.5*a*).

The blastula sheds the fertilization membrane and starts swimming by means of cilia which protrude through the hyaline membrane. There is little division in the next 36 hours or so but there is a lot of cell movement, producing a larva which is quite complex in structure but has only about 2000 cells (about twice as many as the rotifer *Epiphanes*, Fig. 5.1).

The cells of the blastula have processes which protrude into the hyaline membrane, which is presumably why they adhere to the membrane so firmly. They are also attached to their neighbours by desmosomes, close under the membrane. If cells like this become more adhesive they will tend to become taller and thinner so as to have larger areas in contact with each other. If the area of the membrane is more or less fixed this will make it more convex, as shown in Fig. 11.6. Conversely, if the cells become less adhesive they will tend to make the membrane concave.

A dimple appears at the vegetal pole of the blastula (i.e. at the end which is mainly orange in *Paracentrotus*), with a ring around it where the cells are thicker (Fig. 11.5*a*,

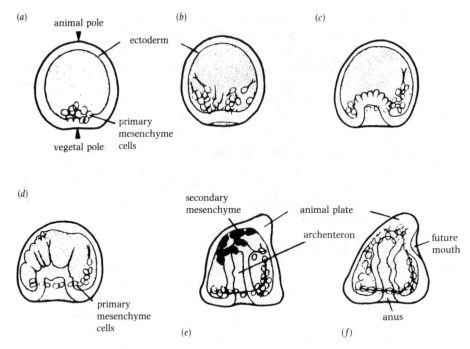

Fig. 11.5. Embryos of *Psammechinus miliaris*, showing stages of development from blastula to gastrula. From T. Gustafson & M. Toneby (1971). *Am. Sci.* 59, 452.

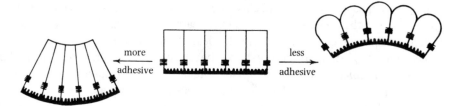

Fig. 11.6. Diagrams of part of a blastula, showing a few cells and part of the hyaline membrane. This diagram is explained further in the text.

b). This may be due to a group of cells becoming less adhesive (producing the dimple) while the cells around them become more adhesive (producing the thick ring). The cells at the centre of the dimple become so much less adhesive that they become detached from the layer of cells. They are the primary mesenchyme cells which will secrete the larval skeleton.

These cells send out fine pseudopodia which seem stiff, and wave around like bristles. The pseudopodia attach at their tips to other cells and then contract, pulling their own cells behind them. Since the pseudopodia are particularly apt to attach to the very adhesive cells of the thick ring this process tends to arrange the primary mesenchyme cells in a ring around the dimple (Fig. 11.5*c*). This is where they start to secrete the skeleton.

At this stage another group of cells on the dimple is extending pseudopodia. Each cell extends several pseudopodia, in different directions. These are the secondary

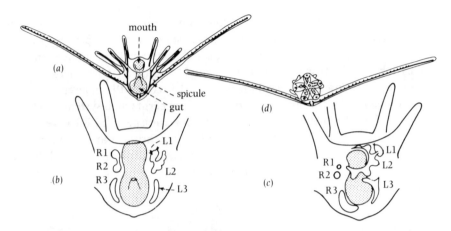

Fig. 11.7.(*a*) **The pluteus larva of a brittle star** (*Ophiothrix fragilis*) **(extreme width about 2.5 mm).** (*b*) **Part of the same, enlarged.** (*c*), (*d*) **Stages in its metamorphosis to the adult form. These diagrams are based mainly on illustrations in E.W. MacBride (1914).** *Textbook of embryology,* **vol. 1. Macmillan, London.**

mesenchyme cells. Unlike the primary mesenchyme cells they do not detach from their original positions so when their pseudopodia contract the dimple is pulled deep into the cavity of the blastula in the process of gastrulation.

Electron microscope sections of secondary mesenchyme cells show that their pseudopodia have microfilaments and microtubules in them. There is a discussion in section 2.8 of the roles of the microfilaments which are present in *Amoeba* and of the microtubules in other protozoans. It is likely that the mechanisms of cell movement are the same in echinoderm embryos as they are in protozoans.

Thus the early stages of sea urchin development seem to depend on quite simple processes involving changes in the adhesiveness of cells and the extension and contraction of pseudopodia. The same processes are probably equally important in the development of other metazoan embryos.

The coelom is formed as a cavity which splits off from the blind end of the gut of the gastrula. A mouth is formed (the original opening of the gastrula becomes the anus). The primary mesenchyme secretes three-pointed spicules (rather like sponge spicules) which push the body wall out into long processes. If skeleton growth is inhibited by sodium fluoride the processes fail to develop, which suggests that they are produced by the stretching action of the growing spicules.

Fig. 11.7(*a*) and (*b*) shows the mature larva of a brittle star. It has long processes stiffened by spicules, like the larvae of sea urchins. There are bands of cilia along the processes. Larvae like this are called pluteus larvae and are characteristic of Echinoidea and Ophiuroidea. The larvae of other classes of echinoderm also have bands of cilia but they lack the processes.

The coelom divides to form (typically) three coelomic cavities on each side of the body. It will be convenient to call these cavities L1, L2, L3 (on the left, numbering from the anterior end) and R1, R2, R3 (on the right). All six cavities are present in the brittle star pluteus shown in Fig. 11.7(*b*), but the larvae of some echinoderms lack R1 and R2. L2 remains connected to L1 and L1 develops an opening to the exterior.

The change from the larva to the adult echinoderm is a drastic metamorphosis. The details differ between the classes. Fig. 11.7 show how it happens in brittle stars. L2 curls round the mouth to form a ring canal from which sprout five radial water canals: it becomes the water vascular system. The connection from it to L1 becomes the stone canal and the external opening of L1 becomes the madreporite. L1 itself becomes the axial sinus but most of the rest of the perihaemal system develops from L3. L3 and R3 form the main body cavity. R1 vanishes and R2 is reduced to a rudiment. These and other changes produce a tiny brittle star in the centre of the pluteus (Fig. 11.7*d*). The long processes of the pluteus fall off leaving a little brittle star which has almost perfect five-fold symmetry although it has developed from a bilaterally symmetrical larva.

Further reading

General

Binyon, J. (1972). *Physiology of echinoderms.* Pergamon, Oxford.

Lawrence, J. (1987). *Functional biology of echinoderms.* Croom Helm, London.

Millot, N. (ed.) (1967). *Echinoderm biology (Symposium of the Zoological Society of London no. 20).* Academic Press, New York.

Nichols, D. (1969). *Echinoderms,* 4th edn. Hutchinson, London.

Development

Gustafson, T. & Wolpert, L. (1967). Cellular movement and contact in sea urchin metamorphosis. *Biol. Rev.* **42**, 442–498.

Horstadius, S. (1973). *Experimental embryology of echinoderms.* Clarendon, Oxford.

Tilney, L.G. & Gibbins, J.R. (1969). Microtubules and filaments in the filopodia of the secondary mesenchyme cells of *Arbacia punctulata* and *Echinarachnius parma.* *J. Cell Sci.* **5**, 195–210.

Yoneda, M. & Dan, K. (1972). Tension at the surface of the dividing sea-urchin egg. *J. exp. Biol.* **57**, 575–588.

12 *Primitive chordates*

Phylum Chordata
    Subphylum Urochordata   (sea squirts etc.)
    Subphylum Cephalochordata   (amphioxus)
    Subphylum Vertebrata
        Class Agnatha,
            Order Petromyzoniformes (lampreys)
            Order Myxiniformes   (hagfishes)
            Order Osteostraci
      and other orders and classes.

12.1.   Sea squirts and amphioxus

The rest of this book is about one phylum, the Chordata. Its best known members are the fish, amphibians, reptiles, birds and mammals, which resemble each other in many striking ways and form an obvious group. They are the vertebrates. The urochordates (the tunicates or sea squirts) and the cephalochordates (amphioxus) are clearly distinct from the vertebrates and from each other, but they have some features which are generally accepted as strong evidence of relationship to the vertebrates. The most obvious of these are the notochord (which will be described), a tubular dorsal nerve cord and perforations in the pharynx comparable to the gill slits of fishes. Many adult vertebrates have no notochord or gill slits, but these structures can nevertheless be found in their embryos.

    The urochordates all live in the sea. The best known of them are the sea squirts, which as adults have very little in common with vertebrates. They are sack-shaped animals, commonly a few centimetres long. They do not move about but remain firmly anchored to a rock or, by root-like structures, to mud. The body is enclosed in a protective tunic which consists of protein and polysaccharide (including cellulose) with only a few cells embedded in it. The only openings in the tunic are the two siphons at the upper end (Fig. 12.1). These are the inlet and outlet for water which the animal pumps through its body and filters to obtain its food. Examination of gut contents shows that unicellular algae (including diatoms) are the main food.

    A large proportion of the space within the tunic is occupied by the pharynx, which has so many small perforations that its wall is in effect a fine network. Water entering the inhalant siphon passes through this network into a surrounding cavity known as the atrium, which leads to the exhalant siphon. It is propelled by cilia. Fig. 12.1(*c*) shows that the bars of the pharyngeal network bear two sets of cilia: the frontal cilia which project into the pharyngeal cavity and the lateral cilia which project across the

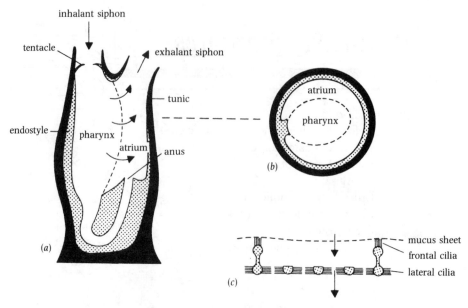

Fig. 12.1. Diagrammatic sections of a typical sea squirt such as *Ciona*. (*a*) Longitudinal section, (*b*) transverse section and (*c*) a greatly magnified section through part of the wall of the pharynx. Arrows indicate flow of water.

perforations. The lateral cilia propel the water through the animal. They beat in such a way as to drive water from the pharynx into the atrium.

A ring of tentacles arranged radially in the inhalant siphon may prevent excessively large particles from entering. Particles that do enter, down to very small dimensions, are retained in mucus on the pharynx wall. The perforations are typically about 50 μm across, but colloidal graphite particles only 1 μm in diameter are retained. Plainly the pharynx wall is not a simple sieve. By careful observation of sea squirts with transparent tunics it can be seen that the perforations are spanned by a sheet of mucus which covers the inner face of the wall. It is in this sheet that the particles are trapped. Parts of the mucus sheet taken from dissected sea squirts have been spread on electron microscope grids. Electron micrographs of them show a rectangular network of fine strands with mesh sizes of the order of 1 μm, but the structure of the sheet in these dried preparations may not have been the same as in the living animal. At least most of the mucus is secreted by gland cells in the endostyle, a structure which lies ventrally in the pharynx. (It is by no means obvious from Fig. 12.1 that this position is ventral, but comparison of the adult with the larva makes it quite clear that it is.) The pressure difference between the pharynx and atrium, which is produced by the lateral cilia, not only drives water through the mucus filter, but also keeps the mucus firmly pressed against the frontal cilia. These move it dorsally until it reaches the dorsal mid-line, where there are other cilia which transport it posteriorly to the digestive parts of the gut. Particles trapped in the mucus are carried with it, and both the mucus and the food particles are digested.

This filter feeding mechanism is very like that of bivalve molluscs (section 6.10), but there is an important difference. Sea squirts use a sheet of mucus to filter the

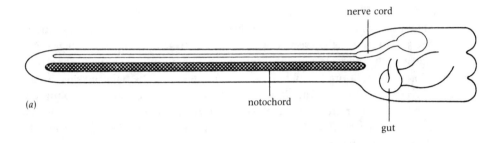

(a)

nerve cord

notochord

gut

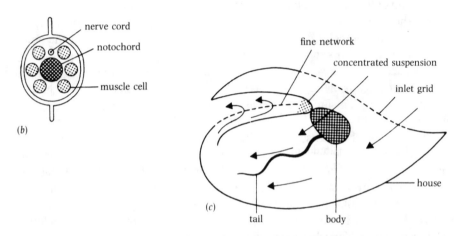

(b)

nerve cord

notochord

muscle cell

(c)

fine network

concentrated suspension

inlet grid

house

tail

body

Fig. 12.2. Diagrams of (*a*), (*b*) a typical urochordate larva and a section through its tail, and (*c*) *Oikopleura*.

water but bivalves seem to catch their food particles on the branched laterofrontal cilia.

Apart from the structures used in feeding, the anatomy of sea squirts is remarkably simple. There is a single nerve ganglion in the body wall between the siphons. There are no special respiratory organs: the large surface area of the pharynx must be ample to enable the animal to take up the oxygen it needs from the feeding current. There seem to be no special organs of excretion or of osmotic and ionic regulation. Ammonia is formed as a waste product of protein metabolism and simply diffuses out of the body.

There is a system of blood vessels, and a heart which is simply a U-shaped tube. Muscular constrictions travel along the heart, driving the blood through. For a few minutes, all the constrictions travel in the same direction and the blood flows one way round the circulation. Then constrictions start travelling in the opposite direction and flow is reversed for a few minutes.

Though the adults of typical sea squirts are stationary, anchored to rocks, they develop from swimming larvae. The larvae of most species are about a millimetre long. They do not feed, but generally settle and start metamorphosis to the adult form after only a few hours.

The larvae look like tiny tadpoles (Fig. 12.2*a*). They swim in the manner of tadpoles, at speeds up to about 3 cm s^{-1}, by tail movements. Running the length of the tail is the notochord, which is a row of 40–42 cylindrical cells with large vacuoles enclosed in a sheath of connective tissue fibres. On either side of it are a few (usually three) rows of muscle cells. These are large cells with myofilaments in their outer layers, and a central core of vacuolated cytoplasm. Swimming must involve alternate contraction of the cells on the left and on the right, bending the tail from side to side. The notochord presumably functions in the same way as the vertebral column of vertebrates, as a structure which is flexible but of fixed length. It cannot shorten because the sheath prevents it from swelling. If it were not there the muscles on both sides of the tail could contract simultaneously, shortening the tail. Because it is there shortening of the muscles on one side lengthens those on the other. The notochord makes the muscles on either side of the tail antagonistic to one another.

Dorsal to the notochord is a tubular nerve cord which is swollen at the anterior end, where it contains two simple organs. These seem from their structure to be a light detector and a sense organ sensitive to tilting. The part of the nerve cord which lies within the trunk of the 'tadpole' is relatively stout and seems to be nervous tissue. The part in the tail is slim, and its cells do not look like nerve cells. It has no obvious function.

The larvae of different ascidians behave in ways which seem likely to take them to suitable sites for settlement, and metamorphosis to the stationary adult. Most swim upwards at first, and later downwards and away from light. This tends to take them to sites such as overhanging rock faces where they can attach themselves firmly and where they are unlikely to get covered by sediment.

Oikopleura (Fig. 12.2*c*) is a member of a peculiar group of urochordates found in marine plankton. They develop from tadpole-like larvae like those of sea squirts but retain the tail as a permanent organ. The body and tail together of *Oikopleura* are only about 5 mm long but it lives in a larger gelatinous 'house' so delicate and transparent as to be extremely difficult to study. It undulates its tail, drawing a current of water through the house and driving the house slowly through the water. Water enters the house through a relatively coarse grid. As it leaves it encounters a large sheet of very much finer mesh, fine enough to stop particles less than 1 μm in diameter. The water itself flows through but the suspended matter accumulates in the position shown in the diagram, close to the animal's mouth.

Oikopleura has only one pair of openings in the wall of its pharynx. It has no atrium, so they open directly to the exterior. Water drawn through them by cilia is filtered through a mucus sheet inside the pharynx in much the same way as in sea squirts. Thus the animal feeds on the suspension that has already been concentrated in the house.

It has been suggested that the vertebrates may have evolved from urochordates by a process of neoteny: that is, by retention of larval features in adult animals. The notochord and tail muscles are found in the larvae of typical sea squirts, but not in the adults. They may have been evolved initially in urochordates, as larval features which disappeared immediately after settlement. Neoteny may have occurred subsequently, producing adult filter-feeding animals which retained the notochord and the tail, with its muscles. Such animals may have been the ancestors of amphioxus (which will be described next) and the vertebrates. *Oikopleura* and its close relatives seem much too peculiar to be ancestors of the vertebrates, but they show that neotenous urochordates are feasible.

The urochordates are quite a large and diverse group but the cephalochordates include only the twenty-odd species of *Branchiostoma* and *Asymmetron*. The familiar name *Amphioxus* has had to be abandoned because the name *Branchiostoma* was used in the original description, but amphioxus (without an initial capital, and not italicized) is often used as the common English name.

Amphioxus are small cigar-shaped animals, up to about 7 cm long. They spend their adult life in the sea bottom, buried in sand or shell gravel. They are very common in some places: a handful of sand from the Lagos area may contain 30–40 *Branchiostoma nigeriense*, and *B. belcheri* is common enough on the Chinese coast to be collected as food.

Careful studies of the distribution of *B. nigeriense* around Lagos showed that it lives only in areas where there is coarse sand with little silt in it. Laboratory experiments were made as well as field observations. Specimens were put in dishes of seawater, with various grades of sand and mud on the bottom. When given mud, they generally rested on the surface. When given fine sand, they buried only the hind part of the body. In either case they were restless, apt to swim if disturbed. This behaviour must tend to make them move away from fine sediments in nature. When given coarse sand they burrowed, penetrating the sand head first and moving through it by eel-like undulations. They either remained under the sand or adopted a position with the mouth, and very little else, above the surface (Fig. 12.3c). Burrowing must make them inconspicuous and inaccessible to predators. They cannot easily penetrate excessively coarse sand.

Amphioxus are filter feeders, using a mechanism very like that of sea squirts. Water enters by the mouth and leaves by the atrial opening, about half way along the length of the body. Feeding is possible in coarse sand even when the animal is completely buried, because water can permeate easily between the grains.

The main structures used in feeding are shown in Fig. 12.3. Two sets of tentacles keep excessively large particles out of the mouth. The wall of the pharynx is not a network as in sea squirts, but is perforated by a series of narrow, sloping slits. The bars between the slits have lateral and frontal cilia, as in sea squirts. The lateral cilia drive water through from the mouth to the atrium. The frontal ones move a sheet of mucus which covers the slits on the inner side.

Small amphioxus are remarkably transparent; if they are fed a suspension of carmine particles, for instance, the particles trapped in the mucus can be seen through the body wall. The movements of the mucus, both in the pharynx and in the gut, have been observed in this way. The mucus is secreted by a ventral endostyle. It is moved dorsally by the frontal cilia to a median groove in the roof of the pharynx, crossing the pharyngeal slits obliquely and trapping food particles from the feeding current which is filtered through it. Cilia in the median groove move the mucus and trapped particles posteriorly to the digestive parts of the gut where a ring of cilia makes the string of mucus rotate.

A deep narrow pouch, the caecum, branches off the gut immediately posterior to the pharynx and runs forward on the right side of the pharynx. It seems to function

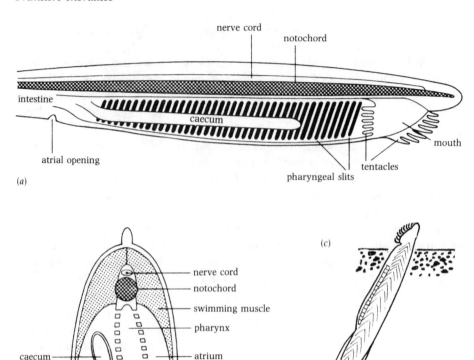

Fig. 12.3.(a), (b) A diagram of an amphioxus and a transverse section through its pharynx. (c) A sketch of an amphioxus in a natural feeding position in coarse sand.

like the digestive glands of molluscs (section 6.10). Cilia beat anteriorly on its roof, and posteriorly on its floor. Small particles become detached from the string of mucus in the gut and move into the caecum. The cells of the caecum secrete digestive enzymes and also ingest fragments of food material which are digested further in vacuoles. Both the liver and pancreas of vertebrates develop as pouches which grow out of the gut in the position of the caecum of amphioxus. The caecum resembles the pancreas in secreting digestive enzymes, but has a blood supply very like the supply to the liver.

Amphioxus has a complicated system of blood vessels, arranged very much as in fish (Fig. 12.4). The sinus venosus is not a heart, but simply a swelling where several vessels meet. Blood flows anteriorly from it in the endostylar artery, then dorsally in the branchial vessels to the dorsal aorta in which it flows posteriorly. It is distributed to the tissues by branches of the dorsal aorta. Blood from the tissues is collected in veins. The veins from all parts of the body except the gut empty into the cardinal veins, on either side of the body, which lead to the sinus venosus. The subintestinal vein collects blood from the network of fine vessels around the digestive part of the gut and delivers it to the caecum where it breaks up into another network before re-forming as the so-called hepatic vein, which goes to the sinus venosus. Hepatic vein is an inappropriate name because the caecum is not a liver, but the subintestinal and

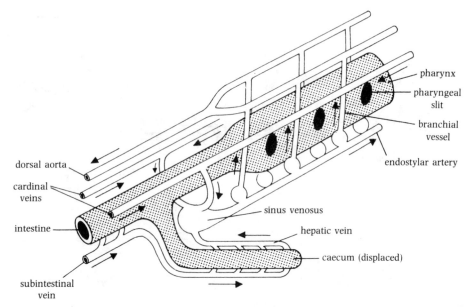

dorsal aorta

cardinal veins

intestine

subintestinal vein

pharynx

pharyngeal slit

branchial vessel

endostylar artery

sinus venosus

hepatic vein

caecum (displaced)

Fig. 12.4. A diagram showing the main features of the blood circulation of amphioxus.

hepatic veins of amphioxus are arranged in just the same way as the hepatic portal and hepatic veins of fish. The blood is propelled by pulsation of the endostylar artery, of the hepatic vein and of the swellings at the bases of the branchial vessels. The blood has no corpuscles, and the layer of cells (endothelium) which encloses the vessels is incomplete.

Amphioxus has nothing comparable to the kidneys of vertebrates, but has protonephridia which presumably work like those of flatworms and rotifers (section 4.8) although their flagella are single instead of forming flame-like bundles.

The notochord consists of a series of discs of muscle with fluid-filled spaces between them, enclosed in a sheath of collagen fibres. The muscle fibres run transversely across it. Notochords dissected from amphioxus become stiffer when stimulated electrically, presumably owing to contraction of the muscle fibres. Stiffening could possibly aid fast swimming by increasing the natural frequency of side-to-side vibration of the animal. Stiffening may also aid burrowing. The notochord extends further forward than in fishes, almost to the tip of the snout. It therefore stiffens the snout, which may aid penetration of sand.

The muscles that are used for swimming and burrowing are divided into myomeres (muscle segments) by partitions of collagenous connective tissue. The myomeres resemble those of fishes but are simpler in shape.

There is a hollow dorsal nerve cord contained in a tube of collagen fibres which runs along the top of the notochord sheath (Fig. 12.3b). Though there are no vertebrae, these collagenous structures enclose the nerve cord and notochord in essentially the same way as do the vertebrae of many fish. There is no swelling of the anterior end of the nerve cord which might be compared with the brains of vertebrates, but the central canal is enlarged at the anterior end. The nerve cord sends a

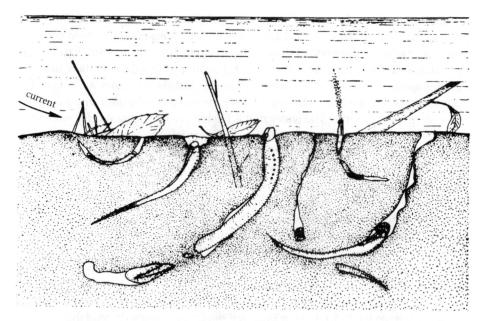

current

Fig. 12.5. Ammocoetes in their natural habitat. From V.C. Applegate (1950). *Spec. scient. Rep. U.S. Fish Wildl. Serv.* 55, 1–237.

motor nerve to each myomere, and gives rise to an equal number of sensory nerves. The motor nerves leave the cord ventrally and the sensory ones dorsally, as in vertebrates.

12.2. Lampreys

The lampreys are the first vertebrates to be described in this book. They are the best known members of the class Agnatha, which consists of primitive fish that have no jaws.

  Most adult vertebrates have a vertebral column that has taken over the function of the notochord. Lampreys, however, have only nodules of cartilage alongside the notochord, and hagfishes (section 12.3) have no vertebral column. The vertebral column is not the most characteristic feature of the vertebrates, despite their name. More characteristic are the brain, the skull, the ears, the kidneys and various other organs, which are basically similar in all vertebrates and quite different from any organs found in invertebrates.

  Lampreys lay their eggs in fresh water. Their larvae (known as ammocoetes) hatch out and grow there, but many species of lamprey spend their adult life in the sea. Both the ammocoete and the adult have slender, eel-like bodies but the adult has a characteristic sucker around its mouth and the larva does not. Ammocoetes are small, up to about 20 cm long, but adults of some species grow to lengths approaching 1 m and masses over 2.5 kg. The larval stage lasts several years, often about five. The duration of the adult stage varies greatly between species but death generally occurs soon after breeding.

Ammocoetes bury themselves in mud on the bottom of slow-flowing streams and seldom leave their burrows (Fig. 12.5). They filter diatoms and other food particles from water which is drawn in at the mouth and passed out through small round gill openings. They live in much finer sediment than amphioxus does, so there must be substantial resistance to flow of water from the gills, but low flow rates suffice because their feeding current is drawn from just above the surface of the mud where diatoms and detritus tend to settle. The water that is filtered probably contains far more food than the bulk of the stream water. The incoming water is strained through a fringe of short tentacles. The large particles that are prevented in this way from entering the mouth are blown clear from time to time by a sort of coughing action (one of the ammocoetes in Fig. 12.5 is doing this).

There are only seven gill openings on each side of the body. Because they are few and small, the pharynx is far less delicate than in amphioxus. It is not protected by an atrium. There are flaps of skin over the openings which act as valves (Fig. 12.6*a*). Water can leave through them but is prevented from entering: any inward movement of water through the openings makes the flaps close over them. Similarly, water can enter through the mouth but is prevented from leaving that way (except in coughing) by the movements of the velum, a pair of flaps just inside the mouth. It can be shown that flow is indeed in one direction only, by releasing drops of dye into the water near an ammocoete. Dye released near the mouth enters the mouth and emerges from the gill openings.

The water is not propelled by cilia as in *Branchiostoma*, but by muscular action. The movements involved have been studied by putting ammocoetes in glass tubes (to simulate a burrow) and watching them through a microscope. They are transparent enough for the velum to be visible through the body wall, and it can be seen to beat rhythmically. Often the water seems to be pumped by the velum alone. Sometimes, however, the whole pharyngeal region of the body contracts in time with the beating velum. Its contractions reduce the volume of the mouth cavity, and also of the parabranchial cavities (Fig. 12.6*a*) which lie between the gills and gill openings. These cavities are enclosed in a framework of cartilage bars, known as the branchial basket, and by a sheet of muscle fibres which run circumferentially round the body (Fig. 12.6*b*). They are compressed when the muscle contracts and water is driven out of them through the gill openings. When they enlarge again water is drawn in through the mouth. There do not seem to be any muscles to produce this enlargement, which is apparently due to elastic recoil of the branchial basket. Note that the cartilage hoops of the branchial basket have inwardly directed kinks in them. When the circumferential muscles contract these kinks become more bent (Fig. 12.6*c*), but when they relax the kinks recoil elastically to their former shape.

The food is captured in a sheet of mucus in essentially the same way as in *Branchiostoma* and the tunicates. The ammocoete has an endostyle but the mucus seems to be secreted elsewhere, by cells in the gills and in the walls of the parabranchial cavities.

The mucus filter and the gills lie between the mouth cavity and the parabranchial cavities. When the cavities are compressed, water is driven through the filter and gills from the contracting mouth cavity. When they expand, water is sucked through the filter and gills by the expanding parabranchial cavities. Thus water may be kept

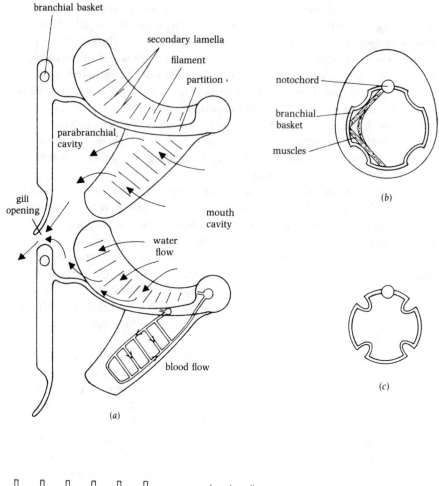

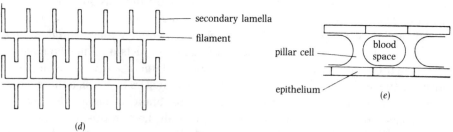

Fig. 12.6. Diagrams of the gills of ammocoetes. (*a*) Two gill arches in horizontal section. Arrows indicate the path of water between the secondary lamellae. The path of the blood is indicated on the filament at the bottom. (*b*) A transverse section of the gill region, with branchial basket expanded. (*c*) The branchial basket compressed. (*d*) A section through two gill filaments, at right angles to the secondary lamellae. (*e*) A section through a secondary lamella.

flowing continuously, like the respiratory currents of cephalopod molluscs (section 6.6). Compare two possible patterns of flow, continuous flow at a steady rate, and intermittent flow at double the rate for only half the time. The average flow rate is the same but the intermittent flow will need more power to drive it. The energy required by a pump is the volume of fluid pumped multiplied by the pressure difference. The power is the rate of flow (volume per unit time) multiplied by the pressure difference. The faster the pump works the more pressure is needed, and there are hydrodynamic reasons for expecting pressure difference to be proportional to flow rate in the ammocoete. Hence power can be expected to be proportional to the square of flow rate. The intermittent flow which is being considered requires four times the power needed for continuous flow, for half the time, so the average power is doubled.

The flow of water over the gills serves for respiration as well as for feeding. In other fish it serves only for respiration but the advantage of continuous flow remains: less energy is needed to pump water over the gills at a given rate, if it is kept flowing continuously, than if it flows intermittently. Continuous flow is also more favourable for exchange of oxygen and carbon dioxide between the water and the blood.

As in molluscs (section 6.6) the gills have a complicated structure which gives them a large surface area. A transverse partition separates each gill slit from the next. From it project horizontal gill filaments, like so many shelves (Fig. 12.6*a, d*). The filaments in turn bear secondary lamellae which project up from their upper surfaces and down from the lower ones. The secondary lamellae are closely spaced so their total surface area is large, and it is through their surfaces that most of the respiratory exchange of gases probably occurs. It has been estimated that their total area in a 1 g ammocoete of *Lampetra planeri* is about 0.7 cm^2, or about one tenth of the whole external area of the body. This is unimpressive in comparison with molluscs, many of which have respiratory surface areas that are larger than the external surface area of the body (section 6.6). It is, however, adequate for the ammocoete's inactive way of life. Teleost fishes of similar size generally have much larger gill areas, for example 7 cm^2 for a 1 g *Micropterus*.

The structure of a secondary lamella is shown in Fig. 12.6(*e*). Its two faces are formed by thin epithelia. If these were not held together in some way they would be pushed apart by the pressure of the blood in the spaces between them, so that the lamellae were not thin and flat-faced but were inflated to a bulbous shape. They are held together by pillar cells which extend (like so many pillars) from one epithelium to the other. Extensions of the pillar cells underlie the whole of the epithelia so that the blood is everywhere separated from the water outside the lamella by two layers of cells. However, the layers are together only about 4 μm thick, so the water and blood are brought very close to each other. The thickness of the lamellae and the distance between pillar cells are just big enough to allow red blood corpuscles (diameter 10 μm) to pass. These cells contain the haemoglobin, the oxygen-carrying pigment of vertebrate blood. Its being enclosed in cells avoids the problem of colloid osmotic pressure discussed, for molluscs, in section 6.6.

The tips of the anterior and posterior filaments of a gill slit almost touch. So do the secondary lamellae of successive filaments. Most of the water passing through the gills must therefore take the route shown in Fig. 12.6(*a*), crossing the filaments in the channels between the secondary lamella and then travelling laterally in the space between the lamellae and the transverse partitions. Thus the water between the lamellae is constantly renewed, and the distances which oxygen must travel by diffusion are small.

Hydrodynamic considerations make it plain that flow in the spaces between the lamellae must be laminar, not turbulent. Since the lamellae are about 30 μm apart, oxygen in the water must diffuse 0–15 μm to reach the nearest lamella. It must then diffuse through the cells of the lamella wall. After this, little diffusion is probably needed: it is probable that the plasma is kept thoroughly stirred as a consequence of the close fit of the red corpuscles in the blood spaces. The average total distance that the oxygen must diffuse, through water and tissue, can be estimated as only about 10μm.

As in molluscs, blood flows through the lamellae in the opposite direction to the flow of water over them. I explained in section 6.6 that this makes for more effective gas exchange between water and blood.

The discussion so far indicates that the gills are constructed in such a way as to allow rapid uptake of oxygen from the water. Their area is large, diffusion distances are small, flow is probably more or less continuous and there is a counter-flow arrangement. How well do they meet the needs of the ammocoete?

When they are resting, 1 g *Ichthyomyzon* ammocoetes use about 50 mm³ oxygen per hour (at 15°C). The measurements were made while the animals were buried in a layer of fine glass beads, because ammocoetes removed from their burrows tend to swim around. We have seen that the gills have an area of 0.7 cm² and that oxygen has to diffuse 10 μm from water to blood. By putting these figures in the diffusion equation (equation 4.1) we find that the difference in partial pressure of oxygen, between the water and the blood, must be about 0.05 atm. (I used a diffusion constant intermediate between those for water and for tissues because the diffusion path is partly through each.) The partial pressure of dissolved oxygen in well-aerated water is 0.2 atm, so the gills are more than adequate for animals resting there.

Adult lampreys have suckers round their mouths, and use them to anchor themselves to stones in streams, or to attach to prey. Some lampreys attach themselves to fish and suck their blood or tear out lumps of flesh. Species that take only blood may remain attached for many days (at least in aquaria) without killing their host, but flesh-eating species soon kill their prey. Other species of lamprey do not feed after becoming adult, but simply breed and die.

The sucker is round when spread open for use, but its lateral margins are generally drawn together when the lamprey swims. The edges of the sucker are studded with tooth-like knobs of keratin which must help to prevent it from slipping when attached to the slippery surface of a fish. In the floor of the mouth cavity is a structure known as the piston which is stiffened by a long central cartilage. Smaller cartilages at the tip of the piston bear horny teeth, which are used to rasp wounds in prey. This is done by the muscles of the piston; alternate contraction of the dorsal and ventral piston muscles makes the teeth rock up and down (Fig. 12.7).

The sucker is muscular, and also contains a ring of cartilage. It seems to work in essentially the same way as a rubber sucker. It is spread open by its muscles, and placed against the surface to which it is to be attached. Any force which tends to enlarge the space between the sucker and the surface will then fix it. Elastic forces which remain after the spreading muscles have relaxed may suffice for this (rubber suckers depend on elasticity), but muscles may also play an active part in maintaining suction. When the lamprey is feeding, blood is passed from the sucker into the gut without detaching the sucker. The mechanism has not been fully explained.

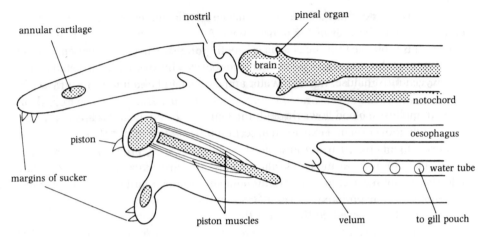

Fig. 12.7. A diagrammatic median section of the head of an adult lamprey.

The blood of the prey is prevented from clotting by the secretion of two large glands which open into the mouth cavity. The secretion also contains enzymes which attack the tissues of the prey.

While the sucker is attached the lamprey cannot breathe in through its mouth. Indeed, adult lampreys do not normally breathe in through their mouths even when they are not attached. This can be demonstrated by releasing a suspension of carmine particles or Indian ink from a pipette near the animal. Particles released near the mouth do not generally enter the mouth but particles released near the gill openings enter the openings as the lamprey breathes in and emerge from them again as it breathes out.

If the gills opened from the pharynx in the same way as in the ammocoete, blood which was being swallowed would get mixed with respiratory water and would be apt to escape through the gill openings. This does not happen because the pharynx of the adult lamprey is divided to form a dorsal oesophagus, which is the passage for food, and a ventral water tube from which the gills open (Fig. 12.7). The opening of the water tube into the oesophagus can be opened and closed. The water tube is narrow, so the openings from it to the gills are necessarily small. The gills are contained in pouches of much larger diameter than their small openings to the exterior and to the water tube. These pouches are compressed when the muscles around them contract, and expanded by the elastic recoil of the branchial basket.

Breathing movements serve also for sniffing. There is a single nostril on top of the head (Fig. 12.7). A tube leads down from it to the olfactory epithelium (where the odour-sensitive cells are) immediately anterior to the brain. It carries on into a sac which lies within the branchial basket and is compressed and expanded by the breathing movements, in the same way as the gill pouches. Water is sucked into the nostril whenever the lamprey breathes in, and expelled when it breathes out. A flap tends to deflect some of this water so that it passes over the olfactory epithelium.

Ammocoetes have rudimentary eyes hidden under the skin, but lampreys have functional eyes. The eyes of lampreys and other fish are very much like the eyes of cephalopods (section 6.8), with spherical lenses that have the refractive index increasing towards the core. Different groups of vertebrates have different arrangements of muscles that focus the eye by moving or distorting the lens.

It will be evident from the above that ammocoetes undergo a remarkable metamorphosis when they change to the adult form. A sucker and a piston develop and the water tube becomes separated from the oesophagus. The eyes move from their original position deep under the skin to the surface of the head. These and other changes fit the animal for a drastic change in feeding habits. There is a great advantage in having them occur rapidly, for while they are in progress the lamprey is not well fitted for either mode of feeding. The change in external appearance takes only about a month but internal changes may take longer and feeding is generally interrupted for several months.

12.3. Other agnathans

As well as the lampreys, the class Agnatha includes the hagfishes (order Myxiniformes) and several orders of extinct fishes known collectively as ostracoderms.

The hagfishes are a small group of bottom-living marine fish. Like lampreys, they have a rather eel-like shape, no jaws, no pectoral or pelvic fins, gill pouches with small external openings and no scales. In a great many other respects they are strikingly different from lampreys.

The North Atlantic species *Myxine glutinosa* is about 60 cm long. It generally lives on the bottom at depths of 100 m or more, so it is hard to observe. A Japanese hagfish, *Eptatretus burgeri*, lives in much shallower water, and a population at 10 m has been studied by scuba diving. The muddy bottom was so soft that the fish could bury themselves, simply by swimming down into the mud. They swam in head first but made U-shaped burrows so that the head protruded again from the mud. They spend the day buried with only their heads visible, but emerged at night to swim around, presumably searching for food.

Fish caught on lines at the bottom in areas where hagfishes are plentiful are often attacked by them. The hagfishes cut a hole through the skin of the fish and then eat out its viscera and flesh, leaving the rest of the skin intact. By the time the fisherman hauls his catch to the surface it may be little more than a skin containing a skeleton and a group of well-fed hagfishes. A study of the gut contents of hagfishes (*Myxine*) caught in the North Sea showed that they had been feeding mainly on shrimps.

Though hagfishes have no jaws they have a flexible plate of cartilage armed with horny (keratinous) teeth. It is normally kept folded in half in the mouth cavity (with the teeth facing inwards) but it can be moved forward (onto the animal's chin, as it were) and opened out (Fig. 12.8). In feeding, this plate moves in and out of the mouth, opening and closing. The teeth rasp at the prey and (since they point posteriorly) tend to pull chunks of prey into the mouth. The hagfish cannot hold on to its prey as it rasps; specimens eating dead fish in an aquarium kept themselves in position by swimming movements.

Like lampreys, hagfishes have only one nostril. It opens on the snout rather than the top of the head and it does not end in a blind sac but opens into the mouth cavity.

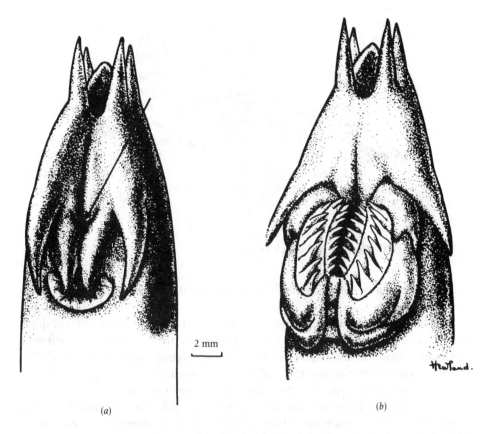

2 mm

(a)   (b)

Fig. 12.8. Ventral views of the head of the hagfish *Myxine* with the toothplates (*a*) withdrawn and (*b*) everted. From A. Brodal & R. Fänge (eds) (1963). *Biology of Myxine*. Oslo University Press, Oslo.

There are five to fifteen pairs of gill pouches, which open directly from the pharynx as in ammocoetes, not from a separate water tube as in adult lampreys. Since hagfishes take solid food, which is relatively unlikely to escape through the gills, there is no need for a separate water tube.

Ostracoderms have been found as fossils in rocks of the Cambrian to Devonian periods, so must have lived 350–570 years ago (Table 1.1). The earliest are also the earliest known vertebrates. Most ostracoderms were quite small: only a few were more than 30 cm long. Unlike lampreys and hagfishes, most of the ostracoderms that have survived as fossils have a covering of thick scales or plates of bone: the name 'ostracoderm' is derived from Greek words meaning 'with skin like earthenware'.

In other respects, the various orders of ostracoderms are very different from each other. The best known is the order Osteostraci, of which *Hemicyclaspis* (Fig. 12.9) is an example. Its head was enclosed in a bony carapace. Its trunk was roughly triangular in section with a dorsal line of scales and a pair of ventrolateral ridges emphasizing the corners of the triangle. The flat belly would help to make the fish stable when it rested on the bottom. The tail was of the type called heterocercal, like the shark tails which will be described and explained in section 13.3. Many

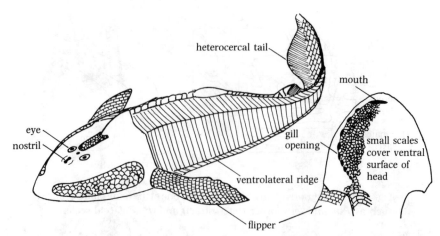

Fig. 12.9. *Hemicyclaspis*, an ostracoderm. Length 20 cm.

Osteostraci had a pair of flippers in the position occupied by pectoral fins in more advanced fishes.

Most of the carapace was a single piece, but it was broken up into a mosaic of small plates in four areas. Three of these were on the dorsal surface and their function is uncertain. The fourth and largest covered most of the ventral surface of the head. The mouth was at the front of this area, and there were ten or so pairs of small, round gill openings around its edge. It is thought that this ventral mosaic area may have provided a sufficiently flexible floor to the mouth and gill cavities, to allow breathing by enlargement and contraction of the cavities. Similar movements may have been made to suck in food.

The internal structure of the heads of Osteostraci has been investigated by serial grinding. The fossil is ground away gradually, so that at each stage a flat section through it is visible. Each section is photographed. When the process is complete the fossil has been reduced to dust, but the investigator has a pile of photographs showing the series of sections through it. More information can often be got from the photographs than could be obtained from an intact fossil. This technique has been particularly successful with Osteostraci because there was apparently a great deal of cartilage in the head, and all the cartilage surfaces were covered with a thin layer of bone. The cartilage has vanished but the layer of bone can often be seen in the sections. This layer even covered the walls of all channels through the cartilage, so the paths of nerves and blood vessels can be traced.

The sections help to explain two holes in the top of the carapace, in the mid-line. One is a single nostril, opening into a structure that seems to have been remarkably like the nasal sacs of lampreys (Fig. 12.7). The other housed the pineal organ, an upward projection from the brain. Sections also show that Osteostraci had only two semicircular canals in each ear (semicircular canals are explained in section 12.6). Lampreys also have two semicircular canals and the hagfishes have only one, but other vertebrates have three.

Though the Osteostraci are so different from lampreys in shape and in being armoured, there are striking similarities. Like lampreys, they had no jaws, they had

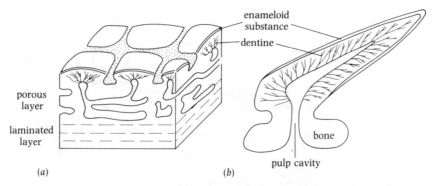

Fig. 12.10. Diagrams of (*a*) a block cut from one of the bony plates of an ostracoderm such as *Psammosteus* (order Heterostraci) and (*b*) a placoid scale.

numerous small, round gill openings, they had only two semicircular canals in each ear and they seem to have had the same peculiar sort of olfactory organ with a single dorsal nostril. These similarities suggest evolutionary relationship.

12.4.   Bone

Bone forms the skeletons of most vertebrates, though not of lampreys and hagfishes. It forms the scales of many fishes, including ostracoderms. It is a composite of inorganic crystals and protein, like mollusc shell, and is stronger than shell though not as stiff (Table 8.1). The inorganic crystals contain calcium, phosphate and hydroxyl ions in about the same proportions as hydroxyapatite ($3 \: Ca_3(PO_4)_2.Ca(OH)_2$). The protein is collagen, in the form of fibres. The crystals are tiny, commonly about 20 nm long, and seem to be firmly attached to the collagen. There are approximately equal volumes of inorganic crystals and collagen fibres but, because the crystals are denser, they make up considerably more than half the mass.

   A distinction is made between true bone and dentine or ivory, which is the main constituent of teeth. Typical bone has cells embedded in it, in tiny cavities with fine branches (canaliculi) reaching out into the surrounding hard tissue. Dentine has no embedded cells. The bodies of the cells that form it remain outside (in the pulp cavity in the case of teeth) but processes of the cells reach into the dentine along fine canals.

   Fig. 12.10(*a*) shows the structure of a bony plate from an ostracoderm. It has three layers: knobs of dentine, resting on porous bony material, which in turn rests on laminated bony material. The bodies of the dentine cells are in the pores of the porous layer. I have written 'bony material' because this is not typical bone. The porous layer has collagen fibres running in all directions but the laminated layer is built like plywood from thin sheets, in each of which all the fibres run parallel.

   The outermost surface of dentine is often covered by a layer of material containing a higher proportion of inorganic crystals. The enamel of mammal teeth is an example. It is harder than dentine, but more brittle. Similar substances found in fish are often called enameloid substances rather than enamel, because of evidence that they may not be formed in the same manner as is the enamel of mammals. It is not always easy to distinguish enameloid substance from dentine in fossils, but it seems to have been present in some ostracoderms.

The scales of sharks (called placoid scales) (Fig. 12.10b) also have an outer layer of enameloid substance with dentine and then bone below. The dentine cell bodies are in a central pulp cavity. There are cells enclosed in the bone when it is first formed, but they disappear later.

12.5.   Water and ions

This section is about how lampreys control their water content, and the concentrations of ions in their body fluids. It explains the workings of the kidney, which is quite different from any invertebrate organ though the 'kidneys' of molluscs are given the same name.

The blood of marine invertebrates has about the same osmotic concentration as seawater, and that of freshwater ones is much more dilute (see section 6.7, on molluscs). The osmotic concentration of lamprey blood plasma has been calculated from measurements of its freezing point and found to be about $0.25 \, mol \, l^{-1}$, whether the animals came from freshwater or from the sea, which has an osmotic concentration of about $1.0 \, mol \, l^{-1}$. Ammocoetes and lampreys in fresh water have plasma more concentrated than the water, so ions must tend to diffuse out from the blood and water must tend to diffuse in. Lampreys in the sea face the reverse situation: ions tend to diffuse in and water to diffuse out.

Consider first an ammocoete or lamprey in fresh water. To maintain the composition of its blood it must get rid of the water that diffuses in and replace the ions that diffuse out. It gets rid of the water as urine. It is not easy to collect and measure the urine produced by an ammocoete, so an indirect method of measuring it has been devised. When the animal is in fresh water the rate at which water diffuses in is matched by the rate of urine production. If it is transferred to a solution of the same osmotic concentration as its own blood water will stop diffusing in immediately, but it will presumably take a little time to halt urine production and during this time the animal will lose mass. It can be assumed that the initial rate of loss of mass after transfer is the rate at which urine is produced in fresh water. It has been found in this way that ammocoetes of *Lampetra planeri* produce about $200 \, cm^3$ urine kg body mass^{-1} day^{-1}, and adults of *L. fluviatilis* (in fresh water) $160 \, cm^3 \, kg^{-1} \, day^{-1}$. Even higher rates have been found when the urine of lampreys has been collected (for instance in a balloon tied round the tail), but these may have been unnaturally high: fish are apt to produce urine faster than usual when they are handled or interfered with.

The urine is much more dilute than the blood plasma, with an osmotic concentration of only about $40 \, mmol \, l^{-1}$, but since so much urine is produced the quantity of salts lost in this way is considerable. This loss is in addition to the loss by diffusion. The losses are apparently made good by uptake of ions through the gills. This has been demonstrated in experiments with ammocoetes which were put in water containing radioactive isotopes of salts. It was shown by autoradiography that ions were taken up rapidly by the gills, but hardly at all by the skin. There are cells in the gill filaments, between the lamellae, which are packed with mitochondria and are believed to be responsible for ion uptake. Adult lampreys living in fresh water have similar cells.

In the sea, lampreys must get rid of excess salts which diffuse into their bodies, and must replace the water that is lost. They are hard to catch in the sea and they unfortunately lose the ability to survive in seawater when they return to fresh water to breed. Experiments to try to find out how lampreys regulate the composition of their blood in the sea have therefore been done on lampreys caught on their way up-river, when they can no longer survive in pure seawater but can still survive in half-strength seawater. This has nearly twice the concentration of the blood plasma so it presents the lamprey with the same problem as seawater, in a less intense form.

The lampreys (*L. fluviatilis*) were anaesthetized. The water was shaken out of their gill cavities and any water in the gut or urine in the kidney ducts was squeezed out by massage. They were weighed. The papilla through which urine is excreted was closed by a ligature and the anus was plugged. The lampreys were allowed to recover from the anaesthetic, and then put into half-strength seawater containing a little of the indicator phenol red. After about 24 hours they were removed and re-weighed. The urine and gut contents were removed, measured and analysed. The net change in the mass of the body (excluding urine and gut contents) was noted.

Most of the animals lost 6% or more of their mass but two kept their mass more or less constant. These two had retained better than the others the ability to regulate their water content in a concentrated environment. They produced no urine, and none of the others produced more than 6 cm³ urine (kg body mass)$^{-1}$ day^{-1}. This is far less than the lampreys produced in fresh water. It was much too little to fill the kidney ducts in the course of the experiment, so the ligature on the urinary papilla presumably did not interfere with urine production.

The purpose of the phenol red was to find out how much water the lampreys drank and absorbed from the gut. Phenol red itself is apparently not absorbed, and it could not be excreted through the plugged anus. Hence the phenol red from all the water the lampreys swallowed remained in the gut. Its concentration in the gut fluid at the end of the experiment was measured by colorimetry, after treating the fluid with alkali to bring out the red colour. (Phenol red is an indicator, and is colourless in neutral and acid solutions.) If it was found, for instance, that the gut of a lamprey contained 1 cm³ fluid with three times the phenol red concentration of the aquarium water, it could be inferred that the lamprey had swallowed 3 cm³ water and absorbed 2 cm³. In this way it was shown that the two lampreys which maintained their mass absorbed 35 and 82 cm³ water kg^{-1} day^{-1}.

Table 12.1 gives more information about one of these lampreys (the other was similar). It shows that, though the blood plasma and the gut fluid had about the same total osmotic concentration, the concentrations of individual ions in them were very different. The gut fluid had very much higher concentrations of magnesium and sulphate ions than either the plasma or the environment. Apparently, these ions are selectively excluded from the body. The phenol red in the gut fluid showed that this particular lamprey had swallowed 3.1 times as much water as was left in the gut. If none of the magnesium and sulphate had been absorbed one would expect their concentrations in the gut fluid to be 3.1 times as high as in the environment: one could expect 84 mmol l^{-1} magnesium and 43 mmol l^{-1} sulphate. The observed

Table 12.1. *Data concerning one of the lampreys which maintained its weight in 50% seawater, in the experiment described in the text*

	Osmotic concentration (mmol l^{-1})	Concentration (mmol l^{-1})			
		Na^+	Mg^{2+}	Cl^-	SO_4^{2-}
Environment (50% seawater)	540	227	27	262	14
Blood plasma	300	155	4	132	3*
Gut fluid	280	64	93	143	63

*Measurement from other animals, in fresh water.
These data are from A.D. Pickering & R. Morris (1970). *J. exp. Biol.* **53**, 231–244.

concentrations are even higher than this: it appears that magnesium and sulphate may actually be excreted into the gut contents. Other ions (mainly sodium and chloride) must have been absorbed with the water, otherwise the total osmotic concentration of the gut fluid would be 3.1 times that of the environment. These ions must presumably have been excreted again, but they cannot have been excreted in urine since no urine was produced.

Some lampreys caught in upstream migration have in their gills mitochondrion-rich cells which are much larger than the ion-uptake cells. Of the fourteen experimental lampreys, only the two which were able to maintain their mass had them. It is believed that these cells excrete the sodium and chloride ions which are absorbed from the gut or which diffuse into other parts of the body.

In more recent experiments, lampreys caught in estuaries were successfully returned to full-strength seawater. They were kept in seawater for at least two weeks, and then urine was collected from them. Its osmotic concentration was 390 mmol l^{-1}, which is a little greater than the osmotic concentration of the blood plasma (310 mmol l^{-1}) but much less than that of the seawater (1040 mmol l^{-1}). Lampreys (and fishes in general) seem unable to produce urine much more concentrated than the plasma. They cannot therefore gain water by absorbing seawater and passing its salts in urine. The kidneys may nevertheless play a part in ionic regulation in the sea by excreting magnesium and sulphate. The urine of the lampreys kept in seawater contained 129 mmol l^{-1} of magnesium ions and 95 mmol l^{-1} of sulphate ions. These are much higher than the concentrations of these ions in the sea, and enormously higher than the concentrations in the blood plasma (Table 12.1).

A kidney consists of a large number of similar nephrons, packed together to form a more or less compact organ. Each nephron is a long slender, tangled tubule. In *Lampetra fluviatilis*, for instance, the nephrons are around 20 mm long. At one end the nephron leads into a branching collecting duct which collects urine from it and from many other nephrons and carries it to the kidney duct (this is known as the ureter in reptiles, birds and mammals, but not in fishes and amphibians, where its embryological origin is different). The other end of the nephron ends blindly, normally as a Bowman's capsule (Fig. 12.11). It is as though the end of the nephron had been blown up into a bulb, and the end of the bulb had then been pushed in to give it a wineglass-like shape. A group of blood capillaries (the glomerulus) fills the

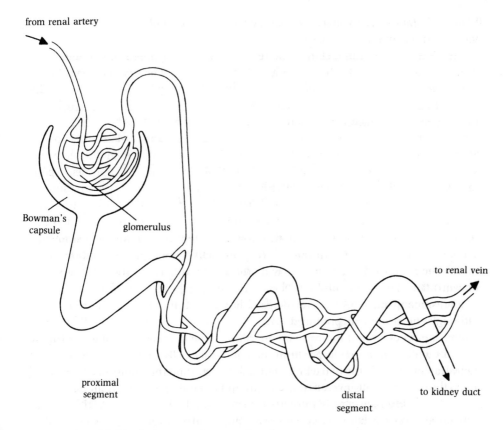

from renal artery

Bowman's
capsule

glomerulus

to renal vein

proximal
segment

distal
segment

to kidney duct

Fig. 12.11. A diagram of a typical nephron.

bowl of the wineglass. The blood in these is separated from the fluid in the nephron
only by the thin capillary walls and the thin wall of the capsule. The blood leaving the
glomerulus travels in an arteriole which breaks up again into a second set of
capillaries, investing the tubule.

Many experiments have been performed to find out how kidneys work. The most
informative ones have involved taking samples of the fluid contained in various parts
of nephrons. This is difficult, because the tubules are very slender, but it is easier in
some vertebrates than others, and has been achieved with the lamprey. The animal
was anaesthetized and its body cavity opened. A nephron was punctured and a tiny
sample of its contents withdrawn, using a glass pipette drawn out to a diameter of 10
μm at the tip. The pipette was held in a micromanipulator so that it could be moved
very accurately, and the operation was performed under a microscope. Many
vertebrates have their nephrons so tangled that it is difficult to discover whether the
part which has been punctured is near the proximal or the distal end of the nephron.
The nephrons of lampreys are arranged in unusually orderly fashion with corresponding parts side by side, and this difficulty does not arise.

Samples from the proximal segment of the nephron (Fig. 12.11) have about the
same osmotic concentration as the blood. This fluid seems to be an ultrafiltrate of the
blood like the pericardial fluid of molluscs (section 6.7). Along the distal segment in

lampreys kept in fresh water, the osmotic concentration falls gradually to the low value of the urine.

Experiments with inulin showed more clearly what was happening. Radioactive inulin was injected into the body cavities of lampreys kept in fresh water. Next day the lampreys were anaesthetized and samples of fluid were taken from the nephrons. The radioactivity of these samples was measured, to determine their inulin content. It was found that the inulin concentration in the proximal segment was the same as in the blood, confirming that inulin passes freely from the glomerulus into the capsule. More distally, the inulin concentration rose by a factor of 1.7. Either inulin was being secreted into the nephron, or water was being reabsorbed from it. In other experiments, radioactive inulin was injected into one of the blood vessels investing a nephron (not a glomerulus). Half this radioactivity appeared in the urine from the injected kidney and half from the other kidney. This suggests that inulin is not secreted into the tubular part of the nephron but enters only at Bowman's capsule. Thus water is apparently reabsorbed. It is probably reabsorbed osmotically, as a more-or-less inevitable consequence of the difference in osmotic concentration between the dilute urine and the blood.

The concentrations of ions in the fluid fell, as it travelled down the nephron, although water was being removed. Hence ions were also being removed. The total concentration of inorganic ions in the blood plasma and in the proximal parts of the nephron is eight times the concentration in the urine. The experiments with inulin indicate that the volume of water entering the nephron at Bowman's capsule is 1.7 times the volume of the final urine. Hence the quantity of ions entering the nephron is $8 \times 1.7 = 14$ times the quantity lost in the urine. About 13/14 or 93% of the ions are reabsorbed from the fluid as it travels down the nephron. This reabsorption must be an active process, unlike the probably passive reabsorption of water.

Similar experiments with lampreys kept in seawater showed that 90% of the small quantity of water entering the nephrons was reabsorbed. This is not enough to account for the magnesium concentration in the urine, which was 50 times the concentration in the plasma. Magnesium must have been being secreted into the nephron. There is evidence that it is secreted into the proximal segment.

It can now be seen how the kidney fulfils its function of getting rid of some constituents of the blood (notably certain ions, metabolic waste products and excess water) while retaining others. Filtration at the glomerulus prevents large molecules from escaping. Small, wanted molecules, such as glucose, and certain ions are recovered by reabsorption. Reabsorption of water, and secretion into the tubule, are both processes which make it possible to excrete unwanted materials at concentrations higher than their concentrations in the blood.

Metabolism of protein produces nitrogenous waste. In terrestrial vertebrates this is produced largely as urea and water, which are excreted by the kidneys. In fish it is mainly ammonia, and excretion of nitrogenous waste is not an important function of the kidneys; most of the ammonia diffuses out of the body at the gills. This has been demonstrated by fixing a fish in a small, divided tank of water, with its body passing through a tight-fitting hole in the partition. Ammonia appears in the water around its head very much faster than in the water around the posterior parts of the body.

12.6.   Lateral lines and ears

Vertebrates in general have ears. Fishes and aquatic amphibians also have lateral line sense organs, sensitive to water movements. Ears and lateral line organs have in common a distinctive type of sensory cell and are referred to together as the acoustico-lateralis system.

The characteristic cells of the acoustico-lateralis system are the hair cells, of which two are shown in Fig. 12.12(*b*). These cells are incorporated in an epithelium and each has a tuft of processes projecting above the epithelial surface. Within the tuft, the longest process is a cilium with the same structure as cilia that beat: it contains filaments arranged in the well-known $9 + 2$ pattern. The rest have a different structure, quite unlike cilia. Sensory neurons synapse with the hair cells; each generally synapses with several hair cells.

Hair cells are found in groups called neuromasts. Typically the tufts of processes of all the cells in the group are embedded in the base of a jelly-like structure called the cupula (Fig. 12.12*a*). There are neuromasts of this sort on the external surfaces of fish (including lampreys) and aquatic amphibians. Fish generally have a line of them running along either side of the trunk: this is the lateral line, and these neuromasts are called lateral line organs. There is a more complicated pattern of lateral line organs on the head. In each neuromast, half the hair cells have the cilium on one side of the bundle of processes and half on the diametrically opposite side. Wherever they lie on the body, lateral line neuromasts are served by cranial nerves, not by spinal nerves.

The physiology of lateral line organs has been studied in experiments both with fishes and with amphibians. Some particularly thorough experiments have been done on the clawed toad (*Xenopus*). The neuromasts were stimulated in various ways and action potentials were recorded from their nerves. When no stimulus was applied the neurons carried a steady succession of action potentials, following each other at a constant frequency. The frequency could be altered by pushing the cupula with a microneedle or by a jet of water from a pipette. The effect depends on the direction of movement. Suppose the neuromast is placed so that all the cilia lie either on the north side or on the south side of their tufts. North–south movements will affect the frequency of action potentials. There are two sensory neurons to the neuromast. One has the frequency increased by northward and decreased by southward movements; the other is affected in the opposite way. Presumably one is connected only to hair cells with north cilia and the other to hair cells with south cilia. East–west movements do not affect the frequency of either.

The lateral line neuromasts also respond to vibration of the water. They can thus detect sounds, at least sounds of low frequency. This does not, however, seem to be an important function. An object vibrating in a fluid has two effects. It pushes the nearby fluid out of its way as it moves, and it sets up sound waves. The first effect (the near-field effect) would occur even if the fluid were totally incompressible. The second (the far-field effect) depends on the compressibility of the fluid: sound waves are alternate compressions and rarefactions of the fluid. Close to the vibrating object, the fluid

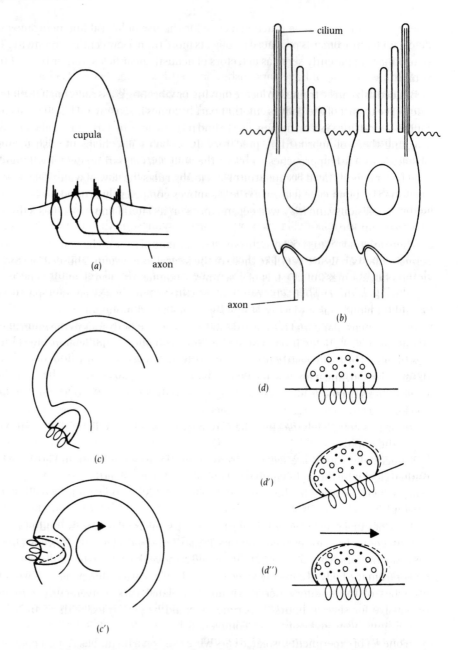

Fig. 12.12. Diagrams of (*a*) a lateral line neuromast; (*b*) two hair cells; (*c*) a semicircular canal, which is rotating in (*c*′); (*d*) an otolith organ, which is tilted in (*d*′) and is accelerating to the right in (*d*″).

movements are mainly due to the near-field effect: far from it, they are mainly due to the far-field effect. The two effects are equal at a distance of about 0.2 wavelengths. Lateral line neuromasts have been shown to respond to vibrations of up to 200 Hz: that is, to frequencies at which the wavelength of sound in water is 7 m or more.

There does not seem to be much evidence for the use of lateral line neuromasts to detect water movements produced by objects more than a few centimetres away. The neuromasts apparently serve as detectors of acoustic near fields, rather than of true sound.

Blinded fish can be trained to locate moving or vibrating glass rods, but if the lateral line nerve to part of the body is cut, that part becomes insensitive. This suggests uses for the lateral line system, helping fish to find moving prey in dark or turbid water and warning them of approaching predators. It probably also helps the fish to avoid obstacles: as a fish approaches a rock or the wall of its aquarium, water will tend to flow faster over its head because the obstacle obstructs the flow of water displaced by the fish. The blind cave fish *Anoptichthys* moves competently around an aquarium without collisions, and may well depend more on its lateral line neuromasts than do fish that can see.

There are two main types of neuromast in the ears of fishes. The neuromasts of the semicircular canals are very like those of the lateral line system. Those of the otolith organs have a mass of crystals of calcium carbonate, or even a single solid mass, embedded in the cupula. The ear is filled with a fluid called endolymph, which resembles blood plasma but contains less protein and more potassium.

Each semicircular canal is a curved tube connected at both ends to the main cavity of the ear (Fig. 12.12c). It has a swelling (ampulla) at one end, which contains the neuromast. The cupula is almost invisible because it has the same refractive index as the endolymph, and it tends to be shrivelled in histological preparations. However, its size has been demonstrated by injecting ink into the endolymph, so that it becomes conspicuous as a clear region in the black fluid. It reaches right across the ampulla, more or less completely blocking it. Any movement of fluid along the canal must move the cupula.

Semicircular canals are sensitive to rotation. If the canal shown in Fig. 12.12c is rotated clockwise the endolymph lags behind, owing to its inertia, so that it is flowing anticlockwise relative to the canal wall (Fig. 12.12c′). The cupula is deflected as shown.

A lot of our knowledge of the physiology of semicircular canals (and of otolith organs) comes from experiments on rays (*Raia*) done by Professor O. Lowenstein and his collaborators. They found that they could remove the part of the skull containing an ear from the body of a freshly killed ray, and still record action potentials from the branches of the auditory nerve within it. Isolated ears conveniently remained responsive for several hours. Experiments could be performed with them, which would have been awkward with complete fish.

In one set of experiments, isolated ears were fixed on a turntable. Action potentials were recorded from the nerve to a semicircular canal. Their frequency was constant while the turntable was stationary. When it was rotated the frequency increased or decreased, according to the direction of rotation.

Nearly all vertebrates have three semicircular canals (Fig. 12.13) in each ear. The exceptions are the lampreys, which have only two canals, and the hagfishes, which have only one (though with two neuromasts). Any rotation can be resolved into components about three axes mutually at right angles. When there are three canals they are set more or less at right angles to each other. Each responds only to

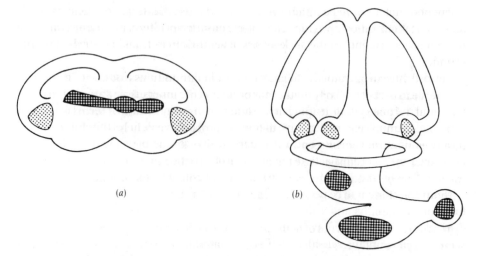

Fig. 12.13. Sketches of the ear, removed from the skull, of (*a*) a lamprey and (*b*) a teleost. The positions of the otoliths and cupulae within them are indicated by coarse and fine stipple, respectively.

components of rotation in its own plane and the three canals can between them provide full information about any rotation.

It is the horizontal canal which is missing in lampreys (Fig. 12.13), and it might be imagined that lampreys would be insensitive to rotation in the horizontal plane. However, it has been shown by recording action potentials from lamprey ears on a turntable that lampreys are, in fact, sensitive to such rotations. The structure of the ear seems to provide an explanation. The semicircular canal neuromasts are not in ampullae but in the main cavity of the ear at the ends of the canals. Each can be displaced by movement of endolymph along its canal (as in other vertebrates), but it can also be displaced (in another direction) by horizontal swirling of the endolymph in the main ear cavity. Other vertebrates have the hair cells of their semicircular canal all oriented in the same way, so as to be sensitive only to movements along the canal. Lampreys have some of the hair cells oriented in this way, but also have some arranged so as to be stimulated by cupula movements at right angles to the canal. These are apparently the hair cells which respond to rotation in the horizontal plane.

Otoliths behave quite differently from semicircular canal neuromasts because they are denser than endolymph. They tend to sink in the endolymph under the influence of gravity, so when the head is tilted their hair cells are affected (Fig. 12.12*d'*). They also tend to lag behind when the head accelerates (Fig. 12.12*d''*). It has been shown in experiments with ray ears that the frequency of action potentials in neurons from the otoliths is affected by tilting, and also by vibration, which is of course acceleration alternately in one direction and in the opposite one. Note that it is to linear, not angular, accelerations that the otolith organs respond. Sensitivity to vibration implies that the otoliths may function in hearing.

There is no way in which an otolith organ can distinguish between tilting and acceleration: the distinction can only be made with the help of other sense organs such as the eyes. This ambiguity seems to have been the cause of many aircraft accidents. A pilot, flying near the ground, accelerates (for instance, on deciding not to land). The acceleration gives him a spurious sensation of tilting, as if the aircraft were tilted nose-up. He moves the stick as if to level the aircraft, and crashes it into the ground.

Teleosts have three otoliths in each ear but lampreys have just one (Fig. 12.13). One surface of this otolith is attached to the floor of the inner ear and another to a side wall, and the hair cells that attach it are oriented in various ways. It seems likely that some of these hair cells are sensitive to up-and-down movements of the otolith, others to fore-and-aft movements and yet others to side-to-side movements. It has been confirmed by recording action potentials from particular groups of axons in the auditory nerve that lamprey ears are indeed sensitive to tilting.

Neuromasts seem to be peculiar to the vertebrates but sense organs with the same functions as otoliths and semicircular canals are found in some invertebrates. For example, some medusae have statocysts, containing a lump of dense material, that serve as sense organs sensitive to tilting. More complicated organs in crabs, also called statocysts, combine the functions of detecting rotation and tilting in a single organ, as in the vertebrate ear.

Further reading

General works on fishes

Alexander, R. McN. (1978). *Functional design in fishes*, 3rd edn. Hutchinson, London.
Bone, Q. & Marshall, N.B. (1982). *Biology of fishes*. Blackie, Glasgow.
Hoar, W.S. & Randall, D.J. (eds.) (1969–78). *Fish physiology* (7 vols). Academic Press, New York.
Love, M.S. & Cailliet, G.M. (eds.) (1979). *Readings in ichthyology*. Goodyear, Sta. Monica.
Webb, P.W. & Weihs, D. (1983). *Fish biomechanics*. Praeger, New York.

Sea squirts and amphioxus

Barrington, E.J.W. & Jefferies, R.P.S. (eds) (1975). Protochordates. *Symp. zool. Soc. Lond.* **36**, 1–361.
Goodbody, I. (1974). The physiology of ascidians. *Adv. mar. Biol.* **12**, 2–149.
Flood, P.R. & Fiala-Medioni, A. (1981). Ultrastructure and histochemistry of the food trapping mucous film in benthic filter-feeders (Ascidians). *Acta zool.* **62**, 53–65.

Lampreys

Gradwell, N. (1972). Hydrostatic pressures and movements of the lamprey, *Petromyzon*, during suction, olfaction and gill ventilation. *Can. J. Zool.* **50**, 1215–1223.
Hardisty, M.W. (1979). *Biology of the cyclostomes*. Chapman & Hall, London.
Mallatt, J. (1981). The suspension feeding mechanism of the larval lamprey *Petromyzon marinus*. *J. Zool., Lond.* **194**, 103–142.
Mallatt, J. (1982). Pumping rates and particle retention efficiencies of the larval lamprey, an unusual suspension feeder. *Biol. Bull.* **163**, 197–210.
Lewis, S.V. & Potter, I.C. (1982). A light and electron microscope study of the gills of larval lampreys (*Geotria australis*) with particular reference to the water-blood pathway. *J. Zool., Lond.* **198**, 157–176.

Rovainen, C.M. & Schieber, M.H. (1975). Ventilation of larval lampreys. *J. comp. Physiol.* **104**, 185–203.

Other agnathans

Brodal, A. & Fänge, R. (eds.) (1963). *Biology of Myxine.* Oslo University Press.
Fernholm, B. (1974). Diurnal variations in the behaviour of the hagfish *Eptatretus burgeri. Mar. Biol.* **27**, 351–356.
Moy-Thomas, J.A. (1971). *Palaeozoic fishes,* 2nd edn (revised by R.S. Miles). Chapman & Hall, London.

Bone

Currey, J.D. (1984). *The mechanical adaptation of bones.* Princeton University Press.
Halstead, L.B. (1974). *Vertebrate hard tissues.* Wykeham, London.
Ørvig, T. (1967). Phylogeny of tooth tissues: evolution of some calcified tissues in early vertebrates. In *Structural and chemical organisation of teeth* (ed. A.E.W. Miles), vol. 1, pp. 45–109. Academic Press, New York.

Water and ions

Logan, A.G., Moriarty, R.J. & Rankin, J.C. (1980). A micropuncture study of kidney function in the river lamprey, *Lampetra fluviatilis,* adapted to fresh water. *J. exp. Biol.* **85**, 137–147.
Logan, A.G., Morris, R. & Rankin, J.C. (1980). A micropuncture study of kidney function in the river lamprey, *Lampetra fluviatilis,* adapted to sea water. *J. exp. Biol.* **88**, 239–247.
McVicar, A.J. & Rankin, J.C. (1985). Dynamics of glomerular filtration in the river lamprey, *Lampetra fluviatilis. Am. J. Physiol.* **249**, F132–F138.
Pickering, A.D. & Morris, R. (1970). Osmoregulation of *Lampetra fluviatilis* L. and *Petromyzon marinus* (Cyclostomata) in hyperosmotic solutions. *J. exp. Biol.* **53**, 231–243.

Lateral lines and ears

Blaxter, J.H.S. (1987). Structure and development of the lateral line. *Biol. Rev.* **62**, 471–514.
Groen, J.J., Lowenstein, O. & Vendrik, A.J.H. (1952). The mechanical analysis of the responses from the end-organs of the horizontal semicircular canal in the isolated elasmobranch labyrinth. *J. Physiol., Lond.* **117**, 329–346.
Lowenstein, O. (1970). The electrophysiological study of the responses of the isolated labyrinth of the lamprey (*Lampetra fluviatilis*) to angular acceleration, tilting and mechanical vibration. *Proc. R. Soc. Lond.* B1**74**, 419–434.

13 *Sharks and some other fishes*

Class Acanthodii   (extinct)
Class Placodermi   (extinct)
Class Selachii   (sharks and rays)

13.1.   Sharks

Fig. 13.1 shows a shark, a member of the class Selachii. As well as median fins like those of lampreys and hagfishes it also has two *pairs* of fins, the pectoral and pelvic fins. It also has jaws.

The jaws are believed to have evolved from the gill skeleton, which is quite different from the branchial basket of ammocoetes and lampreys although the gills themselves are very similar. Selachian gills, like ammocoete ones (Fig. 12.6) have filaments and secondary lamellae. There are parabranchial cavities between the gills and the external openings, which however are tall slits (Fig. 13.1), not circular holes (Fig. 12.5). The selachian gill skeleton consists of bars of cartilage jointed together (Fig. 13.2a) instead of being an unjointed network like the branchial basket of lampreys. Its principal cartilages lie internal to the gills, whereas the branchial basket encloses the gills. Slender gill rays (not illustrated) extend laterally from the principal cartilages in the partitions between the gill slits.

Between each gill slit and the next, in selachians, is a 'gill arch' consisting of four cartilages arranged like the letter W (Fig. 13.2a). The hyomandibular cartilage and the hyoid bar seem to be modified elements of an anterior gill arch (the hyoid arch). They have a full-length gill slit behind them, but in front only a small round opening called the spiracle, which seems to be a vestigial gill slit.

The jaws consist of an upper palatoquadrate cartilage and a lower Meckel's cartilage. Embryological evidence suggests that they have evolved by modification of gill arches. The mesoderm of each segment in a vertebrate embryo differentiates into two main parts, the myotome and the sclerotome (Fig. 13.2b). The myotome becomes muscle and the sclerotome cartilage, which in some vertebrates later becomes bone. As the nerve cord is formed by rolling up of the neural plate, a group of cells known as the neural crest separates from alongside it. Different parts of the neural crest have different fates, forming spinal ganglia, pigment cells and cartilage. The skulls of vertebrates appear initially as cartilage, whether they are to remain cartilaginous as in selachians or to become largely bony. It can be shown by careful anatomical study of embryos, supplemented by experiments in which parts of the embryo are stained or damaged, that this cartilage develops from two sources. The main part of the cranium develops from sclerotome, as do the vertebrae. The gill arches, the jaws and the remainder of the cranium develop from neural crest. There is

323

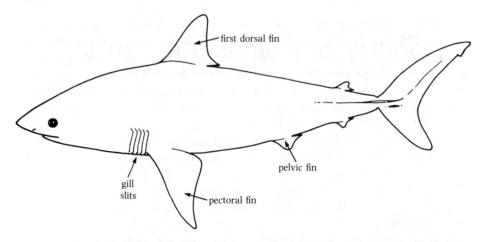

Fig. 13.1. **Porbeagle shark,** *Lamna nasus*(length up to 3 m). A drawing by Valeri Du Heaume from A. Wheeler (1969). *The fishes of the British Isles and North-West Europe.* Macmillan, London.

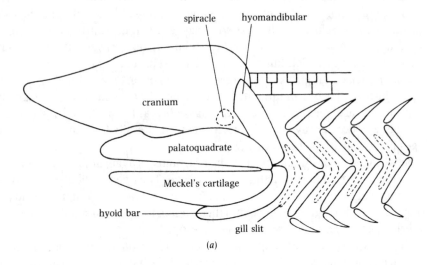

(a)

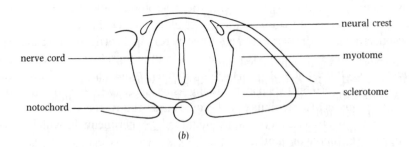

(b)

Fig. 13.2.(*a*) **A diagram of the principal cartilages of a selachian skull and (***b***) a diagrammatic transverse section of a vertebrate embryo.**

evidence in some fossil fishes (members of the Crossopterygii) of an additional rudimentary gill opening, above the palatoquadrate cartilage and anterior to the spiracle. This has led to the suggestion that there were originally two gill arches anterior to the hyoid arch. These are supposed to have formed the jaws. The parts of the cranium which develop from neural crest are supposed to be dorsal elements of these two arches and of the hyoid arch, which have become incorporated in the cranium.

The breathing movements of dogfish (*Scyliorhinus*: a selachian) can be observed when they are resting or swimming slowly in an aquarium. The anterior part of the body gets deeper and wider, then shallower and narrower, as the mouth cavity and parabranchial cavities enlarge and contract. The gill slits, the mouth and the spiracle open and close. The water movements involved can be made visible by introducing a little milk from a pipette. Milk released near the mouth is drawn in through the mouth and expelled through the last three of the five pairs of gill slits. Milk released near a spiracle is drawn in through the spiracle and expelled through the first three gill slits of the same side of the body. Some more active sharks do not make breathing movements as they swim, but keep water flowing over the gills by swimming with their mouths open.

Pressure transducers have been used to measure the pressures involved in breathing by lightly anaesthetized dogfish. They show that the pressure in the mouth cavity is greater than in the parabranchial cavities, both when the cavities are enlarging and when they are contracting. This is probably true also of ammoecoetes (section 12.2), but the pressure recordings that would be needed to demonstrate it for ammocoetes have not been made. Electromyographic recordings show that many muscles are involved in squeezing the gill region when it contracts but that none seem to be active to enlarge it, in gentle breathing. It apparently enlarges by elastic recoil. The gill skeleton is stiff enough to do this, although it is jointed.

The hyomandibular cartilage articulates with the cranium (Fig. 13.2*a*). It and the hyoid bar are attached to the jaws by ligaments. The anterior end of the palatoquadrate rests in a groove in the cranium. This arrangement allows the palatoquadrate to move slightly, relative to the cranium, in dogfish, and to make larger movements in grey sharks (family Carcharinidae).

Fig. 13.3(*a*) shows one of these sharks taking food. The sequence of pictures starts at the top. The cranium tilts up as the lower jaw swings down, so that the mouth opening faces forward. As the mouth closes the upper jaw is protruded from the ventral surface of the head.

Fig. 13.3(*b*) shows the movement more clearly. In the upper picture, the upper jaw is in its resting position. In the lower picture, the palatoquadrates have moved anteriorly and ventrally so that the upper jaw protrudes from the surface of the head. In this position, it is better placed for biting deeply into prey, or for picking up lumps of food from the bottom, but the shark's head is less well streamlined than when the jaws are retracted.

Grey sharks feed largely on fish that are small enough to be swallowed whole, but they also attack larger prey. Pieces of larger fish (including other sharks) and of porpoises have been found in their stomachs. They sometimes take bites out of people, when the opportunity arises. How are bites taken from large prey?

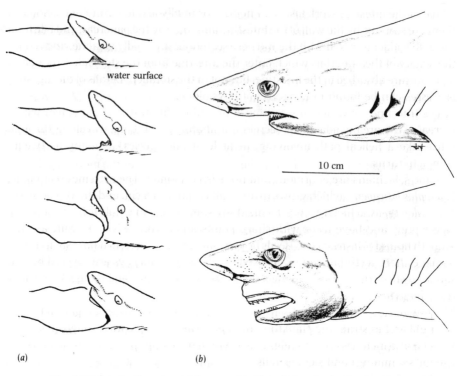

water surface

10 cm

(a)   (b)

Fig. 13.3.(*a*) **Outlines traced from a cine film of the shark** *Carcharinus acronotus* **feeding on a piece of fish which was held at the surface of the water. Similar movements are used in underwater feeding.** (*b*) **The head of a dead shark of the same species with the jaws in two positions. From T.H. Frazzetta & C.D. Prange (1987).** *Copeia* **1987, 979–993.**

The lower teeth of most grey sharks are slender spikes but the upper ones are broad triangles with sharp cutting edges. When the shark has gripped its prey it shakes its head violently. Because of the inertia of the prey this makes the triangular upper teeth move from side to side across the prey, cutting into it like the teeth of a saw. Protrusion enables the upper jaw to extend down into the kerf so that the cut can go considerably deeper than the height of the teeth.

The teeth of sharks have the same structure as the placoid scales that cover the body (Fig. 12.10). Their bases are embedded in the skin that covers the jaws. Rows of new teeth develop posterior to the teeth which are in use, and the skin slides slowly forward over the jaw to bring them into position. Old teeth are shed or destroyed. Teeth of young captive Lemon sharks (*Negaprion*) have been marked by clipping, and their fate followed. It was found that the teeth moved forward one row every eight days or so.

Unlike lampreys and hagfishes, selachians have stomachs. A stomach is generally wider than the intestine which follows it, and receives quite different secretions. Cells in the wall of a stomach secrete hydrochloric acid and the precursors of enzymes of the pepsin type, which attack proteins by breaking them into shorter peptide chains. Pepsins act fastest in acid conditions. By contrast, the many enzymes which are active in the intestine work fastest in neutral or mildly alkaline conditions. Some of them are secreted by the wall of the intestine and some by the pancreas, and bile from the gall bladder is mixed with them. These secretions are mildly alkaline, and reduce the acidity of the materials which enter the intestine from the stomach. Products of digestion are absorbed in the intestine, but not in the stomach. Details of the digestive processes can be found in textbooks of physiology.

The intestines of selachians contain a helical partition (the spiral valve) which increases the area available for secretion and absorption. Food travelling through must take a helical path involving, in at least one species, as many as twenty complete turns.

Some selachians are enormous and none is very small. The extremes seem to be *Squaliolus laticaudus*, which grows to an adult length of 15 cm, and the Whale shark *Rhincodon typus*, which reaches 15 m. Many sharks feed on whole fish and pieces of larger prey, and have teeth with sharp points and cutting edges. In contrast, the Smooth hound (*Mustelus canis*) feeds largely on crabs and lobsters whose shells might damage sharp teeth but can be crushed by blunt ones. Its jaws are covered by flat, square teeth arranged like paving stones. The Whale shark and the Basking shark (*Cetorhinus maximus*, up to 10 m long) feed on planktonic animals such as copepods. They swim slowly with their mouths open, and bristle-like processes (gill rakers) on their gill arches strain the plankton out of the water that passes through. The class Selachii also includes the rays (Fig. 13.6), which have huge pectoral fins (which they use for swimming) and slender tails.

13.2.   Early jawed fishes

The earliest known jawed fishes were not selachians but members of the class Acanthodii, which first appeared in the Ordovician period and became extinct in the Permian. They resembled modern jawed fish and differed from Agnatha not only in having jaws but also in the structure of the gill skeleton, in having pectoral and pelvic fins and in having three semicircular canals in each ear. They seem to be more closely related to the bony fishes (chapter 14) than the selachians: for example, their scales were similar in microscopic structure to those of early bony fishes. Each fin (except the tail fin) had a spine in its front edge, and primitive acanthodians also had a row of spines on each side of the body between the pectoral and pelvic fins (Fig. 13.4*a*). Notice that the lines of spines are comparable in position to the ridges that ran along the triangular-sectioned body of *Hemicyclaspis* (Fig. 12.9), which also had a spine in front of its dorsal fin.

The class Placodermi are another important group of early jawed fishes, found almost exclusively in the Devonian period. Most of them had a rigid shield of bone covering most of the head and another rigid shield enclosing the shoulder region (Fig. 13.4*b*). These were hinged together so that the head could nod up and down, which seems to be the only way in which the animal could have altered the sizes of its mouth and gill cavities to pump water over the gills, or to suck in food. The fins were very like those of early sharks, in internal structure as well as external appearance. It is widely believed that sharks and placoderms are closely related.

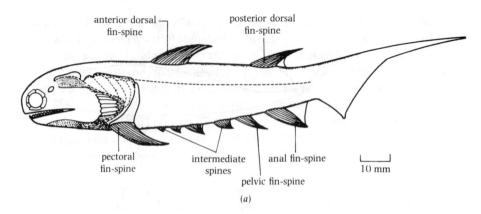

(a)

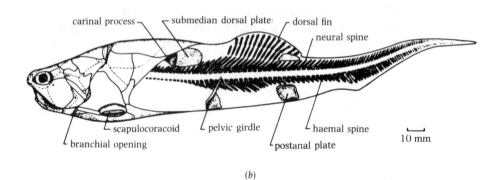

(b)

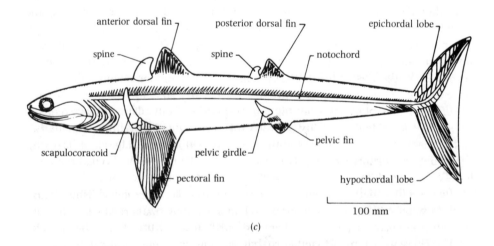

(c)

Fig. 13.4.(a) *Climatius,* an acanthodian; (b) *Coccosteus,* a placoderm; (c) *Cladoselache,* a Devonian shark. From J.A. Moy-Thomas (1971). *Palaeozoic fishes,* 2nd edn (revised by R.S. Miles). Chapman & Hall, London.

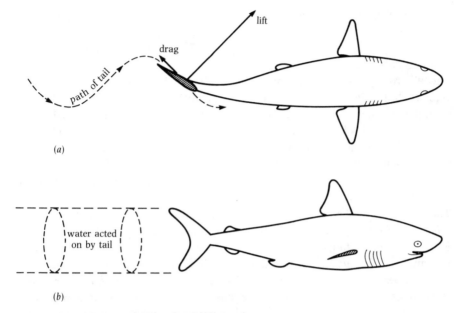

Fig. 13.5. Diagrams of a shark swimming.

The earliest known selachians are fossils such as *Cladoselache* (Fig. 13.4*c*), from the Devonian period. *Cladoselache* must have looked much like modern sharks but its mouth reached the end of the snout instead of being on the undersurface of the head (compare it with *Lamna* in Fig. 13.1). Also, the pectoral fin was triangular, without the narrow base of modern shark pectoral fins. There were spines in front of the dorsal fins. These are differences from most modern sharks but the frilled shark (*Chlamydoselachus*) has a mouth like that of *Cladoselache*, and the spiny dogfish (*Squalus*) has dorsal fin spines.

13.3.   Swimming and buoyancy

Lampreys and hagfishes swim like most other fishes, but I have deferred discussion of swimming to this chapter because it has been studied more thoroughly in selachians.

Fig. 13.5(*a*) shows a shark swimming. It moves its tail from side to side as it swims forward so the tail takes a wavy path through the water. (I have been careful, in drawing the diagram, to make the amplitude and wavelength of the path realistic fractions of body length.) The tail is a hydrofoil, shaped like an aeroplane wing, and is held at an angle of attack to its path (see Fig. 9.7*c*). As it moves to the right, the lift on it acts forward and to the left. As it moves to the left in the return stroke, the lift acts forward and to the right. The components to left and right cancel out and the fish is driven forward through the water. The lift on the tail propels the fish but the drag hinders it, so the larger the ratio of lift to drag the more efficient swimming will be. To get a high lift-to-drag ratio you need a hydrofoil with a high aspect ratio: that is, with a high ratio of span (the distance from tip to tip) to chord (from front edge to back edge). The tails of sharks such as *Lamna* (Fig. 13.1) have reasonably high aspect ratios. These sharks also have well-streamlined bodies, seeming well adapted for efficient swimming.

The tail propels the fish by driving water backwards. It is a reasonable approximation to say that the water it pushes on forms a cylinder, of diameter equal to the span of the tail (Fig. 13.5*b*). The more water it pushes on, the more efficient will swimming be (section 6.5). This leads to the same conclusion as the previous paragraph: a tail with a large span will propel a fish more efficiently than a smaller-span tail of the same area. The energy cost of shark swimming has not been measured but teleost fishes swim much more economically than squids, apparently because they push on more water (Table 6.2).

There is little published information about shark swimming speeds, but 1 m *Carcharhinus* in a large aquarium have been filmed swimming at 0.5–4 m s^{-1}.

Fig. 13.5 seems a fairly realistic representation of swimming by sharks such as *Lamna* and *Carcharhinus*, whose tails have relatively high aspect ratios. It is much less realistic for slender selachians such as dogfish (Fig. 13.10), still less for lampreys and eels. These fish do not have hydrofoil-like tails, and their swimming action does not even approximate to simple side-to-side movement of the tail. They throw themselves into waves which travel backwards along the body. The movements are like the swimming movements of nematode worms (Fig. 5.5*a*) (but the resulting water movements are very different because the Reynolds number is so much higher; see Fig. 2.15).

Rays swim by undulating their huge pectoral fins (Fig. 13.6*c*). Waves are passed backwards along the fin, driving the fish forwards.

The principal muscles of rays are the pectoral fin muscles, but more typical fishes have small fin muscles, and huge axial muscles on either side of the vertebral column. These muscles (the fillets that we eat) may make up more than 50% of the mass of the body. They are mostly white, but most fish have a narrow lateral strip of red muscle. (The colours remain after cooking, so you can see this if you eat fish.) In dogfish (*Scyliorhinus*) 18% of the swimming muscle is red.

The white muscle fibres resemble the white muscle of squid mantle (section 6.3): they have few mitochondria and a relatively poor blood supply. The red fibres have many mitochondria and a good blood supply, like the yellow muscle of squid mantle. Thus the white muscle seems adapted for anaerobic metabolism and the red for aerobic. The red colour is due to myoglobin, a haemoglobin-like compound that seems to be involved in transport of oxygen within the cells.

Long before the distinction between aerobic and anaerobic muscle was known for squid, Dr Quentin Bone demonstrated the distinct functions of red and white muscle in dogfish. He destroyed the fish's brain but kept the body alive by pumping aerated water over the gills. Dogfish prepared in this way still make swimming movements, if fixed by the head with the tail free in a tank of water. The movements continue at a slow, more or less steady rate, but if the tail is pinched the fish swims faster or makes a strong, sustained bend. Bone stuck electrodes into the red and white parts of the swimming muscle and recorded potentials from them. During slow swimming movements potentials were detected in the red muscle, but not in the white. (Dogfish

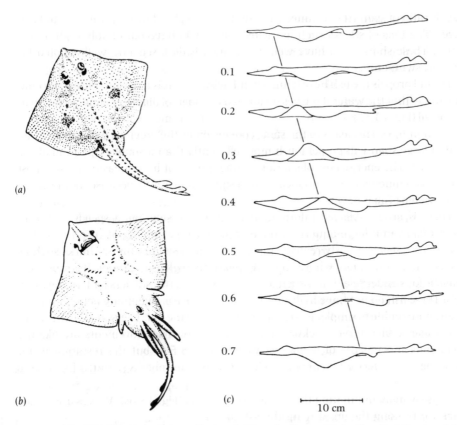

(a)

(b)

(c)

0.1
0.2
0.3
0.4
0.5
0.6
0.7

10 cm

Fig. 13.6.(a) Dorsal view of a female Thornback ray (*Raia clavata*, 0.8 m) and (b) ventral view of a male (0.6 m). From J.R. Norman & P.H. Greenwood (1963). *A history of fishes*, 2nd edn. Benn, London. (c) Outlines at intervals of 0.1 s from a film of *Raia eglanteria* swimming, from T.L. Daniel (1988). *Can. J. Zool.* 66, 630–638.

red muscle does not seem to propagate action potentials, and it is likely that these potentials were action potentials in the nerves rather than the muscles.) When the fish was pinched and so stimulated to make more vigorous movements, muscle action potentials were detected in the white muscle. It is not certain whether the red muscle was active as well because the potentials in the white muscle are much larger than those detected (during slow swimming) in the red. They are large enough to affect the electrode in the thin layer of red muscle and mask any activity in the red muscle itself. These experiments are interpreted as evidence that the red muscle alone is used for slow and sustained swimming but that the much greater bulk of white muscle comes into use for violent movements and bursts of speed. This has been confirmed in more recent experiments in which recordings have been made from the red and white muscles of intact teleost fishes, swimming in water tunnels like the one shown in Fig. 6.19.

Bone also investigated the effects of exercise on the stores of food materials in the red and white muscles. Some fish were kept unexercised for one or two days before they were killed for analysis. Others were killed in a state of exhaustion, after they had been made to swim vigorously for up to ten minutes. Yet others had their brains destroyed and were set swimming in the manner that has been described. They were kept swimming slowly for up to 50 h, and then killed. Exhaustion by a few minutes' exercise halved the glycogen content of the white muscle but had no apparent effect on the composition of the red. Prolonged exercise halved the fat content of the red muscle but had no significant effect on the white. It seems that glycogen fuels the anaerobic metabolism of the white muscle and that fat fuels the aerobic metabolism of the red.

There seems to be a general rule that vertebrate striated muscles are most efficient, and can exert most power, when shortening at about 30% of their maximum rates. The rates at which carp (*Cyprinus*, a teleost) red muscle shortens during swimming have been measured. Films were made of carp swimming at various speeds in a water tunnel to find out how much and how fast they bent. The fish were killed, their bodies were bent into positions like those seen in swimming and they were left until rigor mortis had set in, tightening the muscles. Muscle fibres were dissected out and their sarcomere lengths measured. This made it possible to calculate the rates at which sarcomere lengths changed during swimming. In other experiments, muscle fibres from freshly killed carp were tested (as in Fig. 6.15) to discover their rates of shortening under different loads. The rate of shortening during swimming at the highest speed that the red muscle could sustain was about 30% of the maximum rate. Experiments with carp muscle have also shown that the white fibres (used in fast swimming) have much higher maximum rates of shortening than red fibres from the same fish. Also, the red fibres run parallel to the vertebral column, but the white fibres of teleosts are twisted into helices like the strands of a rope. The consequence of this is that the body bends more, for a given percentage shortening of the white muscles, than for the same shortening of the red ones: it is as though the muscles worked through a gearbox, with the white muscles using a higher gear. Selachians have a different arrangement of white muscle fibres, which gives a much less pronounced gearbox effect.

The swimming muscles are divided into segmental blocks (myomeres) separated by connective tissue partitions. Cooking breaks down the partitions, so cooked fish separates easily into flakes, which are individual myomeres.

Nearly all selachians live in the sea, in water of density about 1026 kg/m³. Many of their tissues are denser, including muscle (about 1060 kg/m³), cartilage (1100 kg/m³) and scales (probably about 2000 kg/m³). Consequently, most selachians are denser than the water they live in. For example, dogfish have a density of about 1075 kg/m³, about the same as the squid *Loligo* (section 6.5).

Such fish must swim perpetually, if they are not to sink. Dogfish spend a lot of time resting on the sea bottom but sharks such as *Lamna* (Fig. 13.1) seem to swim all the time. Dogfish get most of the lift they need to prevent sinking by using their pectoral fins as hydrofoils, but these fins are anterior to the centre of mass (Fig. 6.22b). The moments about the centre of mass are balanced by lift on the tail, which is shaped so as to drive water downwards (producing lift) as well as backwards (producing thrust). The vertebral column bends up into the upper lobe of the tail (the epichordal lobe, Fig. 13.4c) leaving the lower (hypochordal) lobe stiffened only by cartilages called pterygiophores (described below). Such tails are called heterocercal. The forces on the tails of dead selachians have been measured in experiments in which they

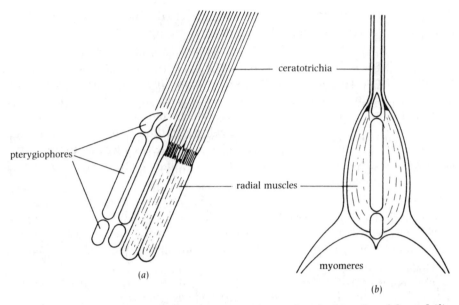

ceratotrichia

pterygiophores

radial muscles

myomeres

(a)

(b)

Fig. 13.7. Diagrams of (*a*) a dissection of part of a selachian dorsal fin and (*b*) a section through the base of the fin.

have been moved sideways through water, at appropriate speeds to imitate swimming. The vertical forces depended on the speed, but had values around those required for equilibrium by the swimming fish.

If the pectoral fins were held always at the same angle of attack, they would produce more lift at higher speeds. The lift that the fish needs is the same at all speeds, because its weight in water is constant. The fish needs muscles to adjust the angle of attack of the fins (and also of the tail) as it changes speed. Fin movements are also used to steer the fish up or down, and rays use them for swimming (Fig. 13.6*b*).

The fins are segmented structures, with muscles which develop in the embryo by separating off the myotomes. The increase in length of the fin as the embryo grows generally fails to keep pace with the growth of the body, so that a dorsal fin derived from fourteen segments (for instance), may only extend over six myotomes in the adult fish. Each segment of the fin contains two or more rods of cartilage (pterygiophores) connected end-to-end (Fig. 13.7*a*). The distal pterygiophores lie between sheets of packed ceratotrichia, which are very thin rods of a form of collagen known as elastoidin. The ceratotrichia extend to the edge of the fin, and stiffen it. There are a great many of them in each segment, but there is only one muscle on each side of the fin in each segment. It is attached by a broad tendon to the sheet of ceratotrichia, and when it contracts it bends the fin over to its side.

In early selachians (Fig. 13.4*c*), and also in rays, the pterygiophores extend near the margins of the fins. In other modern selachians, the pterygiophores are confined to the base of the fin and most of the area of the fin is stiffened only by ceratotrichia. The long-based pectoral fin of *Cladoselache* must have been less mobile than the narrow-based ones of modern sharks.

Not all selachians are substantially denser than seawater. Some, such as the

Portuguese shark (*Centroscymnus*, a deep-sea shark) and the Basking shark (*Cetorhinus*), are very close indeed to the density of seawater. Their pectoral fins do not have to produce appreciable vertical forces except for manoeuvring, and are appropriately small. Their tails are heterocercal but presumably produce more or less horizontal forces. They owe their low density to oil that consists of lipids (density around 920 kg m^{-3}) and the hydrocarbon squalene (density 860 kg m^{-3}). The quantities required are large and the liver, which contains most of the buoyant material, is enormous. In section 6.5 we calculated the volume of gas-filled shell needed to float the mollusc *Nautilus*. A similar calculation shows that to reduce the density of a dogfish (1075 kg m^{-3}) to that of seawater (1026 kg m^{-3}) you would have to add enough squalene to increase its volume by 30%.

I argued in section 6.5 that buoyancy organs are advantageous to the slow-moving cephalopods *Nautilus* and *Helicocranchia*, but that it is better for fast-swimming squid to prevent themselves from sinking by using their fins as hydrofoils. The same argument shows how squalene may be advantageous only to slow-swimming selachians.

Basking sharks swim to filter plankton at about one metre per second, which seems slow for 10 metre fish. The speed of *Centroscymnus* is not known.

13.4.   The heart and the blood

Fig. 13.8 shows the heart of a shark. (The hearts of other fishes, including lampreys, are fairly similar.) The cardinal veins from either side of the body and the hepatic vein from the liver open into the small sinus venosus at the back of the heart. It opens into the large, dorsally placed atrium, which opens in turn into the ventricle which lies ventral to it. Finally, the conus arteriosus joins the ventricle to the ventral aorta, which carries the blood anteriorly towards the gills. All the chambers of the heart from the sinus to the conus have muscular walls, but the walls of the ventricle and conus are much thicker than those of the sinus and atrium. Flap valves at the openings between the chambers prevent blood flowing in the wrong direction. There are four or five valves, each consisting of a ring of flaps, in the conus. The whole heart is enclosed in a chamber called the pericardium, walled partly by the cartilage of the pectoral girdle and partly by sheets of collagenous connective tissue.

Fig. 13.8 also shows results of an investigation of blood flow through the heart. The shark was anaesthetized and dissected so that hypodermic needles connected to pressure transducers could be inserted into the heart, and an electromagnetic flowmeter fitted to the aorta. The records show pressures up to 4000 N m^{-2} in the ventricle and conus, but only low pressures in the atrium. (The highest pressures recorded from the hearts of snails, Fig. 6.23, are about 3500 N m^{-2}.) The pressure in the pericardium fluctuates, but is always negative (i.e. below the pressure in the general body cavity) owing to the elasticity of the pericardium walls. Two main phases of the heartbeat can be distinguished.

(i) The atrium contracts, emptying itself into the ventricle. The pressure is slightly positive in the atrium, and about zero in the ventricle.

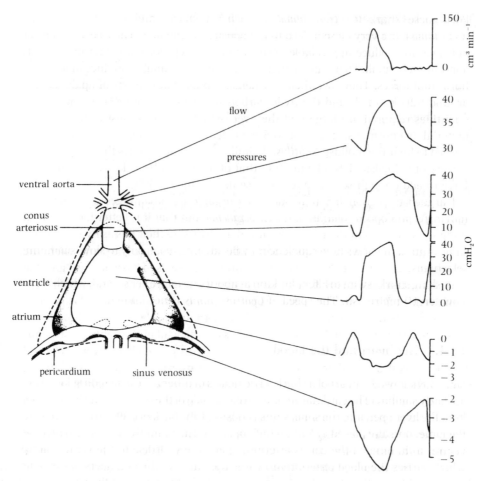

Fig. 13.8. **A ventral view of the heart of the Port Jackson shark,** *Heterodontus,* **and records of blood flow and pressure changes at the positions indicated during a single heart beat. From G.H. Satchell (1971).** *Circulation in fishes.* **Cambridge University Press, London.**

(ii) The ventricle contracts, driving blood out into the conus and aorta at high pressure. This reduces the volume of the heart and so the pressure in the pericardium. The atrium is relaxing and the negative pressure in the pericardium makes it expand, drawing in blood from the sinus venosus and veins.

The conus arteriosus continues to contract after the ventricle has started relaxing, so that high pressure is maintained in it for longer. The pressure in the ventral aorta is remarkably steady, fluctuating only between 3000 and 4000 N m^{-2}. This is partly due to the prolonged contraction of the conus, and partly to the elastic properties of the ventral aorta, which is inflated by blood as the ventricle and conus contract, and then recoils elastically. The recoil keeps the pressure quite high and should keep blood flowing through the gills until the ventricle contracts again. However, the flowmeter records show the flow falling abruptly to zero at the end of the contraction

Table 13.1. *Composition of various fluids from* Squalus acanthias

	Concentrations (mmol l⁻¹)					Osmotic concentration (mmol l⁻¹)
	Na⁺	Cl⁻	Mg²⁺	Urea	TMAO*	
Seawater	440	490	50	0	0	930
Blood plasma	250	240	>1†	350	70	1000
Urine	240	240	40	100	10	800
Rectal gland secretion	500	500	0	18	—	1000

*Trimethylamine oxide.
†Few measurements have been made, but values of 3–4 mmol l⁻¹ seem to be typical.
Data from J. Burger (1967) in P.W. Gilbert, R.F. Mathewson & D.P. Rall (eds) *Sharks, skates and rays*. Johns Hopkins Press, Baltimore.

of the ventricle. Perhaps the elastic part of the aorta is distal to the point where the flowmeter was fitted. Steady flow of blood through the gills, if it occurs, will give the same advantages as steady flow of water over them (section 12.2).

The ventral aorta owes its elastic properties to a high proportion of elastin in its walls. Analysis of the aortas of five sharks of various species showed that elastin constituted an average of 31% of the dry mass of the wall of the ventral aorta, but only 9% in the dorsal aorta. The corresponding figures for collagen were 46% and 69%. There is also smooth muscle in the walls of the arteries: it must account for most of the fraction of the dry weight that is neither elastin nor collagen. Elastin is a rubber-like protein with properties like the resilin of insects (section 9.2): it can be stretched to double its length, and snaps back when released. Collagen is a fibrous protein that cannot be stretched anything like as easily or as far. The Young's modulus of collagen is about a thousand times that of elastin, and the properties of arteries would be dominated by the collagen fibres if they were tight. In fact, the collagen fibres seem to be arranged loosely, so that the elastin can play an important part. The ventral aorta in particular probably has to be distended considerably, to tighten its collagen fibres.

The blood plasma of lampreys and of teleost fishes has an osmotic concentration of about 0.3 mol l⁻¹, almost entirely due to dissolved inorganic salts. That of marine selachians has an osmotic concentration slightly greater than sea water, about 1 mol l⁻¹, but only about half of it is due to inorganic salts. Most of the rest is due to urea and trimethylamine oxide, which are present in remarkably high concentrations (Table 13.1).

Since the osmotic pressure of the body fluids is a little above that of seawater, there is some tendency for water to move into the body, particularly at the gills. Inorganic ions must also tend to diffuse in while urea and trimethylamine oxide tend to diffuse out.

As well as the kidneys there is an organ called the rectal gland, which helps to regulate the composition of the body fluids. It is a small finger-shaped gland which protrudes from the posterior end of the intestine and discharges through a duct into it. Its function was investigated in a series of experiments with spiny dogfish, *Squalus acanthias*. Cannulae were fastened into its duct, and also into the urinary opening.

The dogfish were fastened to a board in some of the experiments but in others they were free to swim in an aquarium, trailing the cannulae behind them. The rectal gland secretion was collected through one cannula and the urine through the other. They flowed at variable rates, but the average for each was 0.5 cm³ kg body mass⁻¹ h⁻¹. Their compositions were quite different, as Table 13.1 shows. The urine contained about the same concentrations of sodium and chloride as the plasma, but much more magnesium (and sulphate and phosphate) and much less urea and trimethylamine oxide. The latter compounds must be reabsorbed from the kidney tubules, so that their rate of loss does not exceed the rate at which they are produced as waste products of protein metabolism. Divalent ions, on the other hand, must be secreted into the tubules. Experiments with inulin (see section 12.5) have shown that the glomerular filtration rate is only about four times the rate of urine production, while the concentration of magnesium (for instance) is more than ten times as high in the urine as in the plasma.

The secretion of the rectal gland (Table 13.1) consists of little else but water and sodium chloride. It contains very little urea or magnesium, but the sodium and chloride concentrations are about twice as high as in the plasma and much the same as in seawater.

There are cells in the gills of selachians which are believed to be capable of excreting salts, but they seem to excrete less than the kidneys and rectal gland.

13.5. Reproduction

The males of most fish do not fertilize the eggs until they have been laid, but selachians copulate and the eggs are fertilized within the female. Males have the posterior ends of their pelvic fins modified to form organs known as claspers (Fig. 13.6b), which serve the function of a penis. Each is cigar-shaped, but with a deep groove along one side, and stiffened by a cartilage. They normally lie flat against the body, pointing posteriorly, but they can be erected individually by muscle action. When the clasper is erect the proximal end of the groove is against the cloaca and semen can be extruded into it. When dogfish copulate, the male wraps his body around the female and inserts one clasper into her cloaca.

Internal fertilization raises two alternative possibilities. The egg may be enclosed in a tough egg case. Since fertilization is internal it can occur before the shell (which would not easily be penetrated by sperm) is formed. Alternatively the egg can be retained inside the body until development has proceeded for some time, before it is laid. In either case the egg is protected in the interval between fertilization and the time when the embryo becomes an active young fish, able to feed and to take action to avoid predators. Both possibilities are exploited by selachians, as they are also by gastropod molluscs (section 6.12).

Dogfish (*Scyliorhinus*) lay each egg in the keratinous envelope known as a mermaid's purse. These are often found near low tide mark with the tendrils that project from their ends tangled in seaweed. The eggs are large, about 13 mm in diameter, and most other selachians lay even larger eggs. In contrast, salmon (*Salmo salar*) eggs are 6 mm in diameter and most teleost eggs are smaller. The larger the egg, the less vulnerable the young when it emerges from the egg case, but fewer eggs can be produced. (See the discussion of molluscs, section 6.12).

Dogfish egg cases are initially closed, filled by a fluid which contains salts and urea in much the same concentrations as the blood plasma. Later, pores open, allowing seawater to circulate through, driven by movements of the embryo's tail. This circulation brings oxygen to the gills, which have filamentous extensions, protruding from the embryo's head.

The Spiny dogfish (*Squalus acanthias*) produces much larger eggs and retains them in its uterus for almost two years until the embryos are ready to swim and feed. While they are in the uterus, their respiration and excretion must depend on diffusion through the uterine fluid. In the course of development, some of the foodstuffs in the body are used for metabolism. The newborn fish contains 40% less organic matter than the egg from which it developed.

The Smooth hound (*Mustelus canis*) also keeps its embryos in the uterus, but not for so long. They are born after 10–11 months, having grown much faster. The new-born young is not only very much heavier than the egg, but it contains about ten times as much organic matter. Foodstuffs are taken up from the mother in the course of development. This is made possible by a placenta.

At first the embryo *Mustelus* depends on the foodstuffs in the yolk. A yolk sac with blood vessels in it grows round the yolk. Blood circulates through, carrying foodstuffs back to the embryo. As the yolk is used, part of the yolk sac becomes firmly attached to the wall of the uterus so that its blood vessels are very close to maternal blood vessels. Foodstuffs, waste products and gases can then diffuse between the blood of the embryo and that of its mother. The arrangement is just like that of the yolk-sac placenta of mammals (Fig. 18.7) and it apparently functions in the same way.

13.6.   Sense organs

The lateral line neuromasts of lampreys and hagfishes all protrude from the surface of the body, but selachians and teleosts have neuromasts enclosed in sunken canals, as well as superficial ones. Fig. 13.9 shows how these are arranged. Neuromasts and openings to the external water alternate along the length of the canal. A difference in pressure between two successive pores causes water to move along the canal between them, deflecting the cupula.

Enclosure in a canal affects the response of the neuromast to stimuli. Superficial neuromasts are very sensitive to water movements parallel to the surface of the body (Fig. 13.9*a*). Canal neuromasts are not, because the pressure differences between successive pores are then small. However, both types of neuromast are sensitive to local water movements at right angles to the body (Fig. 13.9*b*). Water striking the body at right angles is deflected along it, so the superficial neuromasts are stimulated. There is a region of high pressure where the water strikes the body, so water will tend to move along the canals. When a fish is swimming, or resting on the bottom in a current, water is constantly flowing longitudinally over it. The effects of this flow on the superficial neuromast must tend to mask the effects of gentler water movements

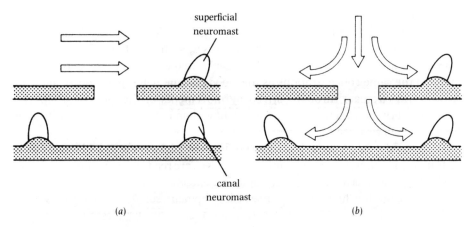

superficial
neuromast

canal
neuromast

(a)   (b)

Fig. 13.9. A diagram showing superficial and canal neuromasts and how they are affected by water movements (a) parallel to and (b) at right angles to the body surface.

produced by wriggling prey or approaching predators. Canal neuromasts can be expected to be less sensitive to it, and so better able to detect movements due to prey and predators, if these have components at right angles to the surface of the body. The main disadvantage of canal organs is that they must tend to be less sensitive than superficial ones to vibrations above a certain frequency, because of the inertia of the water in the canal.

Action potentials in nerves and muscles involve changes of electrical potential at their surfaces. Any aquatic animal which produces action potentials must therefore set up electric fields in the water around itself. These fields are, however, extremely weak. Fig. 13.10 illustrates a series of experiments which demonstrated that dogfish (*Scyliorhinus*) can detect the electric fields generated by potential prey. The dogfish were kept in shallow water in a child's inflatable paddling pool, with sand on the bottom. Young plaice (*Pleuronectes*) or other objects were introduced into the pool while the dogfish were lying inactive on the bottom, as well-fed dogfish do. When the object had been hidden under the sand the dogfish were stimulated to search for food by putting a little fish juice into the water. They were then observed.

Young plaice which were put in the tank buried themselves in the sand with only parts of the head exposed (Fig. 13.10a). A dogfish passing within 15 cm of a plaice would generally uncover and capture it: it would suck off the sand covering the plaice (blowing out the sand through its gill slits) and then seize the plaice and shake it until it tore into pieces small enough to swallow. It could of course have found the plaice by sight, smell or water movements. The next experiment (Fig. 13.10b) was designed to eliminate these possibilities. The plaice was put in a chamber made of agar jelly, of about the same electrical conductivity as seawater. The broken arrows in the diagram indicate tubes used to pass aerated seawater through the cavity, so that the plaice was not asphyxiated. The dogfish responded as before if they swam near the chamber and tried to uncover the plaice. They showed no interest if the chamber was empty. Fig. 13.10(c) shows an experiment in which the living plaice was replaced by pieces of dead fish. The dogfish did not uncover the chamber but tried to find food at

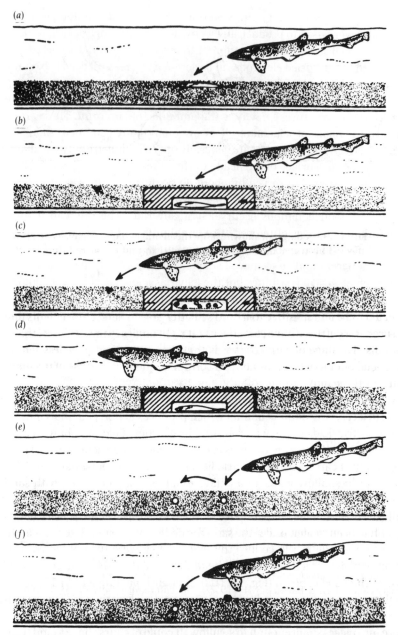

Fig. 13.10. Diagrams illustrating the experiments on the electric sense of dogfish, which are described in the text. From A.J. Kalmijn (1971). *J. exp. Biol.* 55, 371–384.

the end of the outlet tube carrying water from it. They were presumably responding to the odour of the pieces of fish. Fig. 13.10(d) shows that there was no response to a live plaice in the chamber if it was covered by a very thin sheet of a plastic which is a good electrical insulator. More direct evidence that the plaice were detected by an electric sense was obtained by burying electrodes in the sand (Fig. 13.10e). The dogfish showed no interest in the electrodes if no current was flowing, but when the electrodes were used to produce an electric field simulating the field around a living plaice the dogfish responded as though to a real plaice and uncovered them. They responded to the buried electrodes in preference to a piece of dead fish on the surface of the sand (Fig. 13.10*f*).

Other experiments showed that the electric sense could be abolished, by cutting the nerves to the structures known as the ampullae of Lorenzini. These are jelly-filled tubes in the heads of selachians. They are often many centimetres long (and more than a metre in large rays) but are generally only a millimetre or two in diameter. One end opens at the surface of the head but the other is blind, and at this blind end is a group of sensory cells, served by about six axons from the seventh cranial nerve. Electrical recordings have been made from these axons. When conditions are constant, action potentials occur at a steady rate. The rate can be modified by prodding, by changes of temperature or salinity, and by electrical potentials. It is particularly sensitive to electrical potential gradients along the length of the ampullae. The ampullae seem well adapted for detecting such gradients. Their walls have high electrical resistance but the jelly has a low resistance: it contains ions in about the same concentrations as seawater and so has much higher conductivity than the surrounding tissues. Consequently, when there is a potential difference between the ends of the ampulla the electric current tends to be channelled along the ampulla. The current through the sensory cells must be higher than if the jelly had the same conductivity as the tissues. Different ampullae run in different directions, and so are sensitive to potential gradients in different directions.

Further reading

Sharks

Budker, P. (1971). *The life of sharks*. Weidenfield & Nicolson, London.
Frazzetta, T.H. & Prange, C.D. (1987). Movements of cephalic components during feeding in some requiem sharks. *Copeia* 1987, 979–993.
Gilbert, P.W. (1984). Biology and behaviour of sharks. *Endeavour* (n.s.) **8**, 179–187.
Hughes, G.M. & Ballintijn, C.M. (1965). The muscular basis of the respiratory pumps in the dogfish (*Scyliorhinus canicula*). *J. exp. Biol.* **43**, 363–383.
Lineaweaver, T.H. & Backus, R.H. (1979). *The natural history of sharks*. Lippincott, Philadelphia.

Early jawed fishes

Moy-Thomas, J.A. (1971). *Palaeozoic fishes*, 2nd edn (revised by R.S. Miles). Chapman & Hall, London.

Swimming and buoyancy

Alexander, R. McN. (1972). The energetics of vertical migration by fishes. *Symp. Soc. exp. Biol.* **26**, 273–294.
Blake, R.W. (1983). *Fish locomotion*. Cambridge University Press, London.

Bone, Q. (1966). On the function of the two types of myotomal muscle fibre in elasmobranch fish. *J. mar. biol. Ass. U.K.* **46**, 321–349.

Corner, E.D.S., Denton, E.J. & Forster, G.R. (1969). On the buoyancy of some deep-sea sharks. *Proc. R. Soc. Lond.* B171, 415–429.

Daniel, T.L. (1988). Forward flapping flight from flexible fins. *Can. J. Zool.* **66**, 630–638.

Rome, L.C., Funke, R.P., Alexander, R. McN., Lutz, G., Aldridge, H., Scott, F. & Freadman, M. (1988). Why animals have different muscle fibre types. *Nature* **335**, 824–827.

Webb, P.W. & Keyes, R.S. (1982). Swimming kinematics of sharks. *Fishery Bull.* **80**, 803–812.

The heart and the blood

Satchell, G.H. (1971). *Circulation in fishes.* Cambridge University Press, London.

Reproduction

Needham, J. (1950). *Biochemistry and morphogenesis.* Cambridge University Press, London.

Price, K.S. & Daiber, F.C. (1967). Osmotic environments during fetal development of dogfish, *Mustelus canis* (Mitchill) and *Squalus acanthias* Linnaeus, and some comparisons with skates and rays. *Physiol. Zool.* **40**, 248–60.

Sense organs

Kalmijn, A.J. (1971). The electric sense of sharks and rays. *J. exp. Biol.* **55**, 371–384.

Kalmijn, A.J. (1982). Electric and magnetic field detection in elasmobranch fishes. *Science* **218**, 916–918.

14 *Teleosts and their relatives*

Class Osteichthyes,
    Subclass Actinopterygii,
        Infraclass Chondrostei   (palaeoniscoids and sturgeons)
        Infraclass Cladistia   (*Polypterus* etc.)
        Infraclass Neopterygii,
            Division Teleostei,
                Superorder Protacanthopterygii   (salmon etc.)
                Superorder Ostariophysi   (carps etc.)
                Superorder Paracanthopterygii   (cod etc.)
                Superorder Acanthopterygii   (perch etc.)
           and other divisions and superorders

14.1.   Bony fishes

The members of the class Osteichthyes are commonly referred to as the bony fishes. Most of these fish have skeletons which consist mainly of bone, and they are called bony fishes to distinguish them from the selachians which have cartilaginous skeletons. However, even selachians have bone in the bases of their scales and teeth (Fig. 12.10b), and many extinct fish which are not included in the Osteichthyes had a great deal of bone. The thick bony scales and carapaces of many ostracoderms and placoderms, and the bony skeletons of acanthodians, have already been described.

It will be helpful to distinguish between cartilage bone and dermal bone. Structures described as consisting of cartilage bone appear first in embryos as cartilage, which is later replaced by bone. They are derived ultimately from sclerotome or neural crest (Fig. 13.2b), and are homologous with the cartilages of selachian skeletons. Dermal bones are formed directly in the dermis of the skin or the lining of the mouth, without prior formation of cartilage. They are formed in the same way as scales and like them are sometimes coated with outer layers of dentine and enameloid substance (section 12.4). There are no structures in selachians which are homologous with particular dermal bones, though the dermal bones and scales of bony fishes may be regarded as being in a general sort of way homologous with the placoid scales of selachians.

The lungs and swimbladders of Osteichthyes are much more distinctive features than their bone. Most modern Osteichthyes have a lung or a swimbladder but no other fishes, living or extinct, are known to do so.

This chapter is about the members of the subclass Actinopterygii, including the teleosts, which form the great majority of modern fishes. The other subclasses of bony fish, Crossopterygii and Dipnoi, are discussed in chapter 15. I explain there how they differ from Actinopterygii in the structure of their fins and scales.

The early actinopterygians are commonly called palaeoniscoids, but this name has no place in modern schemes of formal classification. *Moythomasia* (Fig. 14.1*a*) is typical of them. Most of them had thick diamond-shaped scales with a thick outer layer of enameloid substance. They also had long jaws, eyes set well forward near the tip of the snout, and heterocercal tails. In all these features they resembled acanthodians (Fig. 13.4*a*). The gill openings of each side of the body were covered by a single large bony operculum, as in the teleosts. They seem to have had a spiracle (which was not covered by the operculum).

In several features, palaeoniscoids resembled selachians more than teleosts. They had heterocercal tails and spiracles. They had stiff fins, apparently incapable of folding. These seem to be primitive features shared by the (unknown) common ancestor of Selachii and Osteichthyes.

The Cladistia, *Polypterus* (Fig. 14.1*b*) and *Erpetoichthys*, are modern African freshwater fishes that have thick scales like those of palaeoniscoids, and spiracles. They have peculiar pectoral fins and symmetrical (not heterocercal) tails. The sturgeons are modern marine and freshwater fish that have spiracles, heterocercal tails and stiff fins like those of palaeoniscoids, but are peculiar in many other ways. Both cladistians and sturgeons have spiral valves in their intestines, like selachians.

Salmon (Fig. 14.1*c*) breed in rivers but spend a large part of their life in the sea. They are primitive teleosts. Their tails are symmetrical, of the shape called homocercal. Their fins are flexible and can be folded like fans. They have no spiracle. Their scales are round, and much thinner than those of typical palaeoniscoids or of cladistians.

Perch (Fig. 14.1*d*) are freshwater fish, and are typical of the Acanthopterygii, a huge superorder of advanced teleosts. The pectoral fins are set higher up on the side of the body than in trout, and the pelvic fins have moved forward to a position below them. The anterior few rays of the dorsal, anal and pelvic fins are not flexible like the remaining rays, but have become stiff spines and presumably help to protect the fish from predators. These spines can pivot at their bases to open or close the fin, but are not easily bent. The upper jaw moves forward as the mouth opens, as will be explained in the next section.

14.2.   Feeding and respiration

Feeding under water, on free-floating prey, presents a severe difficulty. A swimming fish pushes on the water in front of its snout, so as it swims up to potential prey it will tend to push the prey away. This difficulty can be overcome if it enlarges its mouth cavity at the appropriate instant, to suck prey in. Nearly all teleost fish do this. Selachians probably often do the same, but it is only for teleosts that the process of suction feeding has been studied in detail.

Fig. 14.2 shows how the water in front of a fish would move, relative to the fish's body, if it tried to feed in various ways. In (*a*) and (*b*) stationary fish enlarge their mouth cavities to suck in water. These diagrams represent unrealistic extremes: real events would be intermediate between them. In (*a*) the mass of the fish is imagined to

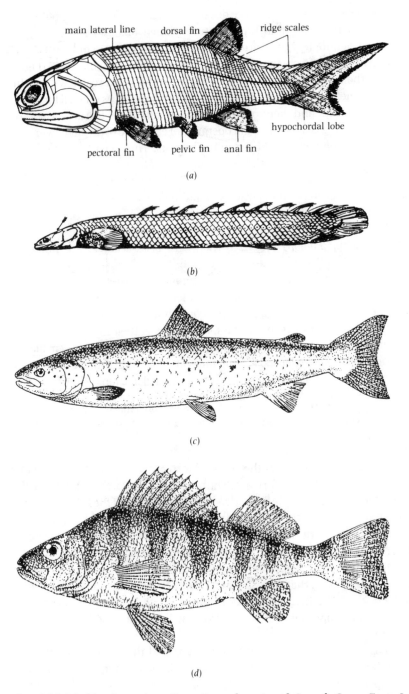

Fig. 14.1.(*a*) *Moythomasia*, a Devonian palaeoniscoid. Length 8 cm. From J.A. Moy-Thomas (1971). *Palaeozoic fishes*, 2nd edn (revised by R.S. Miles). Chapman & Hall, London. (*b*) *Polypterus senegalus*, 0.4 m. The arrow points to the spiracle. From T.W. Bridge (1904). Fishes. In *The Cambridge natural history* (ed. S.F. Harmer & A.E. Shipley) Macmillan, London. (*c*) *Salmo salar*, the Atlantic salmon, about 1.5 m and (*d*) *Perca flavescens*, the Yellow perch, 0.5 m, from A. Wheeler (1975) *Fishes of the world*. Ferndale, London.

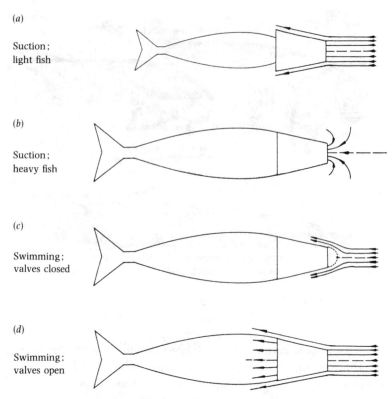

(a)

Suction;
light fish

(b)

Suction;
heavy fish

(c)

Swimming;
valves closed

(d)

Swimming;
valves open

Fig. 14.2. **Diagrams showing how water would move relative to a fish's body, if it tried to feed in various ways.** From M. Muller & J.W.M. Osse (1984). *Trans. zool. Soc. Lond.* 37, 51–135.

be so small, compared with the mass of water it sucks on, that the water hardly moves: the fish sucks itself forward. The movement of water *relative to the fish*, shown in the diagram, is due to the fish moving. In (b) the mass of the fish is imagined to be so large that it remains stationary and only the water moves: it enters the mouth from all directions. In (c) the fish swims towards the food with its gill covers closed, and the water is pushed along in front of the snout (so that its velocity *relative to the fish* is zero) or deflected to either side. In (d) it again swims forward, but with its mouth and gill covers wide open so that the fish can move its mouth over the food without pushing the food away. This diagram shows another unrealistic extreme case: no matter how widely the fish opens its gill covers, there will be some tendency to push the food forward.

Teleosts occasionally feed by swimming round their food, without sucking. For example, anchovies (*Engraulis*) in dense plankton swim around with their mouths wide open, straining out the plankton as the water flows out through their gills. Teleosts picking up food from the bottom may have to rely on suction alone, without swimming. A large proportion of teleost feeding acts, however, use both swimming and suction.

To explain how suction is performed, I must first explain the basic structure of the skull. The skulls of bony fish consist partly of cartilage and cartilage bone, and partly

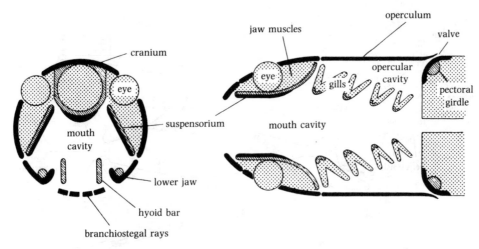

Fig. 14.3. **Diagrammatic transverse and horizontal sections of the head of a typical teleost. Dermal bone is shown black and cartilage bone is hatched.**

of dermal bone. The total number of bones is generally very large (about 135 in the typical case of the herring, *Clupea*) but it is happily unnecessary in this book to describe more than a few of them individually.

Fig. 14.3 shows how cartilage bone and dermal bone combine in the structure of the skull. The diagram is based on teleosts but would be little different if it had been drawn to represent some other group of bony fish. The cranium consists largely of cartilage bone but is roofed by dermal bone and has a ventral covering of dermal bone where it forms the roof of the mouth cavity. The side walls of the mouth cavity are formed by the suspensoria, which each include cartilage bones homologous with the hyomandibular cartilage and parts of the palatoquadrate cartilage of selachians, and also dermal bones formed in the lining of the mouth. The lower jaw consists of a slender core of cartilage and cartilage bone homologous with Meckel's cartilage, invested by dermal bone. The hyoid bars consist of cartilage bones, homologous with the hyoid bars of selachians. The gills are supported by cartilage bones homologous with the corresponding cartilages in selachians. The opercula (gill covers), the branchiostegal rays ventral to them and the bony covering of the cheek consist of dermal bones. At an early stage in the development of the embryo the skull consists exclusively of cartilage and includes cartilages corresponding to all the main parts of the selachian skull. Later, bone develops in these cartilages, and dermal bones are also formed.

Fig. 14.4*a* is a diagram of the skull of a relatively primitive teleost such as a trout (*Salmo*). Fig. 14.4*b* shows the same skull with some of the dermal bones removed. Note that the suspensorium has two articulations with the cranium, one anterior and one posterior to the eye. The latter articulation is formed by the hyomandibular. These articulations act as hinges, allowing the suspensorium to be swung laterally (widening the mouth cavity) or medially (narrowing it). The hyomandibular has a ball and socket joint with the operculum. The quadrate bone (one of the derivatives of the palatoquadrate cartilage) has a hinge joint with the lower jaw. The posterior end of the hyoid bar rests against the inner face of the operculum to which it is attached by

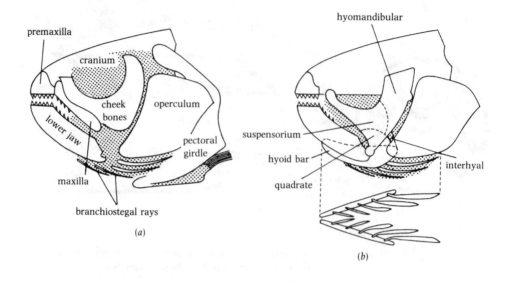

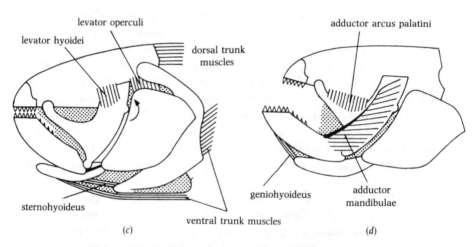

Fig. 14.4. Diagrams of the skull of a typical teleost such as a trout (*Salmo*); (*a*) shows the intact skull and (*b*) shows it after removal of the maxilla and cheek bones. A ventral view of the hyoid bars and branchiostegal rays is shown below (*b*). Parts (*c*) and (*d*) show some of the principal muscles of the head.

ligaments, but is also connected through a small bone (the interhyal) to the hyomandibular. Thus the derivatives of the upper and lower parts of the hyoid arch form an unbroken chain, as in selachians.

If the suspensorium is swung laterally, widening the mouth cavity, the operculum is also carried laterally. Its posterior edge as well as its anterior edge move laterally because a stop on the ball and socket joint between the hyomandibular and the operculum limits movement there; the operculum can hinge outward about a vertical axis through the joint, but inward hinging is limited by the stop. Thus the mouth cavity and the whole of the opercular cavity are widened, and water is drawn in through the mouth. No water can enter under the posterior edge of the operculum

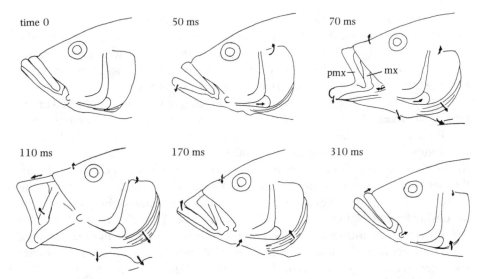

Fig. 14.5. Tracings from a cine film of *Serranochromis,* a cichlid fish, feeding; pmx, premaxilla; mx, maxilla. From K.F. Liem (1978) *J. Morph.* 158, 323–360.

because a strip of flexible tissue along this edge acts as a valve (Fig. 14.3). None can enter the opercular cavity from below because the cavity is floored by a membrane stiffened by the branchiostegal rays. These slender bones are jointed to the hyoid bars, and open like the ribs of a fan when the head is widened. Narrowing the mouth and opercular cavities drives water out under the posterior edge of the operculum, but not through the mouth (unless it is wide open), which is protected by valves.

Since the joint between the hyomandibular and operculum is a ball and socket, it allows other movements as well as the limited hinging about a vertical axis that has been described. It also allows rotation about a transverse axis through the joint. Anticlockwise rotation (as seen in Fig. 14.4c) pulls the mouth open, because a ligament connects the anterior lower corner of the operculum to the lower jaw.

Fig. 14.5 has been drawn from a film of a teleost feeding. The film was taken at 200 frames per second but only a few (irregularly spaced) frames are shown. Before the film was taken the fish had been anaesthetized and electrodes had been placed in five of its head muscles: different combinations of muscles were wired in different experiments and the statements that follow about muscle activity are based on several experiments, each with its accompanying film. The positions of the electrodes were checked by taking X-ray pictures. The wires from the electrodes were glued together to form a cable which was clipped to the dorsal fin so that the fish could not tug the electrodes out of position. The fish swam in its aquarium, trailing the cable, which was connected at its other end to recording equipment.

At time zero, in Fig. 14.5, the fish has its mouth closed and the feeding act has not started. At 50 ms the mouth is just beginning to open, and at 110 ms it is fully open. Between these times, the muscles shown in Fig. 14.4(c) are active. The levator operculi rocks the operculum, as indicated by the arrow, and so opens the mouth. The levator hyoidei helps to swing the suspensorium laterally, but is relatively unimportant: the other muscles that co-operate with it probably do most of the work.

The ventral trunk muscles and the sternohyoideus prevent the tongue from moving forward (some of the trunk muscles hold the pectoral girdle firm against the pull of the sternohyoideus while others run forward with the sternohyoideus to insert directly on the tongue skeleton). The dorsal trunk muscles pull on the back of the cranium, bending the anterior joints of the vertebral column and raising the snout. This moves the posterior ends of the hyoid bars forward but their anterior ends, attached to the tongue, are fixed. The posterior ends can only move forward while the anterior ends stay fixed, if they also move laterally. The combined action of the dorsal and ventral trunk muscles and the sternohyoideus makes the hyoid bars splay apart, widening the head. It also tends to pull the tongue down, producing the bulge in the throat. Further, it helps to open the mouth since backward pressure of the hyoid bar on the operculum tends to rock the operculum. Notice how large a bulk of muscle is involved in sucking food in. Not only are several substantial head muscles involved, but also the anterior trunk muscles. The dorsal trunk muscles seem to do most of the work since they actually shorten, while the sternohyoideus and ventral trunk muscles seem generally to remain more or less constant in length.

If only the relatively small head muscles were involved in feeding, they could do relatively little work in a single contraction, sucking food into the mouth. Most of the work is required to give kinetic energy to the food and the water sucked in with it, so the food could only be sucked in relatively slowly. Involving the large trunk muscles (which are primarily swimming muscles) makes it possible to do far more work in a single suck, and so to suck the food in faster. The faster animal food is sucked in, the less chance it has of escaping.

The fish shown in Fig. 14.5 is an acanthopterygian, with protrusible jaws. The upper jaw is protruded between 70 ms and 110 ms, as will be described later. Between 110 and 170 ms the mouth closes, owing to contraction of the adductor mandibulae muscle (Fig. 14.4d), but the mouth cavity is still expanded and the upper jaw is still protruded. By 170 ms, the other muscles shown in Fig. 14.4(d) are beginning to show electrical activity. The adductor arcus palatini pulls the suspensoria back towards the mid line, narrowing the head, and the geniohyoideus pulls the hyoid bars forward to their original position.

The pressures involved in suction feeding have been recorded, as well as the electrical activity of the muscles. Most recordings have used a long flexible tube inserted surgically into the fish's head, connecting it to a pressure transducer outside the aquarium. It is very difficult to get reliable records in that way because brief pressure pulses are apt to be modified by transmission through flexible tubes. The best records have been made with transducers so tiny that they can be put into the fish, which is then connected only by wires to the apparatus. The record shown in Fig. 14.6 was made with such transducers. A film taken at the same time showed that the mouth began to open at 20 ms, was fully open at 50 ms and was closed by 100 ms. The strongest suction (most negative pressures) were recorded at the instant when the mouth was most widely open: $-10\,000\,\text{N}\,\text{m}^{-2}$ (one metre of water) in the mouth cavity and $-8000\,\text{N}\,\text{m}^{-2}$ in the opercular cavity. The opercular valves opened at 60 ms, and a little later the pressure in the mouth cavity became positive as the water in the expanded mouth was forced out through the gills. These records were obtained from *Amia*, a rather more primitive member of the Neopterygii than are the teleosts, but similar records have been obtained from teleosts.

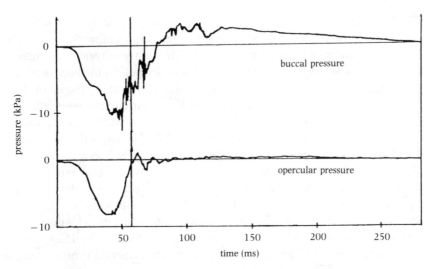

Fig. 14.6. Records of pressure in the mouth and opercular cavities of a bowfin (*Amia*) during a feeding act. From J.L. van Leeuwen & M. Muller (1984). *Trans. zool. Soc. Lond.* 37, 171–227.

The breathing movements of teleosts resemble the movements of suction feeding, but with much smaller amplitude. The trunk muscles and sternohyoideus seem to play no part in normal gentle breathing but the other muscles act in the same sequence as in feeding. The pressures are small, usually within the range ± 500 N m^{-2}. Similar pressures are involved in cephalopod breathing (section 6.6).

Now we will examine the movements of the upper jaw bones, the premaxilla and maxilla. In *Moythomasia* (Fig. 14.1) and other palaeoniscoids they seem to have been rigidly attached to the cranium and the dermal bones of the cheek, so the mouth must have opened as shown in Fig. 14.7(*a*). In trout (*Salmo*) and other teleosts of the superorder Protacanthopterygii, the premaxilla is still fixed to the cranium (Fig. 14.4*a*). The maxilla, however, is free from the cheek bones and attached to the cranium, at its anterior end, by a hinge joint. When the mouth opens, the maxilla is pulled forward and cuts off part of the corner of the mouth (Fig. 14.7*b*). In other teleosts such as herring (*Clupea*) the premaxillae as well as the maxillae are hinged to the cranium and both swing forward as the mouth opens (Fig. 14.7*c*). The corners of the mouth have been entirely eliminated, giving a round mouth opening. The grinning (palaeoniscoid) type of mouth may be best for grabbing prey but the round type may be better for sucking food in, since water can only enter from in front and not through the corners of the mouth. Water entering the corners of the mouth would be most likely to be a problem for fish with large bodies and small heads that suck food in while stationary (Fig. 14.2*b*).

The final refinement in jaw evolution was the protrusible upper jaw, found in acanthopterygians, carps and some other advanced teleosts. The maxillae remain hinged to the cranium, and swing down as the mouth opens, but the premaxillae move bodily forward, giving a round mouth opening at the end of a tube (Fig. 14.5). They do not generally protrude very far, but they protrude fast. For example, one particular film sequence shows a piece of food entering the mouth of a 138 mm lionfish (*Pterois*) at 2.7 m s^{-1}. (This is the speed of the prey relative to the mouth

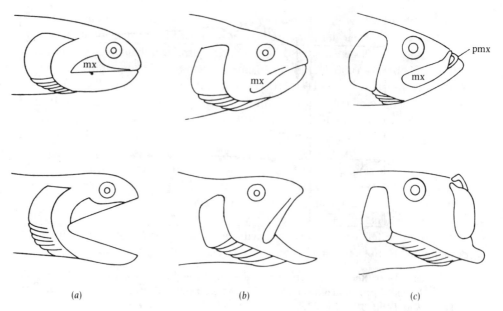

Fig. 14.7. Sketches of (*a*) a palaeoniscoid, (*b*) a typical holostean or primitive teleost, and (*c*) a herring (*Clupea*), showing the mouth closed and open. *pmx*, premaxilla; *mx*, maxilla.

opening.) The jaws were protruded only 5 mm, but they were protruded at 1.5 m s^{-1} (relative to the body of the fish). Their protrusion made a substantial contribution to the relative speed of prey and mouth opening. Protrusion aids feeding in the same way as swimming can do (Fig. 14.2*d*), with little tendency to push water away in front of the mouth. The mouth closes with the jaws still protruded: much of the advantage of protrusion would be lost if the jaws withdrew before closing.

The movements of jaw protrusion are very similar, in acanthopterygians and carps, but the mechanisms that make the jaws protrude as the mouth opens are quite different.

When food has been caught, it has still to be swallowed. Teleosts do this with the help of pharyngeal toothplates. These are plates of dermal bone in the lining of the mouth attached to the underside of the cranium and to the gill skeleton. Typically they bear small, sharp, backward-pointing teeth. The lower toothplates can be moved forward and back by movements of the gill skeleton, so as to drag the food down the throat.

14.3.   Buoyancy

Most teleosts have a swimbladder, a sac of gas in the body cavity which reduces their density. Teleosts without swimbladders, or with the swimbladder deflated, generally have densities between 1060 and 1090 kg m^{-3}, about the same as typical selachians (section 13.3). (The bone of teleost skeletons is much denser than selachian cartilage, but their high urea content must make selachian body fluids denser than teleost ones.) Many teleosts with swimbladders can float almost motionless in mid-water, which shows that their densities are very close indeed to the density of the water.

A 1 kg teleost, of density 1080 kg m^{-3} without a swimbladder, would have a volume of $1/1080 = 0.000926$ m^3 (926 cm^3). A swimbladder containing 74 cm^3 gas would increase its volume to 0.00100 m^3 without appreciably increasing its mass, giving it a density of 1000 kg m^{-3}, the same as fresh water. Similarly, a 49 cm^3 swimbladder would match its density to that of sea water, 1026 kg m^{-3}. These volumes are much smaller than the volumes of squalene that would be needed to have the same effect on its density (section 13.3). A fish with a swimbladder is much less bulky than it would have to be if it depended on squalene for buoyancy, and can be expected to need less energy for swimming at the same speed. However, swimbladders have disadvantages that will soon become apparent.

The swimbladder develops in the embryo as an outgrowth of the gut. Adults of the more primitive teleosts retain a duct (the pneumatic duct) connecting it to the oesophagus. Many of the more advanced teleosts including the Acanthopterygii lose the duct as they develop, so that the swimbladder becomes a closed bag of gas.

Gases have been taken from the swimbladders of fish caught at various depths, and analysed. Swimbladders of fish from shallow water generally contain around 80% nitrogen and 20% oxygen; they contain a mixture of gases very similar to atmospheric air. Swimbladders of fish from greater depths generally contain higher proportions of oxygen. However, trout and their relatives (Salmonoidei) commonly have very high proportions of nitrogen in their swimbladders, no matter how deep they are living. Ciscoes (*Leucichthys* spp.) from depths over 150 m in Lake Huron generally have over 95% nitrogen in their swimbladders.

Most swimbladders have extensible, flexible walls, so the pressure of gas inside matches that of the water around the fish. The pressure in the water (and so in the swimbladder) is 1 atm at the surface and increases by 1 atm for every 10 m of depth, so at D m it is $(1 + 0.1 D)$ atm.

If the water were in equilibrium with the atmosphere, the partial pressures of dissolved oxygen and nitrogen in it would be about 0.2 atm and 0.8 atm, at all depths. About 0.8 atm nitrogen is indeed found at all depths, but the partial pressure of oxygen is always a good deal less than 0.2 atm at depths more than about 200 m. Thus the total of the partial pressures of dissolved gases in the water is 1 atm or less, at all depths. The partial pressures of gases in arterial blood are generally about the same as in the water, because of the exchange of gases that occurs at the gills, so their total is also 1 atm or less, at all depths. However, the pressure in the swimbladder (the total of the partial pressures of the individual gases) is $(1 + 0.1 D)$ atm. Thus at least some of the gases must have higher partial pressures in the swimbladder than in the blood, and will tend to diffuse out. The deeper the fish is swimming, the larger the partial pressure differences and the faster gas will be lost from the swimbladder. Teleosts that have pneumatic ducts and live near the surface may make good the loss by taking an occasional gulp of air at the surface, but others have to secrete new gas, as will be explained.

Secretion of gas from a low partial pressure in the blood to a high partial pressure in the swimbladder is essentially a process of compression, and requires energy. If diffusion losses from the swimbladder can be kept small, energy will be saved. Most swimbladders have sparse blood supplies, except in the specialized regions of secretion and resorption. Many have walls made highly impermeable to diffusion of gases by a layer of extraordinary crystals. These crystals are made of guanine, a purine. They have the form of broad sheets, so thin as to be very flexible. Electron micrographs show that they are only 20 nm thick, and often show them folded into pleats. The crystals have such low diffusion constants that diffusion of gases is effectively limited to the narrow spaces around and between them.

Rates of diffusion of oxygen through swimbladder walls have been measured by means of an oxygen electrode (Fig. 1.6a). The partial pressure of oxygen in an oxygen electrode is effectively zero. Oxygen electrodes are normally used to determine unknown partial pressures of oxygen, by measuring the rate at which oxygen diffuses in through a plastic membrane of known properties. They can, however, be used to determine the diffusion constants of membranes of unknown properties, from the rates at which oxygen diffuses in from a fluid in which it has a known partial pressure. A piece of swimbladder wall was removed from a fish and fastened over the plastic membrane of an oxygen electrode, which was then exposed in turn to moist air and moist oxygen. (Moist gases were used so that the swimbladder wall would not dry out.) The electrode current was measured in each case and the rates of diffusion of oxygen were calculated. The oxygen had to diffuse through the plastic membrane as well as through the swimbladder wall, but account was taken of this in the calculations.

The experiments were performed with the swimbladders of shallow-water marine fish and also of the conger eel (*Conger*) which lives at depths down to 300 m. As might have been expected, the conger swimbladder was the least permeable. Its diffusion constant was about 6×10^{-7} mm^2 atm s^{-1}, which only about one fortieth of the values usually found for other tissues (section 4.3). Some other teleosts live at much greater depths, and many of them have swimbladders. Two species caught at 1400 m and 2800 m had ten times as much guanine, per unit area of swimbladder wall, as the conger.

Since pressure increases with depth, swimbladders are compressed as fish swim deeper and expand when they swim nearer the surface. Fish may need to secrete or resorb gas to compensate for depth changes, as well as secreting to replace diffusion losses.

The process of resorption is straightforward, since diffusion from the swimbladder is in any case inevitable. Many teleosts have means of speeding it up when required. The eel *Anguilla* has a pneumatic duct which could be used to pass gas to the mouth and out of the fish, but which serves as a gas-resorbing organ (Fig. 14.8a). Normally this duct is slender, and though it has plenty of blood vessels little blood flows through them. When gas is to be resorbed it swells and fills with gas from the main cavity of the swimbladder. Its blood vessels dilate, and it has been estimated from measurements of oxygen concentration in them and other vessels that as much as 20% of the blood pumped by the heart may travel through them. Advanced teleosts with no swimbladder duct commonly have a pocket in the swimbladder wall, known as the oval, where resorption occurs (Fig. 14.8b). It has plenty of blood vessels but these are generally constricted so that little blood flows through them and the oval is closed off from the main cavity of the swimbladder by a sphincter. During resorption the blood vessels dilate and the sphincter opens.

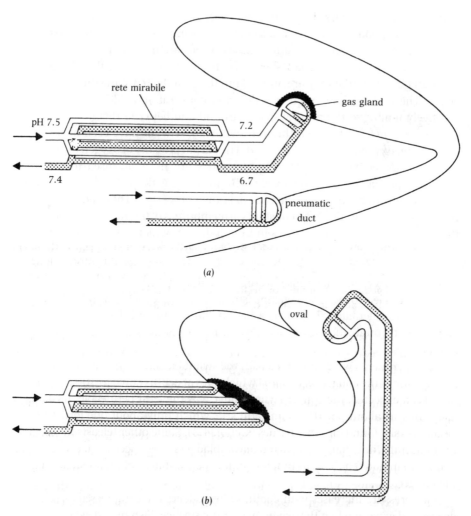

Fig. 14.8. Diagrams of the swimbladder and its blood supply in (*a*) the eel *Anguilla*, and (*b*) a typical advanced teleost with no swimbladder duct. The actual numbers of capillaries are, of course, very much larger than the numbers shown.

Some teleosts have no obvious locality for gas secretion, which probably occurs over large areas of the swimbladder wall. Others have conspicuous red patches known as gas glands. Their function has been demonstrated by injecting cod (*Gadus morhua*) with yohimbine (a drug which stimulates gas secretion). The swimbladders of injected cod were cut open and transparent plastic was laid over the gas gland. Bubbles of gas could be seen collecting under the plastic.

It will be shown that, at least in eels, the process of gas secretion into the swimbladder involves secretion of lactic acid into the blood passing through the gas gland. This lactic acid tends to release gases. Because it is an electrolyte it must reduce the solubility of all gases in blood, just as the salts in seawater make gases less soluble in it than in fresh water. This effect is small, but it is the only known basis for secretion of nitrogen and other inert gases. Because it is an acid, lactic acid tends to release carbon dioxide from bicarbonates in the blood. Also because it is an acid, lactic acid

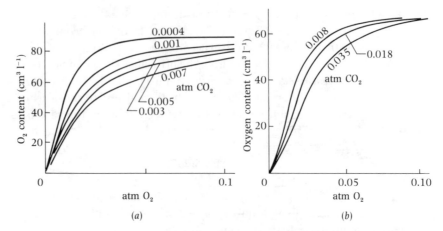

Fig. 14.9. Graphs of oxygen content against partial pressure of oxygen for the blood of (a) rainbow trout (*Salmo gairdneri*) at 15°C and (b) the African lungfish (*Protopterus aethiopicus*) at 25°C. In each case graphs are given for several different partial pressures of carbon dioxide. Re-drawn from F.B. Eddy (1971). *J. exp. Biol.* 55, 695–712; and from C. Lenfant & K. Johansen (1968). *J. exp. Biol.* 49, 437–452.

tends to release oxygen from haemoglobin; Fig. 14.9a shows the effect of carbon dioxide (which reduces the pH) on the amount of oxygen that teleost blood will hold. This sort of effect is not peculiar to teleosts, but also operates in other vertebrates, and a similar effect on snail haemocyanin was shown in Fig. 6.24. The gas secreted into most swimbladders is mainly oxygen, and the effect on haemoglobin is the most important effect of the lactic acid secreted by the gas gland.

Nevertheless, this effect could not by itself release oxygen from the blood at substantial depths. Teleost blood typically contains enough haemoglobin to combine with about 10 cm³ oxygen (measured at atmospheric pressure) per 100 cm³ blood. Additional oxygen can be taken up in physical solution in the blood; if the partial pressure of oxygen is X atm, the amount is $4X$ cm³ per 100 cm³ blood. Oxygen is taken up at the gills from water in which its partial pressure is 0.2 atm (or less), so at a depth of 20 m, where the pressure is 3 atm, the blood is capable of dissolving an additional $4(3.0-0.2)=11$ cm³ oxygen. Even if all the 10 cm³ oxygen per 100 cm³ blood could be released from the haemoglobin it would not pass into an oxygen-filled swimbladder, but would be taken up in physical solution in the blood.

Secretion of oxygen at substantial depths depends on the arrangement of the blood vessels to the gas gland (Fig. 14.8). The artery breaks up into long parallel capillaries which mingle with venous capillaries leading from the gland. The whole group of capillaries is known as a rete mirabile (plural: retia mirabilia). Eels have a slightly different arrangement from other species (compare *a* with *b* in Fig. 14.8); the capillaries join up into a small number of arteries and veins between the rete and the gas gland. This trivial peculiarity has made possible experiments with eels which would otherwise have been a great deal more difficult. Dr J.B. Steen fixed eels (*Anguilla*) in a special holder and opened them to expose these blood vessels. He injected the drug yohimbine to stimulate gas secretion and when this was in progress withdrew for analysis small (40 mm³) samples of blood from the four positions indicated by numbers in Fig. 14.8(*a*). These numbers give the mean pH of samples from each position. Oxygen, carbon dioxide and lactic acid content were also determined, and their values were also in accordance with the theory of gas secretion which will be outlined. It is apparent from the pH values shown in Fig. 14.8(*a*) that acid is secreted into the blood passing through the gas gland. This is lactic acid; the blood entering the gland contains 8 mmol l^{-1} lactic acid, and the blood leaving it 13 mmol l^{-1}. There is also diffusion of ions in the rete, between the venous and the arterial capillaries: this accounts for the fall in pH along the arterial capillaries and the rise along the venous ones.

As blood passes through the gas gland and lactic acid is added to it, the partial pressure of the oxygen in it is increased, mainly through the effect of the acid on haemoglobin. When this blood passes into the venous capillaries of the rete it contains oxygen at a higher partial pressure than in the parallel arterial capillaries. Oxygen diffuses from it into the arterial capillaries, raising the oxygen concentration and partial pressure of the blood there before it arrives at the gas gland. The process is repeated with this blood, and the concentration and partial pressure of oxygen build up higher and higher at the gas gland end of the rete. Eventually the partial pressure exceeds the partial pressure in the swimbladder, and oxygen is released. It can be shown theoretically that the maximum partial pressure that can be built up (and so the depth at which secretion is possible) should increase sharply as the length of the rete increases. In large *Anguilla* the retia (and their capillaries) may be 10 mm long, but teleosts living at depths of 2000 m or more tend to have longer retia. Retia 25 mm long have been found in several deep-sea fishes. These lengths are immensely greater than those of ordinary capillaries, such as the ones in *Anguilla* muscle which are generally 0.5 mm long or less.

The fall in the lactic acid concentration and the consequent rise in the pH of the blood as it travels along the venous capillaries of the rete might be expected to have an adverse effect, partly cancelling out the effect of the rise in concentration in the gas gland. In fact this adverse effect is probably negligible because of a rather surprising property of eel blood. When the pH is decreased, for instance by adding lactic acid, the partial pressure of oxygen is affected very rapidly indeed: the effect has gone half way to completion in 0.05 s. Reversal of the effect by an increase in pH is much slower, and about 15 s is needed for it to go half way. It can be calculated that each blood corpuscle takes a second or thereabouts to travel the length of the rete. This is plenty of time for a fall in pH to take effect, but not for recovery when the pH rises again.

Secretion and resorption are necessarily slow processes, as can be understood by considering the quantities of gas involved. Consider a 1 kg fish with a 50 cm³ swimbladder. Suppose that it moves deeper in the sea, keeping the volume of the swimbladder constant. Every increase in depth of 10 m increases the pressure by 1 atm and requires secretion of a quantity of gas which would have a volume of 50 cm³ at atmospheric pressure. Experiments on brook trout (*Salvelinus*) showed that 1 kg specimens in well-aerated water at 20°C use about 50 cm³ oxygen h^{-1} when resting and up to 150 cm³ h^{-1} when active. The latter is presumably the maximum rate at which the gills can take up oxygen and the blood can transport it to the muscles. Even if one third of the blood pumped by the heart could be directed to the gas gland, it seems most unlikely that oxygen could be supplied fast enough to allow secretion of more than 50 cm^3 h^{-1} (measured at atmospheric pressure). This rate would fill an empty swimbladder in an hour, or maintain the volume of the swimbladder in a descent at the very slow rate of 10 metres per hour.

To find out how fast they could secrete and resorb gases, cod (*Gadus*) were kept in a pressure tank with a viewing window. A pump could raise the pressure in the tank to a maximum of 7.5 atm, corresponding to a depth of 65 m. Fish from shallow water subjected to high pressures were initially denser than the water (because their swimbladders were compressed) but recovered their buoyancy over periods of many hours. The progress of their recovery was monitored by occasional tests taking 3 minutes or less, in which the pressure was adjusted until the fish had the same density as the water and could float with only slight fin movements. The rates of gas secretion were even slower than suggested above, enough only to compensate for depth increases at the rate of one metre per hour. Rates of gas resorption depended on depth: the fish resorbed faster at greater depths, where the differences in partial pressure between swimbladder gases and dissolved gases were greater. At a depth of 5 metres they could compensate for depth reductions only at the rate of 1 metre per hour but at 65 metres they could compensate at 20 metres per hour.

Some teleosts make large depth changes every day. Herring (*Clupea*) spend the night near the surface and the day at depths which vary with locality, but may exceed 150 m. Since swimbladders reflect underwater sound, these movements can be followed by echo-sounding. Even bigger vertical migrations are made by the small lantern fishes (Myctophidae) which are very plentiful in the oceans. Many of them spend the night within 50 m of the surface and the day at 300 m or more: their movements have been observed from deep submersible vehicles (research submarines). The changes of depth at dawn and dusk seem much too fast for compensation by secretion and resorption of swimbladder gases.

When herring and such species of lantern fishes as have gas-filled swimbladders are caught at night at the surface, they have enough gas in the swimbladders to give them about the same density as seawater. It is not known how much gas they have in the swimbladder at their daytime depths, but it seems unlikely that enough extra gas is secreted to compensate for the increased depth. More probably the quantity of gas is kept constant, so that the swimbladder is compressed during the day to a small fraction of its night-time volume. If so, it would be almost totally ineffective by day. The fish would be denser than seawater and would have to depend on the hydrofoil action of fins to avoid sinking further. In these circumstances the advantage of a swimbladder over materials like squalene might disappear. Some lantern fishes have gas-filled swimbladders only as juveniles. As they grow the swimbladder regresses and wax esters are deposited around it. These are quite different chemically from squalene, but have about the same density. Some lantern fishes with no gas in the swimbladder have enough wax ester to give them about the same density as seawater. Their density is virtually unaffected by changes of pressure.

A great many bottom-living teleosts have also lost the swimbladder. The angler fish *Lophius* and the flatfishes (Pleuronectiformes) are among them. A fish which has the same density as water may rest in contact with the bottom but it needs no vertical force to support it, so there are no frictional forces to hold it in place. It will move with every water movement, and·must swim to keep stationary in a current. If it were denser than the water it would use more power when it swam but would not need to swim all the time and so might make a net saving of energy. This is particularly likely to be the case for fish which find their food near bottoms exposed to currents, in rivers and on tidal shores.

Teleosts that have the same density as the water can hang almost motionless in mid-water, and they can execute a remarkable range of manoeuvres by moving their fins. They are not limited to swimming forwards like sharks, but can move backwards, or vertically up and down (with the body horizontal), and they can turn on the spot. Such manoeuvrability is especially useful in cluttered environments, for example among corals. The fins may make rowing movements, or waves may be passed along them. All the fins, including the tail fin, may be used.

Selachian fins are relatively thick and rigid, though the pectoral fins of rays are mobile enough for waves to be passed along them in swimming (Fig. 13.6*c*). Teleost fins are much more mobile. They consist of widely spaced rays joined only by very thin webs of tissue and (unlike selachian fins) can be folded and unfolded like a fan.

Fig. 14.10 shows part of a teleost fin. The pterygiophores are bone. The rays are not simple collagenous rods like the ceratotrichia of selachians but are branched and consist of short lengths of bone joined by collagen fibres. Each consists of two half-rays, one on each side of the fin. Except at the ends of the fin, each ray articulates with a separate group of pterygiophores. The most distal pterygiophore consists of two halves, tightly bound together by collagen fibres. The bases of the two halves of the ray lie on either side of it, attached by ligaments so as to form a rather loose hinge joint. This joint bends when the fin is folded like a fan, and extends as it unfolds (Fig. 14.10*b*). The joint between the distal radial and the next one is also a hinge joint but its axis is longitudinal so that it allows the ray to swing from side to side (Fig. 14.10*d*). This movement occurs when the fin is undulated: each ray rocks from side to side a little out of phase with its neighbours so that the fin is thrown into waves which travel forwards or backwards along it.

Fig. 14.10(*a, c*) shows the muscles which move the rays. The erectors unfold the fin and the depressors fold it. The left inclinator swings the ray to the left and the right one to the right. An inclinator can be assisted by the erector and depressor of its side contracting together, but they also tend to have the effect shown in Fig. 14.10(*e*). The half-ray is pulled basally but cannot simply slide past its mate since the two half-rays are attached quite closely to each other at the edge of the fin. The ray therefore bends to the side. This is particularly apt to happen if the muscles are moving the fin against strong resisting forces.

When they are swimming at all fast teleosts keep their fins (other than the caudal fin) folded flat against the body. Unlike most sharks, they do not rely on spread pectoral fins to provide lift. They spread their fins to brake. Pectoral fins can be spread at right angles to the direction of motion and so make particularly effective brakes, but other fins may help.

If the pectoral fins were simple membranes, they would be bent back at the edges by drag when they were used as brakes. Because of their complex structure, with the muscles attached to half-rays, they tend to bend in the manner shown in Fig. 14.10(*e*). Their edges tend to bend forward, making them more effective as brakes. The forces which can be produced are illustrated by a film sequence of *Taurulus*

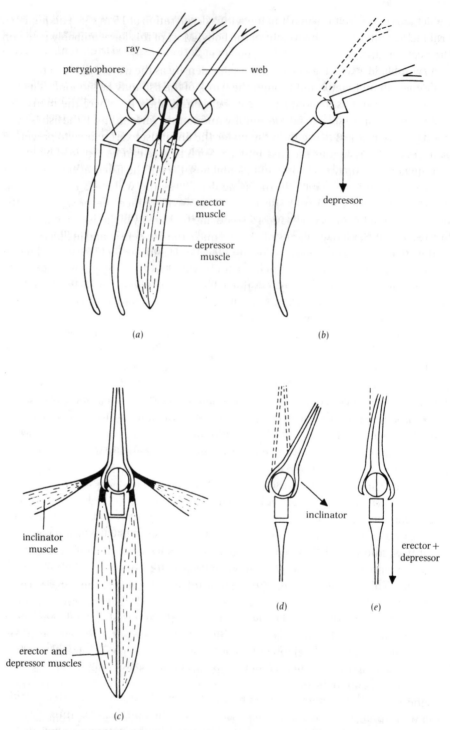

Fig. 14.10. Diagrams of the skeleton and muscles of a teleost fin seen (*a*), (*b*) from the side and (*c*), (*d*), (*e*) in transverse section. This is a dorsal or anal fin: the other fins differ in some details of structure.

(which has large pectoral fins) braking with a deceleration of 15 m s^{-2}. This implies a braking force 1.5 times the weight of the body. Most of this force probably acted on the pectoral fins.

A swimbladder which gives a fish the same density as the water enables it to hover motionless in mid-water, but though the trunk of the fish may be motionless its fins are never motionless for long. This can be seen in aquaria, where the many fish which hover can be seen to be continually undulating various fins. If the fish is even slightly denser or less dense than the water the fins must exert a force to prevent it from sinking or rising, as the case may be. Such a force can be provided by waves passing down or up the pectoral and caudal fins. Even if the fish is at the particular depth at which it has precisely the same density as the water, its equilibrium is unstable. A slight accidental rise in the water will reduce the pressure on the swimbladder, allowing it to expand and reducing the density of the fish so that it tends to go on rising. Similarly, slight sinking will make the fish denser and inclined to sink further. The fish cannot keep at a constant level unless it compensates for accidental vertical movements: it does this by means of its fins. Breathing by pumping water in through the mouth and out through the gills has a jet-propulsion effect, so a teleost hovering in mid water will move slowly forward unless its fins exert a compensating backward force.

14.4.   Silvery shoals

Bottom-living teleosts often have colours which merge with the bottom, and patterns of stripes or blotches which help to make them inconspicuous by distracting attention from the outline of the fish. The teleosts that were cited as examples of bottom dwellers lacking swimbladders (section 14.3) are also examples of this. In contrast, teleosts such as the herring and its close relatives (*Clupea* spp.) which habitually swim in mid-water are often brilliantly silvery. This silveriness is a very effective camouflage. Divers report that silvery fishes, swimming gently, are very difficult to see, except from below.

The problem of camouflage in mid-water is the problem of looking bright against a bright background and dark against a dark background. All the light comes ultimately from above. If a submerged animal in deep water looks upwards towards the sky its eye will receive far more light than if it looks down into the depths. To be inconspicuous from all directions, a fish must match the bright background of the surface water when seen from below, and the dark background of the depths when seen from above. It could of course achieve this by perfect transparency. The paragraphs which follow explain why silveriness is almost as good. Some teleosts are largely transparent but none is completely transparent. Some organs such as eyes and red muscle cannot be made transparent because pigments are essential to their functioning.

Close to the surface on a sunny day, the brightest light comes not from directly overhead but more nearly from the direction of the sun. At depths of 30 m or more in clear water (or less in turbid water) the direction of greatest brightness is much more nearly vertical, whatever the altitude of the sun. The distribution of light is then more

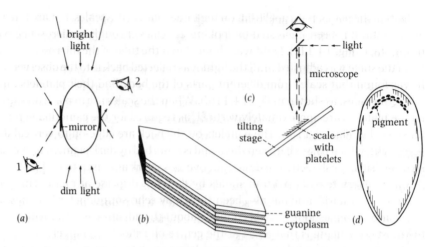

Fig. 14.11. Diagrams illustrating the discussion of silvery fishes. (*a*) How vertical mirrors can make the sides of a fish match their background; (*b*) the structure of a platelet; (*c*) apparatus for measuring the angles of platelets; (*d*) (a diagrammatic transverse section) how the platelets are oriented on different parts of the body.

or less symmetrical about the vertical. In clear ocean water, the vertical downward light is about 200 times as intense as the vertical upward light.

Fig. 14.11(*a*) represents a transverse section through a fish, whose sides are vertical mirrors. An eye at position (1) sees the fish against the bright background of the surface waters, and bright light is reflected into it from the fish's side. To it the fish looks bright, but to an eye at position (2) the side of the fish is dark against a dark background. If the light distribution is symmetrical about the vertical and if the mirrors are perfect, the sides of the fish will match their background perfectly, from whatever direction they are viewed. The horizontal back of the fish can only be seen from above, so if it is to match its dark background it must reflect only a very small fraction of the bright light that falls on it. The horizontal ventral surface can only be seen from below and is silhouetted against a bright background. Herrings and many other silvery fishes have narrow bodies so the strip that cannot be camouflaged is narrow.

The silveriness of teleosts is due to reflecting platelets attached to the inner surfaces of the scales, or in the skin under the scales. If these are scratched off a scale they break up into very thin crystals of guanine (the purine which is also found in swimbladder walls) but it has been shown by electron microscopy that undamaged platelets consist of about five of these crystals, with intervening layers of cytoplasm (Fig. 14.11*b*). The guanine has a much higher refractive index than the cytoplasm so platelets reflect light in the same way as the scales of butterfly wings (section 9.6). Individual platelets reflect particular colours of light, depending on the thicknesses of their layers, but the colour they reflect depends on the angle at which light falls on them. (The same effect makes butterfly scales iridescent.) The light deep in the sea is mainly blue-green but platelets with different layer thicknesses are needed to reflect blue-green light well, whatever direction it comes from. Platelets with different layer thicknesses are superimposed by the overlap of the scales.

Most of the platelets do not lie flat on the undersides of the scales, but are tilted. The angle of tilt has been measured by looking at scales through a microscope with a tilting stage (Fig. 14.11c). Light was shone down the tube of the microscope and the tilt of the stage was adjusted until the light was reflected back into the observer's eyes. It was found that scales from different parts of the body had their platelets tilted at different angles, as shown in Fig. 14.11(d). When the scales are in place on the body, all the platelets are approximately vertical, as required for the camouflage principle illustrated in Fig. 14.11(a). The platelets on the back are edge-on to vertical downward light, which passes between them and is absorbed by dark pigment deeper in the skin, so a silvery fish seen from directly above looks dark against its dark background.

Many silvery teleosts swim in shoals by day, but disperse at night. This can be observed in aquaria, and has also been shown by echo-sounding for herrings in the sea. Various suggestions have been made about the advantages of shoaling. One of the most convincing is that a fish at the centre of a shoal is relatively safe because there are many other fish between it and any predator which approaches. A fish which habitually pushes its way to the centre of a shoal is more likely to survive and leave offspring than one which does not. Constant jockeying for a central position tends to maintain a tight shoal. This is one possible reason why many fish have evolved the habit of shoaling, and why many terrestrial mammals form herds. The open waters where many shoaling fishes live and the open grasslands where many mammals live in herds are both environments where there is no cover – except within a shoal or herd. In support of the idea, minnows (*Phoxinus*) that have been swimming in small groups in an aquarium gather into a single tight shoal when a pike (*Esox*) is introduced.

Another possible reason for shoaling is that members of shoals are likely to get earlier warning of approaching predators. Let the probability that an individual fish will fail to detect a predator in time be P (a fraction less than one). The probability that none of the members of a shoal of n fish will detect it is P^n (a smaller fraction). If any member of a shoal detects the predator, its response may alert the others. To test this idea, groups of one to thirty small teleosts (*Hemigrammus*, a tetra) were alarmed by flashes of light. They responded by brief bursts of swimming. The probability that *no* member of the group would respond diminished as group size increased, but it did not diminish as much as the simple theory predicted. This was possibly because the members of a shoal partly blocked each other's views of the stimulus, but it may have been that the fish in larger shoals spent less time looking around for possible danger. Experiments with other species have shown that fish in larger groups spend a larger fraction of their time feeding.

This leads to a third suggestion about the advantage of shoaling. Fish in shoals can spend more of their time feeding because they may be warned of danger by the movements of others and do not need to spend so much time looking out themselves. Good evidence for a similar suggestion comes from observations of birds, which raise their heads and look around for danger in a very obvious way. Sparrows, ostriches and various other species have been watched while feeding and found to spend less of their time looking around when in a large group, than when in a small one.

A fourth suggested advantage of shoaling has been discounted. Hydrodynamic theory shows that fish could reduce the energy cost of swimming, by swimming very close together in appropriate relative positions. Observations of several marine species showed that their shoals were too loose and too irregular for the fish to be getting any substantial advantage.

14.5. Warm-blooded fishes

There are advantages in being warm. Within limits, higher temperatures enable fish to swim faster. for example, electrodes were placed in the red and white swimming muscles of 18 cm carp (*Cyprinus*), which were made to swim in a water tunnel like the one shown in Fig. 6.19. At 10°C they started using their white muscles when the speed reached 26 cm s^{-1} but at 20°C the red fibres could contract faster and they did not have to start using the white ones until the speed reached 46 cm s^{-1}. In another investigation, 14 cm trout (*Salmo*) were filmed as they accelerated from rest in response to an electrical stimulus. At 5°C their maximum acceleration was only 16 m s^{-2} (1.6 g) but at 15–25°C it was 41 m s^{-2} (4 g).

Most fishes have their whole bodies at the same temperature as the water they live in, but tuna often have the deeper parts of their bodies much warmer. Tuna that have just been caught and killed have been probed with long needles with thermistors at their ends. In a typical case, a Bluefin tuna (*Thunnus thynnus*) taken from water at 19°C had much of its swimming muscle at or above 27°C, and the warmest part was at 31°C. The high temperature presumably improved its swimming performance. The sharks *Lamna* (Fig. 13.1) and *Isurus* also have warm muscles, but the temperature differences between the muscles and the water are generally less than in tuna.

How are the high temperatures maintained? Blood must circulate constantly between the muscles and the gills. If it is warmed in the muscles it must (in a fish with a normal blood circulation) be cooled in the gills, and it will only be possible to maintain a very small temperature difference between the muscles and the water. A quick calculation will demonstrate this.

Let the mean difference in partial pressure of oxygen between the water and the blood at the gills be P. Let the area of the gill surface be A, the distance for diffusion between water and blood d and the diffusion constant for diffusion of oxygen D. Then by equation 4.1 the rate M at which oxygen diffuses from water to blood is given by

$$M = DA \cdot \Delta P / d. \tag{14.1}$$

Similarly, if the mean temperature difference between the water and the blood is ΔT and the thermal conductivity of water and tissue is C, the rate H of loss of heat at the gills is given by

$$H = CA \cdot \Delta T / d. \tag{14.2}$$

From the above two equations,

$$\Delta T = (HD/MC)\Delta P. \tag{14.3}$$

A similar equation can be obtained relating the differences in partial pressure of oxygen ($\Delta P'$) and temperature ($\Delta T'$) between the blood and the muscle. Hence the total temperature difference between the water and the muscle is given by

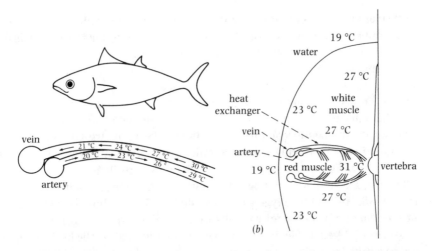

Fig. 14.12.(*a*) A sketch of a tuna (*Thunnus* sp.). (*b*) A diagrammatic transverse section through a tuna showing the blood supply to the swimming muscles and typical temperatures in °C. (*c*) A diagram of the heat exchanger. The temperatures shown in (*b*) are based on actual measurements but those shown in (*c*) are only possible values, chosen to illustrate the principle.

$$(\Delta T + \Delta T') = (HD/MC)\,(\Delta P + \Delta P'). \tag{14.4}$$

Metabolism using 1 cm³ oxygen produces 20 J heat, so if the system is in equilibrium $H/M = 20$ J cm⁻³ or 2×10^7 J m⁻³. The diffusion constant for oxygen diffusing through water is 6×10^{-11} m² atm⁻¹ s⁻¹ and this will be taken as D, although the diffusion is partly through tissue in which the diffusion constant is lower. The thermal conductivity of water is about 0.6 W m⁻¹ K⁻¹, and this will be taken as C. $(\Delta P + \Delta P')$ cannot be greater than 0.2 atm. Hence

$$(\Delta T + \Delta T') \leqslant 2 \times 10^7 \times 6 \times 10^{-11} \times 0.2/0.6$$
$$\leqslant 4 \times 10^{-4} \text{ K}.$$

Taken at its face value, this indicates that the muscles could not be more than a tiny fraction of a degree warmer than the water. However, the main trunk muscles of *Tilapia* (a teleost with quite ordinary blood circulation) have been found to be generally 0.4 K warmer than the water. The discrepancy between theory and observation probably arises because the arteries to the trunk muscles run quite near the veins leaving them. It is by means of a quite different and much more highly developed arrangement of arteries running alongside veins that the high muscle temperatures of tunas are maintained.

This peculiar arrangement of blood vessels is shown in Fig. 14.12. The myomeres do not get their blood from the aorta, but from two arteries running close under the skin on either side of the body. They are not drained by the usual veins but by ones running alongside these arteries. Fine arteries and veins run inwards from these vessels to the red muscle, parallel to each other and in close contact. The blood supply to the white muscle is more sparse but also consists of parallel arteries and veins in contact with each other, extending inwards from larger vessels under the skin. The superficial arteries and veins are cool, close to the temperature of the water. The deeper muscle is warm. Blood in the fine arteries going to the muscle is warmed by blood leaving in the parallel veins. As it leaves again in the veins it is cooled by blood arriving in the arteries. If the parallel arteries and veins are long enough, and in close enough contact, the leaving blood may be cooled very nearly to its original temperature, as Fig. 14.12(c) suggests. Notice that in this diagram there is a small temperature difference between the parallel artery and vein, all along their length, so that heat must be being conducted from the vein to the artery. The same ('countercurrent') principle is used in industrial heat exchangers, which consist of stacks of parallel pipes carrying fluids in opposite directions.

There might seem to be a snag in the system. If heat can be conducted between the arteries and the veins, oxygen can also diffuse. The exchanger which keeps the muscle much warmer than the gills must also be expected to make the partial pressure of oxygen in the muscle fall below that in the gills. The likely extent of this can be worked out in the same way as equation 14.4 was derived. It emerges that a 10°C temperature difference could in principle be obtained for the penalty of only a 0.001 atm drop in partial pressure of oxygen.

The system shown in Fig. 14.12 is a very effective heat exchanger but it is (fortunately for the fish) ineffective as a gas exchanger. We have already in this book met very effective countercurrent gas exchangers, though we have not called them by that name. The rete mirabile of teleost swimbladders is a countercurrent exchanger that maintains large differences in the partial pressures of gases, between the swimbladder and the rest of the body (section 14.3). The ctenidia of molluscs and the gills of fishes are also countercurrent exchangers: the opposite directions of flow of the blood and the water ensure that as much as possible of the oxygen in the water is transferred to the blood (section 6.6).

Further reading

Bony fishes

Greenwood, P.H., Rosen, D.E., Weitzman, S.H. & Myers, G.S. (1966). Phyletic studies of teleostean fishes, with a provisional classification of living forms. *Bull. Am. Mus. nat. Hist.* **131**, 339–456.

Lauder, G.V. & Liem, K.F. (1983). The evolution and interrelationships of the actinopterygian fishes. *Bull. Mus. comp. Zool.* **150**, 95–197.

Patterson, C. (1977). Cartilage bones, dermal bones and membrane bones, or the exoskeleton versus the endoskeleton. In *Problems in vertebrate evolution* (ed. S.M. Andrews, R.S. Miles & A.D. Walker), pp. 77–121. Academic Press, London.

Feeding and respiration

Alexander, R. McN. (1967). The functions and mechanisms of the protrusible upper jaws of some acanthopterygian fish. *J. Zool., Lond.* **151**, 43–64.

Ballintijn, C.M. & Hughes, G.M. (1965). The muscular basis of the respiratory pumps in trout. *J. exp. Biol.* **43**, 349–362.

Lauder, G.V. (1985). Aquatic feeding in lower vertebrates. In *Functional vertebrate morphology* (ed. M. Hildebrand, D.M. Bramble, K.F. Liem & D.B. Wake), pp. 210–229. Harvard University Press, Cambridge, Massachusetts.

Muller, M. & Osse, J.W.M. (1984). Hydrodynamics of suction feeding in fish. *Trans. zool. Soc. Lond.* **37**, 51–135.

Osse, J.W.M. (1969). Functional morphology of the head of the perch (*Perca fluviatilis* L.): an electromyographic study. *Neth. J. Zool.* **19**, 289–392.

van Leeuwen, J.L. (1984). A quantitative study of flow in prey captured by rainbow trout, with general consideration of the actinopterygian feeding mechanism. *Trans. zool. Soc. Lond.* **37**, 171–227.

van Leeuwen, J.L. & Muller, M. (1984). Optimum sucking techniques for predatory fish. *Trans. zool. Soc. Lond.* **37**, 137–169.

Buoyancy

Alexander, R. McN. (1966). Physical aspects of swimbladder function. *Biol. Rev.* **41**, 141–176.

Alexander, R. McN. (1972). The energetics of vertical migration by fishes. *Symp. Soc. exp. Biol.* **26**, 273–294.

Arita, G.S. (1971). A re-examination of the functional morphology of the soft-rays in teleosts. *Copeia* 1971, pp. 691–697.

Berg, T. & Steen, J.B. (1968). The mechanism of oxygen concentration in the swimbladder of the eel. *J. Physiol., Lond.* **195**, 631–638.

Blaxter, J.H.S. & Tytler, P. (1978). Physiology and function of the swimbladder. *Adv. comp. Physiol. Biochem.* **7**, 311–367.

Geerlink, P.J. & Videler, J.T. (1974). Joints and muscles of the dorsal fin of *Tilapia nilotica* L. *Neth. J. Zool.* **24**, 279–290.

Jones, F.R.H. & Scholes, P. (1985). Gas secretion and resorption in the swimbladder of the cod, *Gadus morhua*. *J. comp. Physiol.* B**155**, 319–331.

Lapennas, G.N. & Schmidt-Nielsen, K. (1977). Swimbladder permeability to oxygen. *J. exp. Biol.* **67**, 175–196.

Silvery shoals

Clark, C.W. & Mangel, M. (1986). The evolutionary advantages of group foraging. *Theor. Pop. Biol.* **30**, 45–75.

Denton, E.J. (1970). On the organization of reflecting surfaces in some marine animals. *Phil. Trans. R. Soc. Lond.* B**258**, 285–313.

Godin, J.G.J., Classon, L.J. & Abrahams, M.V. (1988). Group vigilance and shoal size in a small characin fish. *Behaviour* **104**, 29–40.

Hamilton, W.D. (1971). Geometry for the selfish herd. *J. theor. Biol.* **31**, 295–311.

Magurran, A.E. & Pitcher, T.J. (1987). Provenance, shoal size and the sociobiology of predator-evasion behaviour in minnow shoals. *Proc. R. Soc. Lond.* B**229**, 439–465.

Warm-blooded fishes

Carey, F.G. (1982). Warm fish. In *A companion to animal physiology* (ed. C.R. Taylor, K. Johansen & L. Bolis), pp. 216–233. Cambridge University Press.

Rome, L.C., Loughna, P.T. & Goldspink, G. (1984). Muscle fiber activity in carp as a function of swimming speed and temperature. *Am. J. Physiol.* **247**, R272–R279.

Webb, P.W. (1978). Temperature effects on acceleration of rainbow trout, *Salmo gairdneri*. *J. Fish. Res. Bd Can.* **35**, 1417–1422.

15 *Lungfishes and amphibians*

Class Osteichthyes (continued)
    Subclass Crossopterygii,
        Order Rhipidistia (extinct)
        Order Coelacanthini (coelacanths)
    Subclass Dipnoi (lungfishes)
Class Amphibia,
    various extinct subclasses and
    Subclass Lissamphibia,
        Order Salientia (frogs and toads)
        Order Urodela (newts and salamanders)
        Order Apoda

15.1. Coelacanths and lungfishes

Chapter 14 was about the actinopterygian fishes. This section introduces the two other subclasses of the class Osteichthyes: the Crossopterygii and the Dipnoi. All three subclasses first appeared in the Devonian period, about 400 million years ago. The early crossopterygians were members of the order Rhipidistia, like *Holoptychius* (Fig. 15.1a), and the early Dipnoi looked very similar to them. At first sight they may not seem very different from palaeoniscoids such as *Moythomasia* Fig. 14.1a) but some differences are worth noticing. The fins of palaeoniscoids and other actinopterygians generally have the pterygiophores and their muscles embedded in the body: the visible parts of the fins are supported only by rays (Fig. 14.10). In crossopterygians and dipnoans, however, the fins have fleshy lobes at their bases, where the pterygiophores and muscles extend into the fin (Figs. 15.1 and 15.2a; the pectoral fin of *Polypterus* (Fig. 14.1b), is an exceptional case of a fleshy actinopterygian fin). Early members of all three subclasses had thick scales, but the outer layer of enameloid substance was thick in actinopterygians and thin in members of the other subclasses.

Rhipidistians and dipnoans also resemble each other in having internal as well as external nostrils. The nasal sacs of teleosts generally have two openings, both external. Water enters by the anterior nostril and leaves by the posterior one, driven by the fish's movement through the water, by cilia or by the indirect action of breathing movements. Lungfishes have an external nostril connected through the nasal sac to an internal nostril inside the mouth cavity. When a lungfish breathes in water it opens its mouth only slightly and some water must be drawn in through the nasal sac, so breathing aids smelling. (Lungfishes also breathe air, as will be described, but open their mouths widely to do so, so there is probably little air flow through the nasal sacs.)

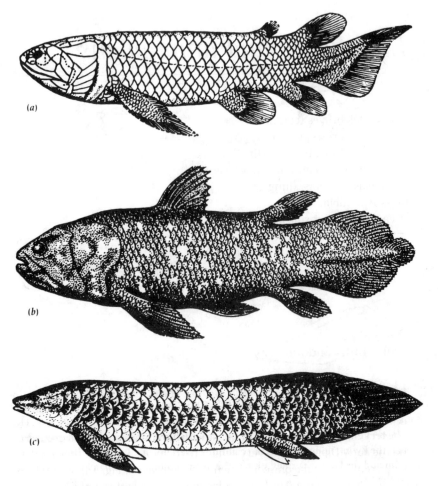

Fig. 15.1.(*a*) *Holoptychius*, a rhipidistian from the Devonian period, typical length 0.8 m. (*b*) *Latimeria*, the modern coelacanth, about 2 m. (*c*) *Neoceratodus*, the Australian lungfish, about 1.5 m. From E. Jarvik (1960). *Théories de l'evolution des vertébrés*. Masson, Paris; and from J.R. Norman & P.H. Greenwood (1963). *A history of fishes*, 2nd edn. Benn, London.

Though rhipidistians and early dipnoans were very similar to each other in many respects, their teeth and jaws were very different. Rhipidistians had pointed teeth along the edges of their jaws and on the palate, which have a characteristic appearance in section because the dentine is deeply pleated (Fig. 15.2*b*). The lungfishes (with a few early exceptions) have no ordinary teeth, and the premaxillae and maxillae (which bear the teeth of the edge of the upper jaw, in other bony fish) are rudimentary or absent. However, the lungfishes have ridged plates of very hard material (similar in chemical composition to enameloid substance) in the roof and floor of the mouth. Ridges in the upper toothplates fit into grooves in the lower ones, making an effective mechanism for chopping or crushing food.

The most peculiar feature of the rhipidistians was a hinge half way along the cranium, that allowed the snout to be tilted up (Fig. 15.2*d*). We will have to examine

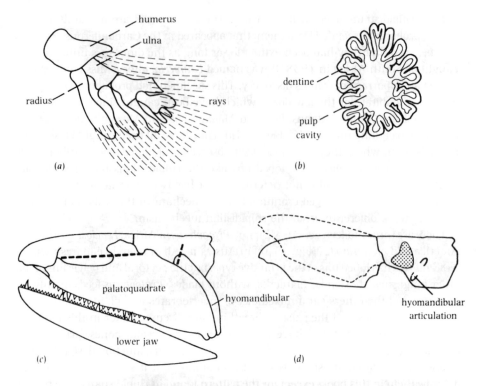

Fig. 15.2. The structure of rhipidistians. (*a*) The skeleton of the pectoral fin of *Eusthenopteron*. (*b*) A section of a tooth of *Eusthenopteron*, showing the pleated dentine. (*c*) The skull of a rhipidistian, with the dermal bones of the cheek removed. The broken lines are the axes of the joints with the braincase of the palatoquadrate and the hyomandibular. (*d*) The cranium alone, showing flexure of the intracranial joint and the facet for articulation of the hyomandibular. The hatching is explained in the text.

other skull structures to understand how it probably worked (Fig. 15.2*c*). The palatoquadrate and hyomandibular were separate, as in selachians (Fig. 13.2*a*), not combined to form a suspensorium as in teleosts (Fig. 14.4*b*). The palatoquadrate was hinged to the anterior part of the cranium and the hyomandibular to the posterior half, so they could be swung laterally to enlarge the mouth cavity to breathe in or to suck in food. However, the axes of the joints were not parallel: the palatoquadrate joint axis was horizontal and the hyomandibular one almost vertical (Fig. 15.2*c*). When the hyomandibulars swung laterally they must have pushed the jaws forward, bending the intracranial joint and tilting up the snout. Teleosts also tilt up their snouts when they feed (Fig. 14.5) but they tilt the whole cranium, not just its anterior part.

   In selachians, the spiracle opens between the palatoquadrate and hyomandibular cartilages. Small toothed plates of bone on the cranium of rhipidistians, indicated by hatching in Fig. 15.2(*d*), have been interpreted as evidence that they had a spiracle in the same position.

Most rhipidistians lived in fresh water. The coelacanths are a mainly marine group, probably descended from them, that appeared in the Carboniferous period. It was believed that they had been extinct for as long as the dinosaurs until one was caught off South Africa in 1938. It was named *Latimeria*, after a Miss Latimer who played an important part in its discovery. This specimen had probably strayed from its normal habitat, for the few dozen which have been caught since have all been caught off the Comoro Islands, between Madagascar and Mozambique. They are large fish, weighing up to 80 kg (Fig. 15.1*b*). They have all been caught close to the rocky bottom, where they have also been observed and filmed from a submersible research vehicle. *Latimeria* has lobed fins like the rhipidistians but its scales are relatively thin and its tail is not heterocercal. It has two external nostrils and no internal one. It has a hinged cranium, but the mechanism that tilts up the snout seems to work differently from the rhipidistian mechanism.

There are three modern genera of Dipnoi, *Protopterus* in Africa, *Lepidosiren* in South America and *Neoceratodus* (Fig. 15.1*c*) in Australia. All of them live in fresh water. *Neoceratodus* has lobed fins of the primitive type, but the pectoral and pelvic fins of the other lungfishes are slender filaments, without blades. *Protopterus* feeds mainly on molluscs and the others eat a mixed diet of invertebrates and plant material. The lungs of lungfishes and their use in breathing are discussed later in this chapter.

The skulls of bony fishes have outer coverings of dermal bones, arranged in complicated patterns which generally have no obvious functional significance. These patterns fascinate specialists in fish anatomy, but it seems unnecessary to describe them in this book, except for the pattern found in rhipidistians. The reason for the exception is that the amphibians seem to have evolved from rhipidistian ancestors. The reptiles evolved from the amphibians and the birds and mammals evolved from them. In the course of this evolution many changes occurred in the skull, many of them of great functional interest. To appreciate them it is necessary to know the names of rather a lot of bones, which are illustrated in Fig. 15.3.

The dermal bones in and around the mouth bear teeth. An outer row of teeth is borne by the premaxillae and maxillae of the upper jaw, and the dentaries of the lower. There is an inner row of big teeth on the vomers, palatines and ectopterygoids. These bones are dermal ones formed in the lining of the mouth rather than on the outer surface of the head (see Fig. 14.3). Other such bones, bearing smaller teeth, are the pterygoids and parasphenoid. The vomers and parasphenoid are borne on the underside of the cranium while the palatines, ectopterygoids and pterygoids cover the inner faces of the palatoquadrates.

Many of the dermal bones of the outer surface of the skull have lateral line canals running through them. The neuromasts of the lateral line system probably played an important part in the development of these bones, just as they seem to influence the development of many dermal bones in modern fishes. For instance, many bones in the skull of the trout first appear as a short length of bony tube in the wall of a lateral line canal around a neuromast, and a little plate of bone beneath. If the epithelium bearing the developing neuromasts is removed surgically at an early stage the bone fails to appear, and additional tubes of bone can be produced by grafting neuromast tissue.

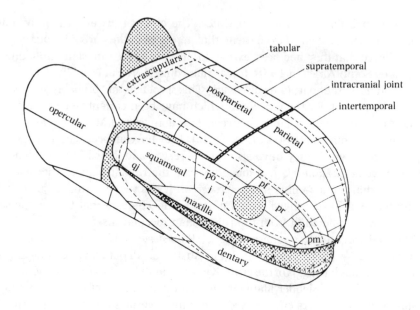

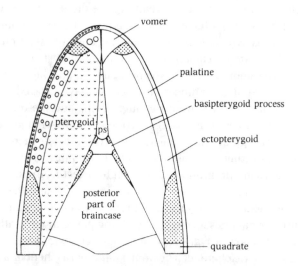

Fig. 15.3. **Diagrams showing the arrangement of dermal bones in the skull of the rhipidistian *Eusthenopteron*. The lower diagram is of the palate seen from below. Abbreviations: j, jugal, l, lacrimal; pf, postfrontal; pm, premaxilla; po, postorbital; pr, prefrontal, ps, parasphenoid; qj, quadratojugal. Broken lines represent lateral line canals.**

15.2. Amphibians

The amphibians, reptiles, birds and mammals are known collectively as the tetrapods. This name means 'four footed' but it is applied to birds and snakes as well as to animals that actually have four feet. The word for animals that walk on four feet is 'quadrupeds'.

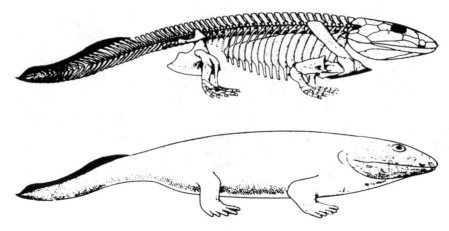

Fig. 15.4. Reconstructions of the skeleton and body of the very early amphibian *Ichthyostega*. From E. Jarvik (1955). *Science Monthly* 80, 141–154. *Ichthyostega* was about 1 m long.

The earliest known tetrapods are some late Devonian amphibians found in Greenland. *Ichthyostega* (Fig. 15.4) is one of them. They must have looked like large newts but they had fish-like features not found in modern amphibians: a fin round the tail and a rudimentary operculum. The fin is not just a fold of skin such as many newts have but a genuine fin with pterygiophores and rays. The operculum is a single small bone and it is not known whether anything remained of the gills. *Ichthyostega* also had small, fish-like scales; the only scales possessed by modern amphibians are the small scales of some of the worm-like Apoda. *Ichthyostega* had lateral line canals in the dermal bones of its skull which suggest that it spent much of its time in water. Lateral line sense organs serve to detect water movements and have no known function out of water. Modern amphibians which live out of water lack lateral line organs, though their aquatic larvae and aquatic adults (such as those of *Xenopus*, the clawed toad) possess them.

As well as generally fish-like features, *Ichthyostega* and its close relatives have features which link them with the rhipidistians. The dentine of their teeth is pleated in just the same way. There is no intracranial joint but there is a suture in the cranium in the corresponding position. The dermal bones of the skull are arranged in a modified version of the pattern found in rhipidistians, which is quite distinct from the patterns found in other bony fishes.

The skulls of *Ichthyostega* and a rhipidistian are compared in Fig. 15.5. It is quite easy to find the homologue in one skull of almost any dermal bone in the other. The parietal bones on either side of the aperture for the pineal organ make a convenient starting point. The remaining bones can be matched with their homologues by studying their positions relative to each other and to the lateral line canals which run through them. The main difference between the two skulls is one of proportions. In the fish, the parts posterior to the eye (including the operculum) are long so that there is room for large gills. In the amphibian they are short, leaving little or no room for gills. However, the snout, anterior to the eye, is relatively longer in the amphibian. There are fewer bones on the roof of the snout in the amphibian: note how a number

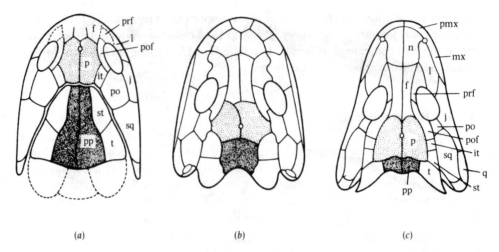

Fig. 15.5. Dorsal views of the skulls of (*a*) a rhipidistian fish (*Osteolepis*), (*b*) the very early amphibian *Ichthyostega*, and (*c*) a slightly later amphibian (*Palaeogyrinus*, of the Carboniferous) to show the differences in proportions. The parietals (p) and post-parietals (pp) are stippled. Names of other bones are given in Fig. 15.4 except f, frontal and n, nasal. From A.S. Romer (1941). *J. Morph.* 69, 141–160.

of small bones (shown in Fig. 15.4) have been replaced by a pair of large frontals and a pair of large nasals. The palatoquadrates formed the side walls of the mouth cavity in rhipidistians (Fig. 15.2*c*) and could be swung laterally to suck in food, but in amphibians they form a mainly horizontal, fixed palate.

There were many amphibians in the Carboniferous, Permian and Triassic periods, some as small as modern newts and others as large as alligators. They were broadly similar to *Ichthyostega* but less fish-like; they had no fins, no trace of an operculum and not even a suture in the position of the rhipidistian cranial hinge.

The modern amphibians belong to three orders, the Salientia (frogs and toads), the Urodela (newts and salamanders, Fig. 15.15) and the Apoda (tropical limbless amphibians). They seem closely enough related to be grouped together in a single subclass, the Lissamphibia, but there has been controversy about this. Many of the similarities between them are associated with their manner of respiration (described in section 15.3): they have moist skins with mucous glands and a good blood supply, small ribs and more or less flattened skulls (giving a large throat to pump air into the lungs). There are other similarities that have no apparent functional relationship to respiration: their teeth are in two parts, a base and a crown separated by uncalcified tissue, and they have a small additional cartilage or bone in the ear (which is called the operculum but is nothing to do with the operculum of fishes).

Despite these similarities, adults of the three modern orders are quite different in shape. Urodeles are shaped much like primitive amphibians but Salientia have long hind legs (used for jumping and swimming) and no tail, and Apoda have no legs.

Terrestrial salientians and urodeles have both evolved long sticky tongues and use them for picking up prey. (Suction feeding as used by fish, section 14.2, will only work under water.) However, the mechanisms of tongue movement are different, indicating that the two orders have evolved them independently. The tongue is attached to the anterior part of the floor of the mouth and at rest is folded back, but is flipped rapidly forward to catch prey. The large toad *Bufo marinus* (about 20 cm from snout to vent) can extend its tongue to a length of 15 cm. The stickiness is due to mucus secreted by glands on the tongue.

15.3. Breathing air

It seems likely that the early members of all three subclasses of Osteichthyes had lungs and got some of their oxygen by breathing air. Modern Dipnoi have lungs and so also do *Polypterus* and some other primitive Actinopterygii. The swimbladders of teleosts probably evolved initially as lungs: a lung inevitably gives buoyancy. *Latimeria* has only a fatty organ that seems to be a degenerate lung or swimbladder but the more primitive Crossopterygii, from which the tetrapods evolved, presumably had lungs.

Mudskippers (*Periophthalmus* species, teleosts) can take up enough oxygen through their skins to survive long periods out of water. They use this ability to leave the water for feeding, crawling on mud flats and mangrove roots on tropical shores. The lungfishes *Protopterus* and *Lepidosiren* survive the dry season in burrows, breathing air, when the swamps that they live in dry up. Most other air-breathing fishes, including the lungfish *Neoceratodus*, spend their whole lives in water. The ability to breathe air enables them to survive in water that contains little dissolved oxygen.

The surface waters of the sea are always well aerated. Very low oxygen concentrations are common at depths of several hundred metres, but lungs would be no use so far below the surface. Low oxygen concentrations are found near the surface in some freshwater habitats, especially in tropical swamps overhung by dense vegetation. During the day these swamps are shaded from the sun. At night, they do not radiate heat to a cold sky but to this same vegetation, which is never much cooler than the swamp and radiates back almost as much heat as it receives from below. Daily temperature fluctuations in the water are therefore small and there is relatively little mixing by convection currents at night, which would distribute oxygen through the whole body of water as the surface water cooled and sank. The water is stagnant, so there is no turbulence to mix it, and the vegetation shields it from disturbance by wind. The vegetation also makes the light so dim that little oxygen can be produced by photosynthesis in the water. Well-aerated fresh water contains about 8 cm^3 oxygen per litre (depending on the temperature) but the concentrations in these swamps are often less than 1 cm^3 per litre except within a few millimetres of the surface. Many of the fish that live in these swamps have evolved air-breathing organs. For example, in South America the swimbladder of the characin *Erythrinus* has re-evolved the capacity to function as a lung, and part of the gut of the catfish *Hoplosternum* has become an air-breathing organ. Most fish that are not air breathers asphyxiate at dissolved oxygen concentrations between about 0.3 and 1.5 cm^3 per litre, but it may be advantageous to breathe air even at slightly higher concentrations, at which a lot of energy would be needed to pump water fast enough over the gills to extract enough oxygen.

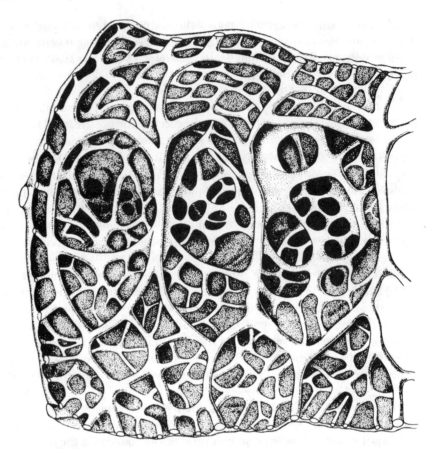

Fig. 15.6. Lung of *Protopterus* opened to show the internal partitions and alveoli. From M. Poll (1962). ***Ann. Reeks Zool. Wetenschap.*** 108, 131–172.

Neoceratodus has a single lung like *Amia*, while *Protopterus* and *Lepidosiren* each have a pair of lungs that merge at the anterior end into a single chamber. If oxygen is to diffuse rapidly into the blood the area available for diffusion must be large and the distance from air to blood short. Complicated partitions in the lungs greatly increase this area (Fig. 15.6). I know no estimates of this area in lungfishes, but the lung areas of many amphibians and reptiles are similar to the gill areas of fishes of the same mass, around 4 cm² (g body mass)$^{-1}$. The lung wall is richly supplied with blood capillaries and the blood is separated from the air only by the walls of these capillaries and the lung epithelium, a distance in *Protopterus* of about 0.5 μm. The corresponding distance in fish gills is commonly over 2 μm but the lungs are in a more protected position and can be more delicate. Also, there is more advantage in having the blood very close to the respiratory surface in lungs than in gills. This is because oxygen diffuses immensely faster through air than through water: its diffusion constant in air is about 3×10^5 times as high as in water. In gills an important part of the resistance to diffusion is due to most of the oxygen having to diffuse through 5 μm or more of water before reaching the gill surface (section 12.2). Shortening the distance from this surface to the blood beyond a certain point has only a marginal effect. In a lungfish lung oxygen might well have to diffuse through 3 mm of air before reaching the respiratory surface, but because of the difference in diffusion constants this is equivalent to only $3/(3 \times 10^5)$mm $= 0.01$ μm water. The layer of tissue between the air and the blood offers the major part of the resistance to diffusion, and there is a great advantage in having it as thin as possible.

Partitions which increase the surface area of a lung could substantially increase the pressure needed to inflate it, owing to the effects of surface tension. The partitions divide the wall of the lung into pockets known as alveoli, of which the smallest in *Protopterus* seem to have radii around 50 μm. If the fluid coating an alveolus of this size had the same surface tension as water, 0.07 N m^{-1}, the pressure ($2 \times$ surface tension/radius) needed to overcome this surface tension as the alveolus was inflated would be $(2 \times 0.07)/(5 \times 10^{-5}) = 3000$ N m^{-2} (30 cm H$_2$O). The actual pressure, which has been measured with pressure transducers in the mouth cavity, is only about half that. The reason is the presence in the lung of a surface-active material (apparently a lipoprotein) which forms a film coating the inner surface of the lung and greatly reduces the surface tension. Similar material is present in the lungs of tetrapods. It is deficient in some human babies, which may die because of the difficulty they have in inflating their lungs. There is, of course, a duct leading from the lungs to the mouth. The glottis, its opening into the floor of the pharynx, has muscles to open and close it.

Neoceratodus makes far less use of air breathing than the other modern lungfishes. It has a full set of gills and it relies almost entirely on them for respiration when it is kept in well-aerated water. It probably breathes little air in its natural habitat. The rivers where it lives seem never to dry up, and never to contain less than about 5 cm^3 oxygen per litre.

Protopterus and *Lepidosiren* commonly live in swamps which are much less well aerated. They have reduced gills, and indeed *Protopterus* has no gill filaments on two of its gill bars. Even in well-aerated water they visit the surface to breathe air every 3–10 min, and they asphyxiate if prevented from reaching the surface. It has been shown that even in well-aerated water *Protopterus* gets only 10% of its oxygen from the water: this was done by keeping the fish in a closed aquarium and following the reduction of partial pressure of oxygen in the water and in the air above.

It might be thought remarkable that a gulp of air every 3–10 min should provide 90% of the oxygen a lungfish needs. Teleosts breathing water make many cycles of breathing movements every minute. For instance, a large trout at 10°C took 82 breaths min^{-1} to obtain 50 cm^3 oxygen kg^{-1} h^{-1}. *Protopterus* at 24°C used 60 cm^3 oxygen kg^{-1} h^{-1} (getting most of this from the lungs) but only took a breath of air about once in 5 min. The trout, obtaining oxygen through the gills at the same rate as the lungfish obtained it from the lungs, took about 400 breaths of water for every breath of air taken by the lungfish. However, the water used by the trout contained only 8 cm^3 oxygen l^{-1} while the air used by the lungfish contained 200 cm^3 l^{-1} or 25 times as much. The trout took only 1.8 cm^3 water (kg body mass)$^{-1}$ in each cycle of respiratory movements while the lungfish took very much larger volumes of air. *Protopterus* normally holds about 35 cm^3 air (kg body mass)$^{-1}$ in its lungs, and most of this is changed at each breath. The air taken in by *Protopterus* at each breath may well have contained 400 times as much oxygen as the water taken in by the trout.

Protopterus can remove a very high proportion of the oxygen from the air in its lungs. This has been demonstrated by collecting the air it breathes out in an inverted water-filled funnel. Analysis of this air shows that if the interval since the preceding breath has been a long one, almost all of the oxygen has been removed. About 60% of the oxygen seems to be removed between breaths at average intervals.

Suppose a lungfish takes in at each breath $10 \, cm^3$ air containing $8 \, cm^3$ nitrogen, $2 \, cm^3$ oxygen and a negligible amount of carbon dioxide. Suppose that $1 \, cm^3$ of oxygen is removed in the lungs. Metabolism which uses $1 \, cm^3$ oxygen generally produces about $1 \, cm^3$ carbon dioxide, so the fish must get rid of about $1 \, cm^3$ carbon dioxide at each breath. If all this leaves through the lungs the air breathed out must contain $8 \, cm^3$ nitrogen, $1 \, cm^3$ oxygen and $1 \, cm^3$ carbon dioxide: that means that the partial pressure of carbon dioxide in it is 0.1 atm. Carbon dioxide could only diffuse into the lungs to this partial pressure if its partial pressure in the blood was even higher. However, the experiments that showed that 90% of the oxygen is taken up in the lungs also showed that 70% of the carbon dioxide is eliminated through the gills. A lot of carbon dioxide can be dissolved in only a little water passing over the gills without giving it a high partial pressure, because carbon dioxide is highly soluble in water. The partial pressure of carbon dioxide in the blood of *Protopterus* in well-aerated water has been found to be about 0.03 atm.

The partial pressure of carbon dioxide may rise very high in the waters of tropical swamps, to values around 0.1 atm. In such conditions, irrigating the gills can do nothing to keep down the partial pressure of carbon dioxide in the blood: rather, it may tend to raise the partial pressure by allowing carbon dioxide from the water to diffuse into the blood. Air-breathing fish tend to stop irrigating their gills when the partial pressure of carbon dioxide in the water is raised. *Protopterus* does this when air containing 0.05 atm carbon dioxide is bubbled through the water. When this is done, or when the fish is removed from water, it takes more frequent breaths and so prevents the partial pressure of carbon dioxide in the lungs and blood from rising too high. The partial pressure in the blood of a *Protopterus* out of water has been found to be 0.06 atm.

Contrast *Protopterus* with an ordinary water-breathing fish in well-aerated water. This fish would probably remove 30–80% of the dissolved oxygen from the water passing over its gills. Suppose it removed 50%, reducing the partial pressure of oxygen in the water by 0.1 atm. Carbon dioxide is 25 times as soluble in water as oxygen, so if carbon dioxide was released into the water as fast as oxygen was removed it would raise the partial pressure of carbon dioxide in the water by only $0.1/25 = 0.004$ atm. The partial pressure in the blood can be correspondingly low, and is generally about 0.005 atm in the venous blood of water-breathing fishes. This is far lower than the values found in *Protopterus*, and a lungfish out of water or in a carbon dioxide-rich swamp would have to take exceedingly frequent breaths to get the partial pressure of carbon dioxide in its blood down to so low a value.

The high partial pressures of carbon dioxide mean that lungfishes and other swamp-living fishes are best served by haemoglobin rather different from that of most other fishes. The effect of carbon dioxide on the ability of haemoglobin to take up oxygen has already been described (section 14.3). In most teleosts the effect is very marked, and a partial pressure of carbon dioxide of 0.03 atm (such as is usual in the

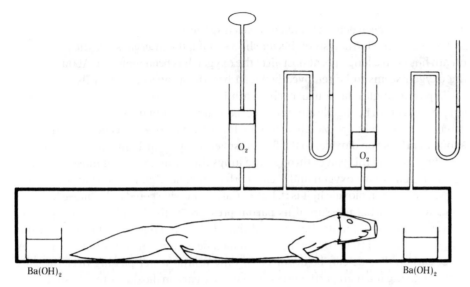

Fig. 15.7. **Apparatus for separate investigation of gas exchange in the lungs and through the skin of amphibians.**

blood of *Protopterus*) may prevent full oxygenation occurring even in the presence of 0.2 atm oxygen. The air in the lung of *Protopterus* will never contain more than 0.2 atm oxygen and will usually contain less. The haemoglobin of *Protopterus* and other air breathers is much less sensitive to carbon dioxide than that of most teleosts (Fig. 14.9).

Adult amphibians have no gills, so cannot get rid of carbon dioxide that way. They lose carbon dioxide largely through the skin, as the experiment shown in Fig. 15.7 demonstrated. A mask formed from a short piece of plastic tube was attached by a few stitches to the skin of the animal's face. The animal was put into a divided airtight container with the mask fitted into a hole in the dividing wall. An airtight seal was made round the mask. All air for the lungs had then to come from the front compartment while nearly all the skin was exposed to the air in the back one. Carbon dioxide given off in each compartment was absorbed by barium hydroxide which was titrated afterwards to find out how much carbon dioxide it had taken up. As oxygen was used up in each compartment the pressure fell, disturbing the manometer. It was replaced by a measured quantity of oxygen from the syringe, just sufficient to return the pressure to atmospheric pressure. Such experiments showed that, at 25°C, various frogs and salamanders got most of their oxygen via the lungs and got rid of most of their carbon dioxide through the skin. At lower temperatures, at which their metabolic rates were lower, most of the oxygen as well as the carbon dioxide travelled through the skin. Some terrestrial salamanders have no lungs and rely on the skin for all their gas exchange.

The skin can get rid of carbon dioxide faster than it can take up oxygen, because the diffusion constant for carbon dioxide in water is about thirty times the value for oxygen. The constants for diffusion through skin are presumably in the same ratio. Suppose that the difference in partial pressure of oxygen between air and blood were

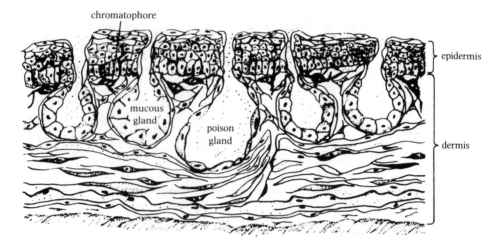

Fig. 15.8. A section through the skin of a frog (*Rana pipiens*). From C.J. & O.B. Goin (1971). *Introduction to herpetology*, 2nd edn. Freeman, San Francisco. Copyright © 1971.

0.2 atm (the maximum possible value). Carbon dioxide could be lost through the skin three times as fast as oxygen was taken up if the difference in partial pressure of carbon dioxide were only 0.02 atm. Carbon dioxide could be eliminated very rapidly by the lungs if the air in the lungs were changed often enough, but this would require much more frequent changes than are needed for oxygen uptake. However much oxygen is removed from the air taken into the lungs at each breath, no more carbon dioxide can be lost there than will make the partial pressure of carbon dioxide in the lungs equal its low value in the blood.

Fig. 15.8 shows the structure of amphibian skin, through which so much respiratory exchange occurs. It consists, as in fishes, of an inner dermis containing the collagen fibres that give the skin its strength, and an outer epidermis. As in fishes the epidermis is formed by cell division at its inner surface and the young cells there are much less flattened than the older cells near the outer surface. However, amphibians and other tetrapods differ from fishes in that keratin accumulates in the epidermal cells. The outer layers of the epidermis consist of keratin-filled cells, which form a thin horny protective covering for the animal. The outermost layer of dead horny cells is shed from time to time as a whole or in pieces. Dead cells are lost individually from the surface of fish skin.

Amphibian skin plays an important part in respiration, but gives little protection against drying up. It has been found that water evaporates from it, in still dry air, about as fast as it would evaporate from a free water surface of the same area. Reptile skin, described in section 16.3, is much more waterproof, but the changes that have made it waterproof have also made it too impermeable to gases to have a role in respiration. There are also a few exceptional frogs that have highly waterproof skin.

Fig. 15.9 shows how a lungfish gets air into its lungs. The pictures were obtained by X-ray cinematography: the air in the mouth and lungs showed up in the pictures because air is more transparent to X-rays than water or tissues are. The fish, still submerged, closes its mouth and drives water from its mouth and opercular cavities

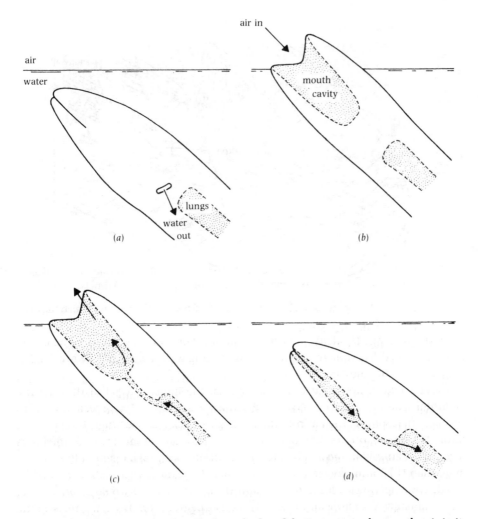

Fig. 15.9. Diagrams showing how the lungfish *Protopterus* changes the air in its lungs, based on X-ray cine films.

out through the opercular openings (Fig. 15.9a). The opercula are then closed by contraction of a muscle which runs transversely across the throat, and they are held closed from now until the air in the lungs has been changed. Next the mouth is thrust above the surface of the water and opened widely, and the mouth cavity is enlarged and filled with air (Fig. 15.9b). The lungs are then nearly emptied, becoming almost indistinguishable in the X-ray pictures (Fig. 15.9c). The mouth is closed and the floor of the mouth is raised. The air inside cannot be driven out through the opercula since these are being held closed, so it is forced into the lungs (Fig. 15.9d). This sometimes completes filling the lungs but a second mouthful of air is sometimes needed.

The pressure inside the lungs has been recorded through a cannula leading to a pressure transducer. It was found that between breaths this pressure was always about 1000–1500 N m^{-2} greater than the pressure in the water around the fish. The lung walls must be taut and the glottis (the opening of the windpipe into the mouth)

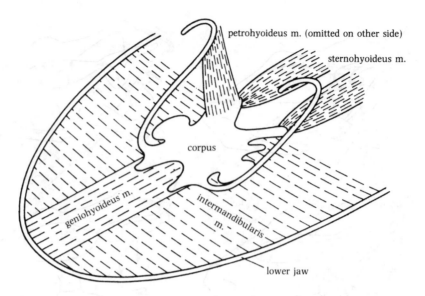

Fig. 15.10. **The hyobranchial apparatus of a frog, and the muscles which move it.**

must be closed. The elasticity of the lung walls must help to empty the lung when the glottis is opened. There is smooth muscle in the lung walls which may be involved as well, as may the muscles of the body wall.

It seems odd at first sight that the air from the lungs is expelled through the mouthful of air that is to be forced into the lungs (Fig. 15.9c). Mixing seems bound to occur, so that some of the air from the lungs goes back into the lungs. It seems that rather less mixing occurs than one might expect, for air withdrawn from the lungs through a cannula immediately after a breath has been taken generally contains more than 0.15 atm oxygen (pure air contains 0.2 atm oxygen). Mixing could be avoided by emptying the lungs before filling the mouth, but this would mean that the fish would have very little air in its body, and be denser than usual, at precisely the moment when it needed to get its mouth to the surface. It might make filling the lungs more difficult.

Frog breathing movements are remarkably like the air-breathing movements of lungfishes, though frogs breathe through their nostrils instead of through the mouth. They have muscles that close the nostrils when required.

Lungfishes pump air into their lungs by movements of their gill skeleton, contracting the mouth cavity to fill the lungs. Tadpoles have a cartilage gill skeleton, but it becomes greatly changed when the tadpole metamorphoses. In the adult frog it consists of a mainly cartilaginous plate (known as the corpus of the hyobranchial apparatus) under the tongue, with a pair of horns curving dorsally to join the cranium (Fig. 15.10). Although these horns reach all the way to the cranium they seem to be homologous with the hyoid bars of fishes: the hyomandibular which should form the upper part of the hyoid arch has become the stapes (section 15.7). Other projections from the corpus seem to be rudiments of other gill arches. Various muscles attach to the corpus. Sternohyoideus muscles run from it to the pectoral girdle. When they contract they pull the corpus posteriorly and ventrally, enlarging

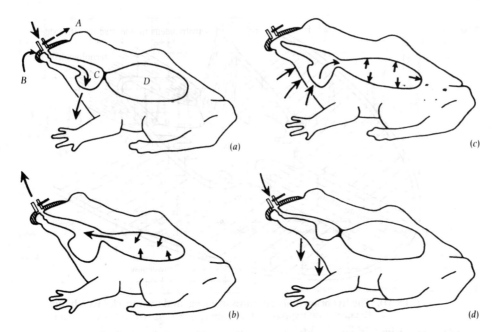

Fig. 15.11. Diagrams, which are explained in the text, showing how frogs change the air in their lungs. *C*, mouth cavity; *D*, lungs. From C. Gans, H.J. de Jongh & J. Farber (1969). *Science* 163, 1223–1225.

the mouth cavity much as the sternohyoideus muscles of fish do. The other muscles shown in Fig. 15.10 co-operate to reduce the mouth cavity by raising the plate and pulling it forward.

Professor Carl Gans and his colleagues have carried out a series of experiments to find out how frogs change the air in their lungs. They used pressure transducers to record the pressures in the lungs and in the mouth cavity, they identified the active muscles by electromyography and they took cinematograph films. Their conclusions are summarized in Fig. 15.11. The air is kept in the lungs under slight pressure, as in lungfish; it is prevented from escaping so long as the glottis is kept closed. To change the air, the mouth cavity is first enlarged by contraction of the sternohyoideus muscles, drawing air in through the open nostrils (Fig. 15.11a). Next the glottis is opened and the air from the lungs is driven out through the full mouth cavity, probably mainly by elastic recoil of the lungs and by contraction of their smooth muscles (Fig. 15.11b). Then the nostrils are closed and the mouth cavity is contracted by the other muscles indicated in Fig. 15.10, driving air from the mouth cavity into the lungs (Fig. 15.11c). Finally the glottis is closed and the nostrils opened again (Fig. 15.11d). This sequence of movements is very like the movements of a lungfish taking a breath of air, but it does not by any means change all the air in the lungs. The capacity of the expanded mouth cavity seems to be only about a quarter of the capacity of the lungs.

This sequence of movements is known as a ventilatory cycle. Between these cycles the frog keeps its glottis closed and its nostrils open and oscillates the floor of the mouth up and down. These oscillatory cycles move air in and out of the mouth cavity

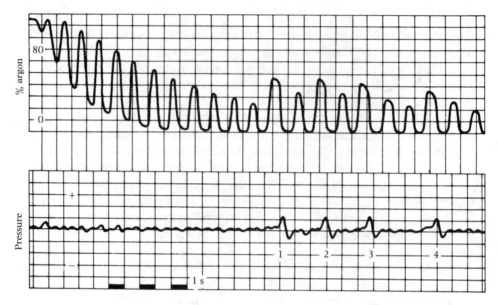

Fig. 15.12. Record of an experiment on frog respiration which is explained in the text. Source as Fig. 15.11.

but do not affect the lungs. The constant rapid movement of the throat is easily seen on a living frog.

Two questions suggest themselves. What function have the oscillatory cycles? In the ventilatory cycles, does not the used air from the lungs get mixed with the fresh air waiting to enter the lungs (Fig. 15.11*b*)? Gans and his colleagues tried to answer both questions by experiments in which they used a mass spectrometer to record the composition of the gas passing in and out of the nostrils. The gas was sampled by means of the probe A (Fig. 15.11*a*) attached to a mask B made of a quick-setting rubber compound.

The most revealing experiments were ones in which the frog was kept for a while in a mixture of 80% argon and 20% oxygen, which was then changed rapidly to air (80% nitrogen and 20% oxygen). Fig. 15.12 shows a record obtained in one of these experiments. The lower line shows the pressure in the mouth cavity, recorded by a pressure transducer. The very small pressure fluctuations mark oscillatory cycles and the large ones (1,2,3,4) ventilatory cycles. The upper line shows the argon content of air passing through the nostrils. The frog made no ventilatory cycles during the period when the gas around the frog was being changed, so at the end of this period the frog was in air containing virtually no argon while its lungs still contained about 80% argon. After the change was completed the record shows alternation between virtually argon-free gas entering through the nostrils and gas with an appreciable argon content passing out through them. The argon content of the gas driven out during the first ventilatory cycle is about twice as high as in the preceding oscillatory cycle, so this gas must have contained a substantial proportion of gas from the lungs. The gas expired in the next oscillatory cycle contains considerably less argon, but nevertheless more than in the oscillatory cycle preceding the ventilatory one. This means that the gas from the lung was not all passed out through the nostrils in the ventilatory cycle, but some of it got mixed with the mouth contents and was only flushed out by subsequent oscillatory cycles. The record after ventilatory cycles 3 and 4 shows gas from the lungs being flushed out of the mouth by a series of oscillatory cycles: note how the percentage of argon in successive expirations diminishes.

It is suspected that most of the early amphibians may have resembled lizards more than frogs in their respiration. Many had scales which would have hindered diffusion of gases through the skin, suggesting that the skin may not have been important in respiration. Most, including *Ichthyostega*, had substantial ribs largely encircling the chest, suggesting that they may have filled their lungs by rib movements (like lizards) rather than by throat movements. Rhipidistians presumably filled their lungs in much the same way as lungfish. They had short ribs or no ribs at all, like frogs and newts. Long ribs, however, are not necessarily used for breathing: many teleosts have them.

It has been shown by electromyography that muscles in the wall of the trunk play a part in the breathing movements of lizards, but the mechanism has not been investigated in detail. The maximum volume of air which a lizard can expel from its lungs and replace in a single cycle of rib movements seems to be about 60 cm^3 (kg body mass)$^{-1}$. The corresponding figures for the throat movements of frogs and *Protopterus* seem to be about 50 cm^3 kg^{-1} and 30 cm^3 kg^{-1}, respectively. (Frogs have relatively broader heads than lungfish and so can take larger mouthfuls of air.)

Reptiles that cannot get rid of carbon dioxide through their skin increase the partial pressure of carbon dioxide in their lungs about as much as they decrease the partial pressure of oxygen. To avoid getting high partial pressures of carbon dioxide in their lungs (and so in the blood) they must change the air in their lungs frequently and remove only a little of the oxygen from it. Lizards (*Lacerta* spp.) have been found to extract only 8% of the oxygen from the air that they breathe.

15.4.   Blood circulation

Lungfishes have lungs as well as gills, and adult amphibians have no gills. Their blood circulations are appropriately different from those of fishes that depend entirely on gills for gas exchange.

Fig. 15.13(*a*) shows the principal arteries of a teleost fish. Very similar arrangements are found in selachians, lampreys and even amphioxus (Fig. 12.4). Blood leaves the heart in the ventral aorta, which branches to form the afferent branchial arteries to the gills. Blood from the gills is collected in efferent branchial arches which join to form the dorsal aorta, which in turn sends branches to all parts of the body. On each circuit round the body, blood goes first to the gills (where it takes up oxygen) and then to the other tissues (where it gives up oxygen).

The diagram shows pressures in some of the blood vessels, and partial pressures of oxygen. To get this information, trout were anaesthetized and opened, cannulae were fixed in the chosen blood vessels and the surgical wounds were sewn up. When the fish had recovered from the anaesthetic they swam in a water tunnel with long

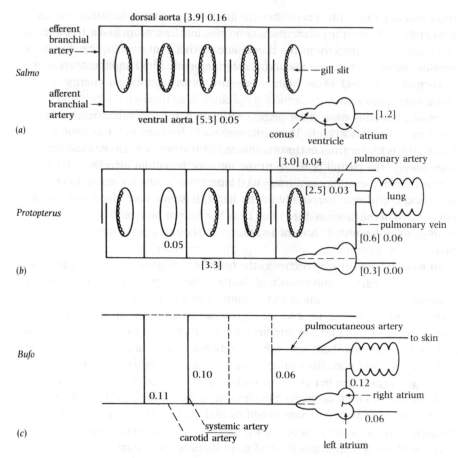

Fig. 15.13. Diagrams of the blood circulation of trout (*Salmo*), the lungfish *Protopterus* and the toad *Bufo*. Peak (systolic) blood pressures in kN m⁻² are shown in square brackets [] and partial pressures of oxygen in atmospheres are shown without brackets, for particular blood vessels. Data from E.D. Stevens & D.J. Randall (1967). *J. exp. Biol.* 46, 307–315; G.F. Holeton & D.J. Randall (1967). *J. exp. Biol.* 46, 317–329; K. Johansen, C. Lenfant & D. Hanson (1968). *Z. vergl. Physiol.* 59, 157–186; and H. Tazawa, M. Mochizuki & J. Piiper (1979). *Resp. Physiol.* 36, 77–95.

tubes connecting the cannulae to pressure transducers. There were also arrangements for withdrawing blood samples through the same tubes. Fig. 15.13(*a*) shows that peak pressure in the ventral aorta (5.3 kN m⁻²) was considerably higher than in the dorsal aorta (3.9 kN m⁻²), which in turn was much higher than in the veins (1.2 kN m⁻²). The first pressure drop drove the blood through the gill lamellae and the second drove it through the capillaries of the body. The partial pressure of oxygen in the water leaving the gills (0.16 atm) was much higher than in the blood arriving at them (0.05 atm). The fish was in well-aerated water, in which the partial pressure of dissolved oxygen must have been about 0.21 atm.

Fig. 15.13(*b*) shows that *Protopterus* has a similar arrangement of blood vessels, with pulmonary arteries (to the lungs) branching off the last efferent branchial arteries. The blood pressures and partial pressures of oxygen were measured in the same way as for the trout, except that the lungfish swam in an aquarium instead of a water tunnel. Two of the gill arches have no gills, and their branchial arteries connect the ventral aorta directly to the dorsal one, without intervening capillaries. The pressure difference between the aortas is consequently small and the heart does not have to pump blood at as high a pressure as in teleosts, so energy is saved. Presumably only a small proportion of the blood flows through the gills that remain. A small gill area is enough for getting rid of carbon dioxide, because its diffusion constant in water is so high. The partial pressure of oxygen in the pulmonary veins (0.06 atm) is higher than in the other veins (0.00 atm), as might be expected. More surprisingly, the partial pressure in the anterior branchial arteries (0.05 atm) is higher than in the pulmonary artery (0.03 atm): you would not expect to find this, if the blood got mixed thoroughly in the heart. It seems that blood that has been oxygenated in the lungs in one circuit from the heart is preferentially directed away from them in the next. This is the beginning of a double circulation like the familiar arrangement in mammals, which have the heart completely divided into left and right halves. Some mixing occurs in the heart of *Protopterus* (otherwise the partial pressure of oxygen would be as high in the anterior branchial arteries as in the pulmonary vein, and as low in the pulmonary artery as in the sinus venosus), but complete mixing is prevented by incomplete partitions in the atrium, ventricle and conus arteriosus. The atrium and ventricle are partially divided into left and right halves. The pulmonary veins bring blood from the lungs directly to the atrium on the left side of the partition. The rest of the blood enters the right side through the sinus venosus. The partition in the conus twists so as to become horizontal at its anterior end, with oxygen-rich blood from the left side of the heart ventral to it and oxygen-poor blood from the right side dorsal to it (Fig. 15.14). The first three pairs of branchial arteries start from below the partition and the last two pairs start from above it. Since the pulmonary arteries are branches of the last efferent branchial arteries they receive mainly oxygen-poor blood.

The partition in the conus is known as the spiral valve (it must not, of course, be confused with the spiral valve of the intestine, which lungfishes also possess). It consists in *Protopterus* and *Lepidosiren* of two ridges rising from opposite walls of the conus and almost meeting in the middle. In *Neoceratodus* only one of the ridges runs the length of the conus and it is obviously formed from a series of valves just like the valves in the conus of selachians.

The advantage of separating the two streams of blood seems plain. The lower the partial pressure of oxygen in the blood entering the lungs, the faster will oxygen diffuse into it. The higher the partial pressure in the blood arriving at other tissues, the faster can oxygen diffuse into the tissues. If mixing is prevented the lungs will receive blood with a low partial pressure of oxygen and the rest of the body will receive blood with a high partial pressure. If complete mixing occurs they will receive blood with the same, intermediate partial pressure and diffusion of oxygen in both will be slower. If complete mixing occurred lungfish would need a bigger lung area, and more capillaries in their tissues, to maintain the same rate of consumption of oxygen. Also, a given volume of blood arriving at the tissues would have less oxygen to give up and a given volume arriving at the lungs would have less capacity for taking up oxygen, so the heart would have to pump blood faster.

In normal water-breathing fishes blood goes first to the gills and then to the rest of

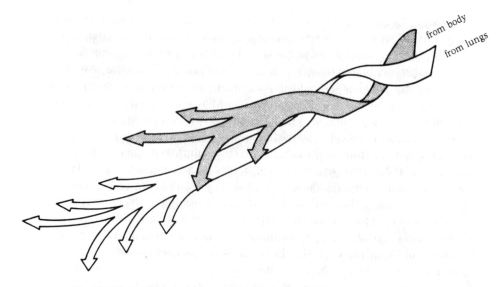

Fig. 15.14. A diagram representing the paths of the two streams of blood through the conus arteriosus and ventral aorta of the lungfish *Protopterus*.

the body before returning to the heart, so there is no mixing of oxygen-rich and oxygen-poor blood. It would presumably have been possible for the lungs of lungfish to have been fitted into the circulation in the same way as gills, so that the blood went from them to the rest of the body before returning to the heart. The disadvantage of such an arrangement would have been that the blood would have to be pumped out of the heart at sufficient pressure to drive it successively through two sets of capillaries, in the lungs and in the rest of the body. There would be a bigger pressure difference between the blood in the lung capillaries and the air in the lungs, so the lung capillaries would have to be stronger to prevent bursting. They would need thicker walls so oxygen would diffuse more slowly into them. (It was explained in section 15.3 why the length of the diffusion path through tissue has much more effect on the rate of uptake of oxygen in lungs than in gills.) A bigger lung area would be needed and more blood would have to be pumped through the lungs.

Tadpoles have arteries arranged more or less as in fish, but adult frogs and toads have a modified arrangement (Fig. 15.13c). As in lungfish the lungs are supplied by the arteries of the last gill arch on each side, but the skin is also supplied from these (pulmocutaneous) arteries. The artery of the next gill arch disappears but those of the two which lack gill lamellae in *Protopterus* remain as a branch of the carotid arteries to the head and as the systemic artery to the trunk and limbs. Some of the arteries that are missing in adult frogs are retained in adult newts and salamanders.

The partial pressures of oxygen shown in Fig. 15.13(c) were measured in samples of blood taken very rapidly (within one to two minutes) after the toads had been killed by destroying their brains. They show almost equal high partial pressures in the pulmonary vein and in the carotid and systemic arteries, and equal low ones in the sinus venosus and in the pulmocutaneous artery. It seems that there are two functionally separate pathways through the heart, but the heart is not completely divided. There are separate left and right atria. The ventricle has no partition but has spongy walls which must tend to prevent blood from swirling around in it and so restrict mixing of the blood from the two atria. The conus is incompletely divided by a spiral valve, and the arteries to the lungs and skin open from one side of the partition while the systemic and carotid arteries open from the other (cf. Fig. 15.14). This arrangement is extremely similar to the arrangement in lungfish and can be expected to make well-oxygenated blood from the lungs tend to go through the carotid and systemic arteries to the general body circulation, while the rest of the blood, which is generally less well oxygenated, tends to be directed to the lungs and skin. If this separation occurs, it has the seeming disadvantage that blood from the skin where it has lost carbon dioxide and taken up oxygen tends to be directed to the skin again, or to the lungs. Some other experiments on toads have indicated more mixing of blood in the heart than did the ones that gave the partial pressures shown in Fig. 15.13(c), but it seems clear that the two streams of blood through the heart are kept largely separate.

15.5.   Aestivation

Many *Protopterus* live in swamps which dry out in the annual dry season. As the water level falls they burrow into the still-soft mud, taking mouthfuls of the mud and spitting it out through the gills. In this way the fish forms a bottle-shaped burrow in which it curls up. It stays in the burrow throughout the dry season, normally 4–6 months, and is said to be aestivating. If it is dug up, it is found to be torpid. Some *Lepidosiren* also aestivate.

   Protopterus have been persuaded to burrow in blocks of mud in laboratories, and to aestivate in them when the mud was allowed to dry out. In these experiments they lost weight much more slowly than lungfish kept starving in an aquarium for the same period. Starving lungfish use about 20 cm^3 oxygen (kg body mass)$^{-1}$ h^{-1}, and the proportion of glycogen in the body falls. Aestivating ones removed from the mud with as little disturbance as possible have been found to use on average only 8 cm^3 oxygen kg^{-1} h^{-1}, and neither the glycogen nor the fat in the body seems to diminish. Not only is the metabolic rate much reduced but metabolism is apparently changed so that virtually nothing but protein is consumed.

   Protein metabolism seems strangely unsuitable in aestivation, since it produces nitrogenous waste which cannot be eliminated from the body. *Protopterus* in water excrete most of their waste nitrogen as ammonia, which would be toxic if it were allowed to accumulate during aestivation. It does not accumulate but urea does, reaching about 2% of the mass of the body in twelve months' aestivation in experiments by H.W. Smith. This is similar to the concentrations found in marine selachians (section 13.4). Would protein metabolism producing this amount of waste be enough to supply the whole of the fish's metabolic needs for a year? We shall work this out very roughly, assuming for simplicity that the protein was synthesized entirely from alanine ($CH_3.CHNH_2.COOH$), an amino acid of moderate molecular mass. The equation for metabolism of this hypothetical protein would be

$$2(-NH.CH(CH_3)CO-)+6\ O_2 = CO(NH_2)_2 + 5\ CO_2 + 3\ H_2O.$$

Metabolism using 6 mol (6×22.4 l) of oxygen would produce 1 mol (60 g) of urea. The urea actually produced was about 2% of the final mass of the body, 20 g (kg body mass)$^{-1}$ or 1/3 mol (kg body mass)$^{-1}$, which would correspond to the consumption in the year of $6 \times 22.4/3 = 45 l O_2$ (kg body mass)$^{-1}$. This is 5 cm^3 O_2 kg^{-1}h^{-1}. This is rather less than the average rate of 8 cm^3 kg^{-1}h^{-1} measured by Smith, but he may have disturbed the fish a little in removing them from the mud and so got too high a value. In any case the correspondence between urea accumulation and oxygen consumption seems to confim that at least most of the metabolism was of protein. The equation indicates that the protein needed to produce 2% urea in the body would be about 5% of the mass of the body. The aestivating fish actually lost on average 27% of their mass: presumably most of this loss was water.

In more recent experiments than Smith's, aestivating lungfish lost weight more slowly and accumulated urea more slowly. This may be due to differences in the conditions in which the fish were kept.

Scaphiopus, the spadefoot toad, is an amphibian that escapes drought by burrowing. It lives in the Arizona Desert where summer thunderstorms leave temporary pools but where there is no substantial rain for the rest of the year. It spends the ten dry months in a burrow, 30–70 cm underground, and can survive loss of water up to almost 50% of body mass. It metabolizes largely fat but apparently also metabolizes protein: the urea concentration in its blood may rise as high as in aestivating *Protopterus*.

15.6.   Locomotion on land

Early amphibians such as *Ichthyostega* were essentially newt-like in shape and might reasonably be expected to have walked on land like newts (Fig. 15.15). Happily we do not have to rely on such comparisons, for there have been preserved in Palaeozoic rocks tracks left by amphibians or reptiles walking over mud. (It is not always possible to decide whether a particular track was made by a reptile or an amphibian.) These indicate newt-like walking with feet well out to the side and generally with the toes pointing forward, though one particularly early track shows the toes pointing laterally. There are seldom marks of the body being dragged along the ground. Apparently early tetrapods normally walked with the belly held clear of the ground like the newt shown in Fig. 15.15. Newts do, however, sometimes crawl with the belly sliding on the ground, for instance on a wet surface.

Walking with the belly off the ground poses a problem of balance. Ideally, the forces on the feet should be kept in equilibrium with the body weight at all stages of the stride. If they are not, the animal's trunk will rise and fall as it walks, or pitch or roll like a ship at sea. These perturbations of the animal's forward motion may not matter if they are reasonably small, but if they are too large the belly will hit the ground at some stage of the stride. They are most likely to be large in slow gaits, which allow

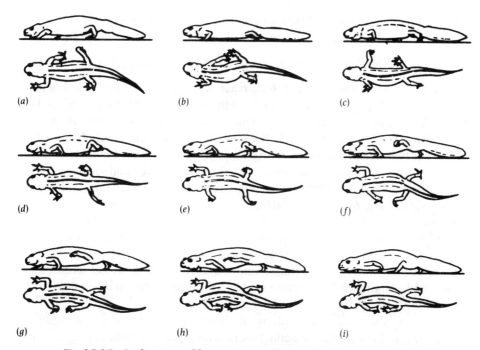

Fig. 15.15. Outlines traced from successive frames of a cinematograph film of a newt (*Triturus pyrogaster*) walking. A mirror was used to obtain dorsal and lateral views simultaneously. From B. Schaeffer (1941). *Bull. Am. Mus. nat. Hist.* 78, 395–472.

ample time for unwanted displacements to occur between one footfall and the next. This problem is particularly serious for very slow walkers such as tortoises.

A person may be in stable equilibrium when standing on two feet or even on only one, but this is because human feet are large and can be placed vertically under the centre of mass. More typical tetrapods with small feet require at least three feet on the ground for stable equilibrium (a three-legged stool is stable but a two-legged one is not). If a four-legged animal is to maintain stable equilibrium as it walks it must move its feet one at a time, so as always to have three on the ground. This implies that each foot must have a duty factor of at least 0.75; that is, it must be on the ground for at least 0.75 of the duration of the stride. Further, a vertical line through the centre of mass must always pass through the triangle of which the corners are the points of contact of the feet with the ground. If the vertical passes outside the triangle the animal will tend to topple over. Fig. 15.16 shows how the vertical can be kept inside the triangle throughout the stride by moving the feet in appropriate sequence. The sequence shown is the only one which can accomplish this for duty factors between 0.75 and 0.83. (Two additional sequences become possible at duty factors above 0.83, but no animal is known to use them in walking.)

Notice that the newt in Fig. 15.15 is moving its feet in the same sequence as the schematic animal of Fig. 15.16. This sequence is also used for walking by lizards and tortoises. However, the schematic animal moves its feet at equal intervals so as always to have three on the ground. The newts and reptiles move diagonally opposite feet almost simultaneously (right hind just after left fore; left hind just after right fore)

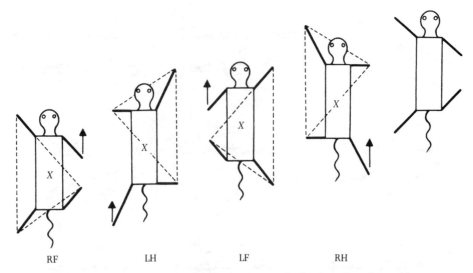

RF   LH   LF   RH

Fig. 15.16. Diagrams showing the order in which a four-footed animal must move its feet if it is to remain stable at all times. Each diagram represents the moment at which the foot marked with the arrow is lifted off the ground. The centre of gravity of the body is marked X. In cases where the centre of gravity lies above one edge of the triangle drawn between the three stationary feet it will move forward over the centre of the triangle before the foot is set down again.

so that though the sequence is as in Fig. 15.16 and the duty factor may be greater than 0.75, there may be times when only two feet are on the ground.

It has been argued that this may actually give a steadier ride, for animals whose muscles cannot develop tension very rapidly. The animal shown in Fig. 15.16 could maintain perfect equilibrium and move absolutely steadily, if its muscles were capable of exerting forces that changed instantaneously, when feet were raised or set down. A mathematical model shows that if they can change forces only rather slowly, the animal may wobble less if it moves diagonally opposite feet almost simultaneously. The diagonal pattern also enables the animal to lengthen its stride by bending from side to side as it walks: notice that in Fig. 15.15 each shoulder or hip points backwards when its foot is raised but swings forward before the foot is set down again. The newt's bending from side to side may suggest the swimming action of a fish, but fish make travelling waves: each bend travels posteriorly along the body. The bends of the newt are standing waves, like the vibrations of a stretched string.

The pectoral and pelvic fins of *Neoceratodus* and *Latimeria* look rather like short legs, but they are not used as legs. Neither fish goes onto dry land, and the fins would not be strong enough to support its weight there. Some of the films of *Latimeria* in its natural environment show it resting on the bottom but none show it using its fins like legs.

No intermediate stage is known between crossopterygian fins (Fig. 15.2a) and the fully developed limbs of early amphibians (Fig. 15.4). It seems fairly obvious that the humerus, radius and ulna are homologous with the pterygiophores indicated in Fig. 15.2, and similar homologies are apparent between hind leg and pelvic fin. It is not at all clear how the remaining bones at the distal ends of the limbs evolved.

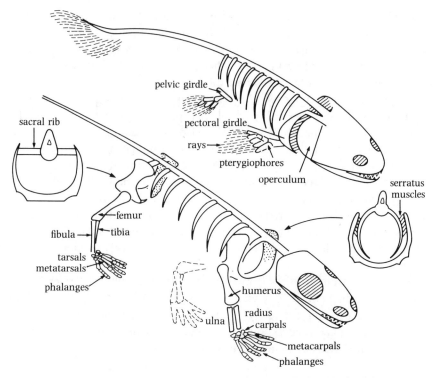

Fig. 15.17. Diagrams of the skeletons of a rhipidistian and an amphibian.

The form of the primitive amphibian limb is obviously suited to the type of locomotion practised by newts. The large number of bones in the foot enables it to accommodate itself to the shape and angle of the surface the animal is walking on and to take advantage of irregularities to obtain a grip. Fig. 15.17 shows how the radius and ulna can cross over each other and so keep the toes pointing forward all the time the foot is on the ground, instead of letting the foot swivel round with the humerus (if the foot did swivel round it would be abraded, and work would have to be done against friction to turn it). X-ray cine films of lizards walking confirm that this twisting happens. The tibia and fibula cross similarly in the hindlimb. It is of course possible for an animal without a separate radius and ulna to walk in amphibian fashion and yet not swivel the foot, provided the elbow is not a simple hinge joint but allows sufficient variety of movement: this is the situation in frogs and toads.

The muscles of the trunk and tail do most of the work when a typical fish swims fast, but the muscles of the limbs do most of the work when a tetrapod runs. There is an appropriate difference in the relative sizes of the muscles. The limb muscles of tetrapods generally make up a very much larger proportion of the mass of the body than do the fin muscles of fishes, and the trunk and tail muscles are correspondingly smaller.

Not only did the origin of the tetrapods involve evolution of limbs from fins and enlargement of their muscles, but it involved modification of the limb girdles. The pectoral girdle of bony fishes consists mainly of a series of dermal bones stiffening the posterior wall of the opercular cavity (Fig. 15.17*a*). The two halves of the girdle meet ventrally and are attached to the cranium dorsally. The girdle provides an anterior attachment for the trunk muscles and also an origin for the sternohyoideus muscle, which is used in feeding and respiration (Fig. 14.4c). Most of the girdle is dermal bone and only a relatively small part attached to its inner side, the scapulocoracoid, is sclerotome bone. It is with this that the fin articulates.

Amphibians have no opercular cavity, so no reinforcement is needed for its posterior wall. The dermal part of the pectoral girdle is reduced but the scapulocoracoid is enlarged (Fig. 15.17b). Many of the limb muscles originate from it. The weight of the animal can best be supported without compressing the viscera and lungs if the pectoral girdle is not simply embedded in the body wall but attached more or less firmly to the axial skeleton. Its attachment to the skull in fishes is not well arranged for this because it is anterior to the fin and linked to it by a series of flexibly connected bones. It has been replaced in tetrapods by an entirely new connection: the rib cage is suspended from the large scapula by short muscles (Fig. 15.17b). Detachment of the girdle from the skull makes the head more freely movable relative to the trunk.

The pelvic girdle in fishes is merely a pair of small plates of sclerotome origin embedded in the ventral body wall. In tetrapods it is larger, giving origin to large muscles and extending dorsally to make a firm attachment with the vertebral column. Fig. 15.17(b) shows how it is sutured to the ends of short stout ribs, which in turn are sutured to one or more vertebrae (the sacral vertebrae).

15.7.   Hearing on land

The hyomandibular bone had two functions in rhipidistians. It played a part in the mechanism of the intracranial joint and it supported the operculum. It is not required for either of these functions in amphibians, which have no intracranial joint and no operculum (apart from the rudiment in *Ichthyostega*). It has acquired a new function in connection with the ear, and is known as the stapes.

In modern amphibians the stapes is a rod of bone connecting the eardrum to the part of the cranium that contains the inner ear (Fig. 15.18). It lies in an air-filled cavity, the middle ear, which is connected to the mouth by the Eustachian tube. The early embryo has a fish-like spiracle and hyomandibular, but the hyomandibular becomes the stapes and the spiracle enlarges and surrounds it instead of remaining wholly anterior to the hyomandibular as in fishes. This arrangement may not have evolved until long after the amphibians first appeared. It is doubtful whether the earliest amphibians had eardrums, and the earliest known stapes are short thick rods that seem unlikely to have worked well as ear ossicles.

The semicircular canals and otolith cavities of amphibians, like those of fishes, are filled with the potassium-rich fluid called endolymph. They are surrounded by perilymph, a fluid that is more like blood plasma. The perilymph cavity reaches the surface of the cranium at two 'windows' where it is bounded only by flexible membranes (Fig. 15.18a). These are the oval window, where the stapes ends on the lateral face of the cranium, and the round window on its posterior face. Most

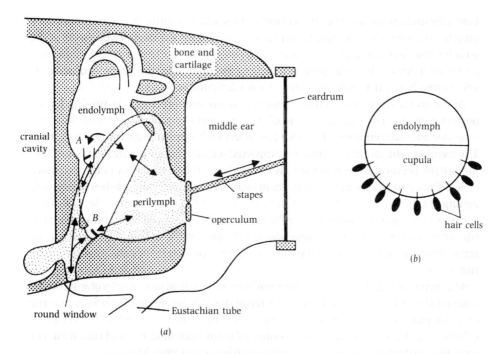

Fig. 15.18.(*a*) A diagram showing the structure of a frog ear. *A*, amphibian papilla;
B, basilar papilla. (*b*) A diagrammatic section through the basilar papilla.

tratrapods have the stapes and nothing else fixed in the oval window. Modern
amphibians have a second structure there, a small bone or cartilage called the
operculum (Fig. 15.18) which is connected to the pectoral girdle by a muscle. Cutting
the muscle (or paralysing it by cutting its nerve) makes the animal less sensitive to
low-frequency sounds (below about 1000 Hz in the tree frog *Hyla*) without affecting
its sensitivity to higher frequencies.

The endolymph cavity contains two neuromast organs not found in fishes. One of
them, the basilar papilla, is present in nearly all tetrapods. The other, the amphibian
papilla, is peculiar to amphibians. Both are papillae which project from the
endolymph cavity and touch the perilymph cavity (Fig. 15.18*a*). The cupula of the
basilar papilla is gelatinous, like the cupulae of semicircular canals and the lateral
line system, but it is not conical. It has the form of a semicircular membrane reaching
half way across the papilla (Fig. 15.18*b*). That of the amphibian papilla (which is
present only in amphibians) has a more complicated shape.

The stapes transmits vibrations from the eardrum to the oval window. The
operculum is also believed to be concerned in the transmission of vibrations from the
pectoral girdle. If the stapes and operculum move inward they push the oval window
in and make the round window bulge out. There is a path entirely in the perilymph
cavity from oval window to round window, but it is slender (Fig. 15.18*a*) and the
membranes separating the perilymph from the endolymph are flexible, so the
endolymph must vibrate as indicated in the figure, through the amphibian and

basilar papillae. Since the papillae are slender the amplitude of vibration will be higher in them than elsewhere. The cupulae will be made to vibrate by the endolymph, and stimulate the hair cells.

Action potentials have been recorded from the parts of the auditory nerve that are believed to serve the two papillae. It appears that low-frequency sounds (below about 1000 Hz in the tree frog *Hyla*) stimulate mainly the amphibian papillae and higher frequencies stimulate mainly the basilar papilla.

The most important high-frequency sounds in the lives of frogs are probably their calls, which are similar in function to the more familiar calls of birds (section 17.5). Some species have distinct mating, territorial and alarm calls. Newts and salamanders do not sing. They have no eardrum, often no stapes and a very small basilar papilla.

The hearing mechanisms of amphibians presumably evolved because the mechanisms of fishes, which work well under water, do not work well in air. An animal in air reflects nearly all the sound energy that strikes it. A satisfactory hearing organ must not reflect all the sound energy that strikes it, but must absorb a reasonable proportion. The same requirement applies to microphones, and we need some understanding of the physics of microphones if we are to understand the role of the eardrum and stapes.

When sound is passing through a material there are fluctuations of velocity in the vibrating material, and fluctuations of pressure. The ratio of the amplitude of the pressure changes to the amplitude of the velocity changes is known as the characteristic acoustic impedance of the material. It is very much higher in dense incompressible materials like water than it is in air. Sound is transmitted well from one material to another if they have similar characteristic acoustic impedances, but hardly at all if they have very different ones. When sound travelling in air strikes water, or vice versa, at least 99.9% of its energy is reflected. When sound travelling in air strikes a microphone some of its energy is used in driving the microphone and some is reflected. The proportions depend on the impedance of the microphone, which is the ratio of the amplitude of the force driving it to the amplitude of the velocity changes of the diaphragm. A substantial proportion of the sound energy is used in driving the microphone only if the impedance per unit area of the microphone diaphragm is close to the characteristic acoustic impedance of air.

The oval window would be a very poor microphone because its impedance per unit area is too high. Adding the stapes and the eardrum must increase the impedance (because they have to be made to vibrate, as well as the fluids of the inner ear), but the area of the eardrum is enormously larger than that of the oval window, so the slightly increased impedance is spread out over a much larger area. The impedance per unit area, measured at the eardrum, is close to the characteristic acoustic impedance of air, as required for the eardrum to work well as a microphone. This was demonstrated by measuring the vibrations of the eardrums of frogs, in response to sounds of different pressure amplitudes. They were measured by an optical interference technique, using laser light reflected from the eardrum.

The frog eardrum is not sensitive to low-frequency sounds because the sound that hits the outer surface of the eardrum also reaches the inner surface via the mouth and

Eustachian tube, cancelling out the effect. Resonance of the mouth cavity makes it very sensitive to a limited range of higher frequencies.

15.8. Reproduction

Most modern amphibians lay numerous, fairly small eggs enclosed in a layer of jelly. These are normally laid in water and develop into aquatic larvae (tadpoles) with external gills. These are tufts projecting from the sides of the head that remain visible in newt tadpoles but become covered, in frogs, by a fold of skin. Traces of similar gills are visible on some early amphibian fossils showing that they too were aquatic larvae. The larvae of the lungfishes *Protopterus* and *Lepidosiren* also have external gills.

The jelly around the eggs is a protein gel. It is secreted by the oviducts as a thin, relatively concentrated layer which swells enormously when it has an opportunity to take up water. It presumably quickly absorbs any water that is present in the oviduct but appreciable swelling does not normally occur until the eggs are laid. The swollen jelly of the common frog (*Rana temporaria*) contains only 0.3% organic matter. Contact with water also makes the surface of the jelly sticky, and the eggs of *Rana* adhere to each other and sometimes to weed or hard surfaces when they are laid. They are usually laid in ponds or ditches, but some salamanders lay their eggs in streams and stick them individually to the undersides of stones so that they are not washed away.

Though the eggs of *Rana* stick together as a bunch of frogspawn they do not form a continuous mass of jelly. Each jelly capsule remains more or less spherical so that though they adhere at their points of contact, channels remain between them. (This can be demonstrated by dropping Indian ink onto submerged frogspawn.) Water can circulate between the eggs, and eggs at the centre of a bunch do not depend for their oxygen on diffusion through the outer parts of the bunch.

The hatched tadpoles do not eat the jelly, so what is its function? It presumably gives some protection from predators and its organic content is apparently too low for it to be itself attractive to predators. It has been suggested that its main function may be to trap heat. The black eggs must absorb radiation from the sun and sky well. Heat gained in this way cannot be lost by direct convection, since convection currents cannot occur in the jelly: the heat must be conducted through the jelly and only removed from its outer surface by convection. Frogspawn in its natural environment has been found to be on average 0.6 K warmer than the surrounding water. Even a slightly raised temperature must lead to faster development which would presumably give a selective advantage.

Not all amphibians lay eggs in water. For example, salamanders of the genus *Plethodon* lays eggs on land, sticking them to the walls of crannies in logs, or of caves. They hatch as miniature salamanders, shaped like adults. *Salamandra atra* does not lay eggs at all but retains the young in its oviduct until after metamorphosis. The walls of the oviduct have a rich blood supply and the larvae have large external gills, so oxygen presumably diffuses from the mother's blood to that of the young.

Further reading

Coelacanths and lungfishes

Bemis, W.E. & Lauder, G.V. (1986). Morphology and function of the feeding apparatus of the lungfish, *Lepidorsiren paradoxa* (Dipnoi). *J. Morph.* **187**, 81–108.

Fricke, H., Reinicke, O., Hofer, H. & Nachtigall, W. (1987). Locomotion of the coelacanth *Latimeria chalumnae* in its natural environment. *Nature* **329**, 331–333.

Panchen, A.L. & Smithson, T.R. (1987). Character diagnosis, fossils and the origin of the tetrapods. *Biol. Rev.* **62**, 341–438.

Thompson, K.S. (1969). The biology of the lobe-finned fishes. *Biol. Rev.* **44**, 91–158.

Amphibians

Bray, A.A. (1985). The evolution of the terrestrial vertebrates: environmental and physiological considerations. *Phil. Trans. R. Soc. Lond.* B**309**, 289–322.

Duellman, W.E. & Trueb, L. (1986). *Biology of amphibians.* McGraw Hill, New York.

Gardiner, B.G. (1983). Gnathostome vertebrae and the classification of the Amphibia. *Zool. J. Linn. Soc.* **79**, 1–59.

Goin, C.J. & Goin, O.B. (1971). *Introduction to herpetology*, 2nd edn. Freeman, San Francisco.

Porter, K.S. (1972). *Herpetology.* Saunders, Philadelphia.

Breathing air

Bishop, I.R. & Foxon, G.E.H. (1968). The mechanism of breathing in the South American lungfish, *Lepidosiren paradoxa*;, a radiological study. *J. Zool., Lond.* **154**, 263–271.

Cox, C.B. (1967). Cutaneous respiration and the origin of the modern Amphibia. *Proc. Linn. Soc. Lond.* **178**, 37–47.

Feder, M.E. & Burggren, W.M. (1985). Cutaneous respiration in vertebrates: design, patterns, control and implications. *Biol. Rev.* **60**, 1–45.

Gans, C., de Jongh, H.J. & Farber, J. (1969). Bullfrog (*Rana catesbiana*) ventilation: how does the frog breathe? *Science* **163**, 1223–1225.

McMahon, B.R. (1969). A functional analysis of the aquatic and aerial respiratory movements of an African lungfish, *Protopterus aethiopicus. J. exp. Biol.* **51**, 407–430.

McMahon, B.R. (1970). Relative efficiency of gaseous exchange across the lungs and gills of an African lungfish, *Protopterus aethiopicus. J. exp. Biol.* **52**, 1–16.

Withers, P.C., Hillman, S.S., Drewes, R.C. & Sokol, O.M. (1982). Water loss and nitrogen excretion in sharp-nosed reed frogs. *J. exp. Biol.* **97**, 335–343.

Blood circulation

Johansen, K., Lenfant, C. & Hanson, D. (1968). Cardiovascular dynamics in the lungfishes. *Z. vergl. Physiol.* **59**, 157–186.

Tazawa, H., Mochizuki, M. & Piiper, J. (1979). Respiratory gas transport by the incompletely separated double circulation in the bullfrog, *Rana catesbiana. Resp. Physiol.* **36**, 77–95.

Aestivation

Janssens, P.A. (1964). The metabolism of the aestivating African lungfish. *Comp. Biochem. Physiol.* **11**, 105–117.

McClanahan, L. (1967). Adaptations of the spadefoot toad, *Scaphiopus couchi*, to desert environments. *Comp. Biochem. Physiol.* **20**, 73–99.

Locomotion on land

Jayes, A.S. & Alexander, R. McN. (1980). The gaits of chelonians: walking techniques for very low speeds. *J. Zool., Lond.* **191**, 353–378.
Jenkins, F.A. & Goslow, G.E. (1983). The functional anatomy of the shoulder of the Savannah monitor lizard (*Varanus exanthematicus*). *J. Morph.* **175**, 195–216.
Warren, J.W. & Wakefield, N.A. (1972). Trackways of tetrapod vertebrates from the upper Devonian of Victoria, Australia. *Nature* **238**, 469–470.

Hearing on land

Clack, J.A. (1983). The stapes of the Coal Measures embolomere *Pholiderpeton* . . . and otic evolution in early tetrapods. *Zool. J. Linn. Soc.* **79**, 121–148.
Hetherington, T.E. (1988). Biomechanics of vibration reception in the bullfrog, *Rana catesbiana*. *J. comp. Physiol.* A**163**, 43–52.
Lambard, R.E. & Straughan, I.R. (1974). Functional aspects of anuran middle ear structures. *J. exp. Biol.* **61**, 71–93.
Pinder, A.C. & Palmer, A.R. (1983). Mechanical properties of the frog ear: vibration measurement under free- and closed-field acoustic conditions. *Proc. R. Soc. Lond.* B**219**, 371–396.

Reproduction

Burggren, W. (1985). Gas exchange, metabolism and 'ventilation' in gelatinous frog egg masses. *Physiol. Zool.* **58**, 503–514.

16 Reptiles

Class Reptilia
    Subclass Anapsida   (primitive reptiles)
    Subclass Chelonomorpha   (tortoises and turtles)
    Subclass Lepidosauria,
        Order Rhynchocephalia   (tuatara)
        Order Squamata   (lizards and snakes)
    Subclass Archosauria,
        Order Saurischia ⎫
                          ⎬ (dinosaurs)
        Order Ornithischia ⎭
        Order Pterosauria   (pterosaurs)
        Order Crocodilia   (crocodiles)
    Subclass Euryapsida   (plesiosaurs etc.)
    Subclass Ichthyopterygia   (ichthyosaurs)
    Subclass Synapsida,
        Order Pelycosauria ⎫
                         ⎬ (mammal-like reptiles see chapter 18)
        Order Therapsida ⎭
        and other orders

16.1.   Introduction

Amphibians produce relatively small eggs without shells, and generally lay them in water (section 15.8). Reptiles produce large shelled eggs and lay them on land. This is regarded as the fundamental difference between the two classes, but it is seldom possible to discover what kind of eggs a fossil laid. The earliest known fossils of reptile eggs are from the Permian period, but fossil eggs are rare (which is not surprising, because eggs are so fragile) and some fossil skeletons from the preceding period, the Carboniferous, are believed to be of reptiles.

Various features of the skeleton have been used in attempts to distinguish fossil amphibians from reptiles. Reptiles generally have two or more sacral vertebrae, each bearing a pair of sacral ribs, but amphibians have only one. There are differences between the classes in the structure of the other vertebrae, and reptiles generally have more joints than amphibians in their toes. However, these features are not reliable. The seymouriamorphs are a group of Permian fossils which in these features are more like reptiles than amphibians, and have sometimes been classed as reptiles, but fossil larvae of seymouriamorphs have been found that show traces of external gills. This establishes beyond doubt that the seymouriamorphs should be classed as amphibians.

Early fossil amphibians and reptiles have the top and sides of the head covered by a

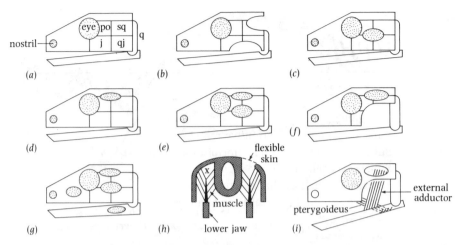

Fig. 16.1.(*a*)–(*g*) Diagrams of reptile skulls showing patterns of fenestration. Abbreviations: po, postorbital; sq, squamosal; j, jugal; qj, quadratojugal; q, quadrate. (*h*) A diagrammatic transverse section through a reptile skull, showing the pennate structure of the jaw adductor muscles. A temporal fenestra is shown on the right side only. (*i*) A diagram of a lizard skull showing the jaw adductor muscles in side view.

sheet of dermal bone (Fig. 15.5). The space between this sheet and the cranium (which contains the brain) is filled largely by the jaw adductor muscles, as it is also in bony fishes (Fig. 14.3). In the course of reptile evolution, holes have appeared in the dermal sheet, so that the jaw muscles are no longer completely covered by it. These holes (known as temporal fenestrae) have appeared in different positions in different groups of reptiles.

The principal patterns of holes are shown in Fig. 16.1. Fig. 16.1(*a*) shows the primitive condition, described as anapsid. The postorbital, squamosal, jugal and quadratojugal bones form a continuous sheet covering the jaw adductor muscles. The quadrate (a cartilage bone) lies a little deeper than these dermal bones and provides the articulation for the lower jaw. Fig. 16.1(*h*) (a diagrammatic section) shows how the jaw muscles fit into the space between the dermal bones and the cranium. The principal adductor is generally pennate, with a central tendon inserting on the lower jaw (compare it with the pennate crustacean muscle in Fig. 8.8). When the muscle fibres shorten to raise the lower jaw, they move the central part of the muscle upwards and squeeze fluid out of the space x (Fig. 16.1*h*). There might be an advantage in having a hole in the dermal bone here, covered by flexible skin that could bulge outwards when the muscles shortened, as shown on the right-hand side of the diagram. Some of the holes that I will describe may have evolved to allow such bulging, and others may have evolved simply to lighten the skull, in places where no rigid attachment for the muscles was needed.

Fig. 16.1(*b*) shows a variant of the anapsid condition, found in many turtles. There are no holes in the dermal bone over the jaw muscles but its posterior and/or ventral edges have been emarginated. Fig. 16.1(*c*) shows the synapsid condition found in the extinct group of reptiles from which the mammals evolved, the subclass Synapsida. There is one fenestra in the sheet of dermal bone on each side of the head, fairly low on

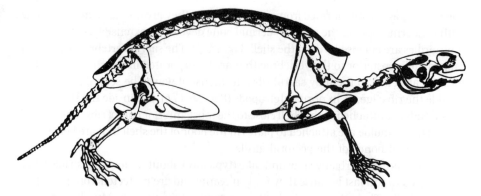

Fig. 16.2. Skeleton of the turtle *Emys*, with the nearer half of the shell cut away. From A.S. Romer (1956). *Osteology of the reptiles*. University of Chicago Press. Copyright © 1956.

the skull: the postorbital and squamosal bones generally meet above it. In the parapsid condition (Fig. 16.1*d*) there is again just one fenestra, but it is higher in the skull, with the postobital and squamosal meeting below it. This condition is found in two groups of extinct marine reptiles, the subclasses Ichthyopterygia (the ichthyosaurs) and Euryapsida (including the plesiosaurs). Diapsid skulls have fenestrae in both the higher and the lower position. The simplest diapsid condition (Fig. 16.1*e*) is found in the tuatara (*Sphenodon*) and other primitive members of the subclass Lepidosauria. Lizards are more advanced lepidosaurs that have lost the quadratojugal bone, leaving the lower fenestra open below (Fig. 16.1*f*). The members of the subclass Archosauria (dinosaurs, crocodiles etc.) are also diapsid but most of them have one or both of the additional holes in the dermal bone of the skull, shown in Fig. 16.1(*g*).

The earliest reptiles had skulls that were anapsid like those of the amphibians from which they evolved. Their bodies were lizard- or crocodile-shaped like those of their amphibian ancestors (Fig. 15.4). Tortoises and turtles also have anapsid skulls but are peculiar in many ways and are best placed in a separate subclass, the Chelonomorpha. They have the body enclosed in a shell which is open at both ends, allowing the head and forelimbs to emerge in front and the hindlimbs and tail behind (Fig. 16.2). Head, limbs and tail can be withdrawn into the shell by many of them (but not by the marine ones). This drives most of the air out of the lungs: the tortoise *Testudo* compresses its lungs to one fifth of their initial volume, when it retires into its shell.

The shells of tortoises and turtles consist of an inner layer of plates of bone and (usually) an outer layer of plates of keratin. Both are formed in the skin, the bone in the dermis and the keratin in the epidermis, which consists largely of keratin even in ordinary reptile skin (section 16.3). The particularly thick keratin plates of the hawksbill turtle, *Eretmochelys*, used to be used for making tortoiseshell combs, etc. The trunk vertebrae and ribs of tortoises and turtles are incorporated in the shell: the dorsal processes (neural spines) of the vertebrae merge with a median row of plates of dermal bone, and the ribs merge with more lateral plates. Vertebrae and ribs are of course cartilage bones, formed from sclerotome, but they grow so as to make contact with the dermis, where the plates of dermal bone develop as extensions of them. The limb girdles are enclosed within the shell (Fig. 16.2). The sacral vertebrae are almost immediately posterior to the vertebrae that are incorporated in the shell. The pelvic girdle is attached to them by sacral ribs, in the usual way. The scapulae do not lie outside the rib cage, as in normal tetrapods (Fig. 15.17*b*), but inside the shell. Their dorsal ends are attached to the shell close to the most anterior pair of ribs. The ventral ends of the scapulae are attached to plates in the floor of the shell that are believed to be the dermal bones of the pectoral girdle.

Tortoises are proverbially slow and take (typically) about 2 s for each stride. They do not need to run fast because they feed on plants and are protected from predators by their shells. Their slowness is due to unusually slow muscles. Muscles removed from the legs of tortoises and stimulated to contract at 0°C took 4 s to develop full tension. They would probably have taken 1 s even at 20°C. Other tetrapods have much faster leg muscles: for instance, frog sartorius muscles develop full tension in 0.08 s at 16°C. The slowness of tortoise muscles confers a positive advantage, for it makes them very economical of energy. Experiments with muscles from various animals have shown as a general rule that slow muscles can maintain tension more economically than fast ones (see section 6.3).

Tortoises and turtles have no teeth but the epidermis of their jaws forms a horny (keratin) beak, which is very effective for biting pieces off plants.

The tuatara (*Sphenodon*) is the most primitive modern member of the subclass Lepidosauria. It is a lizard-like reptile weighing up to a kilogram, found only on a few islands off New Zealand.

Lizards are also lepidosaurians, but have skulls of the type shown in Fig. 16.1(*f*). Loss of the quadratojugal leaves the quadrate free to pivot on the cranium, moving the lower jaw forward and back, but X-ray cine films of lizards feeding show it swinging forward and back through only about 10°. Fig. 16.1(*i*) shows the two principal jaw-closing muscles, which slope in different directions and so can pull the lower jaw forward or back as well as up. The adductor externus attaches to the cranium and the dermal bones of the skull roof, as shown in Fig. 16.1(*h*). The pterygoideus lies median to it, and attaches to a flange of the pterygoid bone, in the palate. At its other end, the pterygoideus curves round the ventral edge of the lower jaw to attach to its lateral surface.

Many lizards have a hinge joint in the roof of the skull, between the frontal and parietal bones. (This is further forward than the hinge in the cranium of crossopterygian fishes (Fig. 15.3).) It has been suggested that this may give extra mobility to the jaws, which may be important in feeding. However, the joint seems to move very little. In a recent experiment, a pointer was screwed to the parietal bones of a monitor lizard (*Varanus*), and a protractor was glued to the skin over its frontal bones. This made it easy to measure changes in the angle of the joint, from films of the lizard feeding. The observed range of movement was only about 10°.

Most lizards feed on insects and other invertebrates but monitor lizards eat larger prey such as small mammals, and a few other lizards eat plants. Most lizards have quite large, fleshy tongues and extend them as frogs also do to pick up small insects: the tongues of chameleons are remarkably long. These tongues are also used to manipulate prey in the mouth, and to aid swallowing. Larger prey are grabbed between the teeth and swallowed by inertial feeding: the mouth is opened momentarily and the head jerked forward to move the prey a little further down the throat before the jaws close on it again. No tongue could be much help in manipulating large prey, and a large tongue might be in the way in swallowing. Monitor lizards do not have fleshy tongues, but slender forked ones like those of snakes.

Many lizards' tongues, whether they are stout or slender, have forked tips. The tips are believed to convey material to a pair of accessory olfactory organs, known as Jacobson's organs. These develop as pockets of the nasal cavity but lose their connection with it and come to open directly into the mouth cavity, anterior to the internal nostrils. Though homologous organs are found in other reptiles they are most highly developed in lizards and snakes and it is only in them that they open into the mouth separately from the nostrils. Snakes and lizards with slender tongues are believed to pass particles to be sensed to Jacobson's organs, by inserting the tips of the tongue into the openings. When they move their tongues repeatedly in and out of their mouths they are probably picking up particles for investigation.

Some lizards have reduced limbs, or no limbs at all. Many of them live in deserts and 'swim' through loose sand much as eels swim through water, by passing waves of bending backwards along the body. A similar action serves limbless species for crawling over the surface of ground and is referred to as serpentine locomotion because it is the most usual manner of crawling by snakes. Fig. 5.5(b) shows a nematode worm crawling in similar fashion.

Snakes probably evolved from limbless burrowing lizards. They have no limbs or limb girdles, apart from rudiments of the hind limb and pelvic girdle in the more primitive groups. They may have evolved this condition as an adaptation to burrowing, but the size of the prey they could swallow would be limited if it had to pass through a pectoral girdle. Few modern snakes burrow, but many snakes eat remarkably large prey. A leopard measuring 1.25 m from snout to rump has been found in a very large python, 5.5 m long.

Fig. 16.3(a) shows the skull of one of the more primitive (non-venomous) snakes. It has evolved beyond the lizard condition (Fig. 16.1f), losing the jugal and squamosal bones. (The bone labelled supratemporal in Fig. 16.3a has sometimes been identified as a squamosal, but it is attached to the parietal in exactly the same way as the supratemporal of lizards and is almost certainly a supratemporal.) There are movable joints between the frontals and prefrontals, between prefrontal and maxilla, between maxilla and ectopterygoid, between pterygoid and quadrate, between quadrate and supratemporals, and elsewhere. These are not simply hinges, but allow rotation about more than one axis, and the ligaments are in many places loose enough to allow some sliding of one bone over another. The left and right maxillae can be moved independently, and the two halves of the lower jaw can be separated widely to enlarge the mouth.

The extraordinary mobility of the head makes the swallowing of large prey possible. The two sides of the mouth are used rather as the left and right hands are used to pull in a rope hand over hand. The prey is held firmly in the left side of the mouth while the right teeth are released and moved forward to a new position, and vice versa. The right maxillary teeth can be released as the right half of the lower jaw

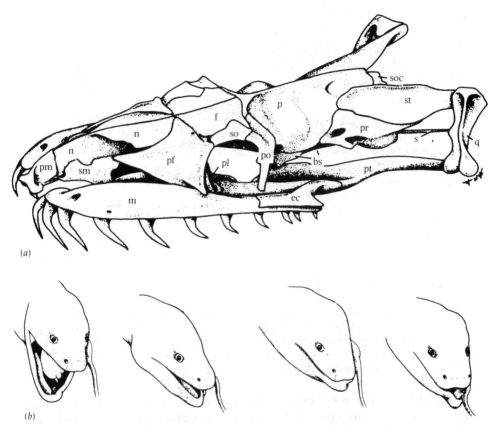

(a)

(b)

Fig. 16.3.(a) **The skull of the African rock python,** *Python sebae.* **Abbreviations: bs,**
basisphenoid; ec, ectopterygoid; f, frontal; m, maxilla; n, nasal; p, parietal; pf,
prefrontal; pl, palatine; pm, premaxilla; po, postorbital; pr, pro-otic; pt, pterygoid; q,
quadrate; s, stapes; sm, septomaxilla; so, supraorbital; soc, supraoccipital; st,
supratemporal. (b) **Tracings from a cine film of an Indian python,** *Python molurus,*
swallowing a mouse. Note the independent movement of the two halves of the lower
jaw. Both after T.H. Frazzetta (1966). *J. Morph.* **118, 217–296.**

is lowered, by tilting up the right side only of the snout. Fig. 16.3(b) shows the two
halves of the lower jaw moving independently. They can be separated widely, if
necessary, when large prey is being swallowed. The glottis can be pushed forward out
of the corner of the mouth so that it is not blocked, and breathing can continue during
swallowing of prey.

Evolution of venom was a further adaptation for dealing with large and possibly
formidable prey. Venom is injected in a quick strike and the snake may withdraw,
avoiding the danger of being wounded by the dying prey. The snake returns later,
using its tongue and Jacobson's organs if necessary to follow the trail of the victim to
the place where it died.

Some venomous snakes have fangs with grooves down which the venom flows. In
more advanced ones the edges of the grooves have met so that the fangs are tubular,
like hypodermic needles. The duct of the venom gland enters the base of the tooth,
and the venom emerges from its tip.

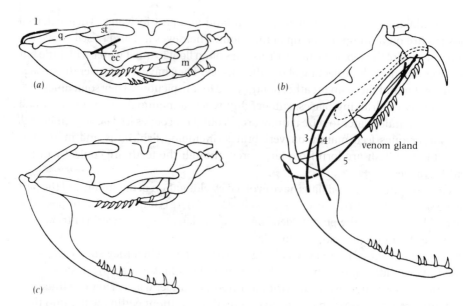

Fig. 16.4. The skull of a viperid snake (*a*) with the mouth closed and fangs retracted, (*b*) with the mouth open and fangs erected, and (*c*) with the pterygoids lowered. Muscle (1) opens the mouth, (2) erects the fangs, (3) closes the mouth, (4) squeezes venom out of the gland and (5) retracts the fangs. Abbreviations as in Fig. 16.3. This diagram is based on sketches of *Agkistrodon* supplied by Dr K.V. Kardong.

Some of the most dangerous snakes, including cobras, have quite short fangs and can close their mouths with the fangs erect. Vipers and rattlesnakes have longer fangs which have to be folded back when the mouth closes. This is possible because the maxillae (which bear the fangs) have become extremely short. Pythons move their palates forward as they strike, and so tilt their relatively long maxillae through a small angle. Vipers move their palates in the same way but the maxillae rotate through a much larger angle, because they are so short (Fig. 16.4).

Vipers keep their fangs folded down while swallowing prey, and use the teeth of the palate and lower jaw. The palatal teeth can be lowered for this purpose, below the level of the fangs (Fig. 16.4*c*). This is possible because the joint between the pterygoid and ectopterygoid, which is fixed in pythons, is movable in vipers.

Lizards and snakes have strips of glandular tissue under the skin of the lips. They discharge into the mouth, through numerous small openings, a mucous secretion which probably helps as a lubricant in swallowing. The venom glands of some snakes seem clearly to be modified parts of the gland in the upper lip, but in other cases their homology is less obvious.

The venoms of snakes are varied and complicated in composition. They include various polypeptides with toxic effects, including in many snakes one with a curare-like action, which causes paralysis by putting neuromuscular junctions out of action. Paralysis of the respiratory muscles results and quickly causes death. Venoms also include a variety of enzymes, including proteases which aid digestion of the prey.

The subclass Archosauria includes the dinosaurs (described in section 16.5) and the extinct flying reptiles (pterosaurs) but the only living archosaurs are the crocodilians.

Crocodilians spend part of their time on land and part in water. Their external nostrils are on a hump on the tip of the snout. Their eyes are also raised above the general level of the top of the head. Crocodiles often float in quiet water with little but the eyes and nostrils projecting above the surface. They have a view over the surface of the water and they can breathe, but they are inconspicuous. They sometimes lurk in this way off drinking places, submerging when a mammal comes to drink and swimming underwater to seize it unawares. Young crocodiles feed largely on insects but larger ones feed mainly on vertebrates, including fish, birds and mammals. Mouthfuls of flesh are torn from large prey, though the teeth are conical without cutting edges.

The long jaws of crocodilians have evolved by elongation of the snout, and the part of the skull posterior to the eyes is relatively short. There is little room for the posterior jaw adductors, and the pterygoideus muscles (which extend forward ventral to the eye) are the main jaw-closing muscles.

Crocodilians have muscles which close their nostrils when they submerge. They also have an adaptation which enables them to breathe through the nostrils while these are above water, even when the mouth is open under water. They can thus hold prey under water to drown it, and go on breathing without getting water into their lungs. The main structure involved is the secondary palate. This is a 'false ceiling' to the mouth cavity, ventral to the primary palate. It is essentially similar to the secondary palate of mammals (see section 18.1) and consists of extensions of the premaxillae, maxillae, palatines and pterygoids. Because it is there, the internal nostrils no longer open into the anterior part of the mouth cavity, but much further back, close to the glottis. Air is conveyed to these openings, in the space between the primary and secondary palates.

16.2. Shelled eggs

I have already mentioned the marked difference between amphibian eggs and reptile eggs. The ova of amphibians (that is, the eggs excluding the jelly) are relatively small. The eggs of reptiles are generally much larger, even if only the ovum (yolk) is considered. For instance, typical (30 g) frogs lay ova of about 2 mm diameter while lizards of similar mass lay eggs of nearly 1 cm diameter (with yolks of perhaps 7 mm diameter). Larger reptiles and birds lay larger eggs and ostrich (*Struthio*) eggs are about 14 cm diameter. Amphibian eggs are usually laid in water and hatch as aquatic larvae. Reptile eggs are laid on land (if they are laid at all, for some reptiles are viviparous) and there is no aquatic larva. Amphibian eggs have no shell but reptile eggs have shells.

A large egg out of water needs a shell (or at least a tough membrane) to maintain its shape. That this is true of hen's eggs will be realized by anyone who has broken one open to fry it. The yolk is more or less spherical in an intact egg (or in a hard-boiled egg) and greatly flattened in an egg broken onto a plate (or in a fried egg). However, if the egg is broken into a jar of water the yolk is largely supported by buoyancy and remains more or less spherical. Consider a spherical egg of radius r and density ρ, enclosed in a taut but flexible membrane. A tension T, acting in the membrane will set up a pressure difference $2T/r$ between the egg and its surroundings. There is a

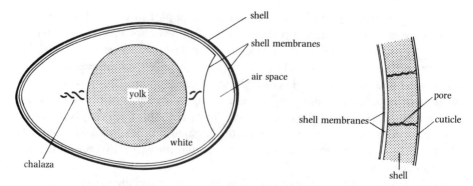

Fig. 16.5. Diagrammatic longitudinal section of a newly laid bird egg and (right) a section of the shell and the shell membranes at a higher magnification.

hydrostatic pressure difference inside the egg, between the top and the bottom, of $2r\rho g$ and if the egg were immersed in a fluid of density ρ there would be an equal hydrostatic pressure difference between the same levels in the fluid. If the egg is in air this is not the case, and it will only remain reasonably nearly spherical if

$$2T/r \gg 2\ r\rho g.$$
$$T \gg r^2\ \rho g.$$

To maintain a particular near-spherical shape as size increased, the tension would have to increase in proportion to the square of the radius. Consequently, a large terrestrial egg would need a relatively thicker membrane than a small one; or a rigid shell. Another important function that can be served by a shell on a terrestrial egg is to restrict water loss by evaporation. This will be discussed later.

Birds, tortoises and crocodilians lay eggs with stiff, brittle shells. Turtles, most lizards and snakes lay eggs with flexible leathery shells. The brittle egg shell of the hen contains only 3% organic matter, with 95% inorganic salts (mainly calcium carbonate) and 2% water. The shell is pierced by pores of about 20 μm diameter, through which oxygen diffuses into the egg while carbon dioxide and water vapour diffuse out. The whole shell is covered by a cuticle 5 μm thick which seems to cover the mouth of the pores. It is too thin to be a serious barrier to diffusion, but it may prevent micro-organisms from entering the pores. Within the shell are two porous shell membranes, each consisting of felted protein fibres. At the blunt end of the egg is the air cell, between the two shell membranes (Fig. 16.5). The stiff shell cannot contract as water evaporates from the egg, but the air cell gets larger. Crocodilian eggs have air cells, but reptile eggs with flexible shells do not.

The white or albumen of the hen's egg contains about 88% water, 11% protein and small quantities of carbohydrate and inorganic salts. Notice that its organic content is immensely greater than that of the jelly of frog eggs (section 15.8). The twisted cords of less fluid albumen known as chalazae, which run lengthwise in bird eggs (Fig. 16.5), are not found in reptile eggs. The yolk of the hen's egg contains only about 50% water, with 16% protein and 32% fat. Only the ovum (yolk) is formed in the ovary. The white, shell membranes and shell are successively laid down around it as it passes down the oviduct, which takes about twenty-four hours.

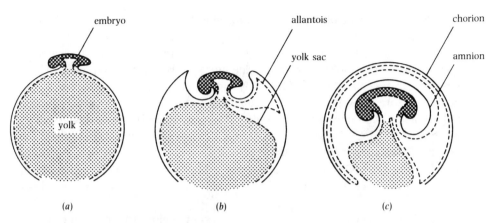

Fig. 16.6. Diagrammatic sections of successive stages in the development of a bird embryo, showing how the embryonic membranes are formed.

As in other large ova such as those of selachians, only part of the ovum of a bird or reptile divides into cells. The embryo develops on the surface of the yolk, which becomes almost completely enclosed by cellular membranes (Fig. 16.6a). An inner membrane which is continuous with the wall of the gut (and is formed of endoderm and mesoderm) invests the yolk closely. It is known as the yolk sac. There is in addition an outer membrane continuous with the body wall (and formed of ectoderm and mesoderm). In selachians this outer membrane is an integral part of the yolk sac but in reptiles and birds it rises in folds around the embryo (Fig. 16.6b). The folds fuse together over the embryo so that it comes to be covered by two membranes, the amnion and chorion (Fig. 16.6c). Mammals also develop an amnion, and the reptiles, birds and mammals are sometimes referred to as the amniotes, to distinguish them from the fish and amphibians which have no amnion. The allantois is yet another membrane (formed of endoderm and mesoderm) which develops in amniotes as an outgrowth of the hindgut. Its main function in birds and reptiles is as an organ of respiration, and it grows so as to cover the whole inner surface of the chorion, to which it becomes attached. A network of blood capillaries develops in the allantochorion so formed, receiving blood from the dorsal aorta and returning it to the heart. Since this network is immediately inside the shell membranes, the blood receives oxygen and gives up carbon dioxide as it passes through. It is very delicate, with a thickness of less than 1 μm of tissue (in chick embryos) separating the blood from the inner shell membrane.

Various suggestions have been made as to why the amnion evolved. The most important reason was probably to allow the allantois to surround the embryo completely, so that the whole of the egg surface could be used for respiration. Another plausible suggestion is that it holds the embryo slightly clear of the shell membranes: temperature fluctuations are likely to be smaller deep in the egg than near its surface. A remark which is often repeated, that the amnion encloses the embryo in a 'private pond', is a pretty metaphor but does not explain anything.

While protecting the egg, the shell and shell membranes must be porous enough to allow respiration. They must allow diffusion of oxygen and carbon dioxide at the rates required for metabolism of the embryo, while the differences of partial pressure of these gases across the shell are reasonably low. The partial pressure of oxygen within the shell must remain high enough to support the respiration of the embryo, and the partial pressure of carbon dioxide must not rise to harmful levels. These partial pressures can be assessed by analysing samples of gas from the air cell. The partial pressure of oxygen is lowest (about 0.14 atm) and that of carbon dioxide highest (about 0.05 atm) towards the end of incubation when the metabolic rate of the embryo is highest. These partial pressures are close to the partial pressures in the air expired by adult fowl, which suggests that a less permeable shell might be harmful. Since the partial pressure of oxygen in air is 0.21 atm, the difference in partial pressure of oxygen between the air outside the egg and the air inside is about 0.07 atm.

A shell restricts evaporation from an egg but cannot altogether prevent evaporation, since it is permeable. A 60 g hen's egg incubated at 38°C in air of relative humidity 60% loses about 7.7 g water in the 21 days of incubation. This is 0.015 g water h^{-1} or 20 cm^3 water vapour h^{-1}. Towards the end of incubation it uses about 25 cm^3 oxygen h^{-1}. Does the ability to take up 25 cm^3 oxygen h^{-1} make the loss of 20 cm^3 water vapour h^{-1} (in these conditions) inevitable? At 38°C and 60% relative humidity, the partial pressure of water vapour is 0.026 atm less than in saturated air at the same temperature. Diffusion of water vapour through the egg shell would be driven by this partial pressure difference, which is 0.026/0.07 of the difference which is needed (as was seen above) to drive the diffusion of oxygen. The diffusion constants (section 4.3) for diffusion of gases in air are inversely proportional to the square roots of their relative molecular masses, so the constant for water vapour (molecular mass 18) is $\sqrt{(32/18)}$ times the constant for oxygen (molecular mass 32). Hence, if oxygen and water vapour are diffusing (in opposite directions) through the same shell, the vapour should diffuse at $(0.026/0.07)\sqrt{(32/18)} = 0.5$ times the rate of the oxygen. Only 12.5 cm^3 vapour h^{-1} need be lost. Actually, as we have seen, 20 cm^3 h^{-1} is lost. The discrepancy is due to the oxygen having to diffuse along a longer path than the water vapour. The shell membranes are moist, and contain about 40% water even at the end of incubation, so water evaporates from their outer surface and has only to diffuse through the shell. Oxygen, however, has to diffuse through both shell and shell membranes. This interpretation has been confirmed by using an oxygen electrode to measure the rate of diffusion of oxygen through eggshells, with and without the shell membranes.

The eggs of fishes and amphibians are laid in water and absorb water osmotically from their surroundings: for instance, an axolotl (*Ambystoma*) ovum increases its mass by about 75% by absorbing water after being laid. The eggs of birds are laid in generally dry places, in nests or on the ground, and lose water by evaporation. Many lizard and snake eggs are laid in fairly dry places (for instance, under stones or logs) and may also lose water during development. Other reptile eggs are laid in damp places and absorb water. (These eggs have leathery shells which allow some swelling.) The Loggerhead turtle, *Caretta*, buries its eggs in damp sand from which they may absorb enough water to increase their mass by 50%. The grass snake, *Tropidonotus*, lays in damp earth or rotting vegetation (such as compost heaps) where its eggs take up water, but if the eggs are kept in drier conditions they lose water and the embryos die. Though a damp place may be necessary, the eggs must not be submerged; oxygen could not diffuse fast enough through the pores in the shell if they were filled with water. Careful excavation of the eggs of a terrapin (*Chrysemys*) showed that they were in underground cavities, so that they lay on damp soil but did not have soil lying on them. The lower surfaces of these eggs took up water osmotically from the soil but the upper surfaces lost water by evaporation.

The amounts of fat, protein and carbohydrate consumed during the development of the embryo can be determined by analysing new-laid eggs and ones that are almost ready to hatch. Such analyses show that protein is the most important source of energy for fish and amphibian embryos. For instance, the material metabolized by frog (*Rana*) embryos has been found to be about 71% protein, 22% fat and 7% carbohydrate. Since fat metabolism uses 2–2.5 times as much oxygen as metabolism of the same mass of protein or carbohydrate, this implies that about 60% of the oxygen used during development is used for the metabolism of protein, and 35% for the metabolism of fat.

Protein metabolism produces nitrogenous waste products. In embryos of fishes and amphibians, as in adults, most of this waste is produced as ammonia or urea. These are soluble materials which diffuse out of the egg into the surrounding water. They diffuse out of turtle eggs which are laid in damp sand, but cannot diffuse in solution from bird eggs, or from reptile eggs which are laid in dry places. Ammonia could diffuse out of such eggs as gas, but not nearly fast enough to prevent a toxic concentration building up if a substantial proportion of the metabolism was of protein, releasing ammonia. (See the discussion of snails in section 6.7.) If urea were produced and accumulated in the egg until it hatched, would it be likely to reach harmful concentrations?

In the course of development a hen's egg uses about 5 l of oxygen. If as high a proportion of protein were metabolized as in the frog egg, about 3 l of this would be used in protein metabolism. A mole of gas occupies 22.4 l (at STP) so this is 3/22.4 mol. A mole of urea results from protein metabolism using about six moles of oxygen (see the chemical equation in section 15.5), so if all the nitrogenous waste were produced as urea, about 0.022 mol would be produced. The contents of a hen's egg, excluding the large air space which is present at the end of incubation, occupy about 45 cm³; 0.022 mol urea dissolved in 45 cm³ fluid would have a concentration of about 0.5 mol l⁻¹. This is higher even than the concentrations found in selachians (section 13.4), aestivating lungfish and spadefoot toads (section 15.5). It would increase very substantially the osmotic pressure of the egg contents.

Analyses show that chick and turtle embryos use a much lower proportion of protein; in the case of the chick, only 6% of the material used in metabolism is protein, 91% is fat and 3% is carbohydrate. (The yolks of the eggs of birds and reptiles are rich in fat.) Only about 3% of the oxygen use by the chick embryo during incubation, and not the 60% supposed above, can actually be used in protein metabolism. If all the nitrogenous waste were retained in the shell as urea, the resulting urea concentration would be only 30 mmol l⁻¹.

Even this concentration is not accumulated, since most of the waste is produced not as urea, but as uric acid, the waste product of terrestrial snails (section 6.7) and insects (section 9.4). Uric acid and its salts are too insoluble to contribute appreciably to the osmotic pressure of the egg contents, even when present as a saturated solution. They are deposited as a precipitate in the cavity of the allantois. Reptile embryos, like bird ones, include a proportion of uric acid in their nitrogenous waste, but the proportion varies greatly between species.

It will be seen later in this chapter that many adult reptiles excrete mainly uric acid or urate crystals, rather than a solution of urea, and that this can enable them to make useful savings of water when water is in short supply.

Internal fertilization is necessary for the production of the shelled, terrestrial eggs of reptiles. *Sphenodon* (the tuatara) has no intromittent organ but the males of other modern reptiles do. The tortoises and turtles and the crocodilians have a grooved penis formed from the ventral wall of the cloaca. It is erected for copulation by engorgement with blood, which also makes the edges of the groove meet to form a tube for the sperm to pass along. A male tortoise places his fore feet on the female's back and curls his tail under hers in copulation. Male lizards and snakes have paired hemipenes which at rest are diverticula of the cloacal cavity. They are erected and protruded from the cloaca by turning inside out: this involves both contraction of a muscle, and engorgement with blood. Only one of the pair of hemipenes is inserted into the cloaca of the female.

Various reptiles are viviparous, but it is probably no coincidence that they include many of the most northerly species. Examples are the viper (*Vipera berus*) and Viviparous lizard (*Lacerta vivipara*), which reach latitudes 67° N and 70° N in Scandinavia. A viviparous species which basks in the sun can keep its body much warmer, during the day, than most hiding places where eggs could be laid. Temperature regulation by basking will be discussed in section 16.4. Sea snakes are also viviparous for a different reason. Reptile eggs are not suitable for laying in water because oxygen could not diffuse in fast enough if the shell were waterlogged. Turtles and crocodilians lay their eggs on land but viviparity enables sea snakes to breed without leaving the water.

16.3.   Skin, water and salts

Reptile skin seems hard and dry in comparison to the soft, moist skin of amphibians. It is generally dry because it lacks the glands found in amphibians (Fig. 15.8). It is relatively hard because its horny outer layer is thicker. Many reptiles have horny scales, which are thicker, less flexible parts of the continuous horny epidermis. Folds where scales overlap make it possible for the skin to stretch, as is necessary when snakes eat large prey. In some cases the horny scales overlay plates of bone in the dermis, which are comparable to the scales of fishes.

Lizards and snakes have a double horny layer. They have an inner sublayer which resembles the whole horny layer of amphibians (Fig. 15.8), consisting of flattened but distinct keratinized cells. They also have an outer sublayer in which cell outlines generally cannot be distinguished. The keratin of the inner sublayer is the same kind (α-keratin) as is found in amphibian and mammal skin. That of the outer sublayer is β-keratin, also found in the feathers of birds. α-keratins have coiled molecules and β-keratins have straighter ones, as can be shown by X-ray diffraction.

The single horny layer of the epidermis of crocodiles and tortoises is apparently added to from within, as in amphibians. It is not shed as a whole, but small flakes are lost from its outer surface. The double horny layer of lizards and snakes could not be maintained in this way, as the outer sublayer would be worn away but only the inner one could be added to. Instead, it is shed as a whole, usually several times a year, and replaced by a complete new double layer. Lizards usually shed it in large flakes but snakes usually shed it complete.

Most reptiles lose water far more slowly than amphibians, when exposed to dry air. This is well illustrated by an observation that a 17 g Garter snake (*Thamnophis*) kept in a desiccator lost 13% of its mass in seven days while a frog (*Rana*) of similar mass in the same conditions lost the same percentage of its mass by evaporation in two to four hours. The difference has often been assumed to be due to the thick horny layers of reptile skin, but it seems to be due more to lipids than to keratin. Water was allowed to evaporate from weighed containers covered with snake skin, into dry air. It evaporated about thirty times faster after the lipids had been extracted from the skin, using a mixture of chloroform and methanol. The principal barrier to evaporation seems to be the outer part of the a-keratin sublayer, where lipid fills the spaces between the keratinized cells like mortar between bricks.

Evaporation of water cannot be cut down below a certain minimum rate, no matter how impermeable the skin. This is because water is also lost from the moist surfaces of the lungs and respiratory tract. The argument applied to a beetle in section 9.4 applies also to reptiles: a lizard in dry air at 40°C must lose almost a milligram of water for every cubic centimetre of oxygen that it uses (a gram for every litre). At lower temperatures it need not lose so much.

The loss is partly compensated for by water formed in metabolism. For instance, when polysaccharide is metabolized according to the equation

$$(C_6H_{10}O_5)_n + 6n\ O_2 = 6n\ CO_2 + 5n\ H_2O,$$

5 mol (90 g) water is formed for every 6 mol (134 l) oxygen used. That is, water is produced at the rate of 0.7 g l^{-1} oxygen. Fairly similar amounts are produced in fat and protein metabolism. This water production could largely compensate for the loss by evaporation from the lungs.

Water losses from the respiratory tract and through the skin have been measured separately for a few reptiles, in experiments in which oxygen consumption was also measured. The animal was weighed and put into a chamber through which dry air was passed slowly. Oxygen consumption was calculated from the difference in oxygen content between the incoming and outgoing air. The animal was weighed again after a period in the chamber, and the total mass of water lost by evaporation was taken to be the loss of mass, minus a correction for the mass of carbon calculated (from the oxygen consumption) to have been lost as carbon dioxide. Urine and faeces were either collected and weighed, in which case their mass was allowed for in the calculations, or they were retained in the body by closing the cloaca with adhesive tape. The experiment was repeated with all but the head of the reptile enclosed by a plastic bag which was fastened closely round its neck. Water could then only be lost from the respiratory tract and from the skin of the head. The loss per unit area through the skin of the head was assumed to be the same as for the rest of the body,

Table 16.1. *Data from experiments described in the text, on evaporative water loss from reptiles weighing about 0.13 kg*

| | Temperature (°C) | Loss from skin (g kg^{-1} day^{-1}) | Respiratory loss | |
			(g kg^{-1} day^{-1})	(g l^{-1}oxygen)
Caiman	23	63.8	8.8	4.8
Iguana	23	9.6	2.3	0.9
Sauromalus	23	2.6	0.6	0.5
	40	6.8	8.1	1.5

Some of the data are presented in a slightly different form from that in which they originally appeared, in P.J. Bentley & K. Schmidt-Nielsen (1966). *Science* **151**, 1547–1549.

and so the respiratory loss could be calculated. Some of the results are displayed in Table 16.1. Note that these results are all for animals of similar size, so that it is reasonable to make direct comparisons between them.

There are several points of interest in the results. First, the three species lost water at 23°C at very different rates. The crocodilian *Caiman*, which lives in and near water, lost water at very roughly one third of the rate which would be expected of an amphibian of similar size. The two lizards lost water much more slowly, but the desert lizard *Sauromalus* lost it even more slowly than the forest lizard *Iguana*. Secondly, *Sauromalus* lost water very much faster at 40°C than at 23°C. Thirdly, although one tends to think of reptile skin as highly impermeable, loss through the skin represented a large proportion of the total loss in every case. Finally, respiratory loss from *Sauromalus* at 40°C was only a little above the minimum (suggested above) of one gram per litre of oxygen.

Reptiles lose water by excretion as well as by evaporation. Some turtles excrete most of their waste nitrogen as a solution of ammonia and urea, but most reptiles excrete mainly urates. Reptiles seem incapable of producing urine of higher osmotic concentration than their blood, so a certain minimum volume of water is required to excrete a given quantity of ammonia or urea. The solubility of sodium urate is only about 7 mmol l^{-1} (which is much less than the osmotic concentration of the blood) so urate can be precipitated and excreted in very little water as a paste (the white droppings of birds are similar urate pastes). How much water can be saved in this way?

A lizard in a warm climate might use about 3 l oxygen (kg body weight)$^{-1}$ day^{-1} (0.13 cm^3 g^{-1} h^{-1}, Fig. 16.7a). Since most lizards feed mainly on insects and other small animals, containing a high proportion of protein, at least 1 l oxygen kg^{-1} day^{-1} would probably be used in protein metabolism. One litre of oxygen is about 0.04 mol, and protein metabolism using it would produce about 0.007 mol urea, if this were the nitrogenous endproduct (the equation in section 15.5 shows that six moles of oxygen are needed to yield one mole of urea). The molar concentration of lizard blood plasma is typically about 0.35 mol l^{-1}, and we will suppose that the urea would be excreted as a solution of this concentration. The volume of water required would be

0.007/0.35 l or about 20 cm³. The lizard could be expected to lose 20 cm³ (20 g) water kg⁻¹ day⁻¹ getting rid of nitrogenous waste. This would be a substantial loss, comparable to losses by evaporation (Table 16.1). It could be avoided by excretion of a urate paste.

Reptiles also lose water in excreting excess salts. Salts must be excreted at the same rate as they are taken in with food. They are normally excreted in the urine, at concentrations not more than the osmotic concentration of the blood. A carnivorous reptile that ate food of about the same salt concentration as its blood would have to use about as much water excreting the salts, as was contained in the food. A herbivorous reptile eating terrestrial plants of lower salt concentration than the blood, on the other hand, need only excrete some of the water from its food and could retain the rest to help compensate for losses by evaporation. If urea were excreted, water needed for this would be additional to the water needed to excrete salts: the sum of the concentrations of urea and salts in the urine would presumably be limited by the concentration of the blood.

Some lizards have glands in their nasal cavities which excrete salts. One of them is *Amblyrhynchus*, which lives in the surf of the Galapagos Islands and feeds on seaweed. Fluid collected from the nostrils of freshly caught specimens has been found to be a concentrated solution of salts, often three times as concentrated as seawater. One cannot be sure that it is secreted at this concentration, since it must tend to be concentrated by evaporation. The secretion is blown out of the nostrils, but an incrustation of salts tends to accumulate around them. There are other herbivorous lizards which have nasal salt glands, including *Sauromalus*, which lives in North American deserts and eats succulent plants.

Of the few marine reptiles all but the Marine crocodile (*Crocodilus porosus*) seem to have salt glands of one sort or another. It is clearly an advantage to them to be able to excrete a salt solution more concentrated than the blood plasma. Like other reptiles, they lose water by evaporation from the lungs. Fresh water which could be drunk to replace it is not available, except on land. Of available foods, fishes have salt concentrations similar to those of the reptiles, while marine invertebrates and algae have higher salt concentrations, about the same as seawater. Marine turtles leave the sea only to lay eggs on the shore. As they crawl up the beach they appear to weep. The tears, which are presumably also produced in the sea, have been collected from *Caretta* and *Lepidochelys*. They were found in each case to be a salt solution considerably more concentrated than seawater. They come from a gland in the orbit which apparently functions in the same way as the nasal glands of *Amblyrhynchus*.

16.4.   Temperature

In reptiles, as in fishes (section 14.5) and other animals, metabolic rates increase with body temperature. This is illustrated by the measurements on *Iguana* shown in Fig. 16.7(*a*). The resting metabolic rate increases with temperature but so does the maximum rate that can be achieved in activity, and the difference between the two is, in this case, greatest at 32°C. This implies that the animal should be able to be most active at this temperature. This was investigated further by experiments on the

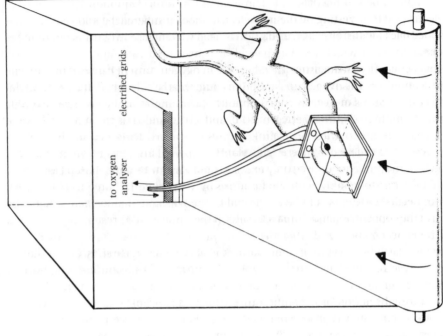

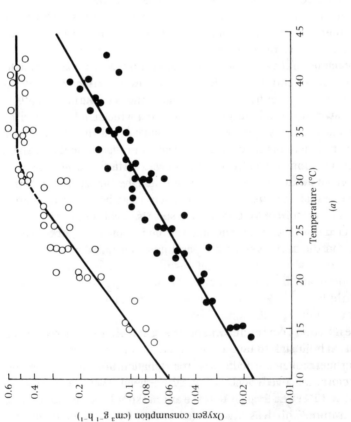

Fig. 16.7. (a) Resting (●) and maximum active (○) metabolic rates at different temperatures of the lizard *Iguana iguana*. The measurements were made on specimens weighing 0.37–1.22 kg. (b) Apparatus used to measure the metabolic rates of lizards running at various speeds. From W.R. Moberly (1968). *Comp. Biochem. Physiol.* **27**, 1–20; 21–32.

treadmill shown in Fig. 16.7(*b*). The animal ran on a continuously moving belt, discouraged from resting by the mild electric shock it suffered if it allowed itself to be carried back against the electrified grid. Its head was enclosed in a transparent plastic helmet through which air was passed at a measured rate, and a paramagnetic analyser was used to measure the difference in oxygen content between the entering and leaving air. The oxygen consumption could thus be calculated. It was found that the cost (in terms of oxygen consumption) of running at any given speed was the same at all temperatures between 20°C and 40°C, and that the maximum speed which could be sustained for several minutes was nearly three times as high at 30–35°C as at 20°C. This confirms what would be expected from Fig. 16.7*a*, that a rise in temperature, at least up to 30°C, enables the lizard to become more active.

Though a reasonably high temperature may be advantageous, too high a one would be lethal. Reptiles become incapable of co-ordinated movement when their body temperatures reach a limit that varies between species, but seems nearly always to be between 39°C and 49°C. They die at slightly higher temperatures. Some enzymes become inactive at temperatures in this range, apparently because of the disordering of molecular structure which is known as denaturation. The harmful effects of high temperatures are probably largely due to this, but it has been suggested that disorganization of cell membranes due to melting of lipids may also be involved.

Terrestrial animals can warm themselves by basking in hot sun, and so obtain the advantages of a limited increase in temperature. Frogs of the genus *Rana* bask in the sun and may become up to about 7 K warmer than the air, but terrestrial salamanders seem to avoid the sun. Amphibian body temperatures, in natural conditions, rarely exceed 35°C and are generally much lower. Reptiles, with their less permeable skins, can bask with far less danger of desiccation. They make much more use of the sun and may attain remarkably high body temperatures, as will be seen. While taking advantage of the sun they have to avoid the danger of overheating.

The problem of temperature regulation is perhaps best appreciated by considering a hypothetical example such as the one illustrated in Fig. 16.8. The reptile is supposed to be standing on level, dark-coloured ground on a clear sunny day, with the sun high in the sky. The air temperature is 30°C but the surface of the ground is at 50°C because it is being heated by the sun. The reptile's body temperature is 37°C. All these suppositions are reasonable ones, as we shall see when we go on to consider a real example. The body temperature of 37°C may seem surprisingly high, for this would be a normal temperature for a mammal, but the bodies of many tropical and subtropical reptiles do, in fact, rise to such temperatures on sunny days. Our reptile exchanges heat with the environment by radiation and convection, heat is produced in its body by metabolism and heat is lost by evaporation of water. The contributions which these make to its heat balance will be estimated in turn. It will be convenient to treat solar radiation and other radiation as separate categories.

Solar radiation   The intensity of the radiation from the sun which reaches the surface of the earth depends on how high the sun is in the sky, because of absorption by the atmosphere. When the sun is near the horizon the radiation takes a much longer path through the atmosphere than when it is directly overhead, and more is absorbed before it reaches the earth. When the sun is high and the sky is clear, solar radiation on a surface set at right angles to the sun's rays amounts to about 850 W m^{-2}. (This

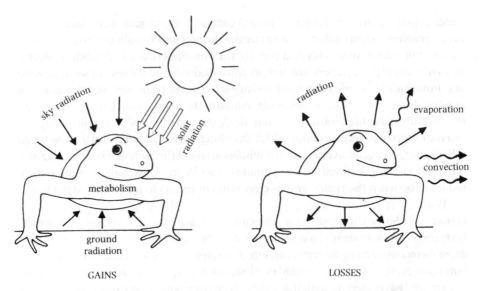

Fig. 16.8. **Diagrams illustrating the heat balance of a terrestrial animal.**

includes radiation scattered by the sky as well as direct rays from the sun.) Only part of the surface of an animal's body can be at right angles to the sun, and the solar radiation averaged over the whole surface of the body will amount to far less than 850 W m^{-2}. A cylinder with its axis at right angles to the sun's rays would receive about $850/\pi = 270$ W m^{-2}, averaged over its whole surface, and this can be taken as an estimate for an animal standing at right angles to the sun. An animal facing the sun would expose less of its area to the sun and receive less energy. If the ground were pale (for instance, sand) our reptile would receive an appreciable amount of solar radiation reflected from it, but since dark ground has been postulated this can be ignored.

Not all the solar radiation reaching the body would be absorbed, for some would be reflected. Pale skin would reflect more than dark skin, but measurements on the skins of various lizards suggest 75% absorption as typical. It can be estimated that our reptile, standing so as to receive as much solar radiation as possible, would absorb about 200 W m^{-2}.

Other radiation   The radiation to be considered now is infra-red of much longer wavelength than solar radiation, because it is emitted by bodies that are much cooler than the sun. Though the surfaces of ground and of animals may be pale in colour and reflect much of the light which falls on them, they generally reflect very little of the long-wave radiation being considered now. In this range of wavelengths they are more or less perfect absorbers and emitters of radiation; in the language of physics, they behave as 'black bodies'.

All bodies emit radiation, at rates depending on the absolute temperatures of their surfaces. The rate per unit area at which radiation is emitted by a black body of surface temperature T_S is $\sigma T_S{}^4$, where the constant σ is 5.7×10^{-8} W m^{-2} K^{-4}. The reptile at 37°C (310K) would therefore emit $5.7 \times 10^{-8} \times (310)^4 = 530$ W m^{-2}. It would also receive long-wave radiation from the ground and from the atmosphere (in addition to the short-wave solar radiation reflected from the ground and scattered by the atmosphere which was considered under the previous heading). Ground at 50°C would emit 620 W m^{-2}. Different parts of the atmosphere are at different temperatures (the outer parts at very low temperatures indeed) and it does not seem possible to estimate atmospheric long-wave radiation from first principles, but it is found in practice that this radiation generally amounts to about 400 W m^{-2} when air temperatures near the ground are about 30°C. The upper half of the reptile's body can thus be expected to receive about 400 W m^{-2} long-wave radiation and the lower half 620 W m^{-2}, giving an average for the whole surface of the body of 510 W m^{-2}. Since radiation is being emitted at an estimated 530 W m^{-2}, exchange of long-wave radiation between the reptile and its environment results in an estimated net loss of 20 W m^{-2}.

Convection   Since the reptile is warmer than the surrounding air, air movements carry heat away from it. Two processes may be involved, free convection which depends on convection currents and forced convection due to wind. Forced convection often predominates in natural outdoor conditions, but if there is no wind only free convection occurs. Calculations of convective heat losses are rather complicated and depend on the speed of the wind and on the size and shape of the body, but a rough calculation for a moderate-sized lizard at 37°C in still air at 30°C gives a loss rate of 40 W m^{-2}.

Metabolism   In the experiment referred to in Table 16.1, the desert lizards (*Sauromalus*, mass 0.12 kg) used about 28 cm^3 oxygen per hour at 40°C. Fig. 16.7(*a*) suggests that they would have used a little less, probably about 20 cm^3 per hour, at 37°C. Metabolism involving 1 cm^3 oxygen produces about 20 J heat so the rate of heat production would be about 400 J h^{-1} or 0.11 W. The surface area of the *Sauromalus* must have been about 0.025 m^2 so this is 4 W m^{-2}: I have expressed it in terms of surface area to correspond with the estimates for radiation and convection.

Evaporation   The 0.12 kg *Sauromalus* at 40°C lost a total of about 2 g water per day (15 g kg^{-1} day^{-1}, Table 16.1). The latent heat of vaporization of water is about 2500 J g^{-1} so this represents a heat loss of 5000 J day^{-1} or 0.06 W, 2 W m^{-2}. Some other reptiles would lose water faster, but water loss would have to be very much faster to affect the heat balance of the reptile substantially. Lizards in laboratory experiments sometimes pant at high temperatures, and so increase evaporative losses, but this does not seem to be a frequent feature of behaviour in nature.

If the body temperature is to be kept constant, heat gains must balance heat losses. The hypothetical reptile we are considering can vary the amount of solar radiation it receives within wide limits, by changing its position. How much solar radiation would be needed to keep its body temperature constant? It has been estimated that exchange of long-wave radiation with the environment results in a net loss from the body of about 20 W m^{-2}. Convection causes a loss of about 40 W m^{-2}. Metabolism gives a gain of about 4 W m^{-2} and evaporation a loss of about 2 W m^{-2}. Hence the solar radiation needed is about $20 + 40 - 4 + 2 = 58$ W m^{-2}. Nearly four times as much would be received if the reptile arranged its body at right angles to the sun's rays. To keep its temperature at about 37°C it would have either to adopt a position in which it received less solar radiation, or spend some of its time in shade.

The ways in which reptiles control their body temperatures are well illustrated by

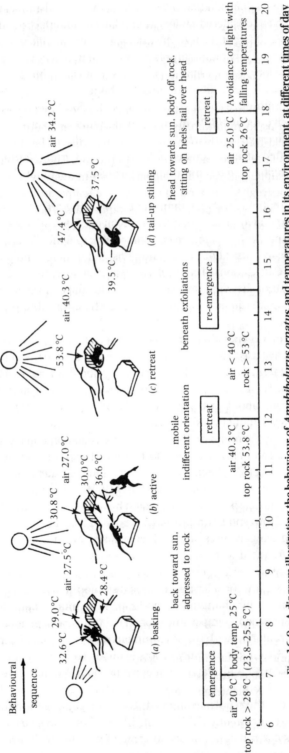

Fig. 16.9. A diagram illustrating the behaviour of *Amphibolurus ornatus*, and temperatures in its environment, at different times of day in summer. From S.D. Bradshaw & A.R. Main (1968). *J. Zool., Lond.* 154, 193–221.

observations on the Australian lizard *Amphibolurus ornatus*. It lives among barren granite outcrops where the only shelter from the sun is provided by the rocks themselves. The surfaces of the rocks become very hot in the midday sun, but very cold at night when they are exchanging long-wave radiation with a cold, clear sky. Fissures in the rocks are exposed neither to the sun nor to the night sky, and vary much less in temperature.

When *Amphibolurus* is kept in a temperature gradient in laboratory experiments, it chooses its position so as to maintain its body temperture at about 37°C. Body temperatures in the field have been investigated by inserting a thermistor in the rectum, with the leads fastened by adhesive tape to the tail. The wires continued through an overhead support to recording apparatus 50 m away which was also connected to thermistors registering air and rock surface temperatures. The lizards were free to run about, and were watched through binoculars. They behaved very much like lizards which had not been fitted with thermistors, except that their wires occasionally got caught in the rocks. Observations on a large monitor lizard (*Varanus*) with a thermistor connected to a miniature radio transmitter attached to its body (so that no wires were needed) gave similar results.

Fig. 16.9 summarizes observations made on a hot summer day. The night was spent in rock fissures, where night temperatures were higher than on the surface. The lizards emerged in the morning with body temperatures of about 25°C, and basked in the sun, retiring again for a while if a cloud passed over the sun or if there was a gust of cold wind. They basked with their bellies in contact with the warm rock, in positions where their bodies were well exposed to the sun's rays (Fig. 16.9*a*). Their body temperatures rose at rates up to 1 K min^{-1}. When they reached about 37°C they became active: feeding, courting and defending their territories (Fig. 16.9*b*). The rock surfaces warmed up more slowly than the lizards at first, but eventually became much hotter. When they reached about 50°C the lizards spent part of their time in the shade, or stood inactive, facing the sun with the belly and often the tail held clear of the ground. Facing the sun reduces the area exposed to it, as has been explained. Holding the body clear of the ground allows free circulation of cooling air around it, and minimizes heat gain by conduction from the rocks. In the hottest part of the day, with air temperatures around 40°C and the rock surfaces over 53°C, the lizards retired to the fissures (Fig. 16.9*c*), but they emerged again for a while in the evening before retiring finally for the night.

Many lizards that live in hot sandy regions behave in much the same way as *Amphibolurus ornatus*, using burrows made by themselves or by rodents instead of rock fissures. Some other reptiles make little use of solar radiation and may be active at much lower body temperatures. For instance, the lizard *Anolis allogus* lives in dense forest in Cuba where hardly any sunlight reaches the ground. It lives in small trees but never climbs high and so has virtually no opportunity to bask. Specimens caught during the day while air temperatures were 26–32°C had body temperatures which were on average about 1 K below air temperature. The legless burrowing lizard *Anniella* lives underground, where its body temperature was found, in a series of measurements in California, to average 21°C. Its temperature must follow closely that of the soil, and high temperatures are avoided by digging deeply in summer; it is found mainly at depths of 0.3 m or less in spring, but at around 1.5 m at mid-summer.

Fig. 16.10. Reconstruction of *Barosaurus*, a quadrupedal saurischian dinosaur. Length (with neck horizontal) about 25 m. From R.T. Bakker (1968). *Discovery, Peabody Museum* 3, 11–22. (Inset) Footprints of a similar dinosaur. Re-drawn from R.T. Bird (1944). *Natural History, N.T.* 53, 60–67.

Sphenodon, the tuatara, spends much of the day in burrows but is active on the surface at night, when body temperatures of 6–13°C have been measured. These temperatures are exceptionally low for an active reptile.

16.5. Dinosaurs

The Mesozoic era was remarkable for its giant reptiles. In the sea there were ichthyosaurs, shaped like sharks or whales and in the same range of sizes: the largest were about 15 m long, which is much larger than an adult killer whale (8 m) but

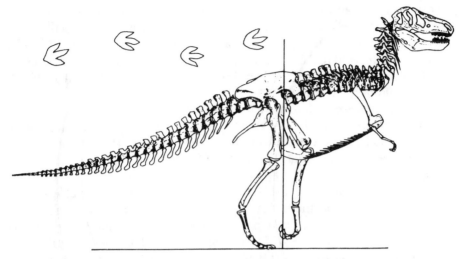

Fig. 16.11. Skeleton of *Tyrannosaurus*, a bipedal saurischian dinosaur, in what is believed to have been the normal walking position. Length about 8.5 m. From B.H. Newman (1970). *Biol. J. Linn. Soc.* 2, 119–123. (Inset) Footprints of a similar dinosaur. Re-drawn from R.T. Bird (1944). *Natural History, N.Y.* 53, 60–67.

smaller than a blue whale (27 m). There were also plesiosaurs with turtle-shaped bodies and (in some cases) very long necks. The longest of them were also about 15 m long. In the air there were pterosaurs, some of them much larger than any modern flying birds. *Pteranodon* had a wing span of 7 m (twice the span of the largest albatross) and fragments have been found of a larger pterosaur that probably had a 12 m span. On land there were the dinosaurs, some of them (with long necks and tails) as much as 30 m long. *Barosaurus* (Fig. 16.10) was neither the longest nor the heaviest. *Brachiosaurus*, a stouter but otherwise similar dinosaur, seems to have had a mass of 50 tonnes (50 000 kg) which is many times as heavy as the heaviest modern land animal (the African elephant, *Loxodonta*, up to 5.5 tonnes) though lighter than the blue whale (110 tonnes). The mass was estimated with the help of a scale model based on careful measurements of fossils, but showing the animal as it probably looked in life. The volume of the model was measured and that of the actual animal calculated from it. The mass was estimated by making the reasonable assumption that dinosaurs (like modern reptiles) had about the same density as water. This section is about the problems of life on land for such enormous animals, but we should remember that there were also small dinosaurs, some of them as small as domestic pigeons.

The form of their teeth seems to show that *Barosaurus* (Fig. 16.10) and many other dinosaurs ate plants, but there were also dinosaurs with teeth like knife blades that seem clearly to have been carnivores. The largest known of them was *Tyrannosaurus* (Fig. 16.11), about 8.5 metres long and with a mass of 7 tonnes. *Barosaurus* and similar dinosaurs obviously walked on all fours but *Tyrannosaurus*, with its tiny forefeet, was obviously bipedal. Many fossil footprints that must have been made by dinosaurs have been found in Mesozoic rocks, most of them made by bipeds (Fig. 16.11, inset) but some by quadrupeds (Fig. 16.10).

There are two groups of dinosaurs that may not be closely related. Both *Barosaurus* and *Tyrannosaurus* belong to the order Saurischia, which includes all the carnivorous dinosaurs and some of the herbivorous ones. Many other herbivorous dinosaurs, including *Triceratops* (Fig. 16.12), belong to the order Ornithischia, which is distinguished by having a distinctively shaped pelvic girdle and an additional, toothless, bone at the front of the lower jaw.

Some long-necked dinosaurs seem to have carried their necks horizontal, but the vertebrae of others fit together best with the neck vertical. *Barosaurus* seems to have carried its head high like a giraffe, and presumably fed like a giraffe on the leaves of trees, but it was twice the height of an adult giraffe. This raises the first of the problems of size that I will discuss.

If *Barosaurus* stood as shown in Fig. 16.10 its brain would have been at least 10 m above its heart. To supply the brain in this position, the heart would have to pump blood at a pressure of more than 10 m water (100 kN m^{-2}). Living reptiles develop blood pressures of only about 5–10 kN m^{-2}, most mammals develop 15–20 kN m^{-2} and even giraffes develop only 40 kN m^{-2}. *Barosaurus* must have had a remarkably strong heart. (Some scientists have argued that blood could be carried up and down giraffe and dinosaur necks by a siphon effect, with no need for high blood pressure, but this is a fallacy. Siphons work only if the tube is rigid enough not to be collapsed by the reduced pressure of the liquid in its higher parts. Veins have flexible walls and would collapse in any attempt to use them as siphons.)

Imagine two animals of the same shape, one twice as long as the other. It is twice as long, twice as wide and twice as high, therefore eight times as heavy. Any surface on the larger animal is twice as long and twice as broad, so has four times the area. As animals increase in size without changing shape, masses increase in proportion to the cube of body length but area only in proportion to the square of length. This is the basis for many of the problems of large size.

For example, our larger animal with eight times the weight, would have to support its weight on feet of only four times the area and would be in more danger of getting bogged down in soft mud. Real large animals are not the same shape as small ones but it is nevertheless a general rule that large animals are more likely to get bogged down. For example, domestic cattle have masses of about 600 kg so their weight (mass multiplied by the gravitational acceleration) is about 6000 N. The undersides of their hooves have a total area of about 0.04 m^2 so the pressure that a standing cow exerts on the ground is about 150 kN m^{-2}. The brontosaur *Apatosaurus* seems to have had a mass of 34 tonnes and footprints made by it or a similar-sized quadrupedal dinosaur have a total area (four feet) of 1.2 m^2, implying a standing pressure of 280 kN m^{-2}.

Brontosaurs were not cow-shaped, but had relatively larger feet. Nevertheless, because they were absolutely much larger than cows, the pressure under their feet was twice as high as for cows, making them more likely to get bogged down in soft ground. Many people used to think that the brontosaurs waded in lakes, but this argument suggests that the soft ground often found around lakes would have been dangerous for them.

Our large animal, eight times as heavy as the small one, has bones of only four times the cross-sectional area, so twice as much stress acts in its bones when it and the small one stand in the same position. The skeletons of very large animals might not be strong enough to support their weight. (This is why some people have imagined brontosaurs wading in lakes where their weight would be supported largely by buoyancy.) How easily could the large dinosaurs have supported their weight, and how athletic could they have been?

People and animals take short strides when walking slowly and long strides when running fast. Footprints made by large dinosaurs (for example, the ones shown in Figs. 16.10 and 16.11) are rather close together, relative to the size of the feet. The animals were travelling slowly, taking short strides. It has been calculated that the footprints shown in Fig. 16.10 were made at a speed of about 1 m s^{-1} (a slow human walking speed) and the ones in Fig. 16.11 at 2 m s^{-1} (a fast human walk). Some footprints of smaller dinosaurs seem to show much higher speeds but there are no known footprints of really large dinosaurs going fast.

This does not prove that large dinosaurs were slow, lumbering monsters. Animals run fast only rarely, so footprints of fast running are unlikely to be preserved. Another argument suggests that some large dinosaurs may have been able to run quite fast. It depends on the assumption that brontosaurs moved like elephants. The proportions of their leg bones are elephant-like, and the fossil footprints show that dinosaurs walked with their feet under the body, as elephants and other mammals do. If brontosaurs had tried to stand with their feet well out to either side of the body like newts (Fig. 15.15) and modern reptiles, the stresses in their leg bones and muscles would have been enormous.

It is much easier to break a stick by bending it than by pushing on its ends, loading it like a pillar. The same is true of leg bones and other long slender structures: forces acting at right angles to their axes, exerting bending moments on them, are much more dangerous to them than equal forces acting along their axes. Veterinarians have performed surgical operations on horses and other mammals, attaching strain gauges to their leg bones. Records from the gauges when the animals ran showed that strains and stresses due to bending moments are much larger than those due to axial loads, in most leg bones. The argument that follows is accordingly based on bending moments rather than axial loads.

The hindlegs of a 34 tonne brontosaur would have had to support about 70% of the animal's mass, 24 tonnes (brontosaurs had heavy hindquarters and tails). Those of a 2.5 tonne elephant have to support 1.1 tonne, as has been shown by standing an elephant with its hind feet only on a weighbridge. If a brontosaur and an elephant ran in similar fashion, all the forces on the brontosaur's bones would be $24/1.1 = 22$ times the forces on the elephant's bones. The brontosaur's femur is 1.8 times as long as the elephant's, so the bending moments at corresponding points on the femur would be $22 \times 1.8 = 40$ times as high as for the elephant. (Bending moments are calculated by multiplying forces by their distances from the cross-sections in question.) Measurements of the dimensions of cross-sections of the brontosaur's femur, and of the femur of a 2.5 tonne elephant, show that to produce equal stresses in the two bones, the bending moment on the brontosaur femur would have to be 32 times as high as for the elephant. Thus if brontosaurs and elephants ran in similar fashion, moving their joints through equal angles and exerting forces proportional to the supported weight, the stresses in the brontosaur femur would be $40/32 = 1.25$ times as large as in the elephant. Similar calculations for the tibia and humerus give ratios

Fig. 16.12. Sketches of *Triceratops* galloping. The positions are copied from a film of a rhinoceros. From R. McN. Alexander (1988). *Dynamics of dinosaurs and other extinct giants*. Columbia University Press, New York.

of 0.67 and 1.27. None of these ratios are very different from 1.0, indicating that similar movements in brontosaurs and elephants would result in roughly equal bone stresses. If brontosaur and elephant bones were equally strong and if their skeletons were built to the same safety factors (section 6.2), brontosaurs could have been as athletic as elephants. Elephants cannot gallop and cannot jump, but can run reasonably fast, at speeds of at least 4 m s^{-1} and probably more. Brontosaurs too could probably have run quite fast. A similar argument suggests that the 6 tonne horned dinosaur *Triceratops* may have been able to gallop, though not as athletically as a buffalo (Fig. 16.12).

When animals of the same shape are heated by the sun, the areas absorbing the radiation are proportional to (body length)^2 but the masses of tissue to be heated are proportional to (body length)^3. The rate of rise of temperature should therefore be proportional to (body length)$^{-1}$ or (mass)$^{-\frac{1}{3}}$. Lizards (*Amphibolurus*) basking in the sun heated at rates up to 1 K min^{-1} (section 16.4). Their body masses were around 0.02 kg. Brontosaurs were not the same shape but we might estimate, very roughly, that a 30 000 kg (3 tonne) brontosaur should heat $(30\,000/0.02)^{-\frac{1}{3}} = 0.009$ times as fast, at about 0.5 K per hour. Its temperature might rise a few degrees during a sunny day but unless it wanted to control its temperature very precisely it would not have to move back and forth between sun and shade like a lizard.

That calculation is misleading. It gives much too high a rate of heating for the brontosaur because it ignores the heat-insulating properties of the skin. The thin skin of small animals has little effect on their heat balance but the thicker skin of large animals has a very important insulating effect. (Once through the skin, heat is circulated round the body by the blood.) The rate of conduction of heat H across a sheet of insulating material of area A, thickness d and thermal conductivity C is

$$H = CA.\Delta T/d, \qquad\qquad \text{(equation 14.2)}$$

where ΔT is the temperature difference across the insulating layer. The surfaces of large animals get hotter in the sun than those of small ones because the heat they receive from the sun is conducted more slowly through the thicker skin, but the surface will not rise above a certain temperature, no matter how thick the skin. Compare animals of different sizes, large enough for their surfaces to heat to about the same temperature. The rates at which heat is conducted into them (when their body tempeatures are the same) will be proportional to A/s. If they are the same shape, their values of A and s will be proportional to (body mass)$^{\frac{2}{3}}$ and (mass)$^{\frac{1}{3}}$, respectively, so their rate of uptake of heat will be proportional to $(\text{mass})^{\frac{2}{3}}$. The heat needed to increase body temperature by a degree is proportional to mass so the rate of rise of temperature will be proportional to $(\text{mass})^{-\frac{1}{3}}$. This argument suggests that the brontosaur should heat up only $(30\,000/0.02)^{-\frac{1}{3}}$ times as fast as the *Amphibolurus*, at about 0.005 K per hour: its temperature would be hardly affected by the sun.

That argument is also misleading because it supposes that the outer surfaces of the two reptiles would heat to the same tempeature. It makes the brontosaur seem to heat too slowly, whereas the first argument made it seem to heat too fast. The true rate of heating would be somewhere between the two estimates, between 0.005 and 0.5 K per hour. We could try to estimate it more precisely, but the important point is clear: large dinosaurs would not have heated up much during a sunny day, nor cooled much in a cold night. Their body temperatures would have remained almost constant, day and night. Birds and mammals also maintain near-constant temperatures, but depend on their very high metabolic rates to keep them warm in cold conditions (section 17.2).

It has been suggested that dinosaurs may have had very high (mammal-like) metabolic rates. Most of the evidence which has been presented is so indirect as to have little value, but there is one part that deserves very serious consideration. Because of their higher metabolic rates, mammals need far more food than lizards of similar size. For instance, it has been estimated from field observations that lions (*Panthera*, about 150 kg) and cheetahs (*Acinonyx*, about 50 kg) kill 40 times their own mass of prey each year. A huge carnivorous lizard, the Komodo dragon (*Varanus komodoensis*, about 50 kg) kills only five times its own mass of prey each year. Hence a given population of prey should be able to support eight times as large a population of carnivorous lizards as of carnivorous mammals. The lion and hyaena populations of Ngorongoro Crater (Tanzania) have a total mass that is 0.01 of the mass of the population of ungulates that they prey on, and similar ratios have been calculated for some other populations of living mammals. If dinosaurs had mammal-like metabolism, we might expect to find similar ratios of population masses for predatory and prey dinosaurs. If they had lizard-like metabolism we might expect to find larger ratios of predators to prey.

An exceptionally rich Cretaceous deposit in Canada has yielded 246 herbivorous dinosaur skeletons (mostly bipedal 'duck-billed' dinosaurs) but only 22 carnivorous dinosaurs (similar to *Tyrannosaurus*, Fig. 16.11). The herbivorous dinosaurs seem to have been the only prey available there for the carnivorous ones. It has been estimated that these herbivorous dinosaurs had an average mass of about 5 tonnes, and the carnivorous ones about 2 tonnes, so the total mass in life of the specimens which have been collected was probably about 1200 tonnes for the herbivores and 44 tonnes for the carnivores, a ratio of predators to prey of 0.04. However, the collectors, faced with so many herbivores and so few carnivores, probably left the less good herbivore specimens behind. The true ratio of predators to prey was probably about 0.02. If the ratio was the same in the living population (which it may not have been) it was higher than for the mammals of Ngorongoro Crater, but not so much higher as to make mammal-like metabolism seem impossible.

It is often said that dinosaurs had very small brains. The brain itself is not preserved in fossils but the volume of the cavity for it in the skull can be measured, and

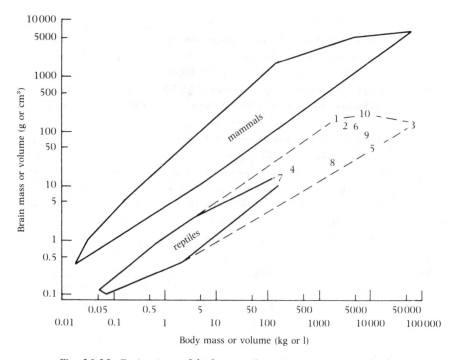

Fig. 16.13. Brain size and body size of reptiles and mammals. The continuous outlines indicate the ranges of variation among living reptiles and mammals. The numerals indicate estimated brain and body sizes of dinosaurs, including (2) *Anatosaurus*, (3) *Brachiosaurus*, (9) *Triceratops* and (10) *Tyrannosaurus*. From H.J. Jerison (1969). *Am. Nat.* 103, 575–588. Copyright © University of Chicago Press, 1969.

comparison with modern reptiles suggests that the brain would have occupied about half this volume. Fig. 16.13 includes estimates made on this basis. It shows that though the brains of large dinosaurs were only a tiny proportion of body mass, their range of sizes was about what might be expected by extrapolation from data for modern reptiles. The notion that they were extraordinarily small is due to the false assumption that brain mass should be proportional to body mass, for related animals of different size. The figure shows that it is more nearly proportional to (body mass)$^{0.65}$. There was a good deal of variation in brain size between dinosaurs of similar size but different habits, but this variation seems to have been no greater than is found in mammals.

The final question about the dinosaurs is, why are they extinct? They died out at the end of the Mesozoic era (though the birds, which survive, are believed to be descended from dinosaurs). At the same time those other magnificent reptiles, the ichthyosaurs, plesiosaurs and pterosaurs, also became extinct, but lizards, snakes, turtles and crocodiles survived. Many invertebrate groups became extinct at the same time as the dinosaurs.

A hint about what may have happened is given by rocks laid down at the extreme end of the Mesozoic. Wherever in the world they are found, these rocks contain an unusually high concentration of iridium, an element that is rare in the earth's crust but commoner deeper in the earth or in extra-terrestrial bodies. One suggestion is that the earth was hit by a large asteroid which exploded on impact, scattering debris all over the earth. This asteroid would have had to have a diameter of about 10 km, to explain the quantity of iridium. Another suggestion is that the earth was hit by a shower of comets, and a third is that there was a period of intense volcanic activity. Any of these events would have spread iridium over the earth. They could also have caused very severe atmospheric pollution, both from the dust they scattered and from acid gases: the high temperatures in the explosion of an impacting asteroid would make nitrogen in the atmosphere combine with oxygen to produce nitrogen oxides and (eventually) nitric acid, and volcanoes emit sulphur dioxide. The acid rain that would have followed might have been far worse than is caused by modern industry and would have affected some groups of animals worse than others.

The extinctions at the end of the Mesozoic were catastrophic, but several other episodes of catastrophic extinction are shown by the fossil record. Also, we are actually experiencing a period of rapid extinction, caused not by asteroids or volcanoes but by human activities. The rain forests that cover 7% of the earth's land surface contain 50% of known animal species, but are rapidly being destroyed.

Further reading

Introduction

Bradshaw, S.D. (1986). *Ecophysiology of desert reptiles*. Academic Press, Sydney.

Condon, K. (1987). A kinematic analysis of mesokinesis in the Nile monitor (*Varanus niloticus*). *Exp. Biol.* **47**, 73–87.

Gans, C. (ed.) (1969–   ). *Biology of the Reptilia*. Academic Press, New York.

Heatwole, H. (1976). *Reptile ecology*. University of Queensland Press, St. Lucia.

Kardong, K.V., Dullemeijer, P. & Fransen, J.A.M. (1986). Feeding mechanism in the rattlesnake, *Crotalus durissus. Amphibia–Reptilia* **7**, 271–302.

Romer, A.S. (1956). *Osteology of the reptiles*. Chicago University Press.

Sinclair, A.G. & Alexander, R.McN. (1987). Estimates of forces exerted by the jaw muscles of some reptiles. *J. Zool., Lond.* **213**, 107–115.

Smith, K.K. (1984). The use of the tongue and hyoid apparatus during feeding in lizards. *J. Zool., Lond.* **202**, 115–143.

Shelled eggs

Burton, F.G. & Tullett, S.G. (1985). Respiration of avian embryos. *Comp. Biochem. Physiol.* **82A**, 735–744.

Packard, G.C. & Packard, M.J. (1986). Nitrogen excretion by embryos of a gallinaceous bird and a reconsideration of the evolutionary origin of uricotely. *Can. J. Zool.* **64**, 691–693.

Rahn, H., Ar, A. & Paganelli, C.V. (1979). How bird eggs breathe. *Scient. Am.* **240**(2), 38–47.

Tracy, C.R., Packard, G.C. & Packard, H.J. (1978). Water relations of chelonian eggs. *Physiol. Zool.* **51**, 378–387.

Skin, water and salts

Bentley, P.J. & Schmidt-Nielsen, K. (1966). Cutaneous water loss in reptiles. *Science* **151**, 1547–1549.

Bradshaw, S.D. & Shoemaker, V.H. (1967). Aspects of water and electrolyte changes in a field population of *Amphibolurus* lizards. *Comp. Biochem. Physiol.* **20**, 855–865.

Dunson, W.A. (1969). Electrolyte excretion by the salt gland of the Galapagos marine iguana. *Am. J. Physiol.* **216**, 995–1002.

Landmann, L., Stolinski, C. & Martin, B. (1981). The permeability barrier in the epidermis of the grass snake during the resting stage of the sloughing cycle. *Cell Tiss. Res.* **215**, 369–382.

Stokes, G.D. & Dunson, W.A. (1982). Permeability and channel structure of reptilian skin. *Am. J. Physiol.* **242**, F681–F689.

Temperature

Avery, R.A. (1979). *Lizards: a study in thermoregulation*. Edward Arnold, London.

Bradshaw, S.D. & Main, A.R. (1968). Behavioural attitudes and regulation of temperature in *Amphibolurus* lizards. *J. Zool., Lond.* **154**, 193–221.

John-Alder, H.B. & Bennett, A.F. (1987). Thermal adaptations in lizard muscle function. *J. comp. Physiol.* B**157**, 241–252.

Porter, W.P. & Gates, D.M. (1969). Thermodynamic equilibria of animals with environment. *Ecol. Monogr.* **39**, 227–244.

Stebbins, R.C. & Barwick, R.E. (1968). Radiotelemetric study of thermoregulation in a lace monitor. *Copeia* 1968, pp. 541–547.

Dinosaurs

Alexander, R.McN. (1985). Mechanics of posture and gait of some large dinosaurs. *Zool. J. Linn. Soc.* **83**, 1–25.

Alexander, R.McN. (1989). *Dynamics of dinosaurs and other extinct giants*. Columbia University Press, New York.

Bakker, R.T. (1986). *The dinosaur heresies*. Morrow, New York.

Charig, A. (1979). *A new look at the dinosaurs*. British Museum (Natural History), London.

Jerison, H.J. (1969). Brain evolution and dinosaur brains. *Am. Nat.* **103**, 575–588.

Norman, D.B. (1985). *The illustrated encyclopaedia of dinosaurs*. Salamander Books, London.

17 Birds

Class Aves,
    Subclass Archaeornithes (*Archaeopteryx*)
    Subclass Neornithes (other birds)

17.1. Introduction

The few fossils of *Archaeopteryx*, from the later part of the Jurassic, are the earliest known birds (Fig. 17.1). They are preserved in a fine-grained limestone that used to be quarried for making lithographic stones, which retains quite detailed impressions of the feathers. These seem to be identical with modern feathers, and the arrangement of the wing feathers is the same as in modern birds.

If the impressions of feathers had not been preserved, *Archaeopteryx* would probably have been identified as a small dinosaur. Its skeleton is so much like those of small saurischian dinosaurs that sceptics have suggested that the feathers are fakes, scratched in cement smeared on genuine dinosaur fossils. (I do not know of any professional palaeontologist who accepts this suggestion but have put references to both sides of the argument in the *Further Reading* list at the end of the chapter.) It has also been argued that *Archaeopteryx* is not particularly closely related to modern birds, but that it and the birds have evolved from the dinosaurs separately, acquiring feathers independently. However, most zoologists regard *Archaeopteryx* as a very early, very primitive bird.

It has teeth like those of saurischian dinosaurs and like them has fenestrae in the skull in front of the eye, and in the lower jaw (Fig. 16.1*g*). No specimen has the back of the skull well enough preserved to show whether it was diapsid. It has only three fingers (as do many dinosaurs) but they were very long. It has a long reptile-like tail of about twenty vertebrae.

Modern birds have no teeth but (like turtles) have horny beaks covering their jaws (Fig. 17.2). The swollen skull contains a large brain, much larger than those of reptiles of equal body mass. If the brain masses of birds were added to Fig. 16.13 they would fall within the area occupied by mammal brains. If those of fishes and amphibians were added, they would lie in and around the reptile area. For animals of similar body mass there is a great difference in brain mass between birds and mammals on the one hand, and lower vertebrates on the other. *Archaeopteryx* seems to have had a brain intermediate in size between those of reptiles and of modern birds: its size is indicated by a natural cast of the cavity that contained it, on one of the fossils.

The principal wing feathers of birds are the primaries, supported by the hand, and the secondaries, supported by the radius and ulna (Fig. 17.3). The hand skeleton has

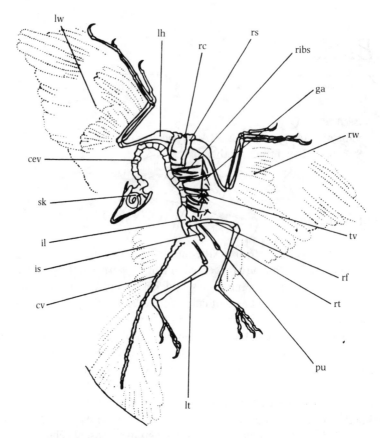

Fig. 17.1. Fossil of *Archaeopteryx lithographica* (the Berlin specimen). Abbreviations: cv, caudal vertebrae; cev, cervical vertebrae; ga, gastralia; il, ilium; is, ischium; lh, left humerus; lt, left tibia; lw, left wing; pu, pubis; rc, right coracoid; rf, right femur; rs, right scapula; rt, right tibia; rw, right wing; sk, skull; tv, trunk vertebrae. From G.R. de Beer (1954). *Archaeopteryx lithographica.* British Museum, London.

three fingers, as in *Archaeopteryx*, but they have been grossly modified. Its largest component is the carpometacarpus, which develops from the metacarpals of these three digits, plus some carpals, fused together as a single bone. A few phalanges of the digits articulate with it. We know that the three digits are the second, third and fourth because vestiges of the first digit (the thumb) and the fifth (the little finger) are found in some embryos. The group of feathers called the alula or bastard wing is attached to the most anterior (second) digit.

The pectoral girdle and sternum are almost as peculiar as the wing itself (Figs. 17.2 and 17.3). To understand them, one must know how the wing muscles are arranged. The largest muscle is the pectoralis, which is responsible for the downstroke of the wings. In reptiles, which stand with the humerus more or less horizontal, pointing laterally, the pectoralis has an important weight-supporting function. It runs from the humerus to the sternum and the ventral part of the pectoral girdle, and tension in

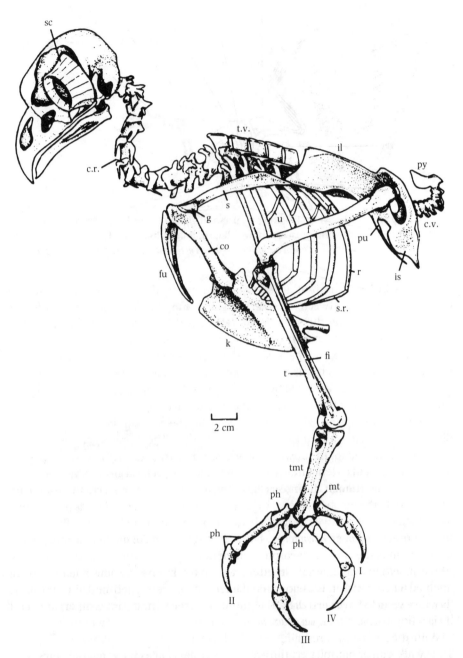

Fig. 17.2. Skeleton, with wings removed, of an eagle owl (*Bubo bubo*). Abbreviations: co, coracoid; c.r., cervical rib; c.v., caudal vertebrae; f, femur; fi, fibula; fu, furcula; g, glenoid; il, ilium; is, ischium; k, keel on ossicles; s.r., sternal rib; t, tibiotarsus; tmt, tarsometatarsus; t.v., thoracic vertebrae; u, uncinate process; I–IV, digits. From A. d'A. Bellairs & C.R. Jenkin (1960). In A.J. Marshall (ed.) *Biology and comparative physiology of birds*, vol. 1. Academic Press, New York.

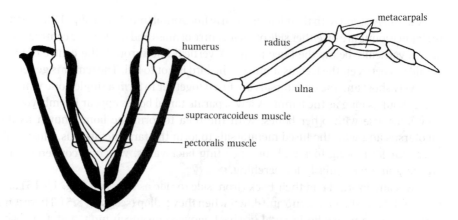

Fig. 17.3. The wing skeleton and pectoral girdle of a bird, seen from in front, with the principal wing muscles shown diagrammatically. From J.R. Hinchliffe & D.R. Johnson (1980). *The development of the vertebrate limb*. Clarendon, Oxford.

it keeps the humerus horizontal and so keeps the animal's chest off the ground. The coracoid, running between the sternum and the shoulder joint, prevents the pectoralis from pulling the shoulders in towards the mid-line. In flying birds, the pectoralis is enormous and the sternum is greatly enlarged, with a deep keel, to provide much of its origin. The coracoids, which prevent the pectoralis from pulling sternum and shoulders towards each other, have become stout pillars of bone. The scapulae are long, but attached by muscle to the ribs as in amphibians and reptiles. The dermal bones of the girdle form the furcula (wishbone).

The pectoralis muscle, which is responsible for the downstroke of the wing, runs directly from the sternum and furcula to the humerus. The supracoracoideus muscle, which is responsible for the upstroke, works less directly (Fig. 11.3). It also originates on the sternum but its tendon runs over a notch in the pectoral girdle which serves as a pulley, so that though the muscle lies ventral to the wings it serves to raise them. The surface of the notch is covered by cartilage, and the sliding tendon is presumably lubricated in the same way as joints between bones. In many birds that fly strongly the pectoralis and supracoracoideus muscles together make up 25% or more of the mass of the body.

Archaeopteryx had a small sternum, without a keel, and some scientists have doubted whether its wing muscles could have been large enough for flight. However, its wings seem to have been as large as those of modern flying birds with similar-sized bodies (for example magpies, *Pica pica*).

With their fore-limbs so highly modified as wings, birds are necessarily bipedal. They walk, run or hop, placing the feet under the body like mammals and dinosaurs, not far out on either side like newts (Fig. 15.15) and lizards. Bipedal dinosaurs seem to have depended on their long tails for balance (Fig. 16.11). *Archaeopteryx* had a dinosaur-like tail (Fig. 17.1) but the flesh-and-bone parts of the tails of modern birds are very short (Fig. 17.2). Some modern birds have long tails but most of the length is made up merely of feathers, too light to give much help with balance. If birds stood with their feet directly below their hip joints, they would be liable to fall on their faces.

They actually stand with their femurs nearly horizontal (much as in Fig. 17.2) so that their knee joints are on either side of their centre of mass and the feet are centred more or less vertically below the knee joints. A vertical line through the centre of mass would fall between the feet, as it must if the bird is not to fall. The femur is generally relatively short and the metatarsals are fused together to form a single bone which is often about as long as the femur. A few separate tarsal bones appear in embryos but they later fuse with other bones, with the tibia to form the bone known as the tibiotarsus and with the fused metatarsals to form the tarsometatarsus. Most birds have four fairly long toes with one pointing backwards, which is convenient for grasping and particularly for perching.

Newts and lizards bend their backs from side to side as they run (Fig. 15.15) and mammals bend their backs up and down when they gallop (section 18.5). There is no obvious way in which birds could use back movements in running, and if the back were flexible it would have to be kept stiff by muscle tension. Quite a long section of the vertebral column has its vertebrae fused to each other and fixed rigidly to the pelvic girdle, and needs no muscle to stiffen it. There is consequently very little muscle between the ilia and the skin. The shape of the pelvic girdle is very peculiar, with hollows on either side of the vertebral column ventral to the ilia, which house the kidneys. *Archaeopteryx* had a much more dinosaur-like pelvic girdle.

The ostrich, emus, rheas, cassowaries and kiwis are often referred to together as the ratites, though they may not be closely related to each other. They have tiny wings and cannot fly, but the bones of their hands are fused together in the same peculiar way as in flying birds, which suggests that they have evolved from flying birds. Their feathers are fluffy, like those of chickens. Their pelvic girdles are rather like that of *Archaeopteryx*. Many of their skull bones are distinct, separated from each other by sutures, instead of being fused together as in the skulls of adult flying birds.

The ratites seem to have evolved by neoteny, by retaining juvenile features in the adults. The chicks of many ground-nesting birds (including domestic fowl) have strong legs and can run well, while their wings are still rudimentary. They have fluffy feathers, and they have sutures between their skull bones instead of having the bones fused together as in most adult birds. *Archaeopteryx*-like pelvic girdles are found in bird embryos. In all these respects, an adult ostrich is like an overgrown chicken. Similarly, *Oikopleura* seems to have evolved by neoteny from more typical urochordates (section 12.1).

The modern birds (subclass Neornithes) are divided into many orders, among which the Passeriformes is much the largest. It consists mainly of small and medium-sized birds including the crows, tits, warblers, thrushes, finches and many others.

17.2.   Feathers and body temperature

The primary and secondary feathers of bird wings, and their tail feathers, have already been mentioned. Smaller feathers cover most of the rest of the body, giving it a streamlined shape and serving as heat insulation. The feet, however, are generally covered by horny scales like those of reptiles (section 16.3).

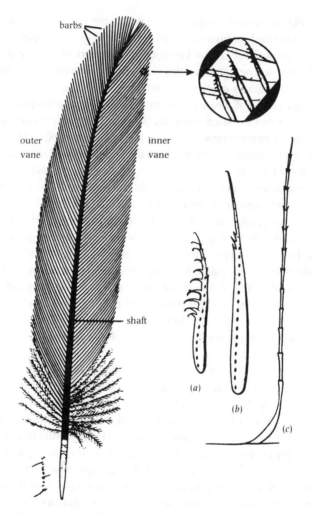

Fig. 17.4. One of the smaller feathers from the wing of a domestic fowl. The enlarged circular inset shows how the barbules interlock. (*a*) A typical distal barbule, (*b*) a typical proximal barbule, and (*c*) a barbule from the fluffy region at the base of the feather. From M.E. Rawles (1960). In *Biology and comparative physiology of birds* (ed. A.J. Marshall), pp. 189–240. Academic Press, New York.

Feathers consist of β-keratin. The base of the shaft (Fig. 17.4) is housed in a follicle in the skin. The barbs are branches on either side of the shaft and they in turn have branches, the barbules, which are of two types. Those on the distal side of each barb have hooks (Fig. 17.4*a*) while those on the proximal side do not (Fig. 17.4*b*). The hooks of the distal barbules catch on the proximal barbules of the next barb (Fig. 17.4, circular inset) so that the barbs form a coherent lamina that is light but not fragile. Rough treatment is less likely to tear it, than to separate some of the barbs. Separated barbs can be interlocked again simply by re-arranging them side by side; this can be done by pulling a feather between the fingers, and birds do it with their bills when they preen.

Many feathers have a fluffy region near the base, where the barbules are all of the type shown in Fig. 17.4c. There are no arrangements for interlocking, and the barbs do not cohere. Chicks and ratites are covered with down feathers in which none of the barbs interlock and similar feathers are present (usually in concealed positions) on adult flying birds.

Barbs which get separated can be interlocked again, but a part which is broken off a feather cannot be replaced. Such damage can only be made good when the feathers are moulted and replaced by entirely new ones, which usually happens at least once a year. Feathers are formed by papillae at the bases of the follicles and as each new one is formed by the same papilla as its predecessor it does not appear until its predecessor has been shed. Moulting generally proceeds gradually so that the bird never has very much less than its full complement of feathers.

Feathers provide heat insulation in the same way as human clothing, by trapping the layer of air next to the body surface and keeping it stationary, preventing convection currents. Little heat can pass across this layer except by conduction, and air is a very poor conductor of heat. The thermal conductivity (equation 14.2) of stationary air is 0.025 W m^{-1} K^{-1}, which is much lower than that of water (0.6 W m^{-1} K^{-1}) and animal tissues. The conductivity of plumage is not quite so low, mainly because heat gets conducted along the feathers as well as through the air. Measurements of rates of conduction through plumage (still in place on pieces of bird skin) gave conductivities of about 0.07 W m^{-1} K^{-1}.

I discussed the advantages of a fairly high body temperature in section 16.4 and explained how many reptiles achieve it, by using solar radiation. They are described as ectotherms, indicating that they depend on a source of heat outside the body. Mammals and birds (and also tuna, section 14.5) depend on metabolism to maintain high body temperatures, and are described as endotherms. Most birds maintain body temperatures of 40–43°C and mammals 36–40°C, and keep their temperatures remarkably constant day and night: a very small deviation of human body temperature from normal is regarded as a symptom of illness. The insulating layer of feathers or fur enables birds and mammals to maintain their body temperatures, with lower metabolic rates than would be needed if they were naked.

The metabolic rate that is needed depends on the temperature of the environment, as an experiment with small birds shows (Fig. 17.5). The birds were resting, and since they had not recently fed they cannot have been using much energy for digestion. Each perched in a five-litre container in a constant temperature cabinet. Air was passed slowly through the cabinet and the rate of oxygen consumption was calculated from its rate of flow and the fall in its oxygen content as it passed through. At temperatures below 18°C the rate of consumption of oxygen was more or less proportional to the difference between body temperature (about 40°C) and the temperature of the surroundings. This is as might be expected since rates of loss of heat are in general proportional to temperature differences. At temperatures between 18°C and 33°C, however, the metabolic rate remained constant at what is presumably the rate required for purposes other than activity or temperature maintenance. This is known as the basal metabolic rate. If it is sufficient to maintain body temperature at 18°C, it is more than sufficient at higher temperatures, at which the bird must make adjustments to avoid overheating. The feathers (which are fluffed out

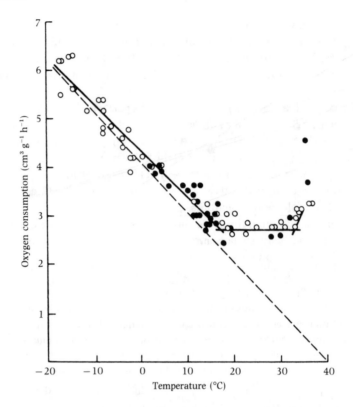

Fig. 17.5. A graph of oxygen consumption against environmental temperature for 22 cardinals (*Richmondena cardinalis*), weighing about 40 g each. ● Birds kept indoors and ○ birds kept out of doors in cold (− 10°C to + 5°C) weather. From W.R. Dawson (1958). *Physiol. Zool.* 31, 37–48. Copyright © University of Chicago Press, 1958.

at low temperatures) are held flatter against the body so that the heat-insulating layer is thinner, and more of the blood is directed to the uninsulated parts of the legs. At very high temperatures, above 33°C, the metabolic rate rises again because the bird's body temperature rises and because it starts panting. Though panting involves muscular activity and so heat production, it results in a net loss of heat because it increases evaporation from the respiratory tract.

Fig. 17.6 shows that the basal metabolic rates of mammals, and of birds other than Passeriformes, are about four times the resting metabolic rates of lizards of equal body mass. The Passeriformes have even higher metabolic rates, about eight times those of lizards of equal mass.

The lizard metabolic rates were measured at a mammal-like body temperature of 37°C. (This would be a normal mid-day temperature for many lizards; see section 16.4.) Therefore, the higher rates for birds and mammals are not simply due to chemical reactions going faster at higher temperatures. Some more fundamental metabolic change seems to have occurred when mammals and birds evolved from reptiles. The difference is between mammals and birds on the one hand, and all the lower vertebrates on the other. At the same body temperature, fishes and lizards of equal mass use oxygen at about the same rate.

When animals run or fly, most of their metabolism is in their muscles, but basal

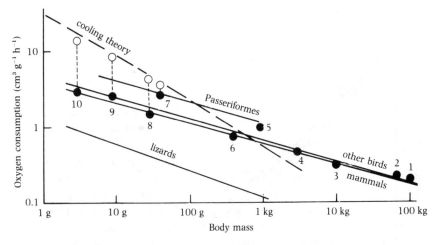

Fig. 17.6. Graphs on logarithmic co-ordinates of oxygen consumption per unit body mass, against body mass. The following regression lines are shown: resting metabolic rates at 37°C of 13 species of lizard; and basal metabolic rates of 35 species of Passeriformes, of 58 other bird species and of 13 species of eutherian mammal. The broken line shows expected oxygen consumptions calculated from heat exchange theory, for birds and mammals with skin temperatures of 37°C in surroundings at 10°C. ●, Basal metabolic rates of a small sample of species and ○, metabolic rates at 10°C. 1, Ostrich (*Struthio*); 2, human; 3, vulture (*Vultur*); 4, domestic cat (*Felis*); 5, raven (*Corvus*); 6, rat (*Rattus*); 7, cardinal (*Richmondena*); 8, deer mouse (*Peromyscus*); 9, harvest mouse (*Reithrodontomys*); 10, various hummingbirds. The raven and cardinal are Passeriformes; other birds listed here are not. The data have been collected from many sources, which are listed in R. McN. Alexander (1975). *The Chordates*, Cambridge University Press, where this figure was first published.

metabolism happens mainly in other tissues, notably the liver, kidneys, brain and heart. Measurements of differences of oxygen content, between arterial blood going to these organs in humans and venous blood coming from them, show that together they account for 70% of basal metabolism. In a comparison of mammals and reptiles of equal body mass it was found that these organs were larger in the mammals and had more densely packed mitochondria. As a result, the mammals had about four times as many mitochondria in these organs as did the reptiles. This agrees well with their having four times the metabolic rate.

Here is a simple argument about bird and mammal heat balance that ignores loss of heat by evaporation. To keep the body temperature constant, the metabolic rate must match the rate at which heat is lost by conduction through the feathers or fur. Compare animals of different sizes with equal body temperatures, at the same environmental temperature. If the surface area of the body is A and the thickness of the insulation (fur or feathers) is d, the rate of loss of heat is proportional to A/d (equation 14.3). If the animals are geometrically similar, A is proportional to (body mass)$^{\frac{2}{3}}$ and d to (body mass)$^{\frac{1}{3}}$, so the rate of loss of heat is proportional to (body mass)$^{\frac{1}{3}}$. (This is essentially the same argument as I used when discussing dinosaurs, in section 16.5.) However, the basal metabolic rates of birds and mammals are about proportional to (body mass)$^{0.7}$, like those of other groups of animals (Fig. 4.9). The metabolic rate needed to maintain constant body temperature is proportional to (body mass)$^{\frac{1}{3}}$ but basal metabolic rates are proportional to (body mass)$^{0.7}$. This suggests that small mammals and birds should cool down and big ones should overheat.

The line labelled 'cooling theory' in Fig. 17.6 was obtained by a more sophisticated version of the same argument, that considered not only how heat was conducted through the feathers or fur, but also how it was lost by convection or radiation from the outer surface. It gives estimates of metabolic rates needed to maintain a body temperature of 37°C in an environment at 10°C. It assumes that the animals have fur as thick, in proportion to the diameters of their bodies, as very furry mammals such as shrews (*Sorex*) and Arctic foxes (*Alopex*). The hollow symbols show the measured metabolic rates of a few species, in environments at 10°C. They lie close to the theoretical line, showing that the theory is realistic. Notice that the theory confirms what a crude argument has already told us. Small endotherms would be unable to maintain their high body temperatures at 10°C, if they did not increase their metabolic rates above their basal levels. Large furry endotherms would overheat, even in such a cool environment. (Notice that the theoretical line falls below the line showing basal rates for birds and mammals, for body masses over 1 kg.) There are two consequences. First, there must be a minimum size, below which it is not feasible to maintain bird- and mammal-like body temperatures by metabolism. A 3 g hummingbird at 10°C consumes oxygen at about four times the basal rate, and a 1 g bird or mammal at the same temperature would have to multiply its basal rate by an even larger factor. The factors would be lower in warmer environments, but even in the tropics there are no birds or mammals lighter than the smallest hummingbirds and shrews, about 2 g. Secondly, large mammals are less furry than small ones and the largest (for example, elephants) have no fur at all. The theoretical line assumed fur as thick, in proportion to body diameter, as in shrews. Only very small mammals, and medium-sized ones from exceptionally cold climates (such as Arctic foxes) have fur as thick as that.

Some birds and mammals conserve energy by allowing the body temperature to fall during the part of the day when they are inactive. As they cool, they become torpid. Hummingbirds in captivity often become torpid at night, and torpid hummingbirds have sometimes been found in nature. Many bats which feed at night become torpid by day.

Adult birds and mammals may be large enough to be able to maintain their own body temperatures, but their embryos are not. Bird embryos are kept close to the adult body temperature by incubation of the eggs, and mammal embryos by being retained in the mother's body (and later in the pouch in marsupials). Even if it were feasible for isolated embryos to maintain a high temperature by their own metabolism it would be wasteful of energy; just as less energy per unit body mass is needed to maintain the body temperature of a large animal than of a small one, less is needed to maintain the temperature of embryos being incubated or in the uterus, than if they were separated from their parent.

I argued in section 9.4 that desert beetles inevitably lose a lot of water vapour to the air from which they obtain their oxygen. Here is a similar argument which shows that birds and mammals lose both water vapour and heat in their breath.

Birds breathe in air containing 210 cm^3 oxygen per litre and breathe out air containing, typically, 140 cm³ oxygen per litre. One litre of air breathed by a bird thus loses about 70 cm³ oxygen, which is used in metabolism yielding about 1400 J. If it left the body at 40°C, saturated with water vapour, it would carry with it 60 mg water vapour. If the air had been cool (say around 10°C) and fairly dry when it was breathed in, nearly all of this water would have been obtained by evaporation within the bird. The latent heat of vaporization of water is about 2.5 J mg⁻¹, so the heat lost in this way would be about $60 \times 2.5 = 150$ J, or around 10% of the metabolic heat production. If the metabolic rate had to be increased accordingly to maintain body temperature, this would be a substantial loss of energy. The loss of water might also be serious, in dry environments.

In fact, neither birds nor mammals breathe air out at body temperature. This has been established by fitting microbead thermistors in their nostrils. These temperature-sensitive devices respond so quickly to changes of temperature that the temperatures of the air entering and leaving the nostrils as the animal breathes in and out are registered separately. Six species of bird breathing in air at 12°C were found to breathe it out again at 14–21°C, and only one (the domestic duck) breathed out at a higher temperature. A litre of air saturated with water at 14–21°C contains only 14–21 mg water vapour, which is much less than the 60 mg at 40°C, so the saving of water and heat is considerable. At higher air temperatures the expired air is warmer and the savings are less dramatic. Similar observations have been made on mammals.

The air must reach body temperature in the lungs but it is cooled again in the nasal cavities. Both in birds and in mammals these cavities contain turbinals, which are thin scrolls of bone or cartilage covered by epithelium. The spaces between the turbinals are narrow, and the total area of the epithelium is large. Incoming air cools the turbinals and is itself warmed. Outgoing air encounters the cooled turbinals, and is cooled again. The principle is the same as that of the counter-current heat exchanger in tuna muscles (section 14.5), though flow in the opposite directions is not simultaneous.

17.3. Flight

Hummingbirds hover in the same way as moths (Fig. 9.7) but most birds make quite different movements when they fly. Fig. 17.7 shows at the top the movements of a pigeon flying fairly slowly. The wings beat forward and down, back and up. They are spread more widely for the downstroke than for the upstroke. The primary feathers are bent during the downstroke, showing that large upward forces are acting on them but there is no obvious bending during the upstroke. It seems that in slow flight like this the downstroke is mainly responsible for producing the forces that support the bird's weight.

The pigeons in Fig. 17.7 had electrodes in their wing muscles. In some of the experiments, long (18 m) wires trailed behind the birds, connecting the electrodes to recording equipment. In others, the birds carried miniature radio transmitters that transmitted the signals from the electrodes. The figure shows that the pectoralis muscles were active at the end of the upstroke and for some of the downstroke: they

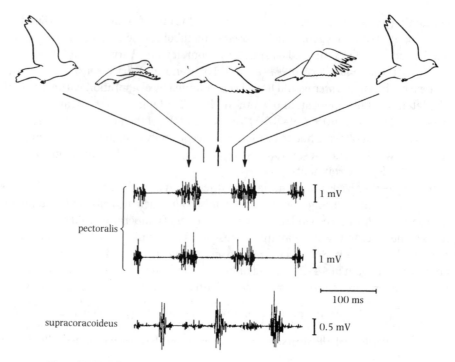

Fig. 17.7. Wing movements of a flying pigeon (*Columba livia*) and electromyographic records from two parts of the pectoralis muscle and the supracoracoideus muscle. From K.P. Dial, S.R. Kaplan, G.E. Goslow & F.A. Jenkins (1988). *J. exp. Biol.* 134, 1–16.

stopped the upstroke and pulled the wings down. The supracoracoideus became active late in the downstroke, stopping the downstroke and accelerating the wings upward. The pectoralis has to produce the large forces needed to support the bird's weight but the supracoracoideus has only to decelerate and re-accelerate the wing (and may be helped in that task by aerodynamic lift on the wing). Accordingly, the pectoralis muscles are much larger than the supracoracoideus muscles.

Flying birds (and aeroplanes) support their weight by pushing air downwards. The air movements involved have been made visible by flying birds through clouds of small soap bubbles. Ordinary soap bubbles are denser than air and sink slowly, but these ones were filled with the low-density gas helium so that the complete bubbles had almost exactly the same density as the air. Thus the movements of the bubbles showed how the air was moving. They were recorded by taking stereoscopic multiple-flash photographs. Each photograph had several images of each bubble, showing where it was when each of the flashes was fired.

Fig. 17.8(*a*) shows the pattern of air movement that was seen when a pigeon flew slowly (at 2.4 m s⁻¹) through the bubbles. Each downstroke of the wings drove a puff of air downwards. This moving air set a surrounding ring of air spinning, forming a vortex ring. Such rings are always formed when a discrete puff of air moves through still air: the smoke rings blown by smokers are vortex rings made visible by the smoke trapped inside them. The upstrokes of the wings produced no obvious air movements.

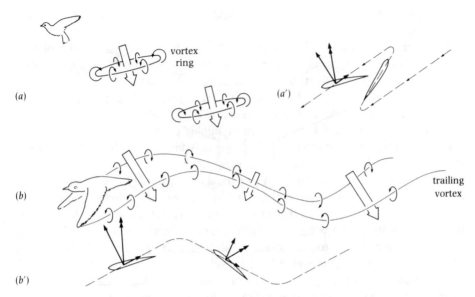

Fig. 17.8. Air movements in the wakes of birds flying (*a*) slowly and (*b*) faster, observed in experiments with soap bubbles; (*a'*) and (*b'*) are diagrams showing the paths through the air of sections of the wings, and the aerodynamic forces that act on them.

This demonstration that air is driven downwards only by the downstroke is consistent with the observation that it is only in the downstroke that the primary feathers are noticeably bent.

Fig. 17.8(*b*) shows a different pattern of air movements that was seen when a kestrel (*Falco tinnunculus*) flew at about 8 m s^{-1}. Instead of discrete puffs the bird leaves behind a continuous band of downward-moving air: the upstroke drives air downwards, producing lift, as well as the downstroke. Vortices form as continuous lines on either side of the downward-moving air, just as wingtip vortices (that sometimes become visible as vapour trails) are formed by aeroplanes. These vortices undulate up and down with the wing beat. They are closer together in the upstroke, when the wings are less widely spread and the band of air that is being driven downwards is narrower than in the downstroke. The smaller quantity of air driven by the upstroke is indicated by a smaller arrow than the ones used for the downstroke.

The wings must produce mainly an upward force, to support the bird's weight, but a small forward thrust is also needed to counteract the drag on the body. Fig. 17.8(*a'*) and (*b'*) show how these forces are produced. The evidence from air movements is interpreted in terms of more conventional aerodynamics.

The pigeon flying slowly apparently holds its wings at an angle of attack during the downstroke (Fig. 17.8*a'*). The resultant force acts upwards and slightly forwards, as required. In the upstroke, the angle of attack is reduced to zero and the aerodynamic forces on the wing are very small. The kestrel flying faster holds its wings at an angle of attack for the upstroke as well as the downstroke (Fig. 17.8*b'*) but the forces must be smaller than in the downstroke: otherwise the backward component of the resultant force produced by the upstroke would cancel out the thrust produced by the downstroke.

The pigeons would only fly slowly in the experiments and the kestrel would only fly fast, but noctule bats (*Nyctalus*) flew at a wider range of speeds. At 3 m s^{-1} they produced air movements like those shown in Fig. 17.8(*a*) but at 7–9 m s^{-1} they made the air move as in Fig. 17.8(*b*). Just as we walk to go slowly and run to go fast, the bats used different gaits at different speeds. It seems likely that birds do the same but so far no bird has demonstrated both gaits in the bubble experiments.

Gliding is an important mode of flight, especially for large birds. A bird gliding in still air inevitably loses height but it can remain airborne or even gain height if the air is rising. Gulls can often be seen soaring on the windy sides of hills, where the wind is deflected upwards by the slope of the ground. Albatrosses and other sea birds use the sloping sides of large waves in the same way and can remain airborne indefinitely, hardly ever beating their wings. Other birds soar in thermals over ground heated by the sun. A patch of hot ground warms the air immediately over it, which rises as a thermal. Vultures and storks gain height by circling in the rising air of a thermal, then glide away (losing height) to another thermal where they circle and rise again. In this way they can cover large distances. Professor Colin Pennycuick, flying a motorized glider in East Africa, once followed a vulture (*Gyps rüppellii*) for a distance of 75 km, from its feeding ground to its nest on a cliff. It covered the distance in 96 minutes, pausing to circle in five thermals but never beating its wings. Thermals often have cumulus clouds at the top, which may help birds to find them.

Pennycuick used the wind tunnel shown in Fig. 17.9 to investigate the gliding ability of pigeons and a fruit bat. A large fan blew a jet of air of almost one metre diameter, at speeds up to 20 m s^{-1}. (A powerful motor was needed.) The 'honeycomb' and the contraction were designed to make the jet as uniform as possible: similar devices were used in the water tunnel shown in Fig. 6.19. The tunnel could be tilted to any desired angle θ. The animals were trained to fly in the jet of air so as to remain stationary near the mouth of the tunnel. (The pigeons were persuaded to do this by offering maple peas on a long spoon that they could reach only when flying in the required position.) If the speed and angle of the jet allowed it, the animals saved energy by gliding; otherwise they had to flap their wings. An animal gliding so as to remain stationary in a jet of air of speed u and angle θ has the same motion (relative to the air) as one gliding at the same speed and angle in stationary air. In the latter case it would lose height at a rate $u \sin \theta$: this is called the sinking speed.

Fig. 17.9(*a*) shows how the lift and drag on the gliding animal balance its weight. The lift acts at right angles to the jet of air, at an angle θ to the vertical, and the drag at angle θ to the horizontal. If the weight is mg the lift must be $mg \cos\theta$ and the drag $mg \sin\theta$, so $\cot\theta$ is the ratio of lift to drag. To glide at shallow angles, an animal must be able to adjust its position so as to get the required lift with very little drag.

In his experiments with the wind tunnel, Pennycuick kept the speed of the jet constant while adjusting its angle, to find the shallowest possible gliding angle for each speed. He calculated the sinking speed and plotted it as in Fig. 17.9(*b*). The maximum lift that the wings can give is proportional to the square of the air speed so below a certain speed (8 m s^{-1} for the pigeon) the animal cannot glide at all. The smallest sinking speeds are obtained at a higher air speed (14 m s^{-1} for the pigeon). The pigeon glides rather badly, with a minimum sinking speed of more than 2 m s^{-1},

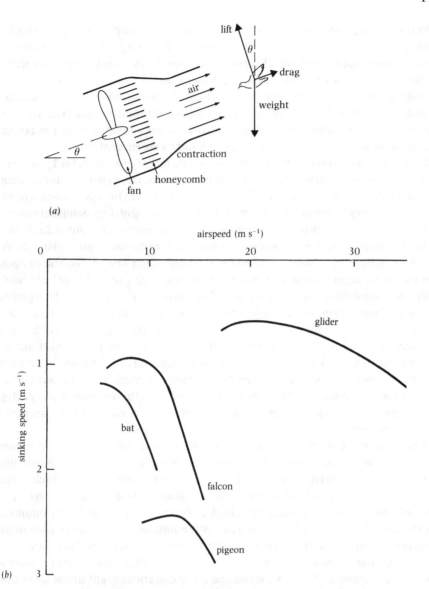

Fig. 17.9.(*a*) A diagram of a wind tunnel used to investigate the gliding performance of birds and bats. The forces acting on the bird are also shown. (*b*) Results of gliding experiments in wind tunnels. Sinking speed (*u* sin *θ*) is plotted against airspeed (*u*). The lines show the performance of a glider aircraft, a pigeon (*Columba livia*, wing span 0.6 m), a falcon (*Falco jugger*, 1.1 m) and a large bat (*Rousettus aegyptiacus*, 0.5 m). Data from V.A. Tucker & G.C. Parrott (970). *J. exp. Biol.* 52, 345–368; and C.J. Pennycuick (1971). *J. exp. Biol.* 55, 833–845.

but the falcon and bat do better, about 1 m s^{-1}. Even lower sinking speeds (typically 0.5 m s^{-1}) are obtained with glider aircraft and models but the comparison is unfair: a gliding bird is burdened by the weight of its flight muscles but a glider carries no engine.

The birds that can glide best (at the shallowest angles) have long, narrow (high aspect ratio) wings. Such wings give the highest possible lift:drag ratio, as explained for fish tails in section 13.3. However, excessively long wings might be awkward, especially when taking off from the ground or from trees. Albatrosses have long, narrow wings but vultures have shorter, broader ones.

The data on gliding can be used to estimate the mechanical power that should be needed for flapping flight. The energy needed to overcome the drag on a gliding bird is supplied by the potential energy that it loses, as it loses height. Thus the power expended against drag is the weight (mg) multiplied by the sinking speed ($u \sin\theta$), and the power per unit mass is $gu \sin\theta$. The animals in Fig. 17.9(b) glided with sinking speeds of $1-3$ m s^{-1}, so the power per unit mass can be calculated by multiplying these by g (about 10 m s^{-2}): it is $10-30$ watts per kilogram. The flight muscles must presumably supply at least this much power, in flapping flight. Physiological experiments with other muscles usually show maximum efficiencies of about 0.25, so the metabolic power requirement is about four times the mechanical power, $40-120$ W kg^{-1}. Metabolism involving 1 cm^3 oxygen releases 20 J energy, so the oxygen requirement for flight should be $2-6$ cm^3 oxygen kg^{-1} s^{-1}, or $7-22$ cm^3 g^{-1} h^{-1}.

Rates of oxygen consumption in flapping flight have been measured for flying birds in horizontal wind tunnels, using a flexible tube and a face mask to collect the air that they breathe out (as in the experiments on running lizards, Fig. 16.7b, but using a better-streamlined mask). The measured rates of oxygen consumption were generally between 10 and 25 cm^3 g^{-1} h^{-1}, in good agreement with the calculation from gliding. In other experiments, the metabolic rates of hummingbirds have been measured by methods like that used for hovering moths (section 9.2). A 3 g hummingbird used oxygen at the rate of 42 cm^3 g^{-1} h^{-1}, close to the rate measured for the similar-sized moth *Manduca*.

These metabolic rates are very high. To make fair comparisons, we must compare animals of similar mass, because larger animals tend to have lower metabolic rates per unit body mass than related smaller animals (Fig. 4.9b). A black duck (*Anas rubripes*) making short flights (not in a wind tunnel) used 14 cm^3 oxygen g^{-1} h^{-1}, but lizards of similar mass (1 kg) running at high body temperatures used a maximum of only 0.5 cm^3 g^{-1} h^{-1} (Fig. 16.7). Even running mammals cannot use oxygen quite as fast as flying birds of equal mass. One kilogram genets (*Genetta*) and rat kangaroos (*Bettongia*), running on a treadmill at the highest speeds they could sustain, used 7 and 11 cm^3 oxygen g^{-1} h^{-1}. However, bats use oxygen in flight about as fast as similar-sized birds.

Birds generally use oxygen ten or more times as fast when they fly, as when they are resting. The resting metabolic rate is enough to maintain body temperature, so birds might be expected to overheat on long flights. However, the air moving past the flying bird must help to cool it, and spreading the wings exposes the flanks which are rather sparsely covered with feathers. Birds do not sweat, and heat loss by evaporation of water is much less important than in mammals. In an experiment in which budgerigars flew in a wind tunnel at $20°$C and 50% relative humidity, evaporation (mainly in the breath) accounted for only 15% of heat loss. If birds were more dependent on heat loss by evaporation they would have to drink more often on long flights, and migrations across broad deserts or seas might not be practicable.

Many small birds migrate across the Sahara desert and have to fly about 2000 km with no opportunity to feed. They set out with very large stores of fat, which are used on the journey. For example, yellow wagtails (*Motacilla flava*) have a mass of 24 g, 7.4 g of it fat, when they leave Nigeria. They arrive north of the Sahara with a body mass of only 15 g, having lost nearly all of the fat and some water. Oxidation of the fat yields an approximately equal mass of water, which helps to compensate for the evaporative losses. (Water produced by metabolism has already been discussed, in section 16.3.)

We will estimate how far a 24 g bird should be able to fly, fuelled by 7.4 g fat. The heat of combustion of fat is 40 MJ kg^{-1}, so the energy available is 0.30 MJ. Metabolism using 1 cm^3 oxygen releases 20 J, so 15 000 cm^3 oxygen would be needed to oxidize the fat, 630 cm^3 per gram of initial body mass. Rates of oxygen consumption of flying wagtails have not been measured so we will have to use data for 35 g budgerigars (*Melanopsittacus*). In experiments in a wind tunnel, these flew furthest for given oxygen consumption at a speed of 40 km h^{-1} (11 m s^{-1}). At this speed they used 24 cm^3 oxygen g^{-1} h^{-1}, so if they started with the same proportion of fat in their bodies as the wagtails they could have flown for 630/24 = 26 hours, travelling 26 × 40 ≈ 1000 km. This is probably an underestimate, because I have taken no account of their getting lighter (so needing less energy for flight) as they used up the fat. Correcting for this would increase our estimate of the range only to about 1200 km so it seems doubtful whether the fat would be sufficient fuel for a budgerigar crossing the Sahara, unless it were helped by a following wind.

17.4. Respiration

Though birds take up oxygen so rapidly their lungs, as distinct from the air sacs connected to them, are relatively small. The lungs proper contain the respiratory surfaces, but change volume relatively little in breathing. The air sacs are much larger and are compressed and expanded greatly as the bird breathes, but their thin walls have a poor blood supply and it can be presumed that little of the oxygen uptake occurs in them. They fill a large part of the body cavity and are connected with the cavities in some of the principal leg and wing bones, which are filled only partly with marrow and largely with air.

I explained the advantages of tubular skeletons in section 8.2. A tube is stronger and stiffer in bending than a solid rod of the same length, made of the same quantity of material. A wide, thin-walled tube can be made lighter than a more slender, relatively thick-walled tube, so long as it is not so very thin-walled as to be liable to buckle like a bent drinking straw. Most of the long bones of tetrapods are hollow, but this saves only a little weight in most cases because the cavities are filled with marrow. The air-filled bones of birds can be much lighter than marrow-filled bones of equal strength.

The lungs of birds contain a very complicated system of air passages (Fig. 17.10). The trachea divides in the usual way into two bronchi, one to each lung. Each bronchus gives rise to two groups of branches, the dorsobronchi and ventrobronchi, which are joined by very large numbers of fine parallel tubes, the parabronchi. There are two groups of air sacs which are connected to the anterior and posterior ends of the lungs.

When the bird breathes it moves its sternum up and down, mainly at the posterior

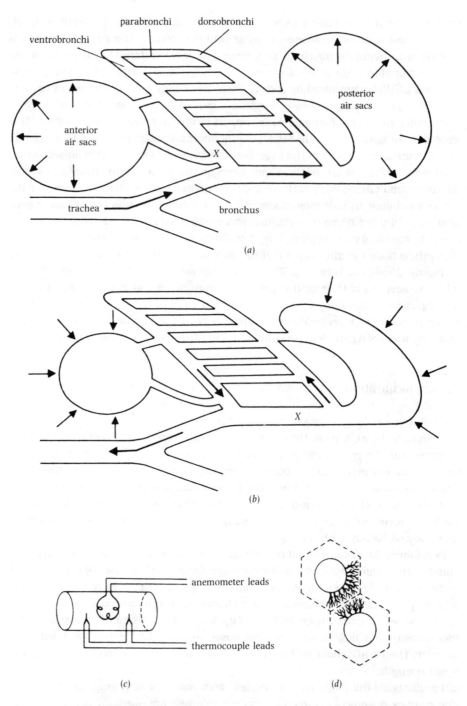

Fig. 17.10.(*a*), (*b*) Diagrams showing the connections between the parts of a bird lung and its air sacs, and the directions of flow determined as described in the text. (*a*) represents inspiration and (*b*) expiration. *X* indicates that there is little or no flow of air. (*c*) A flowmeter used in some of the experiments. (*d*) A diagrammatic section through two parabronchi, showing the air capillaries radiating from them.

end. This compresses and expands the body cavity, pumping air in and out of the air sacs. It is not obvious from anatomy what path the air takes through the lungs, as it travels to and from the air sacs. Various methods have been used in attempts to find out.

Some of the most revealing experiments have used the device shown in Fig. 17.10(c). This is a short tube containing a pair of thermocouples and a hot-wire anemometer. The anemometer is simply an electrically heated filament which is cooled by air flow through the tube. Cooling alters its resistance, so it can be used to give a record of air speed in the tube. It responds in the same way to flow in either direction but, because it heats the air slightly, the air downstream of it is a little warmer than the air upstream of it. The thermocouples therefore indicate the direction of flow. In operations under local anaesthetic, devices like this have been fitted into the dorsobronchi of ducks. Records obtained from them, while the ducks were breathing apparently normally, showed that air flows in the dorsobronchi in the direction indicated in Fig. 17.10 (a,b), both while the bird is breathing in and while it is breathing out.

The dorsobronchi are conveniently placed for these experiments where they can be reached by cutting through skin and muscle without damaging the rest of the respiratory system. Flowmeters of the type described have apparently not, so far, been fitted in other parts of the lungs. A different type of flowmeter has been fixed to the end of a rod and pushed down the tracheae of ducks from an incision in the neck. It was found to be possible to get it into the bronchi and ventrobronchi, as well as the dorsobronchi. These experiments confirmed the information obtained by the other method and provided the additional information on flow which is presented in Fig. 17.10 (a,b).

These experiments show that when ducks (and presumably other birds) breathe in, air is drawn down the bronchi directly to the posterior air sacs. There is little flow at this stage in the ventrobronchi, near their junction with the bronchus, so most of the air going to the anterior air sacs must pass through the dorsobronchi and parabronchi. When the duck is breathing out there is little or no flow in the bronchi, between the dorsobronchi and ventrobronchi. Air from the posterior air sacs must leave mainly through the parabronchi and ventrobronchi while air from the anterior sacs goes directly to the bronchus. It appears that air flows from posterior to anterior through the parabronchi, both while the bird is breathing in and while it is breathing out. There are no valves to prevent it from flowing in the opposite direction, and the one-way flow seems to be due to momentum: air that is already moving in a certain direction tends to continue moving that way, so directions of flow in the lungs depend on the angles at which the various tubes in the lung are connected together. Evidence favouring this idea came from experiments in which the lungs of anaesthetized geese were ventilated artificially. One-way flow as in an actively breathing bird was obtained when the lungs were ventilated with an argon–oxygen mixture (a little denser than air) but not when they were ventilated with a helium–oxygen mixture (much less dense than air). The one-way flow also broke down if the argon was pumped too slowly. Reducing the density of the gas or the rate of flow reduces the momentum of the gas, making the supposed mechanism less effective.

Respiratory exchange between the air and the blood occurs in the parabronchi. Air travels directly to the posterior air sacs but leaves them via the parabronchi. It travels to the anterior air sacs via the parabronchi but leaves them directly. Thus half of the air passes through the parabronchi as the bird breathes in, and the other half as the bird breathes out. The anterior air sacs are filled with air which has already exchanged gases with the blood, but the posterior sacs with largely fresh air (not entirely fresh since the respiratory system is not completely emptied between breaths). This can be inferred from the result of the flowmeter experiments but is also shown by analysis of gas samples from the anterior and posterior sacs.

The respiratory surfaces in the parabronchi are quite unlike those in the lungs of other vertebrates. The walls of a parabronchus are permeated by fine, branching, air-filled tubes, the air capillaries (Fig. 17.10d). These open into the main channel of the parabronchus, and intertwine with the blood capillaries. Air is pumped through the main channel, but movement of gases along the air capillaries presumably depends on diffusion.

The design of the bird lung, with its parabronchi and air capillaries, is very like that of insect wing muscles, with their axial and radial tracheae (Fig. 9.8). A calculation like the one for insect wing muscles in section 9.3 shows that the partial pressure differences needed to drive oxygen and carbon dioxide diffusion in the air capillaries of flying birds would be very small, of the order of 0.002 atm. The very thin layer of tissue between the air capillaries and the blood capillaries is a much more serious barrier to diffusion of gases, than are the air capillaries themselves. A similar conclusion was reached in section 15.3 about diffusion in the (very different) lungs of lungfishes.

Flight requires not only fast uptake of oxygen, but also fast transport of oxygen from lungs to flight muscles. Flying ducks use oxygen about thirty times as fast as can lizards of the same mass (section 17.3). This is made possible partly by having blood which contains a lot of haemoglobin (per unit volume) and so can take up a lot of oxygen, and partly by pumping the blood rapidly round the body. The blood of the lizard in question (*Iguana*), has been found to have a capacity of 8.4 cm³ oxygen per 100 cm³ blood, which is a typical value for reptiles (and also lies in the usual range for fishes). Typical birds (probably including the duck) have capacities of 20–25 cm³ per 100 cm³. The heart of the lizard seems never to beat faster than about 2.5 times per second but that of the duck, immediately after a flight, was found to be making nine beats per second. The duck heart probably also pumps more blood at each beat.

One of the major differences between reptiles and birds is that birds have a completely divided heart. Crocodiles have peculiar hearts with two ventricles that are connected indirectly, but other reptiles resemble amphibians (section 15.4) in having two atria and only one ventricle. Incomplete partitions in the ventricles of reptiles ensure that little mixing of blood occurs there. Deoxygenated blood from the right atrium is directed to lungs and oxygenated blood from the left one to the systemic arteries but in some circumstances (for example, when a crocodile dives) the proportion of blood going to the lungs is reduced and some of the deoxygenated blood goes to the systemics. Birds and mammals have completely separate ventricles so that all the blood from the right atrium and ventricle must go to the lungs and all the blood from the left to the rest of the body. Amphibians and reptiles have two systemic arteries, one on each side of the body, but birds retain only the right systemic artery and mammals only the left.

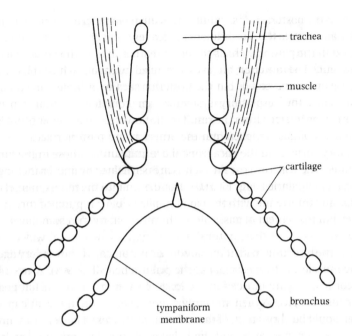

Fig. 17.11. A diagram showing the position and main parts of the syrinx of a typical bird.

While in the egg birds, like reptiles, use the allantois rather than the lungs for respiration. The artery to the allantois branches from the aorta, not from the pulmonary artery. If the blood streams were separate in the embryo, as in the adult, all the blood which returned to the right side of the heart would be pumped through the lungs, and energy would be wasted. In fact the wall between the two atria is incomplete in the embryo, and there are also connections between the systemic and pulmonary arteries. Little blood flows through the lungs until hatching, when they become functional. Thereafter the pathways through the left and right sides of the heart quickly become separate.

17.5. Song

Birds are not the only singing animals. Grasshoppers and cicadas use songs to attract females. Frogs use songs in their mating behaviour and in defending their territories, and some also have alarm calls. Each of these animal groups uses a different mechanism to produce sound. I have chosen to describe only the songs of birds, which include the most elaborate songs and also the ones that are most familiar to those of us who live in temperate climates.

Frogs and mammals have vocal cords, membranes at the top end of the trachea that can be set vibrating to produce sound. Birds have a different structure, the syrinx, at the lower end of the trachea (Fig. 17.11). The walls of the trachea are stiffened by rings of cartilage, and those of the bronchi by incomplete rings. The first few rings of each bronchus have bigger gaps than the rest, so that there is quite a large patch of unstiffened membrane in the wall of each bronchus. This is the tympaniform membrane. It is sometimes exceedingly thin.

The sound-producing mechanism has been demonstrated with excised syrinxes. The syrinx was enclosed in a chamber so that the pressure around it could be varied, and at the same time air could be blown through from the bronchi. So long as the chamber was kept above atmospheric pressure (making the tympaniform membranes cave in) blowing produced a sound. If the pressure was allowed to fall to atmospheric, blowing made the membranes bulge outwards and produced no sound. In intact birds one of the air sacs rests against the membranes. The pressure in the air sac is the same as the pressure in the lungs, which is slightly greater than the pressure in the bronchus when air is being forced out of the lungs. Forceful expulsion of air from the lungs makes the membranes cave in, partly blocking the bronchus.

Two suggestions have been made as to how this may produce sound. The membranes may vibrate at their natural frequency, narrowing and widening the constriction in the bronchus so that the flow of air is not steady, but pulsating. This would produce sound at the frequency of the pulsations. Alternatively, even if the width of the constriction were constant the air downstream of it would form eddies, again making the flow pulsate. The vibrating membrane mechanism would produce tones with harmonics and also partials (sounds at frequencies that are not an integral multiple of the fundamental tone), and could be responsible for many bird calls. The eddy mechanism would produce pure tones (as in some bird songs), tones with harmonics or noise, depending on the speed of flow and the size of the constriction. Both mechanisms may operate. The syrinx has muscles that can distort it, altering the tension in the membranes and so (presumably) the pitch and quality of the sound.

Pressures up to 6000 N m^{-2} (60 cm water) have been recorded in the lungs of cocks while they were crowing. The pressures in normal respiration are far less, about 100 N m^{-2} (1 cm water).

Some birds produce complex songs, modulating both the loudness and the pitch of the notes they produce. They can even tune the two tympaniform membranes to produce sounds of different frequencies: a bird can sing duets with itself.

Birds use numerous calls to co-ordinate the behaviour of members of the same species. Some calls are used to keep flocks together as they move. Others give warning of predators, or are used in courtship. About fifteen distinct calls have been recognized for each of several species of Passeriformes which have been studied carefully. Song is used in establishing and defending territories and often also as part of the sexual display that co-ordinates the breeding behaviour of pairs of birds.

Great tits (*Parus major*) living in a wood near Oxford have been used for a study of the role of song in defence of territory. Great tits feed in flocks in winter, but in spring the males establish territories from which they drive out all other males. They sing regularly in their territories, and each male has a repertoire of up to six types of song. Like most birds, they are monogamous. The male and his mate nest and feed in the territory.

The wood was divided by male Great tits into territories of a little more than 1 ha each. Other Great tits established territories in the fields around the wood, nesting in the hedges. The tits in and around the wood were trapped and coloured rings were put on their legs so that they could be recognized individually. If pairs were removed their territories were taken over within a few hours by pairs from the fields (it seems that territories in the wood are preferred to territories in the fields).

Loudspeakers were placed in some of the territories, and the resident pairs of Great tits were removed. In some of the territories, recordings of a note on a tin whistle were played through the loudspeakers. In others, recordings of Great tit song were played, either a single song played repeatedly or a repertoire of songs played in turn. Nearly all the tin-whistle territories were reoccupied within six hours, but most of the single-song territories remained vacant for 12–18 hours and most of the repertoire ones for 24–30 hours. It seems that Great tit song is rather effective in keeping Great tits out out of a territory, and that a repertoire is better than a single song. The control experiment with the tin whistle showed that intruders were not kept out merely by some general effect of high-pitched musical notes. It is not known why a repertoire is more effective than a single song but it has been suggested that repertoires tend to give intruders an impression of denser population than single songs would do. The larger the intruder's estimate of the resident population the less likely he is, perhaps, to persist in his efforts to establish a territory.

It is not clear what advantage a pair obtain by claiming exclusive use of a territory. They do not seem to be reserving a food supply, for they stop defending the territory when their eggs hatch although their greatest need for food is when they are feeding their nestlings. It has been suggested that a nest is safest in a large territory well away from other nests. Observations of 500 nests over several years showed that predators (mainly weasels, *Mustela nivalis*) raided 23% of Great tit nests which were less than 45 m from their nearest neighbour, but only 11% of more isolated nests.

There is good evidence for some other species of bird that a territory is defended for its stock of food or potential nest sites. For instance an East African sunbird, *Nectarinia reichenowi*, seldom feeds on anything except the nectar of the plant *Leonotis nepetifolia*. The birds in an area of 50 ha were marked with coloured rings on their legs, so that individuals could be identified. It was found that they defended territories of very different sizes (7–2300 m²), excluding all species of sunbird. Though so different in size, most territories contained 1000–2500 *Leonotis* flowers. This seems to be about the number needed to supply the bird's needs.

17.6. Colour and ornament

Some birds have dull colours which seem to serve as camouflage. Others are brightly coloured, and very conspicuous. Often (for instance in peacocks, *Pavo cristatus*, and in many species of duck) the females and juveniles are camouflaged and the adult males are conspicuous. Why have birds evolved in this way?

In many cases, Darwin's theory of sexual selection seems to provide an answer. He suggested that the brightest males, or the ones with the most bizarre plumage, were most attractive to females and so tended to have most offspring. The actual plumage of males was a compromise between sex appeal (favouring conspicuousness) and vulnerability (favouring camouflage).

The best demonstration of sexual selection comes from a species whose males are not bright, but are certainly bizarre. An East African widowbird, *Euplectes progne*, is intermediate in size between the European robin and the (larger) American robin. Females are brown with short tails but males are black with extraordinary tails, about 0.5 m long. Unlike Great tits (section 17.5), they are polygamous. Successful males establish territories and mate with several females each, which build nests on the territory.

In an experiment in Kenya, 36 male widowbirds were trapped at an early stage of the breeding season, when they had on average 1½ nests on their territories. The experimenter cut two thirds off the length off the tail feathers of nine birds and glued the cut pieces to the tails of nine others to give them exceptionally long tails. He cut the tail feathers off another nine birds and glued them back on to check for possible effects of cutting and gluing, and left nine birds intact. He returned all the birds to the wild and counted the nests again 10–14 days later. By then the birds with the elongated tails had acquired on average two new nests but the others had acquired only half to one new nest. It seems that very long tails are attractive to females.

A preference for long tails seems arbitrary, and may have evolved in an arbitrary way. Suppose that genes arose and spread to a significant number of females, that made them prefer long-tailed males. These females would mate with long-tailed males so their offspring would inherit long-tail genes from their fathers and genes for the long-tail preference from their mothers. The two sets of genes would evolve together and mathematical analysis shows that both sets would tend to evolve to extremes: tails would tend to become very long and females would prefer exceedingly long tails. Once the genes were established they would tend to persist because short-tailed mutant males would be unattractive to most females and mutant females that preferred and mated with short-tailed males would tend to have short-tailed, sexually unattractive, sons.

The 'unprofitable prey' hypothesis offers an alternative explanation for bright and bizarre birds. Suppose a hunting predator locates a potential prey animal. It must decide whether to spend time and energy chasing that animal, or to continue searching in the hope of finding prey that is easier to catch. If prey is reasonably plentiful it will catch most, for given time and energy expenditure, by ignoring animals that are hard to catch and concentrating on easier prey. An animal that is hard to catch (and so is unprofitable as prey) may be in little danger, but it will lose time and energy and may be injured if a predator chases it. It will be to its advantage to be quickly and easily distinguishable from profitable prey, if that makes predators more likely to ignore it. Profitable prey tend to be camouflaged, since camouflage reduces the danger that predators will find them. To be as different from them as possible, unprofitable prey should be conspicuous.

It might be thought that profitable prey animals could gain an advantage by evolving bright colours which might make predators mistake them for unprofitable prey. It is doubtful whether they could, since the bright colours would occasionally attract a naïve predator which had not yet learned to distinguish unprofitable from profitable prey.

Juvenile birds are generally easier to catch than adults, so it is generally advantageous for them to be camouflaged. Adults of many species may be unprofitable prey and it might be concluded from the argument so far that both sexes should be conspicuous. However, it is often only the male that is conspicuous. In many such species (for instance, in many ducks) the female cares for the eggs and young. If she were conspicuous she might attract a predator's attention to her vulnerable offspring, so it is probably advantageous for her to be camouflaged. In other species (for instance, in geese) the parents co-operate in caring for their young and tend to have similar plumage.

There is a variant of the unprofitable prey theory which may explain some patches of conspicuous colour. A predator is most likely to succeed if it catches its prey unawares. Hence a prey animal which has noticed an approaching predator may be unprofitable prey, though a similar animal which had not noticed the predator would be profitable prey. It may be to the advantage of a prey animal which has detected a predator, to signal to the predator to this effect. Conspicuous patches on the rump or wings may help to show the predator that the prey, having seen the predator, is turning away and spreading its wings for flight. Most waders have camouflage colours which make them inconspicuous while feeding but some (such as Redshank, *Tringa totanus*) have white rumps and others (such as Knot, *Calidris canuta*) have conspicuous bars on their wings. These bars are hidden until the wings are spread. The same explanation may apply to the white tails of rabbits (*Oryctolagus cuniculus*) and the white rumps of antelopes and of the White-tailed deer (*Odocoileus virginianus*). The alarm calls which some birds give when a predator is near have generally been interpreted as a warning to other members of the species, but may be a means of signalling to the predator that it has been detected.

The red-winged blackbird (*Agelaius phoeniceus*) is common in marshes and fields in North America. The female is speckled brown, and so is inconspicuous, but the adult male is black with brilliant red patches on his shoulders. Up to six females may mate with him, and nest on his territory, but he plays no part in rearing the young. We might try to explain his conspicuousness using either theory, sexual selection or unprofitable prey. However, the red wing patches serve a different (additional?) function in display.

The male sings in his territory, often spreading his wings at the same time. This seems to be his first line of defence against intruders. If another male of the same species trespasses on his territory in spite of his song, he approaches the intruder and spreads his wings widely, displaying the red patches. This visual signal is his second line of defence. The third and final line of defence, which seldom has to be used, is physical attack.

The role of the shoulder patches has been investigated by dyeing them black. Males were trapped in their territories. The shoulder patches of 51 were wiped clean with alcohol and then dyed, but those of another 38 (which served as controls) were merely wiped with alcohol. All the birds were released within 25 minutes of capture and returned quickly to their territories. Twenty-five (49%) of the dyed birds subsequently lost their territories but only three (8%) of the control birds did. This cannot be explained as due to the disturbance of capture and handling since the control birds as well as the dyed ones had suffered this disturbance. It was shown in another experiment that trespassers enter the territories of dyed birds much more often than those of normal birds, and that they seemed to ignore the wing-spreading display of dyed birds. It is concluded that the red patches play a major role in territorial defence. Males which were muted by cutting the nerves to the syrinx succeeded in defending their territories, so song seems less potent than visual display in this particular species.

Dyed males that kept their territories also succeeded in keeping or acquiring mates. It seems that the red patches are not necessary to enable the female to recognize the male's species, nor to make him sexually attractive in the way Darwin supposed. Other observations suggest that the female's most important criterion in choosing a mate is the availability of a good nest site on his territory.

Further reading

Introduction

Campbell, B. & Lack, E. (1985). *A dictionary of birds.* Poyser, Calton.

Charig, A.J., Greenaway, F., Milner, A.C., Walker, C.A. & Whybrown, P.J. (1986). *Archaeopteryx* is not a forgery. *Science* **232**, 622–626.

de Beer, G.R. (1956). The evolution of ratites. *Bull. Brit. Mus. (Nat. Hist.) Zool.* **4**, 59–70.

Farner, D.S. & King, J.R. (eds) (1971–   ). *Avian biology.* Academy Press, New York.

Hoyle, F. & Wickramasinghe, C. (1986). *Archaeopteryx, the primordial bird. A case of fossil forgery.* Davies, Swansea.

King, A.S. & McLelland, J. (1979–   ) *Form and function in birds.* Academic Press, London.

Perrins, C.M. & Middleton, A.L.A. (eds) (1985). *The encyclopaedia of birds.* Allen & Unwin, London.

Thulborn, R.A. (1984). The avian relationship of *Archaeopteryx* and the evolution of birds. *Zool. J. Linn. Soc.* **82**, 119–158.

Feathers and body temperature

Else, P.L. & Hulbert, A.J. (1985). An allometric comparison of the mitochondria of mammalian and reptilian tissues: the implications for the evolution of endothermy. *J. comp. Physiol.* B**156**, 3–11.

Kruger, K., Prinzinger, R. & Schuchmann, K.-L. (1982). Torpor and metabolism in hummingbirds. *Comp. Biochem. Physiol.* **73**A, 679–689.

McNab, B.K. (1983). Energetics, body size and the limits to endothermy. *J. Zool., Lond.* **199**, 1–29.

Schmidt-Nielsen, K., Hainsworth, F.R. & Murrish, D.E. (1970). Counter-current heat exchange in the respiratory passages: effect on water and heat balance. *Resp. Physiol.* **9**, 263–276.

Walsberg, G.E. (1988). Heat flow through avian plumages: the relative importance of conduction, convection and radiation. *J. therm. Biol.* **13**, 89–92.

Flight

Dial, K.P., Kaplan, S.R., Goslow, G.E. & Jenkins, F.A. (1988). A functional analysis of the primary upstroke and downstroke muscles in the domestic pigeon (*Columba livia*) during flight. *J. exp. Biol.* **134**, 1–16.

Jenkins, F.A., Dial, K.P. & Goslow, G.E. (1988). A cineradiographic analysis of bird flight: the wishbone in starlings is a spring. *Science* **241**, 1495–1498.

Pennycuick, C.J. (1983). Thermal soaring compared in three dissimilar tropical bird species. *J. exp. Biol.* **102**, 307–325.

Rayner, J.M.V. (1987). Form and function in avian flight. *Current Ornithology* **5**, 1–66.
Wood, B. (1982). The trans-Saharan spring migration of Yellow wagtails (*Motacilla flava*). *J. Zool., Lond.* **197**, 267–283.

Respiration

Banzett, R.B., Butler, J.P., Nations, C.S., Barnas, G.M., Lehr, J.L. & Jones, J.H. (1987). Inspiratory aerodynamic valving in goose lungs depends on gas density and velocity. *Resp. Physiol.* **70**, 287–300.
Scheid, P. & Piiper, J. (1971). Direct measurement of the pathway of respired gases in duck lungs. *Resp. Physiol.* **11**, 308–314.

Song

Brackenbury, J.H. (1979). Aeroacoustics of the vocal organ of birds. *J. theor. Biol.* **81**, 341–349.
Casey, R.M. & Gaunt, A.S. (1985). Theoretical models of the avian syrinx. *J. theor. Biol.* **116**, 45–64.
Gill, F.B. & Wolf, L.L. (1975). Economics of feeding territoriality in the Golden-winged sunbird. *Ecology* **56**, 333–345.
Greenewalt, C.H. (1968). *Bird song: acoustics and physiology.* Smithsonian Institution Press, Washington.
Krebs, J., Ashcroft, R. & Webber, M. (1978). Song repertoires and territory defence in the great tit. *Nature* **271**, 539–542.
Nowicki, S. (1987). Vocal tract resonances in oscine bird sound production: evidence from bird song in a helium atmosphere. *Nature* **325**, 53–55.

Colour and ornament

Andersson, M. (1982). Female choice selects for extreme tail length in a widowbird. *Nature* **299**, 818–820.
Baker, R.R. & Parker, G.A. (1979). The evolution of bird coloration. *Phil. Trans. R. Soc. Lond.* **B287**, 63–130.
Eckert, C.G. & Weatherhead, P.J. (1987). Owners, floaters and competitive asymmetries among territorial red-winged blackbirds. *Anim. Behav.* **35**, 1317–1323.
Smith, G.D. (1972). The role of the epaulets in the Red-winged blackbird (*Agelaius phoeniceus*) social system. *Behaviour* **41**, 251–268.

18 Mammals and their relatives

Class Reptilia,
  Subclass Synapsida,
    Order Pelycosauria   (extinct)
    Order Therapsida   (extinct)
Class Mammalia,
  Subclass Prototheria
  Subclass Theria,
    Infraclass Metatheria,
      Order Polyprotodonta   (opossums)
      Order Diprotodonta   (kangaroos etc.)
    Infraclass Eutheria,
      Order Insectivora   (insectivores)
      Order Scandentia   (tree shrews)
      Order Chiroptera   (bats)
      Order Primates   (monkey, apes etc.)
      Order Carnivora   (dogs, cats etc.)
      Order Pinnipedia   (seals)
      Order Cetacea   (whales)
      Order Proboscidea   (elephants)
      Order Perissodactyla   (horses etc.)
      Order Artiodactyla   (cattle etc.)
      Order Rodentia   (rodents)
      Order Lagomorpha   (rabbits etc.)

18.1.   From reptiles to mammals

The mammals apparently evolved from a group of reptiles, the class Synapsida. The earliest known synapsid fossils are among the earliest reptiles of any kind, in the later part of the Carboniferous period, but it was in the Permian and Triassic that the subclass flourished. The more primitive synapsids have been found mainly in N. America and are grouped together as the order Pelycosauria. The advanced ones are most plentiful in southern Africa and form the order Therapsida. The therapsids were the dominant land animals immediately before the rise of the dinosaurs.

Pelycosaurs were very like the most primitive (anapsid) reptiles, but had a temporal fenestra in the synapsid position (see Fig. 16.1). They must have looked like lizards, and the structure of the hip and shoulder joints indicates that they stood like lizards and newts with their feet well out on either side of the body (see Fig. 15.15).

458

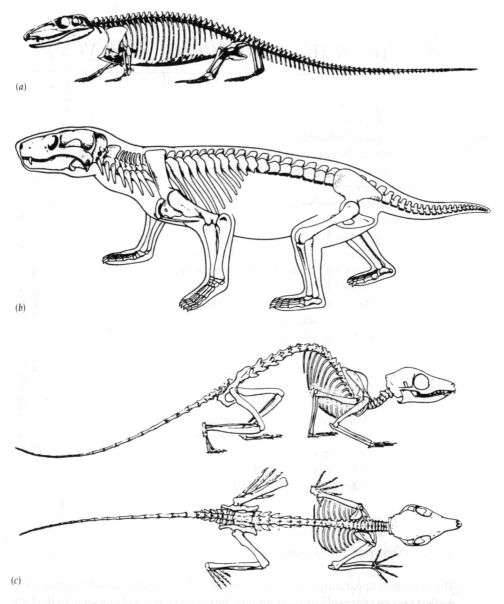

Fig. 18.1.(*a*) *Varanosaurus*, a pelycosaur (length about 1.5 m). From A.S. Romer (1956). *Osteology of the reptiles*. Chicago University Press. (*b*) *Thrinaxodon*, a therapsid (0.4 m). From F.A. Jenkins (1970). *Evolution* 24, 230–252. (*c*) *Tupaia glis*, a mammal (about 16 cm excluding tail). From F.A. Jenkins (1974). *Primate locomotion*. Academic Press, New York.

Varanosaurus (Fig. 18.1*a*) and many others seem to have been carnivorous but some, with blunter teeth, seem to have eaten plants.

The therapsids also included both carnivores and herbivores. *Thrinaxodon* (Fig. 18.1*b*) was a carnivore. Its tail seems to have been short. It had ribs only in the

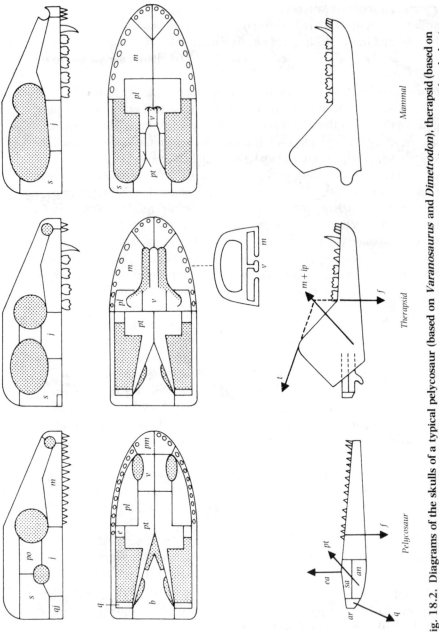

Fig. 18.2. Diagrams of the skulls of a typical pelycosaur (based on *Varanosaurus* and *Dimetrodon*), therapsid (based on *Thrinaxodon* and *Procynosuchus*) and mammal (based on *Didelphis*). Abbreviations: an, angular; ar, articular; b, braincase (basioccipital and basisphenoid); j, jugal; m, maxilla; pl, palatine; pm, premaxilla; po, postorbital; pt, pterygoid; qj, quadratojugal; s, squamosal; sa, surangular; v, vomer. Lower case italic letters indicate forces exerted on the lower jaw by f, the food, m + ip, the masseter and internal pterygoid muscles; ea, the external adductor muscle; pf, the pterygoideus muscle; q, the quadrate articulation; t, the temporalis muscle.

anterior part of the trunk (the thorax), as in mammals (Fig. 18.1c). It had a hole in the pelvic girdle between the ischium and the pubis, again as in mammals. The structure of the shoulder and hip joints show that it stood more like mammals such as *Tupaia* (Fig. 18.1c) than like modern reptiles, with its knees and elbows fairly close to the sides of the body.

The progression from synapsids to mammals also shows clearly in skulls (18.2). Pelycosaurs have small temporal fenestrae with the squamosal and postorbital meeting above them, but therapsids have much larger fenestrae. The fenestrae of mammals are larger still and generally run into the opening for the eye, since the postorbital bone has disappeared. The quadratojugal bone is small in therapsids and missing from mammals. The teeth in different parts of the jaw of a pelycosaur may differ in size but are all more or less the same simple shape. Some therapsids have small conical anterior teeth like mammalian incisors, large conical teeth like canines and many-cusped posterior teeth like premolars or molars.

The secondary palate evolved at the same time as these many-cusped chewing teeth. It provides an air passage from the nostrils to the back of the mouth, by passing the mouth cavity, which may be blocked by food that is being chewed. There is no trace of it in pelycosaurs, whose internal nostrils open into the front of the mouth cavity between the palatine and the vomer (Fig. 18.2). Therapsids have ingrowing shelves from the maxillae and palatines, and a downgrowth from the vomers (now fused to form a single bone). A section in Fig. 18.2 shows how these form a secondary palate, separating an air passage from the mouth cavity, as in crocodiles (section 16.1). In many therapsids, the secondary palate is incomplete, but it is complete in mammals. In the course of these changes, the pterygoid bones became very small.

Pelycosaurs have lower jaws of normal reptilian type (Fig. 18.2). The teeth are on the dentary but there are several other substantial bones. The articular bone at the posterior end of the jaw forms a joint with the quadrate in the skull. In therapsids the dentary became larger and the other lower jaw bones smaller. In mammals the enlarged dentary has made contact with the squamosal to form a new jaw joint, making the articular–quadrate joint redundant. The quadrate has separated from the skull and the articular from the lower jaw, but they remain connected to each other. They have joined the stapes in the middle ear to form the chain of three ear ossicles: stapes, incus (= quadrate) and malleus (= articular) (Fig. 18.5). The angular bone from the lower jaw has become the tympanic bone, which encircles and supports the eardrum and often forms part of a bony wall for the middle ear cavity. The other bones from the posterior part of the lower jaw have disappeared. Embryological studies confirm that the incus, malleus and tympanic bones of mammals are derived from the jaw joint and lower jaw.

It may seem hard to understand how these changes can have occurred. First, it seems odd that the tympanic membrane (eardrum) should be connected directly to the stapes in reptiles (as in frogs, Fig. 15.18) but should be connected through two other ossicles in mammals. How did the additional ossicles insinuate themselves? The answer may be that the common ancestor of mammals and modern reptiles had a stapes but no eardrum, and that the eardrum evolved separately in the two groups.

Secondly, it may seem odd that therapsids had such weak posterior parts to their lower jaws although their large dentary bones and temporal fenestrae imply large, strong jaw muscles. The jaw articulation seems to have got weaker as the jaw muscles got stronger. This paradox can be resolved if account is taken of the directions in which the jaw muscles pulled. Pelycosaurs probably had jaw muscles arranged as in modern reptiles (Fig. 16.1*i*), an upward-pulling external adductor and a forward-sloping pterygoideus. The jaw muscles of a mammal are shown in Fig. 18.3. The internal pterygoid is the homologue of the reptilian pterygoideus. (Another mammalian muscle called the external pterygoid seems to have evolved from a different source and is omitted from Fig. 18.3 because it is very small in *Didelphis*.) The external adductor of reptiles is represented in mammals by two distinct muscles, pulling in different directions. The masseter runs from the zygomatic arch (the bar formed by the jugal and squamosal) to the ventral edge of the lower jaw. The temporalis runs from the braincase to the coronoid process, an upward projecion from the lower jaw. The digastricus opens the mouth and the buccinator is a muscle of the cheek. (Cheeks are a peculiar feature of mammals. They prevent partly chewed food from falling out of the side of the mouth.)

Fig. 18.2 shows at the bottom how these arrangements of muscles help us to understand the evolution of the lower jaw. The arrows represent forces on the jaw during biting with the back teeth. The jaw exerts a force on the food and the food exerts an equal, opposite force *f*, on the jaw. In the pelycosaur, the external adductor and pterygoideus exert forces *ea* and *pt*. For three forces to be in equilibrium the resultant of any two must be equal and opposite to the third, which is possible only if the three forces intersect at a point. The forces *f*, *ea* and *pt* on the pelycosaur jaw obviously do not do that, so there must be a fourth force, a reaction *g* at the jaw joint. The therapsid, however, has a tall coronoid process, apparently for insertion of a mammal-like temporalis muscle which would have exerted a force *t*. There was probably also a masseter-like portion of the external adductor, pulling roughly parallel to the internal pterygoid: these two muscles would have exerted a resultant force $(m + ip)$. Forces *t*, $(m + ip)$ and *f* could have intersected at a point and could have been in equilibrium, so that there was no reaction at the jaw point. Even if they did not intersect exactly, the reaction was probably small compared to the other forces. The evolution of a coronoid process and mammal-like temporalis can explain how therapsids could make do with weaker jaw joints than pelycosaurs, despite having stronger jaw muscles. The posterior bones of the jaw became redundant when the enlarging dentary made contact with the squamosal, forming a new jaw joint lateral to the original one. Fossils are known that have both joints, apparently functioning simultaneously.

The brains of most mammals are ten or more times as heavy as those of reptiles of equal body mass (Fig. 16.13). As the synapsids evolved to become mammals, various bones enlarged to accommodate the larger brain. The frontal and parietal, which were originally limited to the skull roof, extended down the sides of the braincase (Fig. 18.3). Large parts of them came to lie quite deep in the head although they are dermal bones derived from ossifications of the skin (see section 12.4). The squamosal, a dermal bone that was not originally involved in the braincase, came to form a substantial part of its side wall.

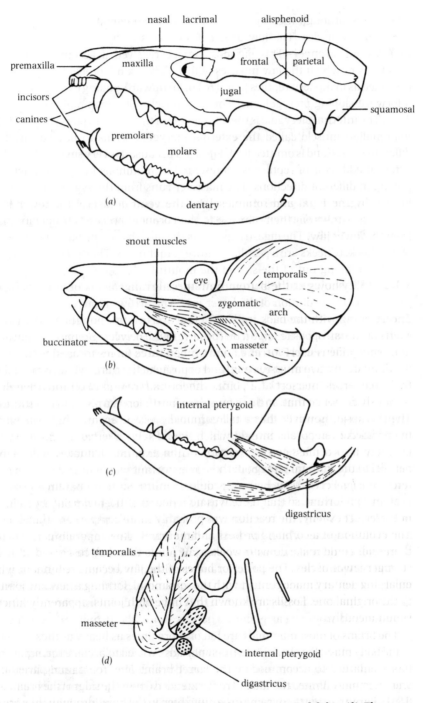

Fig. 18.3.(*a*) Lateral view of the skull of the opossum *Didelphis*. (*b*) The same, showing the positions of muscles. (*c*) Median view of the lower jaw with muscles in position. (*d*) Diagrammatic transverse section through the head of *Didelphis*, posterior to the eye.

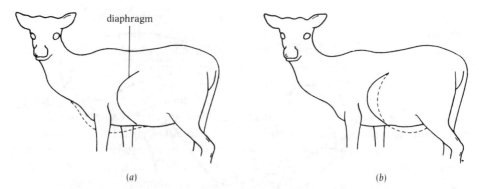

Fig. 18.4. **Diagrams of a mammal showing how (*a*) rib movements and (*b*) contraction of the diaphragm contribute to the volume changes of the thorax involved in breathing.**

18.2. Physiological features of mammals

The evolution of mammals involved other important changes that left no direct evidence in fossils. Mammals are endotherms with much higher metabolic rates than modern reptiles (Fig. 17.6), and body temperatures (in most cases) between 36 and 40°C. Most have a heat-insulating layer of fur, which traps air close to the body just as birds' feathers do (section 17.2). Like feathers, hairs consist of keratin (though of the α rather than the β form) and are formed from the epidermis in follicles in the skin. Mammals use essentially the same mechanisms of temperature control as birds, with the addition of sweating. Sweat is a solution of salts in water, produced by tubular glands in the skin. If it evaporates rather than dripping off it takes most of its latent heat of evaporation from the body. It is an important means of cooling for people, but some other mammals such as dogs depend much more on panting.

Mammals have mammary glands that secrete milk in females but are rudimentary in males. They vary in number from two to more than twenty, and are most numerous in species which give birth to large litters. They may be anterior in position as in women or posterior as in cows, or if numerous may be spread all along the ventral surface of the trunk. Each gland consists of branching tubules opening at a nipple (except in monotremes, which have no nipples). The tubules resemble sweat glands in that they develop from epidermis but grow down into the dermis, and may have evolved from sweat glands. There is more about milk in section 18.3, on reproduction.

Unlike reptiles, mammals have a diaphragm, a muscular partition separating the thorax (containing heart and lungs) from the abdomen (containing liver, stomach and intestines). When the lungs are fairly empty the diaphragm is domed. Contraction of its muscles flattens it, enlarging the thorax and drawing air into the lungs (Fig. 18.4*b*). When it relaxes the process is reversed, largely by the elasticity of the lungs. For this to work well, the wall of the thorax should be relatively stiff but that of the abdomen should be flexible enough to allow the inevitable bulging (Fig. 18.4*b*). We have already seen how the ribs of mammals are limited to the thorax (Fig. 18.1).

Diaphragm movements are not solely responsible for breathing: rib movements like those of lizards are also used (Fig. 18.4a). (Crocodiles have a sheet of connective tissue across the body cavity in the position of the mammal diaphragm, but it is not muscular and is moved, in breathing, by abdominal muscles.)

The additional ear ossicles of mammals are not the only peculiarity of the ears: mammals also have a pinna (external ear) and a cochlea. The pinna acts like the horn of an old phonograph recorder. In such a machine the energy required for indenting the record was provided entirely by the sound, without electrical amplification. The energy was collected from the large area of the mouth of the horn and concentrated on the small diaphragm at its apex which operated the indenting needle. The horn also had a impedance matching effect, making the ratio of pressure amplitude to velocity amplitude greater at its apex than at its mouth (impedance matching and its importance in the ear were explained in section 15.7). The pinna seems to act as an energy collector and impedance matcher, collecting sound from a large area and providing the first stage of impedance matching between the air and the oval window. The eardrum is relatively small and occupies a deep, protected position. It has been shown that the threshold of hearing of cats is raised by about 10 dB when the pinna is removed. The pinna also plays a part in distinguishing the direction from which sound comes.

The cochlea is a coiled structure in the inner ear, evolved from the small basilar papilla of amphibians and reptiles (Fig. 15.18). In these animals, sound vibrations are transmitted from the eardrum to the oval window and thence to the round window, travelling through perilymph, endolymph and perilymph again and being detected by the basilar papilla. In mammals, the basilar papilla, with perilymph on either side, has been drawn out into a long cochlea, coiled into a helix of several turns (Fig. 18.5). (Birds and crocodiles have shorter cochleae, of less than one turn.)

In mammals, as in amphibians, the round and oval windows vibrate together. When one caves in the other must bulge out, and vice versa. The fluids of the inner ear must move with them. These movements could be restricted to the perilymph since there is a pathway, entirely through perilymph, between the windows. However, this pathway is a long one around the tip of the cochlea, and the partitions within the cochlea are flexible. Movements of the windows deflect the partitions, so that endolymph moves as well as perilymph.

Fig. 18.5(b, c) shows how these deflections are detected. A long cupula (known as the tectorial membrane) runs like a shelf along the cochlea, close to one of the partitions. Hair cells in the partition have their sensory 'hairs' embedded in the cupula, and are stimulated by the shearing movement of the cupula relative to the partition, as both are deflected up and down.

Dr G. von Békésy was the first to show how the cochlea distinguishes sounds of different frequencies. He ground small openings in cochleas taken from cadavers so that part of the endolymph–perilymph partition was visible, and sprinkled on silver filings to make it easier to see. He observed it through a microscope while using an electrical device to apply vibrations to the round window. Stroboscopic illumination, by a light flashing at frequencies near that of the vibration, enabled him to see the vibrations of the partition in slow motion. For any frequency of vibration, different parts of the partition vibrated up and down with different amplitudes. The greatest

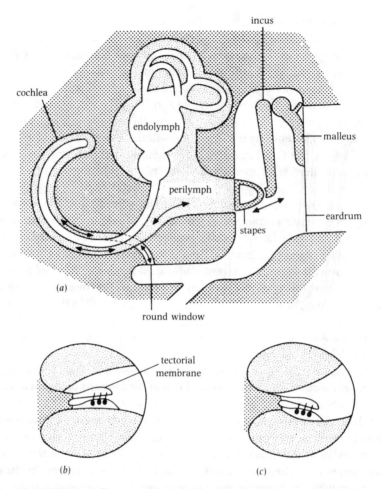

Fig. 18.5.(*a*) A diagram showing the structure of the ear of a mammal. (*b*), (*c*). Diagrammatic sections through the cochlea at two stages in a cycle of vibration of the partitions.

amplitudes were near the base when the frequency was high and further along the cochlea when it was lower. Sounds of different frequencies are probably distinguished by the position in the cochlea of the peak amplitude of vibration, but the peaks are rather broad, which makes it difficult to understand how we perceive small differences of pitch. Experiments with the ears of turtles suggest that hair cells at different positions in the cochlea are tuned to respond selectively to different frequencies.

Another peculiarity of mammals is their ability to produce concentrated urine. Reptiles cannot produce urine more concentrated than the blood (about 0.35 mol l^{-1}). Most of the birds that have been tested produce urine no more concentrated than 0.7 mol l^{-1}, but dogs, cats and rats can all produce urine above 2.5 mol l^{-1}. Desert rodents do even better, up to 9 mol l^{-1} (including 5 mol l^{-1} urea) in the Australian hopping mouse *Notomys*. The ability to produce concentrated urine is particularly important to mammals since they excrete their nitrogenous waste

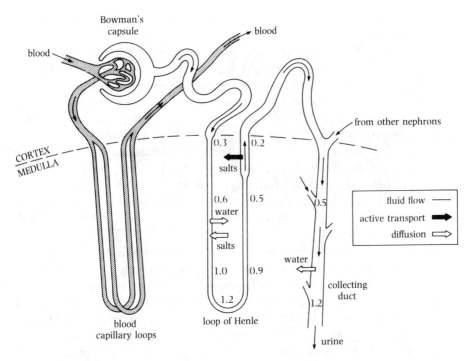

Fig. 18.6. A diagram of one nephron in a mammalian kidney. The numbers represent osmotic concentrations (mol l⁻¹) in various parts of the nephron.

mainly as urea. A reptile or bird can conserve water by passing almost dry urate precipitates but a mammal can only excrete urea in solution.

The shape of typical mammal kidneys is well known. The ureter joins the kidney on its concave inner side. There are two distinct layers of tissue, the medulla on the concave side and the cortex on the outer convex side. The capsules are in the cortex (Fig. 18.6). The structure which enables mammal and (to a lesser degree) bird kidneys to produce concentrated urine is a section known as the loop of Henle, interpolated in each kidney tubule. The tubules do not run straight from the cortex through the medulla to the ureter, but down into the medulla and back before joining one of the collecting ducts which run through the medulla again to the ureter. The section which runs into the medulla and back is the loop of Henle.

Experiments to find out how the loop of Henle works have been performed on rats which had part of the cortex cut away to expose the medulla. The anaesthetized animal's kidney was viewed through a microscope while fine pipettes (about 6 μm diameter) were stuck into loops of Henle. Samples of fluid were withdrawn and their freezing points determined, so that osmotic concentrations could be calculated. Samples were taken at various levels in the medulla, both from the descending limb of a loop of Henle and from the adjacent ascending limb. The osmotic concentrations shown in Fig. 18.6 are based on these data. Notice that the fluid becomes more concentrated as it travels down the descending limb and less concentrated again as it returns up the ascending one. At any particular level, it is about 0.1 mol l⁻¹ more dilute in the ascending limb, and it leaves the loop of Henle more dilute than it entered (which may seem strange, in a structure supposed to be involved in producing highly concentrated urine).

Experiments with small pieces of loop of Henle throw more light on the process. The wall of the descending limb is permeable both to salts and to water. The wall of the ascending limb is impermeable to water, and ions are pumped through it by active transport. Ions are pumped out of the ascending limb but water cannot follow, so the fluid in the tubule becomes more dilute as it travels up this limb. The ions that have been pumped out increase the osmotic concentration in the space between the tubules, so water diffuses out of the descending limb and salts diffuse into it. Thus the fluid becomes more concentrated as it travels down the descending limb. These processes transfer salts from the ascending to the descending limb: salts are trapped near the turn of the loop, and a high osmotic concentration builds up there. This is a process of counter-current multiplication like the one in the rete mirabile in fish swimbladders (section 14.3).

After passing through the loop of Henle the fluid has to travel down through the medulla again, in the collecting duct. It travels down through increasing osmotic concentrations, so there is a tendency for water to be withdrawn from it. How much is withdrawn depends on the permeability of the wall of the collecting duct, which is controlled by a hormone (antidiuretic hormone, ADH). In the presence of ADH the wall is permeable to water and the urine reaches a high osmotic concentration.

The water which diffuses out of the descending limbs and collecting ducts is removed in the blood, but blood flow would make the mechanism ineffective if it resulted in substantial dilution of the interstitial fluid. This is avoided because the blood vessel loops act as counter-current exchangers in essentially the same way as the retia in the blood supply of tuna swimming muscles (section 14.5). Operation of the counter-current principle enables the blood to flow through the medulla where the osmotic concentration is high, without removing substantial quantities of salts, just as it enables tuna blood to pass through warm muscle without removing much heat. Diffusion of salts and water between blood and interstitial fluid concentrates the blood as it travels into the medulla and dilutes it again as it returns.

18.3. Reproduction of mammals

There are three main groups of modern mammals, which reproduce in different ways: three species of monotremes, about 270 of marsupials and about 4000 of eutherian mammals. The monotremes are the platypus (*Ornithorhynchus*) and the echidnas or spiny anteaters (*Tachyglossus* and *Zaglossus*), all of which live in Australasia. The best-known marsupials are the oppossum *Didelphis* of North America and the kangaroos, phalangers etc. of Australasia, but there are many others in South and Central America.

The monotremes lay shelled eggs like reptiles but suckle their young like other mammals. They range in mass from about 2 kg (*Ornithorhynchus*) to 15 kg (*Zaglossus*) but their eggs are less than 2 cm long, smaller than those of reptiles of similar body mass. *Ornithorhynchus* lays its eggs in a burrow in the bank of a stream, and keeps them and subsequently the young warm, by curling round them. *Tachyglossus* lays a single egg and carries it in a temporary pouch which forms on the female's belly at the beginning of the breeding season. The cloaca can reach the pouch if the animal curls up, and it is suspected that the egg is laid directly into the pouch.

Since they come from small eggs, newborn *Tachyglossus* are necessarily small. They weigh about 0.4 g, and are entirely dependent on the mother, whose mammary glands open at two areolae (not raised into nipples) within the pouch. The young can suck milk without leaving the pouch, where it remains until it weighs 170–400 g. By the time it leaves the pouch it is developing the spines, which are a formidable defence for an adult animal but could hardly be tolerated in her pouch by a mother. The temperature in the pouch is about equal to the mother's body temperature but by the time it leaves the pouch the young can maintain its own temperature. The mother leaves it in a burrow and returns periodically to suckle it.

Marsupials and eutherians are viviparous and their ova are tiny, smaller even than those of teleost fish. They range in diameter from 0.1 to 0.25 mm. Like other small chordate ova these divide completely into cells at the beginning of development (only a small part of the large ovum divides in selachians, reptiles, birds and monotremes). Later, however, an amnion, chorion and allantois develop, and a yolk sac which is empty of yolk. These membranes are arranged in essentially the same way as in reptiles and birds, though they sometimes develop quite differently. The reptile ancestors of mammals presumably laid large eggs, in which these membranes served their usual reptilian functions. They presumably had the usual rich blood supply to the yolk sac, responsible for collection of food materials from the yolk, and to the allantochorion, responsible for gaseous exchange through the shell. In mammals (except monotremes) one or both of these membranes become firmly attached to the wall of the uterus, forming a placenta (Fig. 18.7). In most marsupials only the yolk sac forms a placenta, but *Perameles* (the bandicoot) and a few others have chorioallantoic placentae as well. In eutherian ('placental') mammals the chorioallantois forms the main placenta. Yolk-sac placentae also develop in some viviparous selachians (section 13.5) and both yolk-sac and chorioallantoic placentae in various viviparous lizards and snakes.

The blood of mother and foetus come close together in the placenta. In the most primitive arrangement, in which the uterus and embryonic membranes simply interlock (Fig. 18.7b), the two bloodstreams are separated by the foetal capillary wall, connective tissue and epithelium and by the maternal epithelium, connective tissue and capillary wall. This is the situation over most of the area of the placenta in pigs and sheep. In the most advanced arrangement, found for instance in the rabbit (*Oryctolagus*), bare foetal capillaries run through spaces filled with maternal blood (Fig. 18.7c) so that only a single layer of cells separates the bloodstreams. In all cases diffusion between the bloodstreams supplies the foetus with oxygen and foodstuffs, and removes carbon dioxide. Complex interlocking between maternal and foetal tissues ensures that the area available for diffusion is large: it has been estimated as 12 m^2 in the human placenta. The oxygen concentrations in maternal blood arriving at and leaving the placenta have been measured in various species. The results seem to show that the thinner the barrier between maternal and foetal blood, the higher the percentage of the oxygen in the blood that diffuses to the embryo: 70% of the oxygen is given up in the placenta of the rabbit, but only 30% in the placenta of sheep.

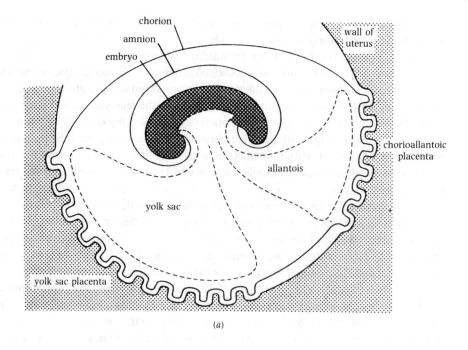

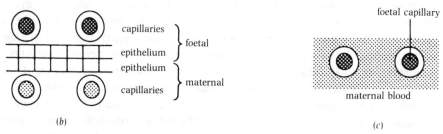

Fig. 18.7. (*a*) Diagrammatic section through a mammal embryo in its mother's uterus, showing how placentae are formed by embryonic membranes. (*b*), (*c*) Diagrammatic sections through two types of placenta.

The foetal blood leaving the placenta necessarily has a lower partial pressure of oxygen than the maternal arterial blood. If its haemoglobin had the same properties as maternal haemoglobin it could never become saturated with oxygen. It does not have the same properties, but becomes saturated at lower partial pressures: the haemoglobin of a late human foetus becomes 50% saturated at about 0.025 atm oxygen, and that of its mother at about 0.035 atm.

In mammal embryos, as in bird ones (section 17.4), the wall between the left and right sides of the heart is incomplete, and there is a connection between the systemic and pulmonary arteries. Little blood flows through the lungs until birth, and it is only after birth that the pathways through the left and right sides of the heart become completely separated. The arteries to the placenta of the mammal embryo, like the arteries to the allantois of the chick, branch from the aorta rather than from the pulmonary arteries.

The females of nearly all marsupials have a pouch. This is permanent, not transitory like the pouches of spiny anteaters. The newborn young are very much smaller than those of eutherians, and though they are in most respects at a very early stage of development they have strong clawed forelimbs, which they use in making their own way to the pouch. The mother adopts a standard posture when giving birth, sitting with the tail extended forward between the legs. She does not seem to assist the young to make its way to the pouch except by licking the fur over which it must travel. The young of the Red kangaroo (*Macropus rufus*) reaches the pouch within a few minutes of birth. It quickly finds and attaches itself to one of the nipples, which lie inside the pouch. The nipple swells inside its mouth, so that it is for the time being firmly attached.

The difference in life history between marsupials and eutherians can be illustrated by comparing two fairly large grazing mammals, the Red kangaroo (adult females weight about 30 kg) and the sheep (70 kg). The young kangaroo is born about one month after copulation, when it weighs about 0.7 g. A sheep embryo (excluding embryonic membranes) has about the same mass at this age but is only born after five months, when its mass is around 6 kg. (Twins are a little lighter than single lambs.) The kangaroo remains in the pouch for about seven months (i.e. to an age of eight months from conception) and at the end of that time weighs about 2 kg. There is a short period when it returns to the pouch occasionally, but thereafter it keeps out of the pouch while continuing to feed partly on milk. The nipple may dangle over the rim of the pouch or the young may put its head into the pouch to suck. Milk is still taken to an age of well over a year (mass 10–15 kg) but grass is also eaten. The sheep also changes gradually from a milk to a grass diet. The half-way point when equal amounts of metabolizable energy are obtained from milk and grass occurs at about three months from birth (eight months from conception) and a mass of over 30 kg.

Milk is an aqueous emulsion of fat globules, ranging in concentration from about 1.5% by weight of fat in the horse to about 20% in the reindeer (*Rangifer*) and in the spiny anteater (*Tachyglossus*, a monotreme) and even more in whales and some seals. The aqueous phase of the emulsion contains sugars, salts and proteins, in solution in water or in colloidal suspension. The sugar in the milk of eutherian mammals is almost entirely lactose, but in marsupial milk other sugars are plentiful as well. The proteins are known as caseins. They are acid proteins containing phosphorus, and are present as calcium salts. Calcium and phosphorus are also present as colloidal calcium phosphate and citrate. These elements are needed by the young animal for building bone, and though they are present in milk in quantities similar to those of sodium and potassium they contribute little to its osmotic pressure, because they are present so largely in protein and colloidal salts. The osmotic pressure of the milk is about equal to that of the animal's blood.

Feeding the young on milk might be expected to be wasteful of energy. Energy from the food used to produce milk is inevitably lost in the processes of digestion and synthesis in the mother. Some of the energy from the milk must in turn be lost in digestion in the young. If the young could feed directly on the food taken by the mother, instead of receiving the energy indirectly as milk, an energy-wasting stage would (seemingly) be eliminated. Would energy really be saved?

The economic importance of cattle has stimulated a great deal of research on their growth and milk production, which makes possible an attempt to answer the question. Experiments have shown that the processes of milk synthesis by cows and of milk utilization by calves are both remarkably efficient. Cattle have been given measured quantities of food of known heat of combustion, so that their total energy intake is known. The heats of combustion of their faeces and urine and of the methane produced by fermentation in the stomach (section 2.3) have been measured. In experiments both with young steers and with lactating cows it has been found that of every 100 J of energy taken in as food, about 30 J is lost in the faeces, 3–4 J in the urine and 7–9 J as methane and as heat released in the fermentation process. This leaves about 60 J available for use in metabolism, growth or milk production, and this 60 J is referred to as the metabolizable energy of the food. The efficiency with which it could be used for growth was determined in experiments with young steers, which were fed varying daily rations of the same food. Suppose that an animal given R J daily grows at such a rate as to incorporate G J daily in its body, and that when fed $(R + \Delta R)$ it incorporates $(G + \Delta G)$. Then if energy is being used at a constant rate for maintenance (as distinct from growth), the efficiency of the growth processes must be $\Delta G / \Delta R$. It was found for steers of a wide range of ages that (within limits) every 100 J of food energy (60 J of metabolizable energy) in excess of the requirements for maintenance could be used to add 30 J to the body by growth. Experiments with adult cows showed similarly that 100 J food energy could be used to produce 42 J as milk, so milk production is a more efficient process than growth.

Milk is also highly digestible. Calves lose in their faeces and urine only about 5% of the energy fed to them as milk. Microbial fermentation is not involved in milk digestion, so the losses associated with it are not incurred. Thus 42 J fed as milk provides 40 J metabolizable energy.

Even so, milk feeding compares poorly with direct feeding as a source of energy. The experiments described above show that 100 J supplied as food to a cow can yield 60 J metabolizable energy to the cow itself, but only 40 J metabolizable energy to a calf drinking the cow's milk.

The comparison is much more favourable to milk feeding if growth is considered. Growth rates of calves fed different daily rations of milk have been compared. In this way it was shown that of every 100 J given as milk in excess of the requirements for maintenance about 75 J was incorporated in the body by growth. Considering the processes of milk production and utilization together, 100 J fed to a cow can yield about 42 J as milk that can be fed to a calf to produce about $0.75 \times 42 = 32$ J growth. Thus 100 J as grass, etc. can produce about the same amount of growth, whether fed directly to growing cattle or to the mothers of unweaned calves.

18.4.   The orders of mammals

The earliest mammal fossils known are from the late Triassic, when the therapsids were declining and the archosaurs were taking over as dominant terrestrial vertebrates. In the Jurassic and Cretaceous, mammals may have been quite numerous but they were small, apparently with much the same size range as modern rodents. Both the marsupials and the eutherians appeared in the Cretaceous but there do not seem to have been any large mammals until the Tertiary, after the dinosaurs had become extinct.

The three genera of monotremes have already been mentioned. They are placed in the subclass Prototheria, separate from the subclass Theria which consists of the infraclasses Metatheria (marsupials) and Eutheria. The earliest therian mammals seem to have been very like the opossums (marsupials of the order Polyprotodonta, Fig. 18.3), the tree shrews (eutherians, order Scandentia, Fig. 18.1c) and members of the eutherian order Insectivora, such as hedgehogs (*Erinaceus* etc.) and the Madagascan tenrecs (*Tenrec, Setifer* etc.). These primitive modern therians are all fairly small (few are heavier than 2 kg). They look rather like rats or squirrels but their teeth resemble those of *Didelphis* (Fig. 18.3) and are very different from the plant-eating teeth of rodents. They eat varied diets including insects and other invertebrates and fruit (and *Didelphis* is a notorious scavenger in dustbins). Opossums, hedgehogs and tenrecs hide by day and are active at night, but tree shrews are active during the day.

Fig. 17.6 shows that typical mammals have resting metabolic rates three or four times as high as reptiles of the same body mass, at the same body temperature. Direct measurements of the resting metabolic rates of hedgehogs and tenrecs have generally given lower values than for typical mammals but the results have been very variable: it is difficult to measure the resting metabolic rate of a restless animal. The rates of oxygen consumption of hedgehogs and tenrecs have also been measured as they walked on a treadmill at various speeds, and resting rates estimated by extrapolating to zero speed. The rates obtained in this way are consistently far lower than for other mammals of the same mass, and about equal to the rates for lizards of the same mass and body temperature. Resting metabolic rates of echidnas (*Tachyglossus*) and opossums (*Didelphis*), estimated in the same way, were much higher than for lizards. The rate for the opossum was about the same as for typical mammals. The rate for the echidna was a little lower than for typical mammals but the difference could be accounted for by a difference in body temperature. It seems that hedgehogs and tenrecs have lizard-like metabolism, but monotremes and marsupials have mammal-like metabolism.

Typical mammals have body temperatures about 38°C, and so do hedgehogs. *Tenrec* and *Didelphis* are a little cooler, at 36°C. *Setifer* and *Tachyglossus* are considerably cooler, about 28°C and 31°C respectively. There is no constant relationship, among these primitive mammals, between normal body temperature and resting metabolic rates.

Many reptiles keep their bodies at fairly constant high temperatures by moving back and forth between sun and shade (section 16.4). They obviously cannot do this at night, and few reptiles are active at night. It has been suggested that mammals evolved endothermy initially as an adaptation to nocturnal life.

It is much easier for an animal to keep itself a little warmer than the environment, than to keep itself cooler than the environment. Warming can be done by metabolism but cooling below the temperature of the environment can only be done by panting or sweating, at the expense of considerable loss of water. An animal trying to keep itself cool might well find it difficult to obtain enough water. An animal which lives in a burrow and emerges only at night is unlikely to encounter temperatures above

Fig. 18.8. Tracing of two images from a multiple-flash photograph of a greater horseshoe bat (*Rhinolophus ferrum-equinum*) catching a moth. From F.A. Webster & D.R. Griffin (1962). *Anim. Behav.* 10, 332–340.

about 25°C, so a body temperature of 25–30°C might be suitable for it. Such a low body temperature might prevent it from being quite as active as typical mammals (this is uncertain) but less energy would be needed to maintain it than to maintain a higher body temperature, in the same environment. The early mammals may have been nocturnal animals with low body temperatures and low metabolic rates, like *Setifer*.

Diurnal mammals are generally likely to be exposed to warm environments and strong sunlight, in which they could not keep their temperatures as low as 30°C except by copious sweating or panting. Typical mammalian body temperatures around 38°C may be the lowest that can conveniently be kept constant in diurnal life. Typical modern mammals may have evolved from nocturnal ancestors, acquiring higher body temperature as an adaptation to diurnal life.

The opossums are placed in the order Polyprotodonta, together with some other marsupials including the (probably extinct) Tasmanian wolf, *Thylacinus*. The remaining marsupials (kangaroos, phalangers etc.) are placed in the order Diprotodonta. Phalangers are fairly small arboreal animals, which feed on leaves and fruit. Kangaroos are generally larger, ground-living animals that eat grass and shrubs. Polyprotodonts have many small lower incisor teeth but diprotodonts (which are all herbivorous) have only a single pair of large, forward-pointing incisors.

I have already introduced members of the two primitive eutherian orders, Insectivora and Scandentia. The bats (Chiroptera) are the only mammals capable of powered flight though flying squirrels (*Petaurista*, etc., Rodentia) and some members of other orders glide quite well. Bat wings consist of skin stretched between the body and the limbs, and between the extraordinarily elongated fingers (Fig. 18.8). They are generally a little larger in span and area than the wings of birds of equal body mass. Bat flying techniques are very like those of birds (section 17.3). Most bats feed on insects which they catch in the air, using echolocation as described in section 18.9. Flying foxes (*Pteropus*, up to about 1 kg) and many other large bats eat fruit, and some bats eat other food.

The order Primates appears next in the list at the start of this chapter, but I will

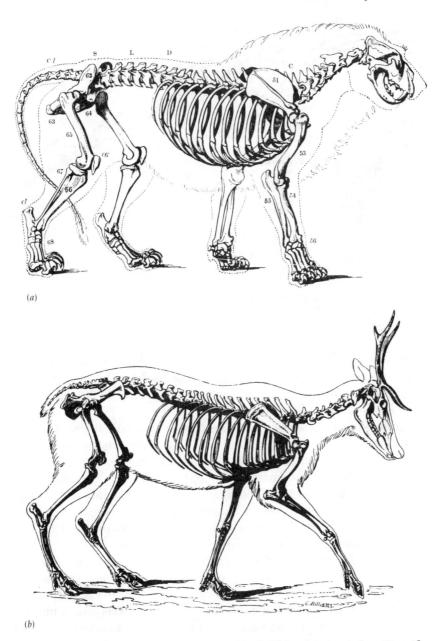

(a)

(b)

Fig. 18.9. Skeletons of (a) a lion (*Panthera leo*) and (b) a female reindeer (*Rangifer tarandus*). From R. Owen (1866–8) *On the anatomy of vertebrates.* Longmans, London.

leave it until later and continue with the Carnivora. This order includes the dogs, cats, bears, weasels, mongooses etc. Some, such as the lion (Fig. 18.9) and other cats feed exclusively on mammals and birds. Others feed largely on invertebrates and even plants: grizzly bears (*Ursus arctos*) feed mainly on plant food, including tubers and berries, but also eat grubs, fish and rodents. The members of the order Carnivora can

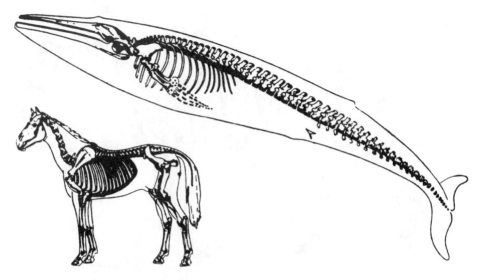

Fig. 18.10. Skeletons of a Blue whale. (*Balaenoptera musculus*), and of a horse (*Equus caballus*). The whale is about 25 m long, and the horse about 1.5 m tall at the shoulder. From E.J. Slijper (1962). *Whales.* Hutchinson, London.

be recognized most easily by their jaws and teeth. Most of them have large canine teeth, but so also do primitive mammals such as *Didelphis* (Fig. 18.3). They also have hinge-like jaw joints, large temporalis muscles and specialized cutting teeth (carnassials) as described in section 18.6.

The Pinnipedia (sealions and seals) are aquatic mammals, that feed mainly on fish and cephalopods. They have short tails and short limbs with large paddle-like hands and feet. Sealions (family Otariidae) swim mainly by movements of their forelimbs. Typical seals (Phocidae) keep their fore flippers against their sides, when swimming fast and straight, and use their hind flippers. Sometimes these are held, soles together, behind the body and used like the tail of a fish: they are swept from side to side by lateral bending of the body. More usually they are used in turn, the left flipper being extended as the pelvis is swung to the right, and vice versa. The fore flippers are used in turning, as hydrofoils or paddles, and sometimes for paddling the seal slowly along.

Seals come on to land (or in many cases ice) to breed. Sealions can turn their hind feet to what is in other mammals the normal position, with the toes pointing forward. Their manner of crawling is not too ungainly. Phocid seals cannot turn their toes forward, and on land can only hitch themselves along. They draw their hindquarters forward, raise the head and thorax on the forelimbs, and push the body forward from behind.

The whales (order Cetacea) are more specialized aquatic animals than seals, and never leave the water. Their forelimbs have become flippers and their hindlimbs are reduced to vestiges that are invisible in the intact animal (Fig. 18.10). Nearly all whales have a rudiment of the pelvic girdle embedded in the ventral body wall with (in some cases) rudiments of the femur and tibia. These are well anterior to the tail flukes, which are not derived from limbs but have evolved as new structures. The tail is thick and muscular. The swimming action is essentially like that of fish with hydrofoil tail fins (Fig. 13.5) except that the movement is up and down, not from side to side. The up and down beat may be a legacy from galloping ancestors: galloping involves up-and-down bending of the back (section 18.5), and the vertebrae and muscles of whale ancestors were probably adapted for this rather than for lateral bending. The body is streamlined, with effectively no neck: the neck vertebrae are greatly shortened and in many species fused together, so that the head merges without interruption into the trunk.

Most of the toothed whales (dolphins etc., suborder Odontoceti) feed on fish and cephalopods. The baleen whales (suborder Mysticeti) have no teeth, but strain zooplankton from the water using closely spaced plates of keratin with fringed edges (baleen) that hang down from the roof of the mouth. Their most important food is krill, crustaceans that are about 3 cm long (excluding appendages). The Blue whale (*Balaenoptera musculus*, Mysticeti) is the largest known animal, living or extinct, with a mean adult body mass of about 100 tonnes.

The remarkable diving ability of seals and whales is discussed in section 18.8.

The order Perissodactyla includes the tapirs, rhinoceroses and horses. The Artiodactyla includes the pigs, camels, deer, antelopes and cattle. The two orders together are referred to as ungulates, indicating that they have hooves, and the elephants (Proboscidea) are sometimes included with them. The illustrations of the horse (Fig. 18.10) and deer (Fig. 18.9*b*) show how they stand on the tips of their toes: the hooves are modified claws. Perissodactyls have digit 3 on each foot stronger than the rest, and in horses it is the only toe. Artiodactyls have digits 3 and 4 about equally strong, and in antelopes and cattle (family Bovidae) only these two toes remain, forming the 'cloven hoof'. Fig. 18.9(*b*) shows that these two are the functional toes of deer but that rudiments of digits 2 and 5 survive.

Ungulates are specialized runners, as shown in more detail in section 18.5. Their feet cannot grasp and manipulate objects, like the paws of more primitive mammals. Advanced ungulates such as horses (Fig. 18.10) and deer (Fig. 18.9*b*) have longer legs than other mammals of equal body mass, owing to elongation of the metacarpal and metatarsal bones (compare them with the lion, Fig. 18.9*a*). Horses have just one metacarpal or metatarsal (known as the cannon bone) in each foot. Advanced artiodactyls have one cannon bone in each foot formed by fusion of the metacarpals or metatarsals of toes 3 and 4.

Pigs eat a varied diet of roots, fruit, insects, etc., but more advanced ungulates feed exclusively on grass or other leaves. These advanced ungulates have evolved specialized grinding teeth, which are described in section 18.6. Bacteria and protozoans in their guts digest the cellulose and hemicellulose in their plant food (section 2.3). These micro-organisms flourish in the stomachs of ruminants (deer, antelopes and cattle), but in the large intestine and the caecum that branches from it in horses. Accordingly, cattle have huge stomachs that account for some 70% of the capacity of the gut, with the large intestine and caecum accounting for only about 12%. In the horse the stomach accounts for a mere 10% of gut capacity, and the large intestine and caecum for 60%.

The ungulates are large herbivores. The smallest seems to be the Royal antelope (*Neotragus pygmaeus*, adult mass about 3 kg) and the largest is the African elephant (*Loxodonta africana*; males grow to about 5.5 tonnes). In contrast, rodents are generally small herbivores. The capybara (*Hydrochoerus*, from South America) attains masses of 45 kg or more but other rodents are much smaller, most of them less than one kilogram. The order Rodentia includes the squirrels, rats, mice, porcupines, cavies etc. and has many more species than any other order of mammals. They can be recognized by their chisel-like incisor teeth, one pair in each jaw. I describe how they gnaw and chew in section 18.6. Their limbs and feet are like those of tree shrews (Fig. 18.1*c*) and other primitive therian mammals.

The Lagomorpha (rabbits etc.) resemble rodents but have a small second pair of upper incisors.

The order Primates includes the lemurs, lorises, monkeys, apes and ourselves. Many primates spend much of their time in trees but others (for example baboons, *Papio*) live mainly on the ground. Primates have prehensile (grasping) hands and feet, and long limbs, that seem to be adaptations for climbing.

It is difficult to identify the features of the hands and feet that enable primates to grasp and manipulate objects so dexterously. Compare the human hand with the hand of the primitive mammal *Didelphis*. They have the same number of bones, and their joints allow movement through similar ranges of angles. They each have about 44 muscles. (To get this number, I have counted separately any parts of muscles that seem capable of acting independently.) One important difference is that people and most other primates have thumbs that can be separated widely from the other fingers, so that they can grasp an object between the thumb on one side of it, and the fingers on the other. Another difference is that primates have unusually large hand and foot muscles.

Fig. 18.11 gives information about the flexor muscles of the wrist and fingers (important muscles for grasping). The primates represented in it are a 0.6 kg bushbaby (*Galago*) and four species of monkey. It compares them with carnivores, which have relatively unspecialized feet, and with antelopes (Bovidae), which are specialized runners. Fig. 18.11(*a*) shows that these muscles are much larger in primates than in other mammals of the same body mass. Fig. 18.11(*b*) shows that this difference is due to the primates' having remarkably long muscle fibres. The primate muscles are not unusually strong (the total of the cross-sectional areas of their muscle fibres is about the same as in other mammals of equal mass) but the long fibres should enable them to move the joints through large angles while exerting substantial forces (section 8.3). An important difference between climbing trees and running on the ground is that climbing animals must be able to exert large forces with their feet in a wide variety of positions.

Fig. 18.11 also shows that antelopes have unusually small flexor muscles of the hand and wrist, with very short fibres. I will discuss this in section 18.5.

Graphs of bone lengths against body mass (similar to the graphs in Fig. 18.11) show that the femur, tibia, humerus and ulna of primates are typically 20–50% longer than in other mammals of the same body mass, but that the metatarsals and metacarpals (which are exceptionally long in ungulates) are about the same length in primates as in other mammals. Primates and ungulates both have long limbs, but they have acquired them by elongating different bones.

Further comparisons show that primates have large brains, typically 50% heavier than in other mammals. Humans have exceptionally large brains even for primates,

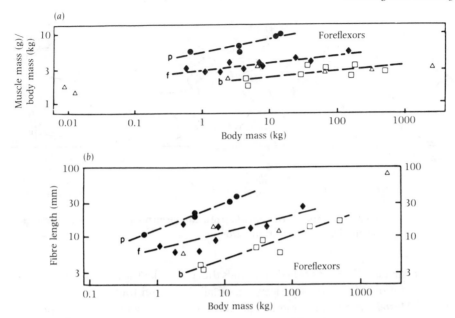

Fig. 18.11. Dimensions of the flexor muscles of the wrist and hand plotted against body mass on logarithmic coordinates: (*a*) shows the total mass of these muscles and (*b*) the harmonic mean length of their fibres in ●, primates; ◆, carnivores; □, antelopes and △, other mammals. From R. McN. Alexander, A.S. Jayes, G.M.O. Maloiy & E.M. Wathuta (1981). *J. Zool., Lond.* 194, 539–552.

about five times as heavy as in typical mammals of the same body mass. However, we are not unique. Dolphins of similar mass to ourselves have brains of about the same size.

Primates feed largely on leaves, fruit and other plant materials but many also eat insects and other small animals. Monkeys and apes have teeth like ours: they have incisors with straight cutting edges, suitable for biting pieces from plant food, and square crushing molars.

18.5.   Walking and running

Fig. 18.1(*c*) is based on X-ray cine films of *Tupaia*. These show that it stands and walks on strongly bent legs, with the humerus and femur more nearly horizontal than vertical. Also, these bones are inclined a little laterally, giving the animal a rather bow-legged stance. The left and right footprints do not form a single line but two parallel lines, a short distance apart.

X-ray cinematography would be little use for studying the leg movements of large animals such as lions and reindeer (Fig. 18.9) because the field of view is very restricted, but light cine films show that they walk with relatively straight legs, with the humerus and femur more nearly vertical than horizontal. The stance is not bow-legged, and the lines of footprints formed by the left and right feet are relatively close together.

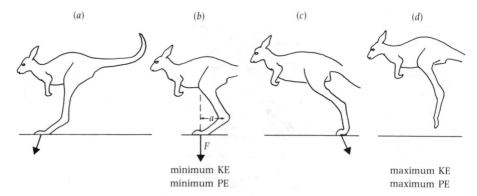

(a)   (b)   (c)   (d)

minimum KE   maximum KE
minimum PE   maximum PE

Fig. 18.12. **Outlines traced from a film of a kangaroo hopping. The arrows represent forces exerted by the feet on the ground.**

Professor Farish Jenkins called mammals that move like *Tupaia* 'non-cursorial' and ones that move like lions and deer 'cursorial'. He found that opossums, rats and ferrets (*Mustela*) were non-cursorial and that dogs and cats were cursorial. Indeed, all mammals of less than 1 kg body mass are non-cursorial and almost all of more than 5 kg are cursorial. (The domestic cat is one of the smallest cursorial mammals.)

Some mammals place more of the foot on the ground than others. *Tupaia* (Fig. 18.1c) sets down the whole of each foot, right back to the wrist or heel, and is described as plantigrade. Lions (Fig. 18.9a) place only the toes on the ground, as far back as the distal ends of the metacarpals or metatarsals, and are described as digitigrade. Deer (Fig. 18.9b) stand on the tips of their toes with only their hooves on the ground and are described as unguligrade. The unguligrade stance is peculiar to the ungulate mammals. Most other large mammals are digitigrade and small mammals are plantigrade. The only large plantigrade mammals are primates (including ourselves) and bears.

The hopping of kangaroos is a conveniently simple gait to study, because only two legs are used and both do the same thing. I will use it to introduce principles that apply also to other mammal gaits, including the bipedal running of humans and the quadrupedal running of most other mammals.

Fig. 18.12 shows a kangaroo hopping. It travels in a series of bounds so it rises and falls: its potential energy is lowest at stage (b) and highest at stage (d). The forces exerted on the ground have been recorded by having kangaroos hop across force plates, force-sensitive plates set into the floor. The records show that the forces act in the directions shown by the arrows, sloping at every stage of the step so as to be more or less in line with the leg. (Only the vertical component of the ground force is needed to support body weight, but if the force were kept vertical it would exert very large moments about the hip joints at some stages of the stride, and much larger forces would be needed in muscles.) At stage (a) the kangaroo decelerates itself by pushing forward on the ground and at stage (c) it accelerates itself by pushing backwards, so it is travelling most slowly at stage (b), and has minimum kinetic energy then, but is travelling fastest and has maximum kinetic energy at stage (d).

Thus the kangaroo has minimum (kinetic plus potential) energy at stage (b) and

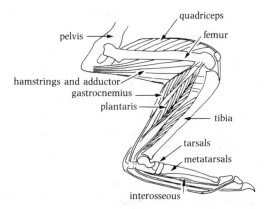

Fig. 18.13. The skeleton, and some of the principal leg muscles, of a typical mammal. From R. McN. Alexander (1984). *Am. Zool.* 24, 85–94.

maximum (kinetic plus potential) energy at stage (*d*). This suggests that the leg muscles must act as brakes at stage (*a*) of the stride, removing mechanical energy from the body and degrading it to heat; and that they must do work at stage (*c*), restoring the mechanical energy of the body. The quantities of work involved can be calculated from the force plate records. However, the need to do this work could be avoided if there were springs in the legs that stored up mechanical energy at stage (*a*) and returned it in an elastic recoil at stage (*c*): the animal would bounce along like a rubber ball and its leg muscles would have to do only enough work to make up for frictional and viscous losses.

Fig. 18.13(*a*) shows how the principal muscles in the legs of kangaroos (and other mammals) are arranged. Notice that the gastrocnemius and plantaris muscles, and some others, have long tendons. The gastrocnemius tendon ends on the heel but the plantaris tendon wraps round the heel and goes right to the toes. These are the springs that are important in hopping.

To assess their importance as springs it was necessary to find out how much energy they stored (as elastic strain energy) at the stage of the hop shown in Fig. 18.12(*b*). To do that, it was necessary to discover the forces in the tendons at that stage, and to measure their elastic properties. The forces were estimated from force plate records and measurements on films. In Fig. 18.12(*b*), if the force plate registers a force $2F$ we can assume that force F acts on each foot. If the line of action of the force is at a distance a from the ankle joint, its moment about the joint is Fa and the ankle muscles must exert an equal, opposite moment. This moment must be supplied mainly by the gastrocnemius and plantaris muscles because the other muscles that could assist them run too close to the ankle joint to exert large moments, but it could have been supplied by the gastrocnemius alone, by the plantaris alone or by both acting together. My colleagues and I, who did the experiment, assumed that both muscles were active, exerting equal stresses (force per unit cross-sectional area). We found out how much the calculated forces would stretch the tendons, and how much elastic strain energy they would store, by dissecting the tendons from carcasses and stretching them in a tensile testing machine of the kind used by engineers. The tests showed that tendon has a fairly high Young's modulus, in the usual range for fibres (section 3.5), so quite large stresses stretch it only a little. The forces involved in hopping would have stretched the gastrocnemius and plantaris tendons by about 3% of their lengths, and the graphs of force against extension told us how much strain energy would be stored. The tests also showed that tendon is an excellent elastic material, returning in its elastic recoil 93% of the work done stretching it.

From the results of these experiments we calculated that when an 18 kg kangaroo hopped slowly, it lost and regained 39 J (kinetic plus potential) energy in each step. If its tendons were inextensible it would have had to use its muscles as brakes and then do 49 J work with its muscles, in each step (49 rather than 39 J because during part of the step muscles do work aginst each other). However, the elastic properties of the tendons enable them to store and return 16 J strain energy, reducing by about one third the work that the muscles have to do. The work would be reduced by larger fractions in faster hopping.

This presumably saves metabolic energy but it is uncertain how much. Even if the tendons had ideal elastic properties, stretching and shortening by so much that the muscles remained constant in length and so did no work, the muscles would still have to exert tension. (Remember that a force does not work unless it moves its point of application.) Muscles use metabolic energy whenever they develop tension, even if they do no work.

However, a lot of metabolic energy is presumably saved by similar mechanisms in advanced ungulates such as deer and horses. Experiments on them show that the tendons in the lower parts of the legs have near-ideal elastic properties, so that their muscles have to make only very small length changes during running. The muscles have responded by evolving extremely short muscle fibres, so that the forces that are needed can be exerted by much smaller volumes of muscle. (Fig. 18.11 shows how this has happened in the forelegs of antelopes.) Less muscle has to be activated in each step, so less metabolic energy is needed. The extreme adaptation is found in camels, in which the muscle fibres of the plantaris have almost vanished. The plantaris remains as a substantial tendon running all the way from knees to toes, that can only serve as a passive spring. (Though the musles of the lower part of the leg are small there are large muscles in the thigh, capable of providing the power needed for acceleration.)

Most mammals walk at low speeds and run to go faster. Walking gaits are ones in which the left and right feet of a pair move alternately, and each is on the ground for more than half a stride. Thus there are stages of the stride when both feet are on the ground. In running, each foot is on the ground for less than half a stride and there are stages when both are off the ground. Energy savings by tendon elasticity seem to be important in running, but relatively unimportant in walking.

Quadrupedal mammals walk at low speeds, moving their feet in the order shown in Fig. 15.16. At moderate speeds most of them trot: the trot is a running gait in which the left forefoot and the right hindfoot move together, and the right forefoot and the left hindfoot together. A trotting quadruped is like two people running one behind the other, half a cycle out of step. At high speeds quadrupedal mammals gallop, setting down the two hindfeet and then the two forefeet (Fig. 18.14).

To understand why mammals gallop, we have to consider a component of the body's kinetic energy that has not so far been mentioned. The forces between the feet and the ground decelerate and re-accelerate the body as a whole, causing fluctu-

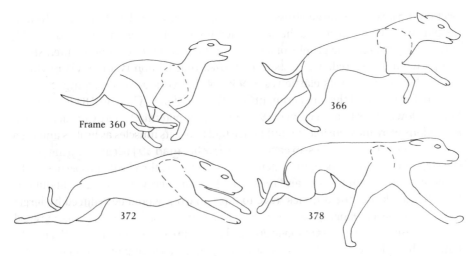

Fig. 18.14. Outlines traced from film, showing four stages of a stride of a greyhound galloping at 15 m/s. From A.S. Jayes & R. McN. Alexander (1982). *J. Zool., Lond.* 198, 315–328.

ations of kinetic energy due to the legs being decelerated and re-accelerated, relative to the trunk, as they swing backwards and forwards. These fluctuations increase rapidly as speed increases, but galloping seems to use an elastic mechanism to save some of the energy. At frame 360 of the film (Fig. 18.14) the hindlegs have finished their forward swing and are about to swing back, and the forelegs are finishing their backward swing and are about to swing forward. Both pairs of legs are temporarily halted, so the kinetic energy of the body falls. However, the back is bent and its extensor muscles are taut. A sheet of tendon by which these muscles attach seems to serve as a spring, storing elastic strain energy as the legs are halted and returning it in the elastic recoil to start them moving the other way.

Human walking and running are very different from the gaits of other mammals, but we are not unique in being bipedal. Kangaroos and various other mammals hop bipedally. Among the primates, captive gibbons (*Hylobates*) walk bipedally when they are on the ground, but wild gibbons seldom leave the trees. Wild chimpanzees (*Pan*) sometimes travel bipedally in long grass, or when carrying a load. However, we are unique in walking with our trunks erect, and keeping our legs straight while the foot is on the ground. Chimpanzees and other primates walk bipedally with the trunk sloping forward and the knees bent (Fig. 18.15). With the legs bent like this, the line of the ground force must inevitably be a long way from some of the leg joints, so large moments act about some of the joints and large forces are required in muscles. When we walk with straight legs, keeping the ground force more or less in line with the legs, the moments about the leg joints are small and small muscle forces suffice. Metabolic energy is needed to generate tension in muscles, so our straight-legged style of walking should reduce metabolic energy costs.

The arguments in this section suggest that horses and antelopes should be able to run very economically, because the elastic properties of their leg tendons have made it possible for some major leg muscles to be small and short-fibred: and that people should be able to walk very economically because we keep our legs straight.

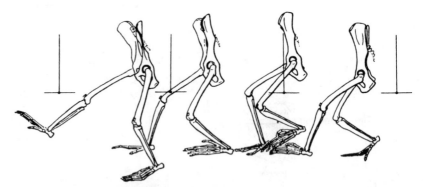

Fig. 18.15. Outlines from X-ray cine pictures of a chimpanzee walking bipedally. The dots with straight lines through them represent a fixed point in the background. From F.A. Jenkins (1972). *Science* 178, 877–879.

Unfortunately, measurements of oxygen consumption have failed to confirm these expectations. The metabolic energy cost of locomotion for horses, antelopes and people are little different from the costs for other species of the same body mass, travelling at the same speed.

18.6.   Teeth and chewing

The opossum *Didelphis* will be taken again as an example of a primitive mammal. Its teeth are very like those of some early mammal fossils, and its chewing movements have been carefully studied. It eats a wide variety of foods including fruit, insects and small vertebrates. The general arrangement of its teeth is shown in Fig. 18.3. The most anterior are the incisors, which are small, peg-like teeth probably used mainly for picking up and manipulating food. The lower incisors are not vertical, but tilt forwards, and so can more easily be slid under objects which are to be picked up. Posterior to the incisors are the much larger canines, used by other mammals and presumably by the opossum for tearing at prey. They are appropriately sharp and curved. Next come the premolars, which are perhaps most useful for piercing food in the early stages of chewing. Finally, there are the molars, which are much the most complicated teeth in the mouth and much the most interesting. They are used for chewing the food before it is swallowed. By breaking the food up into small pieces they speed the action of digestive enzymes on it. Fast digestion is important to animals with high metabolic rates. Lizards chew their food to some extent, but mammals break their food up much more thoroughly.

Fig. 18.16*A*, *B* show the form of opossum molar teeth. Note that the upper and lower molars are quite different. The upper ones are triangular in plan with an inner relatively low part (stippled) and an outer generally higher part. Three main cusps (labelled *a*, *b* and *c*) rise above these levels, and sharp ridges run between the cusps and along the edges of the tooth. Each lower molar consists of a relatively high triangle and a generally lower squarish heel. Again there are cusps (labelled with Greek letters) and ridges. All the cusps have accepted scientific names, but as the

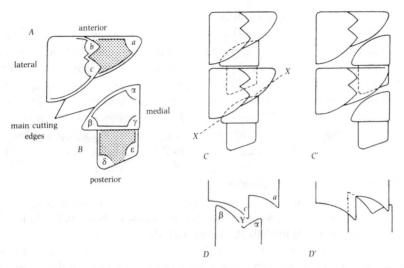

Fig. 18.16. *A*, *B* Diagrams showing the main cusps and ridges on upper and lower molar teeth, respectively, of a primitive therian mammal such as *Didelphis*. *C*, *C'* Diagrammatic plans showing the relative positions of upper and lower molars at the beginning *C* and end *C'* of the effective stroke in the cycle of chewing movements. *D*, *D'* Diagrams representing thick vertical sections along *XX*, showing upper and lower molars in the same two positions as *C*, *C'*. Lettering is explained in the text.

names are long and not particularly memorable, letters seem more appropriate in this book.

The chewing movements in which the opossum uses these teeth have been observed and recorded by X-ray cinematography. Lateral views show that chewing a mouthful of food occurs in two stages. First come a series of bites which pierce and crush the food without actually cutting through it, so the upper and lower teeth do not meet. The cusps are important in this part of the chewing process, and it is probably largely as a result of it that their points become worn. Later, when this first process has softened the mouthful of food, bites pass right through it, and upper and lower molars make contact. The ridges rather than the cusps are important in this chopping phase of chewing. Ridges on the upper and lower teeth slice past each other like the blades of shears, and their vertical faces become worn.

The lower jaw is less wide than the upper one, so it has to be moved to one side for chewing, which cannot proceed simultaneously on both sides of the mouth. X-ray cinematographs in vertical view show this. The lower jaw is lowered, shifted to one side and raised again, so that the lower molars of that side take up a position directly below the upper ones. Pressure is then applied to the food as the jaw moves back towards the median position. It is first applied with the lower molars in the position shown in Fig. 18.16*C* and is continued as they move to the position shown in Fig. 18.16*C'*. In the latter position the high triangles of the lower molars occupy the triangular gaps between the upper molars, and the low heels of the lower molars (stippled in Fig. 18.16*B*) rest against the low inner corners of the upper molars (stippled in Fig. 18.16*A*) and food may be crushed between them.

The main cutting edges are indicated in Fig. 18.16*A*, *B* and shown in the sectional view in Fig. 18.16*D*. This shows the position when they are beginning to slice past each other. Notice that because the cutting edges are concave, food trapped in the space Y cannot easily slip out of position and escape being cut as the lower molar moves to the position shown in Fig. 18.16*D'*. The same principle is applied in the design of garden shears with concave cutting edges, which will cut branches that would slip out of straight-bladed shears. The edges indicated as the main ones are probably the most important, but every edge apparently slices past another at some stage in the movement from the position of Fig. 18.16*C* to that of Fig. 18.16*C'*.

In the life of a reptile, the teeth are shed and replaced many times. Waves of replacement travel backwards along the jaw in such a way that alternate teeth are always very different in age. A young tooth, or an old one which is being resorbed at the base prior to being shed, is flanked on either side by mature, firmly fixed teeth. Thus at least half of the teeth in any given part of the jaw are always fully functional. In mammals, however, very little replacement occurs. In eutherians the original ('milk') incisors, canines and premolars are replaced only once, and the molars not at all. In marsupials only one tooth in each jaw (the last premolar) is replaced.

The teeth of a lizard can perform their functions of holding and piercing provided only that there is always a reasonable proportion of mature teeth. The chewing action of mammal molars such as those of the opossum depends on a precise fit being maintained between each upper tooth and a pair of adjacent lower teeth. This fit could not be maintained while alternate teeth were being replaced, especially if the animal was growing and the new teeth were bigger than their predecessors. Continual replacement in the reptilian manner would make the molars much less effective. 'Milk' premolars are generally much more like molars than adult premolars, and are used in chewing by young animals. They are not shed and replaced until there are molars to take over their chewing function.

Reptiles feed in the adult manner using their teeth, as soon as they emerge from the egg. If mammals took solid food and needed teeth while still so small a fraction of the adult size, their teeth would surely have to be replaced many times during growth to maintain a reasonable proportion between tooth size and head size. Young mammals are born quite large and subsequently suckled, so that only quite a modest amount of head growth occurs while teeth are in use. Suckling may have made possible the mammalian pattern of (very limited) tooth replacement.

Different parts of the molar teeth of *Didelphis* serve to cut food or to crush it, but in more advanced mammals with more specialized teeth one function or the other has been emphasized. In Carnivora, the last upper premolar (pm^4) and the first lower molar (m_1) have evolved to become the carnassial teeth: when a dog gnaws at a bone with the side of its mouth it is using these teeth to scrape off flesh. Fig. 18.17(*a*) shows that in the mongoose, the main cutting edges (see Fig. 18.16*a*) of pm^4 and m_1 are enlarged, but the teeth are not very different from those of *Didelphis*. In cats (Fig. 18.17*b*) the exaggeration has gone to extremes, and the only substantial cusps that remain are the ones involved in the cutting edges. The same diagram shows that the first upper molar (m^1) has lost its cutting edge and become a specialized crushing tooth in the mongoose (it crushes food against the heel of m_1). The same tooth has been reduced to a rudiment in the cat (which has also lost the heel of m_1). The

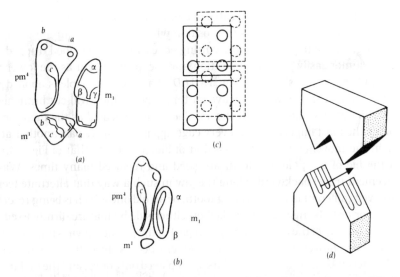

Fig. 18.17.*A*, *B* Diagrammatic plan views of teeth from the posterior part of the jaw of two carnivores, (*a*) a mongoose (*Herpestes*) and (*b*) a cat (*Felis*). (*c*) A diagrammatic plan showing how the upper and lower molars of monkeys fit together in the closed mouth. (*d*) A diagram of upper and lower molars of a ruminant mammal.

mongoose retains the ability to crush food although it has evolved carnassial teeth, but the cat, which is more strictly carnivorous, has no crushing teeth.

The jaw joints of carnivores are formed as hinges that allow the jaw to open and close, rotating about a transverse axis, but do not allow it to swing from side to side. However, they allow the slight lateral movements of the jaw (sliding along the axis of the hinge) that are needed to ensure that the cutting edges of the carnassial teeth pass close to each other. Their cutting action depends on this, just as the effectiveness of a pair of scissors depends on the tightness of the rivet.

The main difference between plant and animal tissues, as food, is that plant cells are enclosed in cell walls of cellulose, hemicellulose and lignin, substances that are not digested by vertebrate enzymes. The dry matter of wheat grains consists of about 2% salts, 12% of these cell wall materials and 86% fat, protein and easily digestible carbohydrate (mainly starch). The dry matter of mature ryegrass (*Lolium*) is about 5% salts, 60% cellulose and hemicellulose, 11% lignin and less than 25% easily digestible matter.

Not only are cell wall materials indigestible to animals lacking appropriate enzymes, but they may protect the contents of the cells from digestion. The digestible materials are not accessible until the cells have been broken open. Grain, fruit and other plant materials, containing a large proportion of digestible material enclosed in relatively thin cell walls, may only need crushing to make most of their food content available. Materials such as grass which contain relatively little digestible material, and that enclosed in thick cell walls, may require fine grinding. Though vertebrates do not themselves produce enzymes capable of digesting cell wall materials, many herbivorous mammals have in their guts microorganisms capable of digesting cellulose and hemicellulose (sections 2.3 and 18.4). These could digest the cell walls

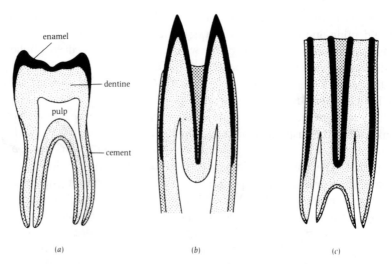

Fig. 18.18. Diagrammatic transverse sections of molar teeth of herbivorous mammals. (*a*) A bunodont molar; (*b*) an unworn hypsodont molar and (*c*) a worn hypsodont molar.

of unground food and (eventually) make the cell contents available, but grinding can make cell contents immediately available.

The molar teeth of pigs, bears and primates have lost their cutting edges, but their crushing surfaces have been enlarged. Fig. 18.17(*c*) shows their form in monkeys, apes and people. Both the upper and the lower molars are square, with four cusps. Three of the cusps on the upper molars seem to be homologous with cusps *a*, *b* and *c* of *Didelphis* (Fig. 18.16*a*), but the fourth is new. The cusps of the lower molars seem to be β, γ, δ and ε: α has been lost. When the mouth closes, cusps of the upper molars fit between cusps of the lower ones, and vice versa.

These teeth are of the bunodont types, shown in Fig. 18.18(*a*). More advanced, hypsodont teeth are found in horses and ruminants. Initially they have high cusps (Fig. 18.18*b*) but their points soon wear away, and the fissures between the cusps are largely filled by cement. This is a form of bone which is present in small quantities only, around the roots, in bunodont teeth. Hypsodont teeth are adapted for grinding rather than mere crushing. They are particularly suitable for grinding grasses, which tend to be more abrasive than other foods. Most grasses have abrasive silica crystals in their leaves, and they are very apt to have grit on their surfaces because they grow so near the ground. An indication that the external grit may abrade the teeth more than the internal silica in the grass is provided by the observation that the teeth of horses kept on sandy pastures get particularly severely worn. The remarkably complete fossil record of the evolution of the horses shows that the early ancestors of modern horses had bunodont teeth but that horses with hypsodont teeth appeared in the late Tertiary, at about the time when grasses first became widespread and abundant.

As a hypsodont tooth wears away its constituent layers of enamel, dentine and cement become exposed edge-on (Fig. 18.18*c*). Enamel has a higher inorganic content than the other constituents and is accordingly more resistant to wear. As wear proceeds, the enamel comes to stand in ridges slightly above the worn surfaces of the dentine and cement. This leaves it rather more exposed to wear than the dentine and cement, so the latter do not wear into progressively deeper hollows. Once the difference in level has been established all the materials wear more or less evenly and the difference remains more or less constant. The rough file-like surface which is so effective for grinding is maintained.

The hypsodont molars of ruminants have four cusps, like the bunodont molars of monkeys, but the cusps are elongated, making more effective file-like surfaces than round cusps would. Fig. 18.17(*d*) shows how their lower molars move across the upper ones. The cusps are elongated longitudinally, and the lower jaw moves transversely to grind food. Notice that there are tall ridges running across each tooth from cusp to cusp, as well as the much lower ridges of enamel running along the tooth. It is the low ridges of enamel which have the file-like action. Horses have molars which wear much flatter, with nothing like the tall ridges of ruminants. Their cusps have complicated shapes so that the crown of the worn tooth has very long convoluted lines of enamel.

There are marked differences in the jaw muscles as well as in the teeth, between carnivorous and herbivorous mammals. In carnivores, the temporalis muscle is the largest but in ungulates it is relatively small, and the masseter and pterygoideus muscles are larger. For example, the composition by mass of dog jaw muscles was found to be: temporalis, 67%; masseter, 23%; pterygoideus, 10%. That of bison jaw muscles was: temporalis, 10%; masseter, 60%; pterygoideus, 30%.

The temporalis, which is large in carnivores, is well arranged for use when the animal is tugging at prey with its canine teeth. The force exerted by the food on the lower jaw is then a forward and somewhat downwardly directed one which is likely to be more or less in line with the force exerted by the temporalis. If the masseter muscles alone were used to hold the mouth shut while the animal tugged, the whole force of the pull would have to be taken by the ligaments of the jaw articulation and there might be a danger of dislocating the jaw.

Herbivores seldom need to pull forcibly with their teeth, and most have no canines, so they seem to have no need for large temporalis muscles. The masseter and internal pterygoids, however, are well placed for use when grinding food. Ruminants and horses use a side-to-side jaw movement, essentially like the chewing action of *Didelphis*. When chewing food on the left side of the mouth they open the jaws and move the lower jaw to the left. They then close the left lower teeth against the upper ones and move them, in the effective grinding stroke, towards the mid-line. The jaw muscles must press the left teeth together and at the same time pull the lower jaw towards the right. The pull to the right can only be provided effectively by the left internal pterygoid muscle (see Fig. 18.3*d*). Chewing on the right side would similarly demand a pull to the left, exerted by the right internal pterygoid. Thus the internal pterygoids apparently play a major role in chewing, and it is not surprising that they are large. Since the jaw is shaped in such a way as to make room for large internal pterygoids on its inner side, there is room also for large masseters on the outer sides. The masseters pull more or less vertically and can contribute to the vertical component of the force exerted in the effective grinding stroke, but not to the transverse component.

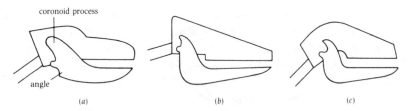

Fig. 18.19. **Diagrams of the skulls of three mammals. (*a*) is a carnivore, and (*b*) and (*c*) are herbivores. Further explanation is given in the text.**

There is a striking difference in shape between the lower jaws of typical carnivorous mammals, and typical herbivorous ones (Fig. 18.19). In carnivores the large temporalis muscles have a tall coronoid process to attach to, but the angle of the jaw, where the masseter and pterygoid muscles insert, is shallow. In herbivores the angle of the jaw is deep, making room for the large masseter and internal pterygoid muscles, while the coronoid process is small. There is a corresponding difference in the skull. Carnivores generally have the jaw articulation approximately in the plane of the secondary palate, but herbivores have it higher.

Fig. 18.19 also shows a difference between two types of herbivore. Pigs, rats etc. have stout necks and keep them more or less horizontal. They have relatively straight skulls as in Fig. 18.19(*b*). Sheep, horses, rabbits etc. have longer, more slender necks and generally hold them at a steep angle, carrying their heads high and so getting a better view of possible danger. The back of the skull bends down to join the neck as in Fig. 18.19(*c*).

Many rodents including rats and squirrels have bunodont molars, but grazing rodents such as voles have hypsodont molars. All of them, however, have the characteristic rodent arrangement of teeth at the front of the jaw: a single pair of chisel-shaped incisors in each jaw, with no canines, and a toothless gap (the diastema) between incisors and premolars. They also have peculiar jaw joints: the condyles of the lower jaw are narrow, and they rest in longitudinal grooves in the squamosals, so the lower jaw can slide forward and back along the grooves as well as opening and closing.

The movements of rat jaws have been studied by X-ray cinematography. The rats were fed biscuit impregnated with barium sulphate (which is opaque to X-rays) so that the food as well as the bones and teeth could be distinguished in the films. Two main jaw actions were observed, one used for biting and gnawing with the incisors and the other for chewing with the premolars and molars (Fig. 18.20). The jaw slides forward for biting and gnawing. In this position, the upper and lower incisors meet edge-to-edge without bringing the premolars and molars into contact (Fig. 18.20, upper). It moves backward for chewing so that the lower incisors are posterior to the upper ones and the premolars and molars can make contact (Fig. 18.20, lower). Similarly, if you bring your incisors edge-to-edge to bite, your premolars and molars cannot meet. You must move your lower incisors behind the upper ones to chew. Fig. 18.20 (upper) shows the cycle of jaw movements made by a rat biting off a piece of food. The incisors meet edge-to-edge, and then the lower ones pass posterior to the upper ones. Each has a thick layer of enamel on its anterior face and none on the

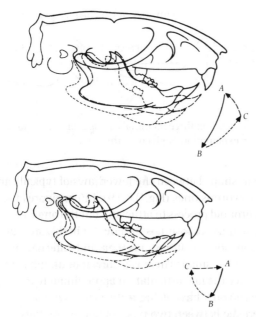

Fig. 18.20. Outlines of the skull drawn from X-ray cinematograph films of rats (*Rattus norvegicus*) feeding. The upper outline shows the sequence of jaw positions involved in biting with the incisors. The lower one shows the sequence of positions involved in chewing. From K. Hiiemae (1971). *Zool. J. Linn. Soc.* 50, 111–134.

posterior face. The cutting edges of the teeth are formed by the layer of hard enamel. The lower teeth scrape the upper ones as they pass from position *C* to position *A* in gnawing (Fig. 18.20, upper), and so sharpen them.

The toothless diastema enables rodents to gnaw without ingesting the shavings, since the lips can be drawn in through it, behind the incisors, to close off the rear part of the mouth. A squirrel can gnaw through the shell of a nut but ingest nothing until it reaches the kernel, and a beaver can fell a tree without filling its mouth with wood shavings. The diastema also enables grazing mammals to eat long lengths of grass. When a rabbit (*Oryctolagus*), for instance, eats long grass it bites it off near the ground, allowing the long ends to protrude from the side of the mouth through the diastema. These ends are drawn into the mouth by the tongue as chewing proceeds.

Rat chewing is not a side-to-side movement like the chewing of *Didelphis* and of ruminants. Instead it is mainly a forward-and-back movement (Fig. 18.20, lower). The jaw moves anteriorly during the active phase of the cycle, while it is actually grinding the food, and posteriorly in the recovery phase. The main chewing muscles are the masseters, which pull the jaw upwards and forwards.

18.7. Conflict and co-operation

Two families of artiodactyls are particularly successful, the Bovidae (antelopes, cattle etc.) and the Cervidae (deer). Both have weapons, the horns of Bovidae and the antlers of Cervidae. Female antelopes generally have rather smaller horns than males, or no horns at all. Female deer have no antlers, with the sole exception of the reindeer (Fig. 18.9*b*).

Horns are permanent and unbranched. Antlers are shed annually, and are often branched. They are generally shed in the autumn and regrown in spring, and tend to be larger and have more branches each successive season in the life of the deer. A horn is a bony projection of the skull with a sheath of keratin (the material known as horn). An antler is also a bony projection of the skull, but it is covered during growth by soft skin with a rich blood supply. Later in the season the skin is shed and the bare bone exposed.

Both horns and antlers may grow enormous. The horns of a wild American ram may be as much as 12% of body mass. A moose (*Alces*) 2.3 m high at the shoulder may have antlers with a span of 1.8 m. The extinct Irish elk (*Cervus giganteus*) was smaller in the body than a moose but had antlers up to 3.5 m across.

The use of antlers is illustrated by observations of Red deer (*Cervus elaphus*) on the Isle of Rhum (Scotland). Mating occurs in autumn. Each mature male tries to gather a group of females around him at this time, and defends them against other males so that only he can mate with them. A typical harem consists of about six females.

Antlers are used in contests between males, one or both of which has a harem. The winner takes over part or all of the loser's harem.

The contest starts with the two males roaring at each other. One may concede the contest and retire at this stage. Otherwise, the males approach each other and walk side by side, apparently displaying antlers to each other. At this stage again, one may concede defeat. Otherwise, the males interlock antlers. They push and wrestle until one is pushed rapidly backwards, breaks contact and runs off. The winner pursues the loser for only a few yards and neither may get hurt. However, if one slips and falls the other will try to horn it. Even if neither falls, there may be serious injuries. It has been estimated that 23% of the mature males are injured in tournaments in each mating season, including 6% with injuries which have lasting effects such as blindness or lameness. Antlers are quite often broken, although they are made of bone that is unusually flexible and so is hard to break by impact.

Such contests result in gains and losses. Females may be gained by the victor and lost by the loser. Both contestants may be injured and both will lose time which could otherwise have been used for feeding and mating. These gains and losses influence the contestants' chances of passing on their genes to the next generation so if the strategies used in the contests are inherited they will be subject to natural selection.

We will consider just three of the many possible strategies, which we will call Hawk, Dove and Assessor (you will soon see why).

A Hawk fights at every opportunity, winning sometimes and losing sometimes. When he wins he gains females, or retains females that he already holds. When he loses he is likely to be injured.

A Dove always runs away, never gaining anything and never getting hurt.

An Assessor sizes up his antagonist. He fights if he thinks he can win but retires if he is sure he would lose. Thus he probably wins as often as a Hawk would, but he generally avoids getting hurt.

It seems fairly obvious that if you start with a mixed population of Hawks, Doves and Assessors, the Assessors are likely to leave most offspring. If the strategy used by an animal is determined genetically, successive generations will include bigger proportions of Assessors until the other strategies are eliminated from the population. Red deer stags seem to be Assessors, and the elaborate ritual before fighting seems to be a means of assessing each other's strength. In particular, the roaring matches seem to test the stags: the one that can make more roars per minute is probably the stronger. These procedures seem to ensure that only fairly well-matched stags fight. Young stags have lower body masses than older ones and do badly in fights against them, and seldom fight with antagonists more than a year older than themselves.

Hawk, Dove and Assessor are not by any means the only possible strategies for stags. Various others have been considered, and their relative merits were sometimes not clear until they had been analysed mathematically. The appropriate mathematics was the theory of games, the same theory as you would have to use if you wanted to work out the best strategy for a game such as poker. The conclusion stands, however, that Assessor seems to be a better strategy than any other that has so far been thought of.

Stags are fierce rivals, but male lions cooperate in small groups. A pride consists (typically) of two or three adult male lions, five to ten adult lionesses, and their cubs. When a lioness is about to come into oestrus the nearest male claims her and stays with her, copulating frequently until (several days later) oestrus is past. Males sometimes fight for females but the rule is generally respected: the male who first claims a female is allowed to keep her.

This seemed puzzling. One of the males in the pride is presumably the strongest and could capture females from the others, or drive the others out of the pride. Why have lions not evolved an Assessor strategy? Why do they not sieze females from males that are weaker than themselves?

A possible answer comes from long-term studies of lion prides in Tanzania, which have been observed for many years and whose members are known individually: they can be recognized by scars, irregular teeth, spots on their muzzles and such-like features. These studies show that the female members of a pride have been born in the pride and have spent all their life in it. Most female cubs remain in the pride but male cubs are ejected while still juveniles, and live in small all-male groups for a few years until they are able to capture a pride, driving out the males who previously held it and taking their place. The males of a pride have frequently spent their whole lives together, growing up together in a different pride and wandering together as young bachelors, before taking over their present pride.

There is a strong selective advantage to the males in working in groups, because a group is better able to capture and hold a pride than is a lone male. Only four of 23 lone males held prides, but 23 of 39 pairs held prides and so did 20 of 21 groups of three or more. Pairs of males held prides for an average of 18 months but groups of three held them for an average of at least 40 months. Companions greatly increase the number of offspring a male is likely to have, even though he has to share females with them.

The penalty of having to share is reduced by the companions being related. The females of a pride have all been born in the pride and are likely to be relatives. The males of a pride have generally been born together in another pride. In some cases, they are brothers. In others, they are more distant relatives. It has been calculated that two males in a pride share on average almost one quarter of their genes: that is to say, they are almost as closely related as half brothers. A male who allows his companion to breed with a female pride member may be missing an opportunity to breed himself, but some of the genes that his companion passes on will be genes that the two lions share. This is the same kind of argument as one in section 9.7, in the discussion of social insects.

Similar thinking seems to explain another aspect of lion behaviour. The mothers in a pride suckle each other's cubs, as well as their own. Since the mothers are related, they increase the number of genes matching their own that are passed to later generations, by helping each other's cubs to survive.

Selective advantages may be gained by helping related animals, but there is generally no advantage in helping unrelated ones. New males taking over a pride may kill the cubs that they find in it. These cubs are not related to them, and if the cubs are killed the females will be ready sooner to breed with the new males.

18.8.   Diving

Seals and whales are remarkable divers. The longest recorded dives have been made by whales (sperm whales, *Physeter*, have been known to stay submerged for over an hour), but most physiological research has been done on seals.

It is not easy to follow a seal under water, and it may surface some distance from where it dived. This difficulty was overcome in a study of the Weddell seal, *Leptonychotes weddelli*, which spends the winter under the Antarctic ice sheet. It breathes at holes in the ice: these probably originate as natural breaks but it uses its teeth to keep them open. Captured Weddell seals were taken to a man-made hole in an ice sheet which was apparently otherwise unbroken over a wide area. They were released and dived, but had to return to the same hole to breathe. Instruments were attached to them that made records of pressure changes (and so depth changes) and could be recovered when the seal returned. Records of 381 dives by 27 seals were obtained. Most dives were to less than 100 m and most lasted less than 5 min, but the deepest was to 600 m (this was about the depth of the bottom) and the longest lasted 43 min. Deep dives were generally fairly brief and seem to have been hunting forays. The longest dives were relatively shallow, and it is thought that the seals may have been searching for other holes in the ice. Information on the diving habits of other seals is sparse, but several species have been hooked on lines at 180 m or more.

How does a seal survive a 40 min dive without asphyxiating? The quantity of oxygen it can take with it is strictly limited. Typical seals have lungs which occupy about 10% of the volume of the body, when full. Seals generally breathe out before they dive (the probable advantage of this will be considered later), but even when the lungs are full of air containing 20% oxygen, they can only contain about 20 cm^3 oxygen (kg body mass)$^{-1}$. More oxygen can be carried in the blood. Various seals which have been investigated contained about 120 cm^3 blood (kg body mass)$^{-1}$. This is more than is usual in terrestrial mammals: the blood volumes of people and dogs are about 70 and 90 cm^3 kg^{-1}, respectively. The oxygen capacity of the blood of various species of seals has been measured, and values of around 300 cm^3 l^{-1} have been found, 1.5–2 times as high as in typical terrestrial mammals. (Such very high values are made possible by the high haemoglobin content of seal blood: about 20%.) Hence when all the blood is saturated it contains about 40 cm³ oxygen (kg body mass)⁻¹. Oxygen can also be stored in the muscles, combined with myoglobin. Seal muscle contains about 80 g myoglobin kg⁻¹ (eight times as much as beef) and is consequently a dark blue-black colour. It can be calculated from this that the myoglobin of a seal containing 35% of muscle must be able to combine with about 40 cm³ oxygen (kg body mass)⁻¹. Thus the lungs, haemoglobin and myoglobin between them can take up about 100 cm³ oxygen (kg body mass)⁻¹; far more oxygen than a terrestrial mammal could take with it on a dive, but still not enough to support metabolism for long. The basal metabolic rates of various seals have been measured, and found to be considerably higher than Fig. 17.6 would suggest. As an example, consider *Phoca vitulina*, which is known as the Common seal in Britain and as the Harbor seal in the United States. A 30 kg specimen was found to use 400 cm³ oxygen (kg body mass)⁻¹ h⁻¹, so the estimated oxygen store of 100 cm³ oxygen kg⁻¹ would only last for about 15 min if the seal was resting, or less if it was active. In fact this species has survived 28 min submergence in a laboratory experiment.

Long dives are apparently made possible by anaerobic metabolism. In one series of experiments seals were submerged in a bathtub, and samples were taken periodically from their back muscles. When the samples were analysed it was found that the oxygen (combined with myoglobin) in the muscle declined almost to zero in the first ten minutes or so of submergence, and that the lactic acid concentration rose thereafter.

The partial pressure of dissolved nitrogen in the blood of animals breathing air at the surface is 0.8 atm, the same as the partial pressure of nitrogen in the atmosphere. If the animal dives the air in its lungs is compressed, the partial pressure of nitrogen in it rises, and more nitrogen dissolves in the blood and tissues. Two of the hazards of diving, for people, result from this. The first is nitrogen narcosis: high partial pressures of dissolved nitrogen in the blood have an anaesthetic effect. Human divers, breathing compressed air, suffer only slight effects at 30 m but many become incapable of useful work at about 80 m. For deeper dives, oxygen–helium mixtures are used instead of air. The other hazard is decompression sickness, which does not occur during the dive but afterwards, when the diver returns to the surface. It is due to the additional dissolved nitrogen coming out of solution, as the pressure falls, forming bubbles in the blood and nervous system. Pain and even death follow. The severity of the symptoms depends on the duration of the dive as well as its depth, but symptoms are rare even if the dive is long if the depth is 13 m or less. Decompression sickness is avoided by returning to the surface in stages, pausing to allow the release of excess nitrogen. In experiments on cats, bubbles were only found after the partial pressure of dissolved nitrogen in the blood had been allowed to rise to 3.3 atm. This is the partial pressure which would be reached in the course of a long dive, using compressed air, to 31 m. (The pressure at 31 m is 4.1 atm, and the partial pressure of nitrogen in air compressed to 4.1 atm is $0.8 \times 4.1 = 3.3$ atm.)

Seals dive to depths far greater than 30 m, and so might be expected to suffer from both nitrogen narcosis and decompression sickness. How do they avoid these hazards? Four Weddell seals were released at an ice hole, with equipment attached to them that would take small samples of their blood during their dives, at known depths. Analysis of the samples showed that the partial pressure of dissolved nitrogen in the blood reached about 3 atm when the seal had dived to 50 m, but rose no higher: indeed, it tended to fall as the seal dived deeper. The higher partial pressures that would cause trouble were not reached, although the seals dived as deep as 200 m.

Part of the explanation seems to be that the seal breathes out before diving and so starts with rather little air (and so nitrogen) in the lungs. This is a major difference from human divers using compressed air, who have a virtually unlimited supply of nitrogen. However, there is enough nitrogen in the seal's part-filled lungs to cause decompression sickness, if it were allowed to dissolve in the blood. The other part of the explanation is that seals have their bronchi reinforced by rings of cartilage that are much more rigid than those of mammals. When the pressure increases in a dive, the lungs collapse but the bronchi do not, so the small quantity of air in the lungs is forced into the bronchi, from where it is much less easily absorbed than if it had remained in the alveoli. Evidence of this comes from X-ray pictures of seals in a compression chamber at pressures up to 31 atm (equivalent to 300 m depth). These show that the bronchi were hardly compressed, even at the highest pressures. (The bronchi would not normally have shown up well in the X-ray pictures, but powdered tantalum had been blown into them to coat their walls.)

18.9.   Echolocation

Most bats are active at night and do not depend on sight for detecting obstacles and finding food, but more on echolocation. They get information by emitting sounds and listening to the echoes from their surroundings. Bats blindfolded by covering their eyes with black collodion still fly normally around obstacles. However, if their ears are tightly plugged they collide with obstacles apparently at random, even in daylight with their eyes uncovered.

The sounds which most bats use for echolocation are inaudible to human ears, because of very high frequency. They can, of course, be detected by suitable microphones and displayed as wave forms, for instance on a cathode-ray oscilloscope. They vary considerably between groups of bats. Bats of the family Vespertilionidae, which includes *Myotis* and most of the other bats of Britain and the northern United States, make very brief but intense chirps. (A sound inaudible to people on account of high frequency may nevertheless be intense in terms of energy flux.)

Bats chirp more frequently as they approach an obstacle or insect prey, and this gives an indication of the distance at which objects are detected. Many observations have been made of bats flying along a room, steering between wires stretched from floor to ceiling. *Myotis* started increasing their chirping rate while still 2 m from wires 1 mm in diameter. They must have been able to detect the wires at *at least* that distance.

The chirps become shorter as a bat approaches an insect. *Myotis* have been filmed and their chirps recorded, as they chased mealworms thrown into the air. Each chirp lasted about 1.5 ms while the bat was 0.5 m from the mealworm (as shown in the film), but only 0.3 ms at 0.1 m. Sound travels at 330 m s^{-1} in air, so in 1.5 ms it travels 0.5 m and in 0.3 ms, 0.1 m. In each case the first sound waves must have been reaching the mealworm as the last were emitted by the bat. If chirps 1.5 ms long were still being emitted when the bat was only 0.1 m from the mealworm, the echo would return to the bat before the chirp was over. The relatively weak echo might not be easily heard, during the intense chirp.

There is rather surprising evidence that their echoes from large objects may actually sound louder to bats than the chirps themselves. Electrodes have been inserted into the ears of a Mexican bat (*Tadarida*) to record electrical potentials (cochlear microphonics) produced by sounds. These electrodes were connected to long, light leads so that the potentials could be recorded while the bats crawled and even flew around the laboratory. It was found that echoes of chirps from a wire window screen, over 1 m from the bat, produced microphonics almost as large as did the original chirps. This is probably partly due to the sound being directed forward from the mouth, not back towards the pinnae. However, there is another effect. Microphonics were recorded as before while a sound of constant amplitude was emitted by a loudspeaker. The size of the microphonics due to this sound diminished immediately before each chirp, and returned to normal immediately after it. The ear was apparently made less sensitive to sound while the chirp was being emitted, and allowed to return to normal before the echo returned. There is evidence that this is achieved by brief contraction of muscles attached to the ear ossicles.

High-frequency sounds are most appropriate for locating small objects. This is because the intensity of sound reflected or scattered by an object depends on the ratio between the dimensions of the object and the wavelength of the sound. Objects which are large relative to the wavelength reflect sound in particular directions, as a mirror reflects light. Those that are small scatter sound in all directions and the intensity of the scattered sound falls off very rapidly indeed as the diameter falls below about 0.1 wavelength. The frequency falls in the course of a chirp but may be 100 kHz or even more at the beginning. Sound of this frequency, in air, has a wavelength of only 3 mm, and so is scattered reasonably effectively by insects down to the size even of gnats.

How much information can a bat obtain from an echo? It seems obvious from the facility with which insects are caught and obstacles avoided, that it can distinguish the direction from which the echoes come. A vespertilionid bat with one ear blocked is nearly as helpless as with both blocked, so the sense of direction presumably depends largely on comparisons between the two ears (as in other mammals). It also seems clear that distance can be judged. This has been confirmed by training *Eptesicus* to fly for food to whichever of two similar platforms was the nearer. It could make the correct choice even with one platform at a distance of 60 cm and the other at 58.5 cm. In other experiments it was shown that the bats could distinguish between triangular plates differing in area by 17% or more, and between triangles of the same area but different proportions.

That account of echolocation is based mainly on Vespertilionidae. Bats of other families make rather different sounds and in some cases apparently use distinctly different mechanisms. For instance, horseshoe bats (Rhinolophidae, Fig. 18.8) emit rather long pulses of sound which overlap with the returning echo, and their ability to avoid obstacles is not seriously impaired by blocking one ear.

Dolphins are believed to use echolocation, much as bats do. A captive *Tursiops* was persuaded by patient training to allow an experimenter to fit opaque plastic suction cups over her eyes. She swam without hesitation round the tank immediately after blindfolding. She could navigate blindfold between obstacles, find and pick up food without touching nearby obstacles, and perform tricks such as pressing an electric bell. Echolocation has also been demonstrated convincingly for porpoises (*Phocaena*) and killer whales (*Orcinus*) but not, so far, for the larger whales. The sounds which are apparently used in echolocation are high-frequency clicks that seem to be produced by structures in the nostril.

Echolocation will only work if the animal can emit sound in narrow beams, or judge the direction of returning echoes, or both. Echolocation by dolphins probably depends partly on directional emission, and partly on directional hearing. It has been shown by sound measurement in the water around captive dolphins that at any given distance the echolocation sounds are much more intense ahead of the animal than in other directions. Speculations as to how this is achieved involve the idea that the sound may be focused into a beam by the concave upper surfaces of the bones of the snout, which may serve as a concave mirror, and by the lens-shaped mass of fat which gives the bulbous shape to a dolphin's 'forehead'. Sound travels slower in fat than in water so a submerged convex mass of fat acts as a converging lens for sound. Possible mechanisms of directional hearing have also been suggested. There is no pinna to aid directional hearing, and indeed a pinna would not work in water in the way it does in air, because there is little contrast in acoustic properties between water and tissues. For the same reason the head would not shield the left ear from sounds coming from the right, were it not that the ears are surrounded (except on the lateral side) by foam. Sound approaching from the same side can enter freely but sound from other directions tends to be reflected by the gas in the foam.

Further reading

From reptiles to mammals

Crompton, A.W. (1963). On the lower jaw of *Diarthrognathus* and the evolution of the mammalian lower jaw. *Proc. zool. Soc. Lond.* **140**, 697–753.

Kemp, T.S. (1978). Stance and gait in the hind limb of a therocephalian mammal-like reptile. *J. Zool., Lond.* **186**, 143–161.

Presley, R. (1984). Lizards, mammals and the primitive tetrapod tympanic membrane. *Symp. Zool. Soc. Lond.* **52**, 127–152.

Physiological features of mammals

Coles, R.B. & Guppy, A. (1986). Biophysical aspects of directional hearing in the Tammar wallaby, *Macropus eugenii. J. exp. Biol.* **121**, 371–394.

Holmes, M.H. (1987). Frequency discrimination in the mammalian cochlea: theory versus experiment. *J. acoust. Soc. Am.* **81**, 103–114.

Jamison, R.L., Bennett, C.M. & Berliner, R.W. (1967). Countercurrent multiplication by the thin loops of Henle. *Am. J. Physiol.* **212**, 357–366.

Lote, C.J. (1987). *Principles of renal physiology*, 2nd edn. Croom Helm, London.

Reproduction of mammals

Blaxter, K.L. (1962). *The energy metabolism of ruminants.* Hutchinson, London.
Pond, C.M. (1977). The significance of lactation in the evolution of mammals. *Evolution* **31**, 177–199.
Tyndale-Biscoe, H. & Renfree, M. (1987). *Reproductive physiology of marsupials.* Cambridge University Press.

The orders of mammals

Alexander, R.McN. (1985). Body size and limb design in primates and other mammals. In *Size and scaling in primate biology* (ed. W.L. Jungers), pp. 337–343. Plenum, New York.
Corbet, G.B. & Hill, J.E. (1986). *A world list of mammalian species,* 2nd edn. British Museum (Natural History), London.
Crompton, A.W., Taylor, C.R. & Jagger, J.A. (1978). Evolution of homeothermy in mammals. *Nature* **272**, 333–336.
Griffiths, M. (1978). *The biology of the monotremes.* Academic Press, New York.
Kielan-Jaworowska, Z., Crompton, A.W. & Jenkins, F.A. (1987). The origin of egg-laying mammals. *Nature* **326**, 871.
Lee, A.K. & Cockburn, A. (1985). *Evolutionary ecology of marsupials.* Cambridge University Press.
Macdonald, D. (ed.) (1984). *The encyclopaedia of mammals.* Allen & Unwin, London.
McNab, B.K. (1986). Food habits, energetics and the reproduction of marsupials. *J. Zool., Lond.* A**208**, 595–614.
Martin, R.D. (1981). Relative brain size and basal metabolic rate in terrestrial vertebrates. *Nature* **293**, 57–60.
Richard, A.F. (1985). *Primates in nature.* Freeman, New York.
Southern, H.N. (1964). *The handbook of British mammals.* Blackwell, Oxford.

Walking and running

Alexander, R.McN. (1984). Walking and running. *Am. Sci.* **72**, 348–354.
Alexander, R.McN. (1988). *Elastic mechanisms in animal movement.* Cambridge University Press.
Cavagna, G.A., Heglund, N.C. & Taylor, C.R. (1978). Mechanical work in terrestrial locomotion: two basic mechanisms for minimizing energy expenditure. *Am. J. Physiol.* **233**, R243–R261.
Jenkins, F.A. (1971). Limb posture and locomotion in the Virginia opossum and in other non-cursorial mammals. *J. Zool., Lond.* **165**, 303–315.
Jenkins, F.A. (1972). Chimpanzee bipedalism: cineradiographic analysis and implications for the evolution of gait. *Science* **178**, 877–879.
Ker, R.F., Dimery, N.J. & Alexander, R.McN. (1986). The role of tendon elasticity in hopping in a wallaby. *J. Zool., Lond.* A**208**, 417–428.

Teeth and chewing

Crompton, A.W. & Hiiemae, K. (1970). Molar occlusion and mandibular movements during occlusion in the American opossum, *Didelphis marsupialis* L. *Zool. J. Linn. Soc.* **49**, 21–47.
Hiiemae, K.M. & Ardran, G.M. (1968). A cinefluorographic study of mandibular movement during feeding in the rat (*Rattus norvegicus*). *J. Zool., Lond.* **154**, 139–154.

Hiiemae, K. & Jenkins, F.A. (1969). The anatomy and internal architecture of the muscles of mastication in *Didelphis marsupialis. Postilla* **140**, 1–49.

Smith, J.M. & Savage, R.J.G. (1959). The mechanics of mammalian jaws. *School Sci. Rev.* **40**, 289–301.

Conflict and co-operation

Bygott, J.D., Bertram, B.C.R. & Hanby, J.P. (1979). Male lions in large coalitions gain reproductive advantage. *Nature* **282**, 839–841.

Clutton-Brock, T.H., Guinness, F.E. & Albon, S.D. (1982). *Red deer. Behaviour and ecology of the sexes.* Edinburgh University Press.

Packer, C. (1986). The ecology of sociality in felids. In *Ecological aspects of social evolution* (ed. D.I. Rubinstein & R.W. Wrangham), pp. 429–451. Princeton University Press.

Schaller, G.B. (1972). *The Serengeti lion. A study of predator–prey relations.* University of Chicago Press.

Smith, J. Maynard (1982). *Evolution and the theory of games.* Cambridge University Press.

Diving

Blix, A.S., Elsner, L. & Kjekshus, J.K. (1983). Cardiac output and its distribution through capillaries and A–V shunts in diving seals. *Acta physiol. scand.* **188**, 109–116.

Falke, K.J. and eight others (1985). Seal lungs collapse during free diving: evidence from arterial nitrogen tension. *Science* **229**, 556–557.

Ridgeway, S.H. & Howard, R. (1979). Dolphin lung collapse and intramuscular circulation during free diving. *Science* **206**, 1182–1183.

Echolocation

Amundin, M. & Andersen, S.H. (1983). Bony nares pressure and nasal plug muscle activity during click production in the harbour porpoise and bottlenosed dolphin. *J. exp. Biol.* **105**, 275–282.

Cahlander, D.A., McCue, J.J.G. & Webster, F.A. (1964). The determination of distance by echolocating bats. *Nature* **201**, 544–546.

Kick, S.A. (1982). Target-detection by the echo-locating bat, *Eptesicus fuscus. J. comp. Physiol.* A**145**, 431–435.

Simmons, J.A. & Vernon, J.A. (1971). Echolocation: discrimination of targets by the bat *Eptesicus fuscus. J. exp. Zool.* **176**, 315–328.

Index